Jim Gri

To Jim -

may we all live to see our problems with silviculture get successful solutions.

T. W. Daniel

Principles of Silviculture

McGraw-Hill Series in Forest Resources

AVERY Natural Resource Measurements
BOYCE Forest Pathology
BROCKMAN and MERRIAM Recreational Use of Wild Lands
BROWN and DAVIS Forest Fire: Control and Use
CHAPMAN and MEYER Forest Mensuration
DANA Forest and Range Policy
DANIEL, HELMS, and BAKER Principles of Silviculture
DAVIS Forest Management
DAVIS Land Use
DUERR Fundamentals of Forestry Economics
GRAHAM and KNIGHT Forest Entomology
GUISE The Management of Farm Woodlands
HARLOW, HARRAR, and WHITE Textbook of Dendrology
HEADY Rangeland Management
PANSHIN and DE ZEEUW Textbook of Wood Technology
Volume I—Structure, Identification, Uses, and Properties of the Commercial Woods of the United States
PANSHIN, HARRAR, BETHEL, and BAKER Forest Products
RICH Marketing of Forest Products: Text and Cases
SHARPE, HENDEE, and ALLEN An Introduction to Forestry
SHIRLEY Forestry and Its Career Opportunities
STODDART, SMITH, and BOX Range Management
TRIPPENSEE Wildlife Management
Volume I—Upland Game and General Principles
Volume II—Fur Bearers, Waterfowl, and Fish
WACKERMAN, HAGENSTEIN, and MICHELL Harvesting Timber Crops
WORRELL Principles of Forest Policy

Walter Mulford was Consulting Editor of this series from its inception in 1931 until January 1, 1952.

Henry J. Vaux was Consulting Editor of this series from January 1, 1952, until July 1, 1976.

Principles of Silviculture

Second Edition

Theodore W. Daniel
Utah State University

John A. Helms
University of California

Frederick S. Baker
Late Professor of Forestry
University of California

McGraw-Hill Book Company

New York St. Louis San Francisco Auckland Bogotá Düsseldorf
London Madrid Mexico Montreal New Delhi
Panama Paris São Paulo Singapore Sydney Tokyo Toronto

PRINCIPLES OF SILVICULTURE

90 DODO 898

This book was set in Times Roman by Automated Composition Service, Inc. The editors were Jean Smith, Marian D. Provenzano, and Stephen Wagley; the production supervisor was Dominick Petrellese. New drawings were done by J & R Services, Inc.
R. R. Donnelley & Sons Company was printer and binder.

Library of Congress Cataloging in Publication Data

Daniel, Theodore W.
Principles of silviculture.

(McGraw-Hill series in forest resources)
Bibliography: p.
Includes index.
1. Forests and forestry. 2. Forest management.
3. Trees. I. Helms, John A., joint author.
II. Baker, Frederick Storrs, date joint
author. III. Title.
SD391.B32 1979 634.9'5 78-12570
ISBN 0-07-015297-7

DEDICATION

This book is dedicated to F. S. Baker, who is known as "Bake" to generations of foresters—his students, his friends, and the profession. He was able to distill the essence of the nature of the interaction between trees and their environment into words. In that transition from nature to words there was a twist that was peculiarly Bake. Bake did very well in transmitting his knowledge of trees, but in this revision of his work there was no way to incorporate his style into the changing emphasis and enlarged volume of research that this book had to digest.

F. S. Baker became Dean of the School of Forestry at Berkeley, but this book is dedicated to the teacher who raised questions rather than give the common bromides that served as answers, to the humorist who could make you laugh at his skits, and to a very human person who boiled the complexities of silviculture down to an understanding of the fundamentals of a plant's interactions with its environment.

Contents

2
THE TREE

3
FOREST SITE

4
STANDS

* uneven-aged mgt. Q factors, etc.

* Langsaeters Curve, p. 318

5
FOREST MANIPULATION

Preface

This second edition is based on Frederick S. Baker's text *The Principles of Silviculture* which was published in 1950. Baker's text has served for many years as the only book written for North American forestry that deals with the basic principles underlying the practice of forestry. Since 1950 there has been a tremendous increase in knowledge of forest science. The basic silvicultural principles are the same, but today we have much more information which places these principles on a firmer foundation.

In this edition, a few sections are much the same as in Baker's text. The principles he elucidated are as valid today as they were in 1950. Other sections are completely new, reflecting the advanced state of knowledge and the profession's changed orientation toward forest management.

A large part of this edition is devoted to considerations of basic plant-environment interactions. This follows Baker's emphasis, and to quote Baker, "a sufficiently wise and flexible silvicultural art can be developed on the ground only by practitioners who understand the forest as a biological entity."

ACKNOWLEDGMENT

The authors acknowledge the help, advice, and criticism of many associates over the years this edition was incubating. Few of our professional friends have not been called on to provide expertise, and to list those who helped would be to call the roll. We can only express our heartfelt appreciation of their efforts and hope for an opportunity to reciprocate in the future.

Theodore W. Daniel
John A. Helms

Part One

The Forest

Chapter 1

Introduction

In introducing the *Principles of Silviculture* it is both appropriate and informative to consider two excerpts from the introduction given by F. S. Baker to his 1950 text of the same title. Baker has this to say:

> "Forestry," says a recent definition, is "the scientific management of forests for the continuous production of goods and services." . . . This compressed definition implies many things. Forestry is more than the utilization of such products as the wild and unmanaged forest may furnish, for it assumes that there is a skillful planning to assure a certain and continuous production and that "goods and services" will be precisely that—useful products and services necessary to our present-day economic life.
>
> Forestry as thus defined can fail at two points. It may either fail to assure continuous production by rank mismanagement, or it can defeat its own ends by overelaborate management that will cost more to carry out than the "goods and services" warrant. In private forest enterprises this is always a bogey, perhaps sometimes fanciful but nevertheless effective in deterring the operator from spending money for forestry. But even in public

forestry, where a strict dollars-and-cents accounting for all tangible and intangible "goods and services" produced by the forest is neither possible nor desirable, there must be some balance in the long run between the cost of forestry and its returns.

Forestry must be sound both biologically and economically if it is really to work. To do this in America means that silviculture must be developed in such a way as to get the maximum of effectiveness coupled with low costs. As Buttrick (1943) points out, silviculturists, like all technicians engaged in the production of goods, work for the highest attainable output of the highest possible quality and, given the opportunity, will tend to ignore the factor of diminishing economic returns.

In the never-ending conflict between the ambitious silviculturist and those charged with financial management, having an eye to profits, there can be no well-defined, sound determination of what constitutes the point of maximum net profit because, on the one hand, the silvicultural measures do not guarantee perfect results and, on the other hand, the income from the forest crop is never determinable until the year of harvesting. . . .

A knowledge of the nature of forests and forest trees, how they grow, reproduce, and respond to changes in their environment, makes up the broad field of forestry called *silvics*. This is practically equivalent to the *forest ecology* of some writers. The methods of handling the forest in view of its silvics—modified in practice by economic factors—is *silviculture*. Silvics is therefore the basis of silviculture on the biological side. The present book is essentially a book on silvics—considering first the forest as a static affair (what it is) and then the natural laws of its development (how it works).

In the last three decades, considerable changes have occurred in the evolution of the forestry profession. These are primarily due to changes in society, of which forestry is an integral part. Foresters and society as a whole have entered a new era of general concern for conservation, which means "wise use," of our natural resources. In addition, society is rightly expecting its forests to produce more than just wood products. Sustained and efficient wood production is still the prime objective in forestry, but this must be achieved simultaneously with increased outputs of high-quality water, wildlife, recreation, and aesthetics, and without environmental degradation. This precipitates conflicts in land use, raises questions of compatibility of multiple objectives, and has introduced demands for public involvement in decision making. Suddenly we are confronted with the need for land-use planning, which requires the ability to quantify the "carrying capacity" of forests for all the diverse goods and services required.

Public concern with forest management is reflected in the recent burgeoning of federal and state legislation regulating forestry. These incentives and constraints, coupled with the real-world environment of costs and taxes, provide a totally new context for the practice of forestry with vastly increased restrictions on operations. This situation calls for a new generation of silviculturists who are knowledgeable and versatile, who can accurately predict the likely outcomes from alternative stand treatments, and who can develop silvicultural prescriptions compatible with physiological, ecological, managerial, and social constraints. This can be done only by people well grounded in the principles of silviculture.

With these increased complexities and expanded expectations we are surely entering the most challenging and stimulating period in forestry.

Chapter 2

The Role of Silviculture in Forest and Wildland Management

The practice of forestry in any one country or region is in a continuous state of evolution which must parallel that of the local society. Where wood and land are abundant relative to societal demands, forestry practices are necessarily minimal and extensive. This was the situation in much of North America up to the 1940s. The need for close silvicultural control of stand development up to this period was generally minimal at best.

As society develops in size and complexity, demands on resources increase in both intensity and diversity. Land-management practices must continually change accordingly. It is common to find, for example, that practices which are normal and acceptable in one decade are clearly unacceptable, and in some cases illegal, in the next. Problems arise when practices are introduced which are either ahead of or behind their times. In either case, the practices are socially unrealistic. The aim of the forester is to utilize practices which are managerially sound, socially acceptable, and anticipate future needs.

In many parts of the world, society has rapidly developed to the point where there are competing potential uses for land, and where there are concerns regarding the finiteness of natural resources and the capacity to maintain timber supply. Many countries are net importers of wood, and the Food and Agriculture Organization (FAO) estimates that on a worldwide basis some $30 billion per year was spent in 1974 on the import of wood products. The obvious impact on national trade balances of more self-sufficiency in wood supplies on the part of these countries is being expressed in a worldwide expansion of reforestation programs and efforts to increase growth and yield.

In addition, worldwide concerns for conservation and the protection of the environment from deterioration and pollution are increasing. These concerns should be the cornerstone of good silviculture, and indeed most forestry agencies and industries are improving management practices both because of the need to become more efficient and productive and because of a genuine desire to provide a better forest environment for the future. To stimulate the general movement in this direction, there has been an outburst of federal, state, and local legislation and regulation aimed at conservation and protection. For example, in the United States, the National Forest Management Act of 1976 calls for reforestation of all understocked land by 1985. Public Law 92-500 requires the use of "best possible" practices in land management by 1985 and for "zero discharge" of sediments into streams from silvicultural practices. In addition, many states have initiated or revised forest-practice acts which specify minimum standards for forestry practices and stocking on private land. All these factors are creating an environment which calls for a rapid upgrading of land-management practices.

It is this rapid rate of change and increased expectations from the forestry profession that have opened up new and exciting challenges for silviculturists. Indeed a new generation of silviculturists is needed which has attitudes and capabilities more suited to carrying out the new, complex tasks of today's forest management. To explore these new challenges, we must first identify the nature of the lands being managed, the way in which land-management objectives have changed, and hence the changed function of the silviculturist in a more intensive management program.

FORESTS AND WILDLANDS

The kinds of land on which foresters and silviculturists have managerial concerns and responsibilities are very broad. They include such diverse types as forests, brush, ranges, fresh-water aquatic areas, and wilderness areas. Products from management may be correspondingly diverse. Because of this diversity, the single term *wildlands* is often used to describe this range of vegetation types involved in the practice of forestry. This term has come into common acceptance despite the difficulty that it includes such nonwild areas as managed forests, plantations, and urban forests.

Forests and wildlands are a major resource, forests alone covering one-third the surface of both the United States and of the world in general. The importance of wise management of this resource is apparent when it is recognized that even in today's industrialized societies, forests are still indispensable to human existence and well-being. Some of the ways in which forests influence our lives include:

1 Development and support of a favorable atmosphere with a stable component of oxygen
2 Provision of our current coal and oil deposits which have resulted from forest growth that occurred in the geologic past
3 Development and protection of a soil mantle which we now use to support forest and agricultural production
4 Production of clean water and protection of watersheds from erosion
5 Provision of habitat and food for animals, insects, pathogens, fish, and birds
6 Supply of building materials, fuel, and multitudinous forest products
7 Provision of essential intangible benefits through aesthetic values, recreation, wilderness, and parks

All these values, except the production of fossil fuels, relate to forest management. Their relative importance depends on the size and nature of the forest and the magnitude and array of societal needs.

In the United States, forests and wildlands are managed by federal agencies such as the Forest Service, the National Park Service, the Bureau of Land Management, and the Bureau of Indian Affairs; by large private corporations and industries; and by small private owners. Concerns of foresters and silviculturists have traditionally been directed toward larger landowners since they have been the prime source of natural resource products and the employer of most foresters. In the future, however, the contribution of products from the collectively large areas owned by the many small owners will become increasingly important. As this comes about, silvicultural information and expertise will increasingly have to be made available to these owners to ensure the sound management of the nation's resource base.

The objectives of management have changed over time as interests have broadened from forests to wildlands. In the past, silviculturists were concerned solely with timber production, since this was commonly the sole objective of management. Currently, however, this traditional and important objective must be placed within the context of other outputs. Forests and wildlands produce wood, water, fish, game, livestock, and recreational products. Determinations must be made of the extent to which these uses should be single or multiple, and whether certain combinations are mutually exclusive or compatible. Land-management objectives have rapidly become more complex and more intensive. With this has come requirements that outputs be more highly quantified and predictable.

ROLE OF SILVICULTURE

Silviculture is concerned with controlling the establishment, growth, composition, and quality of forest vegetation. This can only be done in any given forest cover and locality if there is a clearly defined management objective that describes what is to be achieved. Each management objective, then, must be interpreted in terms of the kind of stand structure that is most suitable. Objectives varying among production of wood, water, wildlife, and recreation will obviously require forests of quite different structure; that is, the presence, absence, or relative abundance of litter, grass, brush, reproduction, saplings, poles, mature trees, and overmature trees should vary in a managed stand depending on the particular management objective. Furthermore, each objective will influence the mixture of species and age classes present and the extent to which vegetation components are layered or grouped both vertically and horizontally in relatively uniform aggregations or mosaics within a stand.

Simple objectives such as wood production commonly require a simple and efficient structure. Complex objectives involving multiple uses commonly require a correspondingly complex and diverse structure with probably some loss in efficiency for any one particular use. Formulation of silvicultural strategy is, therefore, dependent on recognition of the overall objective of stand management. As objectives change, so will the silvicultural prescription, since stands of different structure will probably be required.

Silvicultural control of structure requires a prescription that blends biological, managerial, and economic knowledge. This prescription must then fit within a socially acceptable framework. Thus there is no such thing as inherently "good" silviculture which does not at the same time make good practical managerial and social sense.

The development of silvicultural prescriptions involves a thorough understanding of basic *principles* and their translation into recommended *practices*. These two components are distinct and yet interdependent. The basic principles are universal and provide the foundation for practices which tend to be regional and vary with the forest cover type. The principles of silviculture revolve around a thorough understanding of plant-environment interactions. Manipulation of establishment, growth, composition, and quality of forests to meet particular objectives requires the silviculturist to appreciate the interrelationship between the growth of forest vegetation and the physical and biological components of the operational environment. Therefore, knowledge of the potential growth of individual trees, brush, and grass species provides a basis for controlling the growth of stands. Figure 2-1 illustrates this basic concept. The silviculturist's first task is to fully understand and describe the current structure and interactions of the stand in question in terms of soils, vegetation, pests, microclimate, and ecological interactions. Second, by knowledge of stand history and ecological interpretation, the silviculturist can appreciate past stand structure and

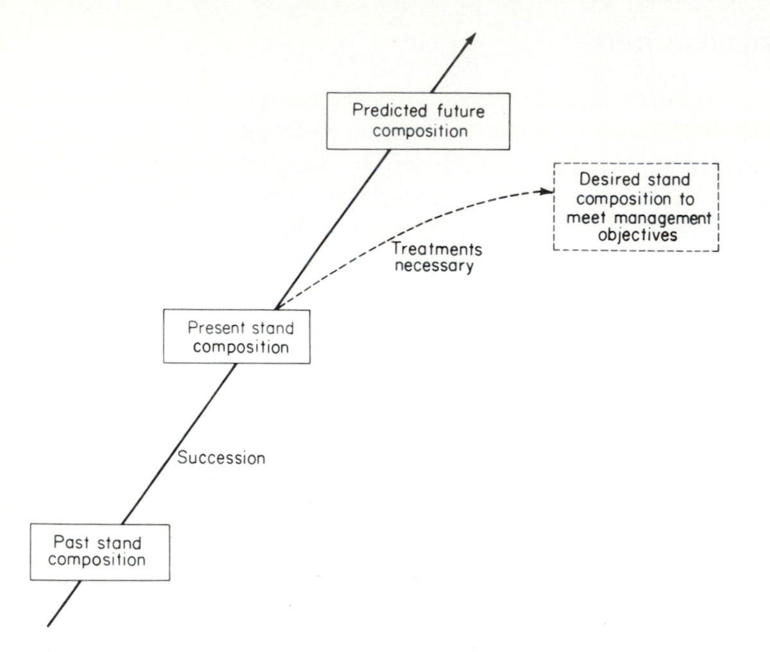

Figure 2-1 Predicted change in stand composition (or aggregations within a stand) and the deliberate modification of composition through silvicultural treatment.

the nature and rate of ecosystem change. This permits a projection of what the nature of the unmanaged stand structure will be for each relatively uniform aggregation type some time into the future. If the rate of natural change and the natural development of stand structure suit the management objective, then little silvicultural action is called for. If, however, natural stand development is not likely to be suitable, then some treatment or sequence of treatments is needed to produce the desired structure. The silviculturist then needs to specify the kind, intensity, timing, and cost of treatments; justify the recommendation; and predict the likely consequences of the treatments in silvicultural and environmental terms.

This basic concept is also illustrated in Fig. 2-2, which emphasizes the point that proper choice of silvicultural treatment depends on an understanding of the extent to which genotype-environment interactions control physiological development. The control of stand structure implies the control of growth of both desirable and undesirable plants. Since the growth of any plant is controlled by genotype-environment interaction, the silviculturist must appreciate that *all* treatments, including harvesting, thinning, site preparation, and fertilization, directly influence this interaction. Success in meeting management objectives, therefore, depends on the ability of the silviculturist to predict the effects of alternative treatments in creating suitable environments in which the desired plants can grow.

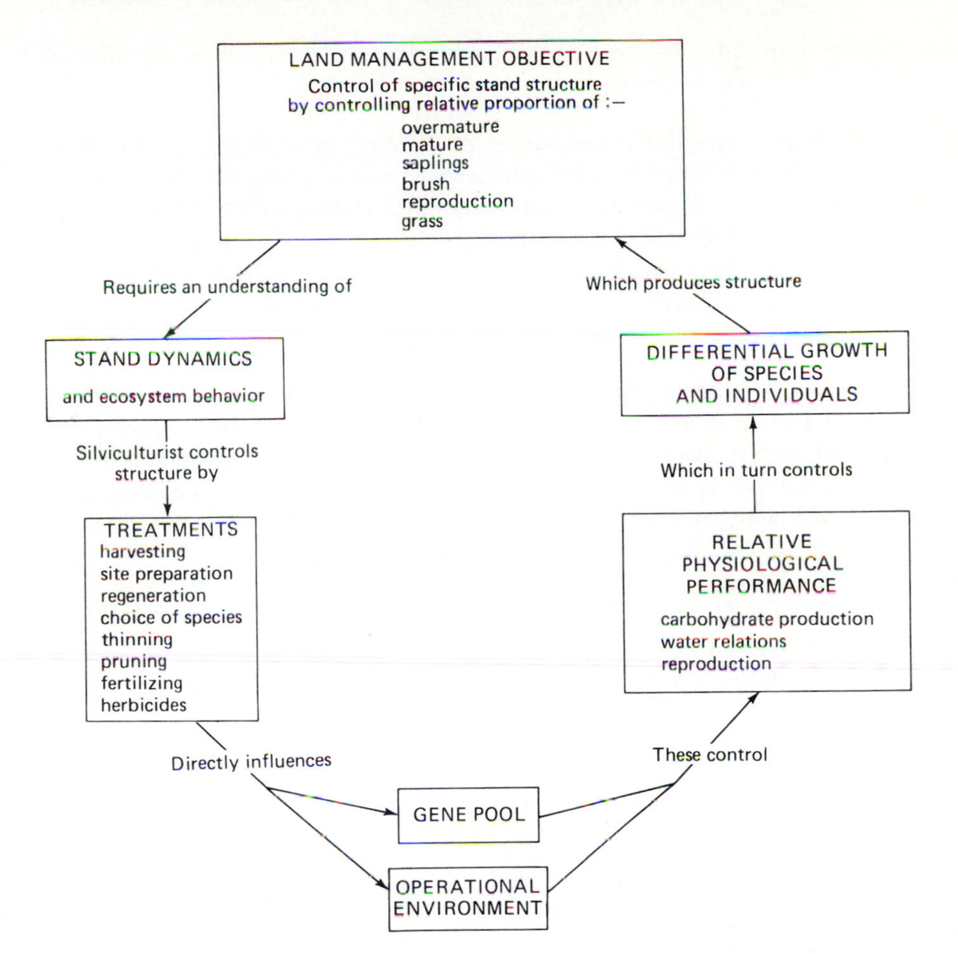

Figure 2-2 Control of stand structure is obtained by differentially controlling growth of species and individual plants through silvicultural manipulation of gene pools and operational environment.

CONSTRAINTS TO MANAGEMENT

Choice of silvicultural treatments is always constrained by ecological, managerial, and social considerations. These constitute the real-world environment in which the practicing forester must operate.

Ecological Constraints

Site Quality Inherent site quality or potential for productivity is a dominating factor. The fertility of the soil; soil depth, elevation, aspect, and slope; and other site factors will strongly influence the range of silvicultural opportunities. Generally, the more productive the site, the greater the array of treatments one is able to consider.

Existing Vegetation The nature of existing aggregations of vegetation within each stand must be taken into account and capitalized on. The presence of particular genotypes and their relative physiological characteristics will influence the choice of silvicultural treatment. This is so because certain treatments will be aimed at retarding some components of the vegetation mix, while others will be deliberately aimed at release. The capacities of plants to respond in either of these alternate ways must be predicted.

Microenvironment Since plant growth is strongly influenced by environment, the specific microenvironment in terms of light intensity, temperature, evaporative stress, and soil moisture availability within each aggregation of uniform vegetation in the stand or within each habitat type must be recognized. These microenvironments must be taken into account in selecting treatments to either enhance existing plant growth or establish new regeneration.

Pests The current or potential presence of animals, insects, diseases, or competing vegetation should be evaluated and incorporated into the prescription. Mistletoe, *Fomes*, gophers, deer, bear, *Dendroctonus*, *Scolytus*, and sprouting hardwoods will all constrain the choice of treatment.

Managerial Constraints

Technical These include such factors as operational requirements or limitations of particular equipment, requirements of the management plan which may limit the allowable cut or specify particular procedures, and economic considerations.

Policy Operations may be constrained by policy decisions which prescribe particular practices on areas adjacent to highways, i.e., visually important areas, or other practices such as the Forest Service's policy on even-aged management.

Social Constraints

Legal These may be federal, such as the National Forest Management Act of 1976 which regulates the management of public lands; state, such as state forest practice legislation, fish and game requirements, pollution control, and especially taxes; or local, such as county ordinances which impose standards for forest practices within a county.

Social Pressure This develops through the activities of conservation groups, hunting lobbies, recreation groups, summer-home development, and concerned activists.

Ecological, managerial, and social forces thus provide the framework, either through incentives or constraints, within which all silvicultural recommendations must fit. Problems in the past which have created public and professional concern have developed largely because foresters have paid disproportionate and sometimes insufficient attention to the managerial aspects of decision making, as well as inadequate attention to the ecological and/or social components of the system. In other words, the evolution of changes in forestry practices has not kept pace with the rate of change in the social environment within which forestry operates. Examples of these problems are areas that have unexpectedly failed to regenerate satisfactorily, have developed high erosion characteristics, or have become visually unacceptable. Where these have occurred, legitimate concern has been expressed regarding the adequacy of current forest-management practices.

WHAT IS EXPECTED OF THE SILVICULTURIST?

This rapid increase in both internal and external pressures and incentives to enhance forest-management practices is making special demands on the silviculturist. In particular, there is a marked worldwide increase in the need, and expectation, for silviculturists to be able to accurately predict the likely consequences of their prescribed treatments. Surprisingly, little specific information is available on the likely biological and economic benefits or gains as opposed to the expected costs or losses resulting from various treatments. This kind of information is necessary for every proposed treatment for each species mix, age class, and site quality. Such information is essential if silvicultural decision making is to be put on a rational basis. Predictions of this kind, however, are difficult and can be made only after thorough analyses.

A final point concerns the manner in which land-management decisions are made. In the past, it was common for sequential treatments to be prescribed incrementally without integration or forward planning. Treatments were often prescribed by different people and at different times whenever the next treatment was thought due. Sound management cannot occur in this manner. What is needed is the development of a silvicultural prescription that integrates a sequence of treatments into a coordinated plan aimed at meeting a particular management objective. The prescription also justifies the treatment choices on specific ecological, managerial, and social grounds. Use of this systems approach enables the silviculturist to predict the likely treatment outcomes and minimizes the possibility of undesirable surprises. This approach of developing stand prescriptions has been put into practice in some regions of the United States through the Forest Service's program of silviculturist certification and will contribute substantially to enhancing forest-management practices.

The outlook for the further development of sound silvicultural practices has never been brighter. Incentives from the manager and general public to increase overall wildland productivity while at the same time securing environmental quality will ensure that silviculturists become more proficient in controlling establishment, growth and yield, stand structure, and overall quality of the world's wildlands.

Chapter 3

Classification of Forests

With the diversity of the vegetational formations that exist on the earth and the increasingly available data on the environmental conditions under which each develops, there has been considerable effort devoted to understanding or establishing causal relations between plants and their environment in terms of temperature and moisture. In light of all the variables that influence the successful establishment of a plant, temperature is shown to be a powerful influence on plant distribution when zonations of vegetation based solely on temperature demonstrate such significant correlations. Each attempt at forest classification makes a contribution to our understanding of the plant-environment interaction, and each draws on the efforts of the pioneers until a workable system emerges. This chapter presents the current status of these alternative approaches.

MAJOR DIVISIONS OF THE WORLD'S FORESTS

It has been estimated that about 22 percent of the land area of the earth, exclusive of the polar regions, is forested (Zon and Sparhawk, 1923). This area, as indicated in the maps (Figs. 3-1 and 3-2), consists of three broad classes of

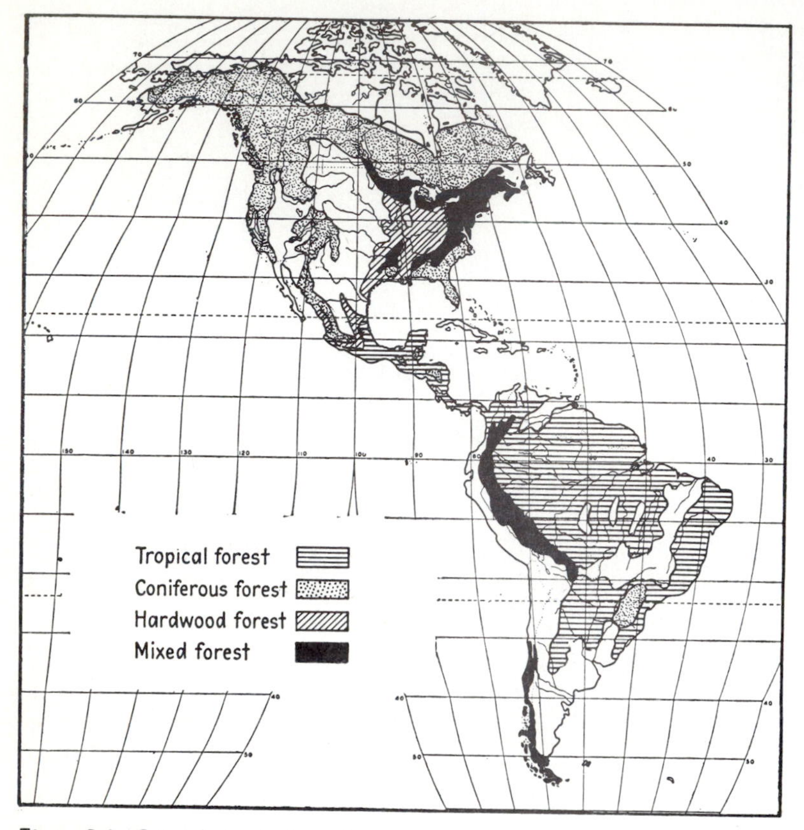

Figure 3-1 General distribution of forests upon the western continents. (*Redrawn from Zon and Sparhawk, 1923.*)

forest, two of which are frequently closely intermingled: (1) the tropical hardwood forest, which comprises approximately 50 percent of the entire forest area of the world; (2) the temperate hardwood forest (15 percent); and (3) the coniferous forest (35 percent). These broad classes are made up of many smaller elements which may be extremely variable in ecological character. A brief review of the more important of these, following the general classification of Schimper (1903), will show the complexity of the forests and their silvical characteristics.

The Tropical Hardwood Forest

The tropical hardwood forest (Beard, 1944; Burtt-Davey, 1938; Champion, 1936) ranges all the way from the dense forests of the jungle to thin, open thorn forests and savannas where groups of trees stand in the grassland depending on the amount and seasonal distribution of the rains. The *tropical rain forest* is the most highly developed and complex of all these forms (Fig. 3-3). This is an

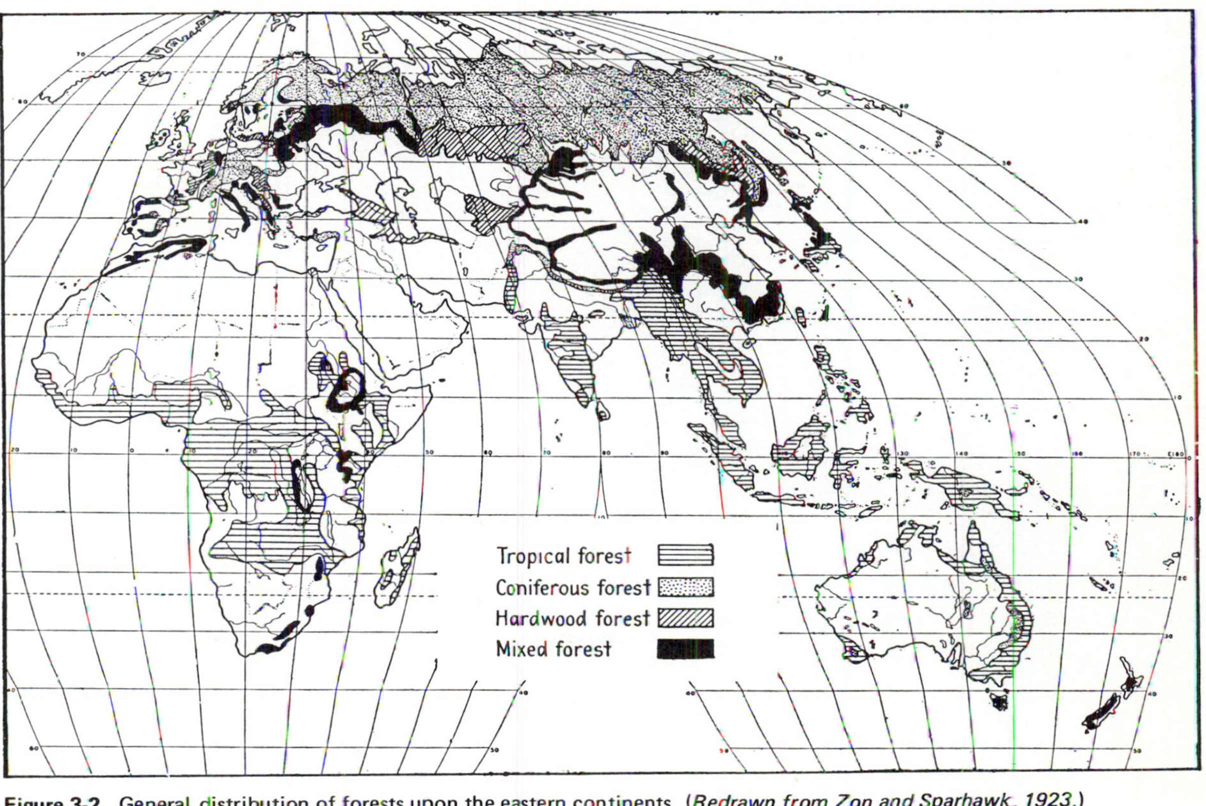

Figure 3-2 General distribution of forests upon the eastern continents. (*Redrawn from Zon and Sparhawk, 1923.*)

Figure 3-3 Profile diagram of a 60-m (200-ft) strip of tropical rain forest in southern Nigeria, showing all trees over 4.6 m (15 ft) in height. Fourteen species of trees appear in this transect. (*From Richards, 1939.*)

evergreen broad-leaved forest of majestic proportions and great density which is much less widely distributed than is popularly supposed. It occurs where temperatures do not fall below 5°C at any time during the year and where the annual rainfall is well distributed and amounts to at least 1800 to 2000 mm. Humidity is always high—usually 80 percent or more. The tropical rain forest is found chiefly along the coasts and rivers of equatorial Africa, in Burma, Thailand, Kalimantan, and East and West Irian, and in the Amazon and Orinoco basins of South America. The driest months are insufficient to affect the development of the forest to any degree.

The forest is composed of several crown layers, the chief canopy being, as a rule, the second layer, made up of tall, slender trees rising to heights of 30 to 45 m (100 to 150 ft), above which rise scattered "outstanding" trees of the uppermost layer—45 to 60 m (150 to 200 ft) tall (Fig. 3-3). The crowns are small and irregular with few branch orders, and the leaves are usually compound. The thin main canopy allows much greenish-tinged light to filter through, permitting the development of subsidiary layers of highly tolerant species with tall, slender stems and small crowns. Lianas are not particularly common, and the epiphytes which are so much a feature of the tropical forest are on such high branches that they are hardly noticeable. On the forest floor there may be small palms, creeping bamboos, ferns, or almost no vegetation. The forest has repeatedly reminded travelers of Gothic cathedral aisles lighted with a dim greenish light. The timber stands are not remarkably heavy, since many trees are below merchantable sizes. The stand of commercially important timbers is usually very light, since there may be as many as 80 species of trees per acre, only a few of which have any commercial value. Competition is very sharp; dead trees and debris decay very rapidly. The net growth of the forest is not remarkably high owing to the heavy mortality.[1] After cutting such a forest, reproduction is prompt, but it usually consists of very different species than comprised the bulk of the original forest.

In tropical mountains, the rain forest may give way to the *montane rain forest* which, owing perhaps to shallower soils and steeper slopes, is less dense and majestic than the true lowland rain forest. Still higher may appear the *cloud forest*, dripping and sodden nearly all the time and composed of small and poorly formed trees and enormous numbers of epiphytic plants. Along tropical coasts in brackish waters run the extensive *mangrove forests*, while inland in smaller ecological niches are other specialized forms.

While the rain forest is virtually unaffected by the dry season, *the seasonal forest*, or *monsoon forest* as it is often called (Schimper, 1903), is more or less modified by a drought period of sufficient length and intensity to impress itself upon the forest. Almost always some to nearly all of the species will be deciduous during the dry season. The forest is not so tall, hardly exceeding 30 m (100 ft) to the top of the main canopy; the crowns are somewhat more fully developed; and the stems are shorter and stouter than in the rain forest. The impression is often of huge trees in the midst of smaller growth rather than the columned aisles of the rain forest. The leaves are largely compound, particularly those of the upper canopy. Much light filters through, and as a result, there are more lianas and epiphytes than in the rain forest. Several crown layers are usually evident, but as the climate becomes drier, the layering becomes less complex, the trees are shorter, and the proportion of deciduous species increases. It finally often breaks into a *thorn forest* of small trees mostly belonging to the pea family (Leguminosae), which occupies large areas in Africa and India between the seasonal forests and the deserts. Where the rainfall is highly seasonal and the dry season windy, and perhaps where other factors are also involved (the matter is very controversial), the seasonal forests give way to *savanna* as the dryness increases. This transition between the forest and grassland formations is very widespread in Africa and consists typically of scattered trees or clumps of trees in grassland. Both savanna and thorn forests cover great areas but yield few important forest products.

[1] An annual increment of 4.0 m^3 ha^{-1} (57 ft^3 per acre) per year has been computed for the dipterocarp forests of the Philippines (Brown and Matthews, 1914); it is an amount that is no greater than is produced by fully stocked stands of most American species on average and better sites. In 100 years, teak in India is as tall as Douglas-fir on good sites, and the tree diameters are greater, but since there are only about one-fourth as many living trees per hectare, the volume of timber is only about half as great. At the same time, a sprout stand of ipil-ipil in the Philippines (Matthews, 1914) has been found to produce as much as 42 m^3 ha^{-1} (600 ft^3 per acre) per year at 5 years of age. This is about double the maximum for our most productive forests.

The monsoon forest is very widespread and variable—depending on the character of the rainfall climate—and is frequently subdivided by ecologists on the basis of its "evergreenness." It is the source of many valuable tropical hardwoods and, in forestry, is noted for the celebrated forests of teak in India, which have long been under formal forest management. In the monsoon forest it is easy to get reproduction after logging, but it is very hard to control species composition. For example, teak regenerates easily after clearcutting from the hard-shelled nuts that lie dormant on the forest floor, but the regeneration is accompanied by so much competing woody vegetation that it is considered wiser to clear the whole area and plant the teak in lines so that it can be cultivated (Troup, 1928).

These forests occupy regions where the temperatures never fall below 5°C and the mean annual temperature is around 27°C. Rainfall may be as low as 1000 mm per year and as high as 4000 mm, but there is always a well-defined dry season during which the rainfall will not exceed 50 mm per month for 3 to 8 months.

The tropical rain forest and the seasonal forest together make up nearly half the forest area of the world. Silviculturally they are best known in India and Burma, but their exploitation in the many tropical countries is expanding their silvicultural base. Their problems under management center around fire, competition between valuable and weed species, tree diseases, biotic factors, and the problems of soil fertility in consequence of heavy leaching and overrapid decay of organic materials.

The Temperate Hardwood Forest

As far south as the hill region of India, North Africa, and the mountains of Central America appear the southern fringes of the temperate hardwood forest. This differs from the tropical hardwood forest in having typically one well-defined canopy level, winter leaf fall, and few lianas and epiphytes (see Fig. 3-4). The leaves are generally simple, more or less oval, and of a bright green color.

Figure 3-4 Profile diagram of a 60-m (200-ft) strip of northern hardwood forest composed of red oak, red maple, black birch, yellow birch, beech, and white ash.

The family Fagaceae is prominent, with oaks, chestnuts, and beeches in the Northern Hemisphere and the South American "beeches" (*Nothofagus*), *Eucalyptus*, and *Acacia* below the equator. In this forest the crowns are generally well developed, and mature trees reach heights of from 23 to 30 m (75 to 100 ft). Branches are often stout, and there are from six to eight branch orders. In composition this forest is relatively clannish, with only a few species per hectare, although, on the other hand, pure stands are rare. In the mature forest the stands appear open and empty compared to the tropical forest because of the absence of intermediate crown levels and lianas. Long, clear stems are rare except in the best of the managed forests or in high-class virgin stands, for in understocked stands the branching tends to be heavy and the stems may also be crooked. The net growth rate is good and the yields per hectare are fair, especially where the stand is composed chiefly of merchantable species.

The hardwood forest is characteristic of the warmer portion of the temperate zone having well-distributed rainfall in excess of about 600 mm per year. It is quite sensitive to differences in soil quality and does best on rich, well-drained, deep soils. While the hardwood forest may occupy areas of poor and shallow soils within its native climatic province, the trees are likely to be very short, branchy, slow growing, and of little commercial importance. As a result, many areas once covered with the best type of hardwood forest have been cleared for agriculture, and the remnants are in small areas incidental to farm holdings. The largest remaining areas exist in the Eastern and Central United States, Central Europe, parts of the U.S.S.R., and the more remote sections of China.

A rather distinct form of the hardwood forest is the *broad-leaved sclerophyll forest*, or Mediterranean forest, which occupies regions of moderate annual rainfall having well-defined summer droughts and winters that are wet and not very cold. These forests tend to be irregular and open, with low-crowned, branchy trees whose leaves are predominantly small, simple, hard or leathery, and often shiny on the upper surface. There is often an undergrowth of shrubby species having similar sclerophyllous (leathery) leaves. In very dry places and on thin soils, or after forest destruction, these shrubs may replace the forests for a long period of time. As a forest formation, this Mediterranean forest is of no great importance, but it contains a great number of well-known species because our civilization developed in such a region. The olives, evergreen oaks, laurel, and shrubs used for ornamental planting are typical of this forest. It occurs in the region for which it is named and also on the South Pacific Coast of the United States, in Chile, southern Africa, and parts of Australia. There are also a few conifers characteristically mingled with these hardwoods; in the Northern Hemisphere they are usually pines, junipers, or the true cypresses (*Cupressus*).

Silviculturally, the temperate hardwood forest is rather easily reproduced by sprouts or seeds and it makes a tough, persistent forest. Intensive management in some areas is difficult, for it can be hard to get desirable species composition without herbicides, intermediate cuts, or planting.

The Coniferous Forest

As far as usefulness is concerned and as far as forestry is involved, the coniferous forest is the most important of all forest formations. Although it occupies only about one-third the total forested area of the earth, it has always been of outstanding importance because of the nature of the wood, the form of the forest, and its accessibility to modern civilizations.

As shown by Fig. 3-5, a fully developed coniferous forest consists of a stand of characteristically tall, straight boles topped with relatively small crowns whose tops are often at a level of about 30 m (100 ft) above ground. The canopy is dense and even, usually single-storied (unless of several age classes), and the thick, dark needles make an almost unbroken shade within the forest. As a result of the darkness and the moderate rainfall, the forest floor is commonly clean and strewn with needles. The coniferous forest consists of pure stands or mixtures of a relatively small number of species, all of which are generally merchantable. While the rates of growth of the individual trees are not phenomenal, there are so many stems to the hectare and there is so little mortality that net increment is excellent. Thus these forests are the most productive of the earth in terms of sawlog volumes.

Coniferous forests are typical of the colder parts of the north temperate zone, running northward to latitudes where the mean July temperatures fall to about 10°C. These forests are poorly represented south of the equator but are widespread in the continental climates of the Northern Hemisphere, especially

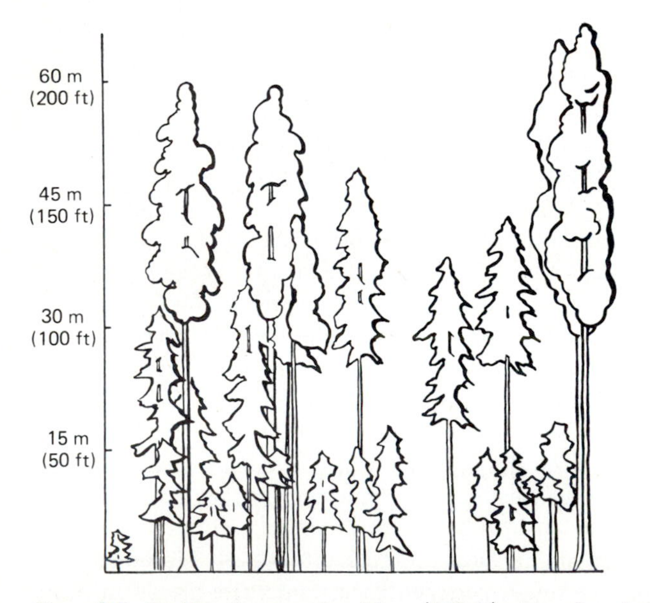

Figure 3-5 Profile diagram of a 60-m (200-ft) strip of a virgin uneven-aged coniferous forest of the Sitka-spruce–western-hemlock type near Quinault, Washington.

where winters are cold. They occur in Central and Northern Europe and sweep east through Siberia and the higher mountains of Asia to the Pacific Ocean; in North America, they dominate the West, wherever rainfall is adequate, and occur across Canada and down the eastern mountains and coastal plains.

Owing to the location of these forests and the heavy use of their timbers by civilizations, forest management has developed in these regions. Their silvics are the best known, their management problems have been under study for over a century, and most silvicultural textbooks automatically concentrate on this particular kind of forest. Growing in regions where climate is not too favorable or competition too serious, these forests ordinarily face problems of reproduction and maintenance of a rapid growth of high-class trees.

FOREST REGIONS AND CLIMATIC PROVINCES

A forester recognizes that the broad climax formations are associated with climatic factors and that temperature and precipitation are primarily involved. Various attempts at correlating natural vegetation zones with climate have used temperatures and moisture variations. Each has shown some interesting delineations, but most have lacked the capacity to define the boundaries of the various vegetational provinces.

Merriam's Life Zones

Merriam's (1898) classification of life zones was the first developed in the United States. It was based on temperature summations above 6°C for the colder boundary of each zone and the sum of the means for the warmest 6 weeks of the year for the warmer boundary. He characterized each of his zones by the plants and animals that commonly occupy it. His zones are not well integrated with some of the forest regions of the East (Fig. 3-6), but in the West the relationship is much clearer, especially in the areas where he worked—the Southwest and California. Some of the reasons for the sharper definition of life zones in the West are rapid changes in temperature over short distances (because of the steep topography) and the excellent correlation between decreasing temperature summations and increasing rainfall. The West was made up of arid zones, except the Pacific transition, while the East had humid zones. His summations of temperatures are no longer of interest, but his associated plants and animals in the West define the boundaries of the zones as follows:

Lower Sonoran The true desert of the Southwest with cacti, creosote bush, and other characteristic plants. The equivalent zone in the East is Austroriparian.

Upper Sonoran The zone of sagebrush, piñon pines, and desert junipers; and in California, digger pine, several species of oak, and chaparral. Equivalent zone in the East is Carolinian.

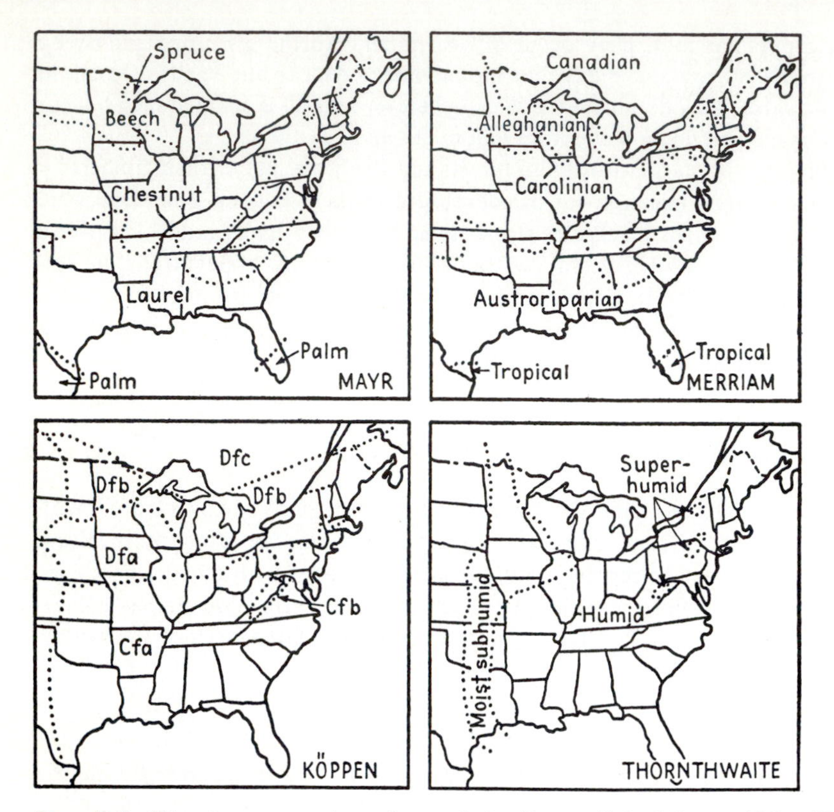

Figure 3-6 Climatic zones and provinces of the Eastern United States. *(After Merriam, 1898; Mayr, 1909; Köppen, 1923; Thornthwaite, 1931.)*

Transition (arid) The zone of ponderosa pine primarily. Pacific transition (humid)—the Douglas-fir region of the North Pacific Coast. Equivalent zone in the East is Alleghenian.

Canadian Douglas-fir in the Rocky Mountains and lodgepole pine; in northern Idaho, western white pine; on the Pacific Coast, noble fir and red fir. Equivalent zone in the East is Canadian.

Hudsonian High-altitude pines, spruces, mountain hemlock, alpine fir, and mountain larch. Equivalent zone in the East is Hudsonian.

Mayr's Zonal Classification

Mayr's (1909) classification was the first that treated forests in any detail, and it divided the Northern Hemisphere into six narrow belts defined by the mean temperature of the 4 months, May, June, July, and August. Each zone was named after a characteristic tree. It has not been utilized in this country, but it is recognized as useful in Italy and other Mediterranean countries where the zones are particularly meaningful. Its application to Eastern United States is illustrated in Fig. 3-6. Mayr's zones and the associated trees are given in Table 3-1.

Table 3-1 Mayr's Zonal Classification of Forests with Zone Names and Their Temperature Boundaries

Zone	Temperature, °C (4 mo)	Characteristic regions and trees
Palm	over 26°	Tropics with a wealth of hardwood species, palms
Laurel	22 to 26°	Mediterranean region, Gulf states: live oaks, laurel, and other Lauraceae, as California bay, sassafras; hard pines on poorer soils.
Chestnut	18 to 22°	Central Japan, China, Europe, and North America: deciduous oaks, chestnuts, ash, and other hardwoods.
Beech	14 to 18°	North Japan, Manchuria, Northern Europe, and the Lake states and Northeast in America: beech, maples, pines (both hard and soft) on poorer soils, some species of birch, and northern oaks.
Spruce-fir	10 to 14°	Siberia, Northern Russia, Scandinavia, Canada: spruces, firs and larches, northern birches, and poplars of the aspen group. The limit of forest is at about the line 10°C summer mean temperature both poleward and altitudinally in the mountains.
Alpine-polar	6 to 10°	Scattered stunted trees on the edge of the tundra or high in the mountains: Asia, Europe, North America: scrub willows and birches, spruces, larch, and high-elevation white pines.

Source: Mayr, 1909.

Köppen's Climatic Provinces

Köppen (1923) departed from a simple reliance on temperature and developed a complex plan delimiting the boundaries of his provinces. To simplify his method for illustration, the temperature zones were from *A* (warmest) to *E* (coldest), with zones *C* and *D* in Eastern United States (Fig. 3-6). Then there was a subdivision based on "winter dry," "summer dry," or "moist," with the *f*, or moist, subdivision covering all of Eastern United States. Then a third division was based on summer warmth, with *a* warmest, *b* moderate, and *c* coolest. The West had a very intricate pattern because of the desert to alpine conditions found in so many localities, as illustrated by the 11 provinces found in California (Russell, 1926).

Thornthwaite's Classification

The use of index numbers derived from combinations of temperature and rainfall (or humidity) data had been explored for a long time by European workers. Thornthwaite (1931), in a similar effort, devised a complex formula for recognizing five primary climatic regions by their P/E (precipitation/evaporation) indices. His five regions are a series from arid to superhumid, which places the East almost entirely in the one category "humid" (Fig. 3-6) and provides the poorest means of distinguishing the forest provinces.

Holdridge's Life-Zone System

Holdridge's system of ecological analysis was developed in the tropics, but it provides a basis for a comparative analysis of environments that is applicable globally. The system assumes the macroclimate as the primary and independent regulator of the earth's ecosystems; i.e., the effects of terrain, the soil complexes, and the plant communities with their fauna component are fundamentally subordinate to the macroclimate. Thus the structure, life forms, and growth habits of a plant community are considered to reflect the climate prevailing in its locality; as a corollary, it is postulated that the life zones may be derived from an analysis of the vegetation as well as from basic climatic data. The system divides the earth's macroclimates into over 100 ecologically equivalent units. Each unit represents one sector of the climatic continuum and supports a distinctive *set* of plant associations. The variety of associations will depend on the soil and topographic conditions within the climatic unit.

The system has three levels of classification: the climatically defined life zones which form the primary level, the subdivision of each life zone into associations which are differentiated on the basis of local environmental conditions, and finally, the subdivision of the associations on the basis of actual vegetation cover or land use (Holdridge et al., 1971). Holdridge defines *association* as "a unique ecosystem, i.e., a distinctive habitat or physical environment complete with an evolved naturally adapted community of plants and animals." Holdridge's system has four basic types of associations: climatic, edaphic, hydric, and atmospheric. Thus a climatic association is attuned to the macroclimate of the life zone and grows on a zonal soil, so there can be only one such association in each life zone. Holdridge uses the climatic association for naming the life zone.

The three-dimensional model that relates the life zone to the critical climatic factors is presented in Fig. 3-7. Holdridge's critical factors are biotemperature, total precipitation, and humidity, where the humidity province is determined by the potential evapotranspiration ratio. A unique feature of Holdridge's model is the logarithmic scale to the base 2 on which the climatic data are plotted. It recognizes the implication of Mitscherlich's law of the minimum, where a small increment of *any* factor at a critical level has a disproportionately large effect as compared to the effect of the same increment when the factor is less critical. It is the logarithmic scale as well as the choice of factors which gives the system its global applicability (Costa Rica and Thailand: Holdridge et al., 1971; Mediterranean and North Africa: Steila, 1966; Colorado: Thompson, 1966; and Eastern United States: Sawyer and Lindsey, 1964). Biotemperature for a station is the

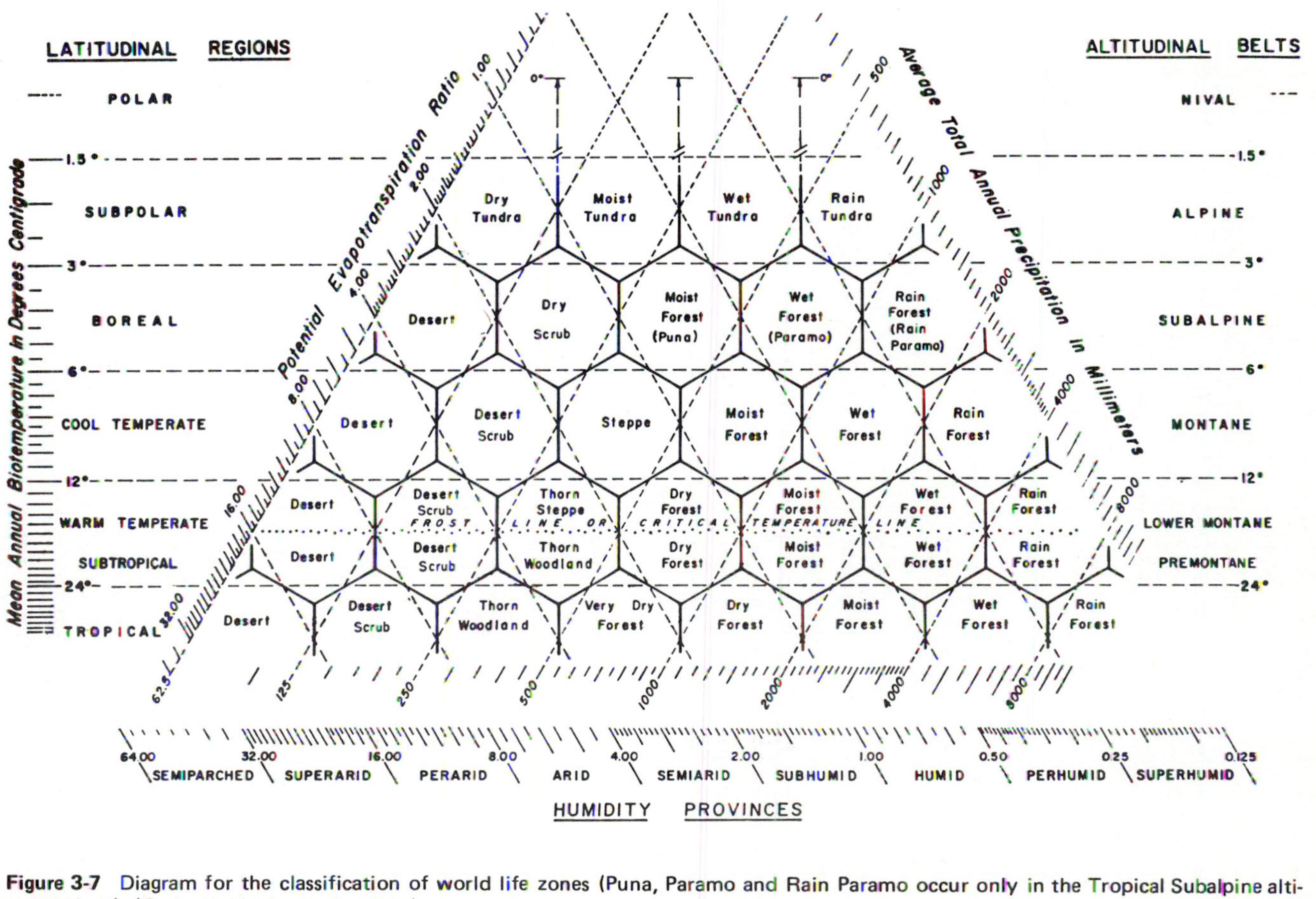

Figure 3-7 Diagram for the classification of world life zones (Puna, Paramo and Rain Paramo occur only in the Tropical Subalpine altitudinal belt). (*From Holdridge et al., 1971.*)

mean positive annual temperature, which is determined by the summation of the mean positive daily temperatures (i.e., if min = -12°C and max = +8°C, then mean positive temperature for the day is 4°C) divided by 365 (Tosi, 1964). Temperatures below 0°C are assumed to have little effect because plants are dormant. The exception is the division of the latitudinal region or altitudinal belt 12 to 24°C into two parts based on the occurrence of a killing frost within 2 or 3 years. Mean annual precipitation for a station is expressed in millimeters. The potential evapotranspiration ratio is the potential evapotranspiration expressed in equivalent millimeters divided by mean annual precipitation (E/P ratio). Holdridge derives the potential evapotranspiration by multiplying the biotemperature by 58.93—a figure obtained experimentally in the process of developing his first model (Holdridge, 1947).

Biotemperature determines the latitudinal region or altitudinal belt of the life zone, not the latitude in degrees or the altitude in meters. In the Tropical region, the full range of altitudinal belts may exist, so an Alpine belt can be found at the Equator at the higher elevations of the Andes Mountains. In the Boreal region, the potential number of altitudinal belts have been reduced to three (Fig. 3-7). The mean annual biotemperature establishes the altitudinal belt, but the latitudinal region is determined by correcting the altitudinal biotemperature to the corresponding sea-level biotemperature. Thus, if the mean annual temperature is 2.9°C at the top of the Andes in Ecuador, the station would be in the Alpine belt, but a sea-level mean annual biotemperature 24.5°C places the station in the Tropical region. If the rainfall at the station is 900 mm, then the life zone would be the Tropical Alpine Rain Tundra or, expressed in three dimensions, the Tropical Alpine Superhumid Tundra. Where the mean annual biotemperature's altitudinal belt is the same as the latitudinal region, then the latitudinal region alone is used; i.e., a station with a mean annual biotemperature 7°C and 700 mm of precipitation is in the Cool Temperature Moist Forest Life Zone if the sea-level biotemperature is 10.5°C.

In applying Holdridge's system to Eastern United States, Sawyer and Lindsey (1964) mapped the life zones illustrated in Fig. 3-8. There are 11 of the life zones shown on the map, but others were also present, such as Cool Temperate Alpine Semisaturated Tundra on Mount Washington in New Hampshire. (The high rainfall gives a life zone not shown in Fig. 3-7 but indicated to the right of Cool Temperate Alpine Rain Tundra Life Zone.) Sawyer and Lindsey, in their discussion of how well Holdridge's system was able to delimit the vegetation regions, point out that a comparison with vegetational maps would not necessarily make a valid test of Holdridge's accuracy in defining bioclimatic formations since the maps differ among themselves in theoretical and classification bases (Bailey, 1976; Küchler, 1964; Braun, 1950).

Lindsey and Sawyer (1971), in expanding their work on the application of Holdridge's system to Eastern United States, used Braun's (1950) map of the forest regions. The weather stations used in the distribution map of Holdridge's

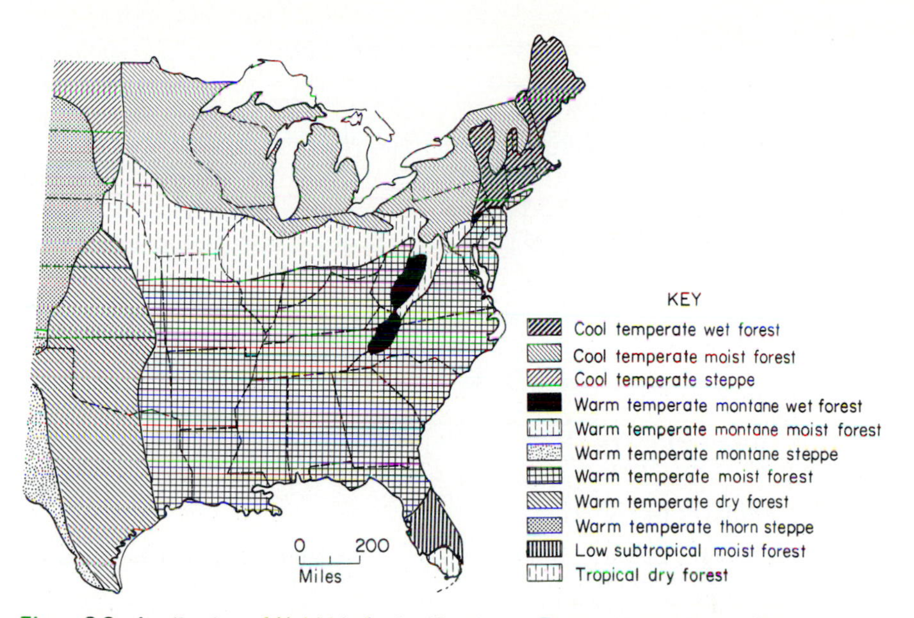

Figure 3-8 Application of Holdridge's classification to Eastern United States. (*From Sawyer and Lindsey, 1964.*)

life zones (Fig. 3-8) were classified according to their location within eight of the forest regions recognized by Braun. (The oak-hickory type, the most westerly region of hardwoods extending from Minnesota to Texas, was omitted because its western boundary was not clearly determinable.) The values of each station were plotted on Holdridge's three-axis graph, and the area of concentration of the points for each forest region is shown in Fig. 3-9. The central location of Braun's mixed-mesophytic type (MM), from which, she proposes, all the other forest types of her Deciduous Forest Formation have been derived, indicates that it is climatically as well as geographically central. The points corresponding to the oak-pine region had no area of concentration, and its diffused dot pattern had little graph area not shared by some other forest region, principally by SE, WM, and OC (see Fig. 3-9).

The mean climatic values for the various Braun forest regions are given in Table 3-2. The relative north-south position of the forest regions corresponds rather well with the increasing mean annual biotemperatures. However, while the actual amount of precipitation tends to increase from north to south, the effective moisture is greater for the hemlock-hardwoods region than for the southeastern-evergreen region.

The realistic stratification of Braun's forest regions (Fig. 3-9) indicates how well Holdridge's life-zone model corresponds to variations in biological responses, although actual testing of the biological aspects of the model against external

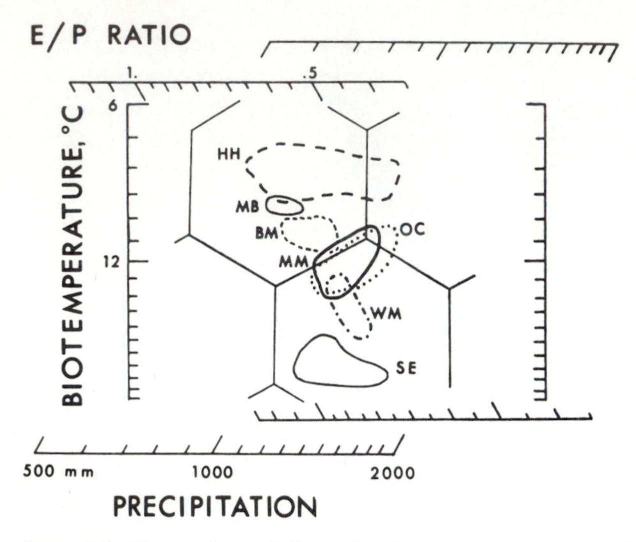

Figure 3-9 Three-axis graph from plotting the most concentrated 80 percent of the weather station data in each forest region to show the most typical climatic conditions in each region. The symbols are: HH, hemlock-hardwoods; MB, maple-basswood; BM, beech-maple; MM, mixed mesophytic; OC, oak-chestnut; WM, western mesophytic; and SE, southeastern evergreen. The upper left hexagon delimits the Cool Temperate Moist Forest of Holdridge, the upper right shows the Cool Temperate Wet Forest, and the lower hexagon depicts the Warm Temperate Moist Forest. (*From Lindsey and Sawyer, 1971.*)

criteria is weak for lack of such external criteria. In terms of the physical data used by Holdridge, the total annual precipitation and mean annual biotemperature are available in absolute form from any weather station, but the calculation of the potential evapotranspiration ratio depends on the estimate of the potential

Table 3-2 Mean Climatic Values for Vegetation Regions of the Eastern United States

(Values Are Listed in Descending Order of Effective Moisture from Top to Bottom of the Last Column, which Gives Potential Evapotranspiration/Precipitation Ratio, the Inverse of Effective Moisture)

	Biotemperature, °C	Precipitation, mm	E/P ratio
Hemlock-hardwoods	8.5	899	0.556
Oak-chestnut	12.0	1138	0.622
Mixed mesophytic	12.2	1117	0.642
Beech-maple	10.7	889	0.706
Western mesophytic	14.3	1186	0.710
Maple-basswood	9.5	739	0.756
Oak-pine	16.3	1196	0.802
Southeastern evergreen	18.9	1297	0.859

Source: Lindsey and Sawyer, 1971.

evapotranspiration, for which the different formulas vary greatly in their accuracy. Thornthwaite's formula, which serves as a "standard" in the United States, is based solely on temperature, with a complex correction applied (also derived from the temperature) that varies with locality (i.e., no correction for variation in radiation, humidity, or wind). Holdridge has also based his estimate of the evapotranspiration on temperature, but he applies a constant to the mean annual biotemperature. In comparing Thornthwaite's and Holdridge's estimates of evapotranspiration for 300 weather stations scattered over the 50 states, Lindsey (1978) finds, on average, only a 1.65 percent lower estimate and a somewhat larger coefficient of variability by Holdridge as compared with Thornthwaite. In northern stations, Holdridge underestimates the values derived from Thornthwaite, and in southern stations, he overestimates them. In view of the lack of data to provide more accurate estimates of evapotranspiration, Holdridge presents a simple, realistic estimate of potential evapotranspiration.

LAND CLASSIFICATION FOR MANAGEMENT INFORMATION SYSTEMS

Classifications have the purpose of organizing information to render it easily available for answering questions or solving certain problems. Each classification has a purpose for which it was designed. This purpose was frequently served by choosing a classification with management implications applicable to a certain location. Other classifications used some integrated or partially integrated system like land type, ecoregion, or biogeocoenose. However, Holdridge's classification provides for recognition of life zones by use of component factors, and it makes no pretense of being able to describe the boundaries of particular cover types or even formations in an area. While he uses components of the climate that are independent of value judgments, the question his classification answers is too narrow to help in management decisions except in a broad land-use context.

Among methods that have been developed to serve as frameworks for information systems for use by managers of terrestrial wildlands and water resources is a successful one called *ECOSYM* (Davis and Henderson, 1976). ECOSYM is a classification and information system designed to help answer wildland-management questions involving allocation of resources considering physical, biotic, social, and economic potential of an area. As a result of 2 years of testing, it will be applied to the management of three national forests and the Olympic National Park. In order to be useful on a local, state, or national level, the classification system stratifies ecosystems into classes about which meaningful generalizations can be made, allows for aggregation and disaggregation of information from local to national levels, and satisfies the preceding two criteria in a cost-effective manner. The characteristics of the classifications used are:

1 They are based on components of the ecosystem such as vegetation, soil, and climate.

2 They are hierarchical in their structure, i.e., for each component there are different levels of generalization and resolution. This means that each level in the hierarchy is broken down into finer and mutually exclusive subsets. Thus in soils classified in soil taxonomy (Soil Survey Staff, 1975) the hierarchy goes from bottom to top: series, family, group, suborder, and order. If the series is known, then the whole hierarchy up to the order is known.

3 They are objective, i.e., the delineation of classes at any level in the hierarchy is based on objective and quantifiable criteria where possible.

The major advantage of a component classification that includes the various physical, biological, social, and economic factors is that it allows for greater flexibility and a more universal applicability. It does not restrict a user to any particular component or set of components. The manager has the freedom to choose what works best in the particular area. The component system allows integration and interpretations by virtue of the number of combinations and permutations of components. Thus it has many more "types" and more closely approaches the real-world-continuum situation.

FOREST REGIONS OF THE UNITED STATES

There is no standard classification of the forest regions of the United States, except as administrative divisions have certain regional designations. Küchler (1964) developed a map of the potential natural vegetation of the United States which incorporates some refinements in classification by indicating climax and major seral species. Shantz and Zon's (1924) map makes no attempt to fit any ecological concept but represents the vegetation that was on the ground when the first settlers saw it. It also represents the potential uses under today's management. The map (Fig. 3-10) shows an area as though it were totally covered by a forest of limited composition, whereas most areas have not only a mosaic of various forest cover types but also a great deal of admixture of agricultural lands and cities. The area and proportion of each forest region described by Shantz and Zon for the 48 states are given in Table 3-3.

Forest Cover Types of the United States

Forest cover types are not adaptable to a "natural" system of classification like the Linnean system in botany. They belong to a group that requires an "artificial" system of classification. The types are not separate and discrete entities either, they flow and merge into each other in many cases like the colors of the spectrum. Sometimes, it is true, the types are very clear-cut and are readily recognized–such as the pure forests of lodgepole pine that come up after a fire or the similar aspen and birch forests of the East. But in the variable mixture of hardwoods in the Mississippi delta or in the complex coniferous forest of the Pacific Northwest, the situation is not so easy to classify. In such situations the forester will see patterns well enough, but different people will see different

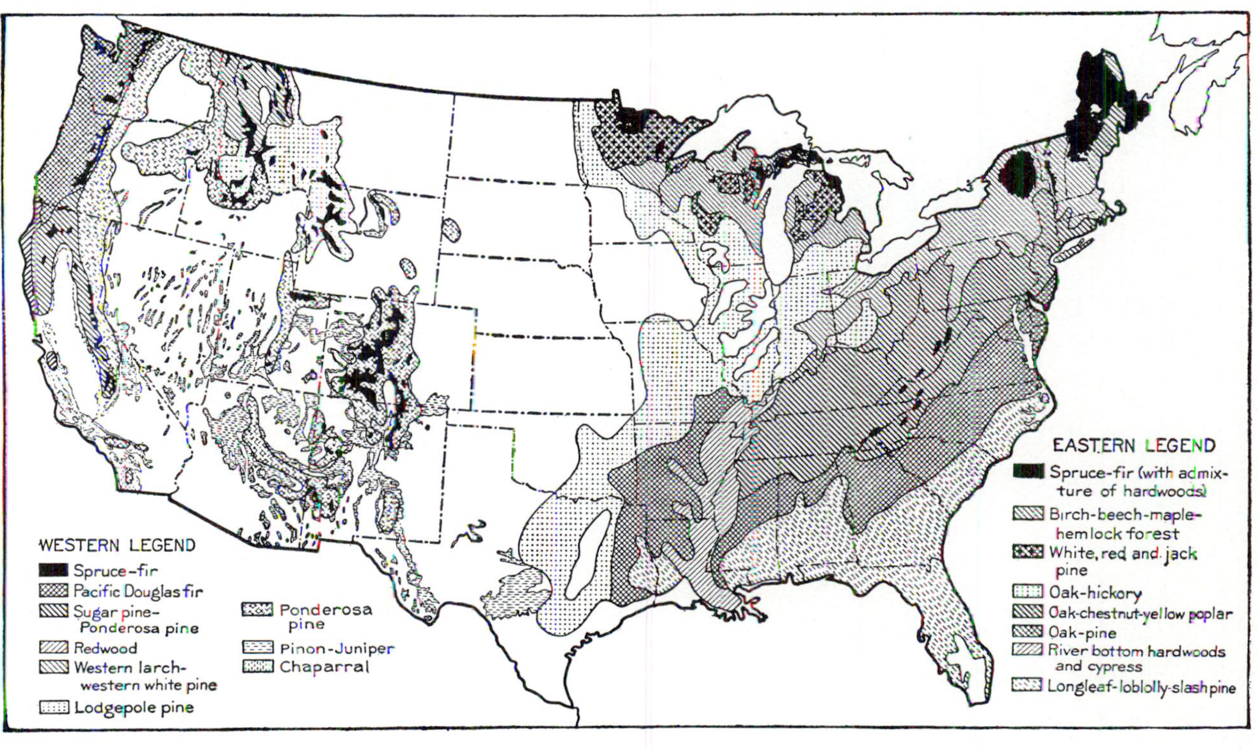

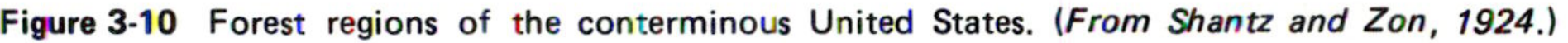

Figure 3-10 Forest regions of the conterminous United States. *(From Shantz and Zon, 1924.)*

Table 3-3 Forest Regions of the Conterminous United States

Region	Area, km^2	Percent total forest	Character
Coniferous forest:			
Spruce-fir	152,130	5.67	Southern fringe of circumpolar spruce-fir forest that covers great areas in Siberia, Northern Europe, and Canada
White, red, jack pine	63,860	2.38	Occupies immature sandy soils along southern edge of the spruce-fir forest region in Lake states and Northeast
Longleaf–loblolly–slash pine–shortleaf	228,630	8.52	Generally subclimax forest of four "southern" pines on sandy soils or where hardwoods are held in check by fire or grazing
Western larch–western white pine	29,240	1.09	Area marked by forests of western white pine and larch in which fire is a very important ecological factor.
Lodgepole pine	73,590	2.74	Region marked by forests of lodgepole pine, largely of fire origin, occupying a climatic zone somewhat warmer and drier than spruce fir
Ponderosa pine	223,640	8.33	Climax forests in which ponderosa pine or Douglas-fir predominate. The lowest of the timber types in the interior western mountains
Piñon–juniper	231,270	8.62	An open woodland type fringing the deserts of the interior West, developed mainly on immature soils
Pacific Douglas-fir	75,540	2.81	Forest occupying a humid region of mild winters in the Pacific Northwest, composition greatly affected by fire history
Redwood	7,110	0.26	A climax forest region marked by a humid coastal climate and much summer fog
Sugar pine–ponderosa pine	49,900	1.86	A region marked by climax forest of ponderosa pine, sugar pine with Douglas-fir, white fir, and incense cedar in a region of wet winters and dry summers
Hardwood forest:			
Birch–beech–maple–hemlock	227,970	8.49	Region of cool summers in Lake states, Northeast, and Appalachian Mountains; a climax forest of the named species

Table 3-3 Forest Regions of the Conterminous United States (*Continued*)

Region	Area, km²	Percent total forest	Character
Oak-chestnut-yellow poplar	341,540	12.73	Climax forest of variable composition occupying humid, moderately warm, usually hilly or mountainous country in East
Oak-hickory	535,430	19.96	Climax forest occupying region of moderate rainfall and usually giving way to grass types on deep, fine-textured, mature soils on level topography
Conifer-hardwood forest:			
Oak-pine	277,530	10.34	Characteristic of Piedmont and hilly rolling lands of the Southeast, a variable mixture in which the pine is a persistent element if not a dominating species
Cypress-tupelo	120,020	4.47	Forest of bottom lands subject to flooding and swamps of lower Mississippi and the Southeast
Chaparral	44,870	1.67	Brushland or woodland of sclerophyll (hard leaved) broad-leaved species generally greatly modified by fire but showing evidences of conifer-hardwood climax under long protection

Source: Shantz and Zon, 1924.

patterns. One may consider the whole broad mixture as the "type"; others may break the complex down into more restricted mixtures and may even choose to recognize small pure types as locally significant. There is no basic reason why one viewpoint is more "correct" than another here. As a result, there can be no single "perfect" classification of types. There can be more useful and less useful classifications, however, and it is clear that it is highly desirable to have an accepted standard classification based on a uniform philosophy and system of nomenclature and an equal degree of "splitting" in all parts of the country.

The Society of American Foresters published a forest cover type classification for North America (exclusive of Mexico) in 1954. The committees (one for the East and one for the West) limited their classification of cover types to those which met their three qualifications: (1) the cover type had to be distinctive and easily separated from other cover types that closely resemble it; (2) it had to occupy an aggregate of hundreds of thousands of acres but did not have to cover any single large area in a solid stand; and (3) any forest cover that met the two previous conditions had to be named. In naming the types, no name longer than a trinomial was used, though holding to the trinomial limit required the occasional usage of generic or general designations. The name was based on species predominance in the type, in that those species named had to constitute more than 50 percent of the dominant and codominant trees. Species listed in the name are in order of their importance or indicator value within the type. Thus the yellow birch-red spruce type has more yellow birch than red spruce among the dominants and codominants of a stand. An indicator species had to be characteristic and indicative of the type but did not have to predominate such as with western white pine, redwood, and Port Orford cedar in the West.

The classification takes the middle ground between the minute divisions of an area for research purposes and the broad groupings of inaccessible areas for administration purposes. The cover types are listed so that they can be expanded or contracted to fit the needs of various uses. There are 106 named forest cover types in eastern North America and 50 in the West, exclusive of Mexico. The East is divided into four forest regions: Boreal, Northern, Central, and Southern; and each region has forest cover types associated with dry, fresh to moist, or wet sites. The low relief gave meaning to a temperature gradient as a basis for designating the regions.

The West, with its rugged topography and high plateaus, made a regional division on temperature unsuitable, so the regions were designated partly by geography and partly by elevation. The regions are Northern Interior; High Elevations in the Mountains; Middle Elevations, Interior; North Pacific; Low Elevations, Interior; and South Pacific, except for the high mountains.

In the East hardwoods form a part of 72 forest cover types and conifers a part of 64, while in the West there are 10 hardwood forest cover types and 40 conifer cover types.

Chapter 4

Forest Composition, Stand Structure, and Tree Classification

To the uninitiated, forests seem to be nothing but wide expanses of trees. However, the closer forests are examined, the more they differ, and their differences may be expressed in a great number of ways, depending on the viewpoint that is adopted. A *forest* is defined as an ecosystem characterized by more or less dense and extensive tree cover (Ford-Robinson, 1971). Dengler (1944) pointed out that a collection of trees becomes a forest only when it is sufficiently dense and covers a large enough area to develop a set of local climatic and ecological conditions that are distinct from those outside. There must be significant change in temperature, moisture, light, wind, humidity, flora, and fauna, as well as the character of the upper soil layers.

Forests and units within forests can be divided into groups in a number of ways:

1 Origin Forests that develop from seed are classically called *high forests*, while those which reproduce vegetatively by sprouts or suckers are called *coppice forests.* A forest can be composed of a mixture of coppice and *standards*, where standards are trees of seed origin or sprouts allowed to grow for two or more coppice rotations and are usually too large to efficiently sprout when cut.

This is called a *coppice-with-standards forest.* All coniferous forests are high forests, though rare exceptions may occur in a strong sprouter such as the coast redwood. Hardwood forests, on the other hand, usually contain many trees of sprout origin; the proportion may be especially high in cutover or burned lands.

2 **Age** Forests are differentiated by age into old-growth and young-growth. *Old-growth* is primarily represented by the rapidly dwindling area of virgin forest, but it still constitutes the bulk of the harvesting in Western United States. *Young-growth* consists of the forests that after logging or fires are in many stages of development.

3 **Composition** Forests can be classed as consisting of *pure* or *mixed* species. Since truly pure stands are rare except in the West, where ponderosa pine, lodgepole, true fir, and aspen have large pure areas, an arbitrary 90 percent of one species has been selected as separating pure from mixed stands. Even with a 90 percent level for pure stands, most stands are mixtures of two or more species. Pure stands also occur in plantations or on special sites like a black-spruce bog.

4 **Stands** Stands are usually the management units which make up a forest. A *stand* can be defined as a reasonably homogeneous unit that can be clearly differentiated from surrounding stands by its age, composition, structure, site quality, or geography. There is no precise area implied by *stand*, and size may change as management intensifies. In addition, a stand's particular conditions may be repeated many times within a forest or working circle.

5 **Structure** *Structure* of a stand or forest refers to the distribution of age and/or diameter classes, and of crown classes.

6 **Age-Class Distribution** A forest can have an even-aged or uneven-aged structure. *Even-aged forests* are composed of stands in which all the trees in a stand are approximately the same age, but the component stands may be of all different ages. *Uneven-aged forests* have stands that are uneven-aged, where each stand may have at least three or four well-spaced age classes. A forest of northern hardwoods or ponderosa pine may have a mixture of stands, some even-aged and some uneven-aged.

7 **Working Circle** A *working circle* is the forested area needed to support the wood industry of a given community without transport from excessive distances. There is the implication in the use of *working circle* of the capability of maintaining an even flow of forest products through the community in perpetuity —in other words, a managed unit. A forest could be a working circle or made up of several working circles. Thus a National Forest composed of a million hectares could be divided into several working circles for management purposes or a big pulp mill could have its own working circle.

PURE AND MIXED STANDS

Stands are designated as *pure* if 90 percent or more of the dominant or codominant trees are of the same species. A stand may have an understory of other species without changing the pure designation. A small area within a stand might be pure, but it is the management unit as a whole which is critical.

Pure stands owe their existence to one or more of the following basic causes.

1 Climatic conditions are so severe that few tree species can exist in an area. Severe climates dominated by cold, such as the upper timberline in the mountains and areas near the arctic tundra, favor resistant species like spruce, whitebark pine, and limber pine, so they form pure stands under such conditions. At the lower timberline, lack of rainfall or unbalanced distribution of rainfall limits the number of species that can survive to such as ponderosa pine and junipers.

2 Edaphic conditions control the tree species that can survive. Only a few species, like bald cypress and water tupelo, are adapted to prolonged periods of inundation. Bogs, with their highly acidic organic soils and cold, wet conditions, restrict species to tamarack, black spruce, or white cedar. The high magnesium content of soils developed on serpentine rock favors incense cedar.

3 A catastrophe (fire, wind, insect epidemic, or flooding) occurs and only a few species are capable of rapid invasion of the site or of surviving the high exposure of an open site. Fire is the most prevalent catastrophe. Douglas-fir can invade from the side and lodgepole and jack pine have serotinous cones which remain closed on trees for many years and can provide seed after being opened by heat. Longleaf pine has the ability to endure fire as a mature tree or in the grass stage.

4 An aggressive, very tolerant species, such as hemlock, beech, sugar maple, or red fir, may be able to shade out every competitor even on a good site.

5 Plantations are usually made with single species because the interrelationships of the species are so complex and vary so from site to site that most planted mixtures result in the dominance of one species.

In forestry there are certain well-defined advantages of pure stands and equally marked advantages of mixed stands. In Europe there has been an argument as to the relative superiority of each form for over 200 years, and it will continue in North America because there are social and technical considerations basic to these alternatives.

The mixed stand has had many biological and economic advantages attributed to it, while the proponents of pure stands minimize the potential biological weaknesses and often exaggerate the economic advantages. In the United States, the multiplicity of valuable species and the tremendous range and variety of climates, site conditions, and natural catastrophes have produced great expanses of pure stands in primeval forests, and great effort is concentrated on maintaining these stands even against the consequences of normal succession. There is also an effort to expand the ranges of the best of the fast-growing species like red and slash pine. However, serious efforts of this kind in this country are less than 50 years old; we have yet to see the consequences of efforts to hold an area at a seral[1] level through several rotations for species whose ranges have been stretched. Of course, these potential consequences may never materialize because the intensity of silivicultural practices applied to most forest areas normally is not aimed at eradication of competing species. In addition, many of our species such as ponderosa and lodgepole pine which have maintained pure stands naturally for many generations do not appear to have created adverse effects on site quality. Their greatest hazards have been insect epidemics and diseases like dwarf mistletoe which have developed in overstocked, overmature, and unmanaged stands. By management, the insect and disease problems in these same species can be reduced, but management practices and direction of natural succession may well be opposed.

However, there is value in reviewing the differences between pure and mixed stands for the potential of such differences in making apparent or ameliorating problems in the woods. Each site dictates what can be done and each stand has its own problems. If mixed composition is considered desirable in even-aged stands, a real Pandora's box of problems opens: species composition, percent by species, single-tree mixing, small- or large-group mixing, ad infinitum. Almost any single species is prone to some natural weakness. A species may be shallow-rooted, very subject to snow break or damage by a certain insect or disease, or its litter may decompose slowly. This Achilles' heel of such pure stands may be protected by maintaining an admixture of one or more appropriate species which will offset the weakness.

Köstler (1950) lists the biological advantages of mixed stands:

1 Roots utilize the soil profile better and increase wind resistance of a stand when shallow- and deep-rooted species are mixed, not to mention the possible improvement resulting from differences in demand for particular nutrients.

2 Mineral cycling is improved by the more rapid breakdown where the 0 horizon is a mixture of leaves.

3 Crown space is better utilized and closure is better, especially where the mixture is one of tolerant and intolerant species.

4 The interior climate of a mixed stand exhibits fewer fluctuations, even frost damage can be reduced for sensitive species.

5 Mixed stands are healthier and less susceptible to various damaging agents. Pine and spruce in hardwood stands are less subject to heart and root rots. Insect damage is not as likely or as disastrous in a mixed stand, although spruce with a high content of balsam-fir has an increased risk of spruce budworm. However, eastern white pine grown in mixture with hardwoods is much less susceptible to the deformations caused by white-pine weevil. Red- and jack-pine plantations are subject to *Scleroderris* canker in Wisconsin and Michigan, while natural stands seem to escape. Black-locust borers riddle pure stands of black

[1] *Sere* is the change in vegetation on a site from the earliest pioneer stage to the climax stage, so a *seral stage* is the vegetation on an area at a particular time, usually below the climax.

locust. Species like black walnut and cherrybark oak are allelopathic, that is, they produce inhibitory substances that discourage growth of regeneration and other competing vegetation.

Economic and administrative advantages have been the prime justification for growing pure stands. The arguments are as follows:

1 The entire forest can be devoted to the most valuable species compatible with various site conditions.

2 Management of the stands is relatively simple and inexpensive.

3 Harvesting and marketing costs are reduced.

4 It is simpler to reproduce the *desired species*, especially today when planting has become a common regeneration practice and may assume a *dominant* role in introducing and perpetuating genetically improved stock in the forest. On the other hand, it is easier to secure reproduction *of some sort* in stands of mixed species, but it is difficult to secure the desired proportions of the component species unless intensive silviculture is applied to stands after establishment.

5 Where range values are high, pure stands of intolerant species allow for greater forage production.

Economic advantages are not all on the side of pure stands, particularly if the social costs are weighed in the decisions.

1 Pure stands lack flexibility in meeting changing market demands, while mixed stands permit changes in emphasis among the species. Some valuable species are unlikely to change their desirability with time, but demand can change a species' status, such as has happened with white birch in New England, lodgepole pine in the Rocky Mountains, and white fir and hemlock on the West Coast.

2 Multiple use has modified fiber objectives on public lands and influences the decisions of the private owner as well. Wildlife needs dictate some management decisions, such as leaving isolation strips around elk wallows and snags for raptor birds. Mixed stands provide the variety of shelter and food necessary for a broad spectrum of wildlife, while pure stands support little variety. Watershed protection is enhanced by mixed stands through their effects on porosity of soil profile and the maintenance of a receptive soil surface.

3 Aesthetic and recreational values are generally favored by mixed stands.

4 Yield superiority for mixed stands on some sites under some conditions is possible. Data supporting this depend on European sources and show the importance of the measurement units. In terms of sawn volume, pure spruce outyields pure beech by 54 percent, but the difference is reduced to 4 percent on the basis of cubic-meter volume. If total weight were used, then beech would produce more than spruce. Burger (1928) reworked the data for some mixed stands and found the following relative production for pure and mixed stands, using a value of 100 for pure beech stands as a basis for comparison (Table 4-1). If the average specific gravity is 0.63 for beech, 0.54 for larch, and 0.40 for spruce, then by weight, given 100 for spruce, larch-beech equals 163.

Table 4-1 Comparisons of Sawn Volumes and Cubic Volumes for Pure and Mixed Stands of European Species

	Sawn volume, %	Cubic-meter volume, %
Pure beech	100	100
Beech-larch	131	103
Larch-beech	142	117
Pure spruce	154	104

Source: Burger, 1928.

For even-aged mixtures, Dengler (1944) specified the following requirements for success:

1 The best mixture is a tolerant with an intolerant species.
2 The species mixture should have a favorable effect on the soil.
3 Growth response on the particular site should be relatively the same.
4 If both species have the same tolerance, the slower starter should have an earlier start.
5 Both species should mature within the expected rotation.
6 Single-tree mixtures are the best but the most difficult to handle, so group mixtures provide the best balance between costs and results.
7 Favorable mixtures at maturity can be produced by removing the competition when only a few of one species among the several species making the final crop are in the young stand.

EVEN- AND UNEVEN-AGED STANDS

Stands are classified on the basis of age-class composition. Strictly defined, an *even-aged stand* is one in which all the trees are of one year or having been planted at one time. *All-aged stands*, on the other hand, theoretically contain trees of every age—from seedlings of the current year to mature veterans. These rigid definitions mark the two theoretical end points, but in nature the two forms rarely exist. Even-aged plantations are, of course, common. Even-aged stands in common field use can have a range in ages of about 20 percent of the *rotation age* (the age at which the stand is to be harvested and regenerated). *Uneven-aged stands* have at least three distinct age classes present and usually have gaps in the age-class distribution.

Practically speaking, these terms refer more to the form of the stand than to its actual composition by age groups. Even-aged stands are marked by an even canopy. The smallest trees are tall spindly members of the stand that have fallen behind their associates. The greatest number of stems are in a diameter class represented by the average of the stand; there are fewer trees in the classes both above and below this mean. On the other hand, the uneven-aged stand presents a broken and uneven canopy in which the smallest trees represent young seedlings and saplings that have come into the stand where some veteran has recently disappeared or, under managed conditions, in the holes created by harvesting. The largest number of stems is in the smallest diameter class; the number decreases more or less regularly with increasing size, so that at last there is only a mere scattering of the largest-sized trees. It happens that a stand with trees from 1 to 50 years of age would be considered as typically uneven-aged, but a stand from 200 to 250 years old would be decidedly even-aged in appearance and behavior. An uneven-aged stand has reproduction coming in waves since openings, seed, and climatic conditions are in conjunction. Because the trees in any one age class grow at different rates according to the light and nutrient supply, an uneven-aged stand frequently acquires the characteristic all-diameter distribution that an all-aged stand is presumed to have. Some even-aged stands of tolerant[2] trees may appear uneven-aged because the distribution of their diameter classes conforms to the inverted-J-shaped curve normally associated with the distribution of diameter classes in a regular uneven-aged stand.

Between the two typical forms there are many intermediate forms of stands, some of which are well enough marked to be recognized by name. They are illustrated, together with the classical even- and uneven-aged stands, in Fig. 4-1.

Storied stands have usually originated when an old even-aged stand starts to break up rather suddenly, or when a tolerant species forms a subsidiary layer beneath an intolerant overwood. This form is very common in the tropical rain forest, where two-, three-, or even four-storied forests may exist. In the ponderosa pine of the Black Hills of North Dakota, a managed two-storied stand is produced by a modified shelterwood method which does not remove the overstory until the regeneration is in the pole stage.

Irregular stands develop where single trees or groups of trees happen to be logged or naturally killed in even-aged stands. The reproduction is confined to these small openings instead of existing as an understory. The frequency distribution of trees by diameter classes may be very similar to that in the preceding case.

Even-aged group stands are much more common than is generally appreciated. Most virgin forests are of this type, especially in the West and in other regions where reproduction occurs only irregularly, or where the overmature stands tend to break up suddenly, as they do under some types of bark-beetle attacks to which the old trees are very susceptible. The stand tends to be composed of successive waves of reproduction, like a series of superimposed even-aged stands. With three or more waves of such reproduction, the stand takes the form of an uneven-aged condition. Group selection under managed conditions would take this form.

Reserve-form stands are frequently produced intentionally in silivicultural

[2]*Tolerance* is the forestry term for expressing the relative capacity of a tree to compete under low-light and high-root competition.

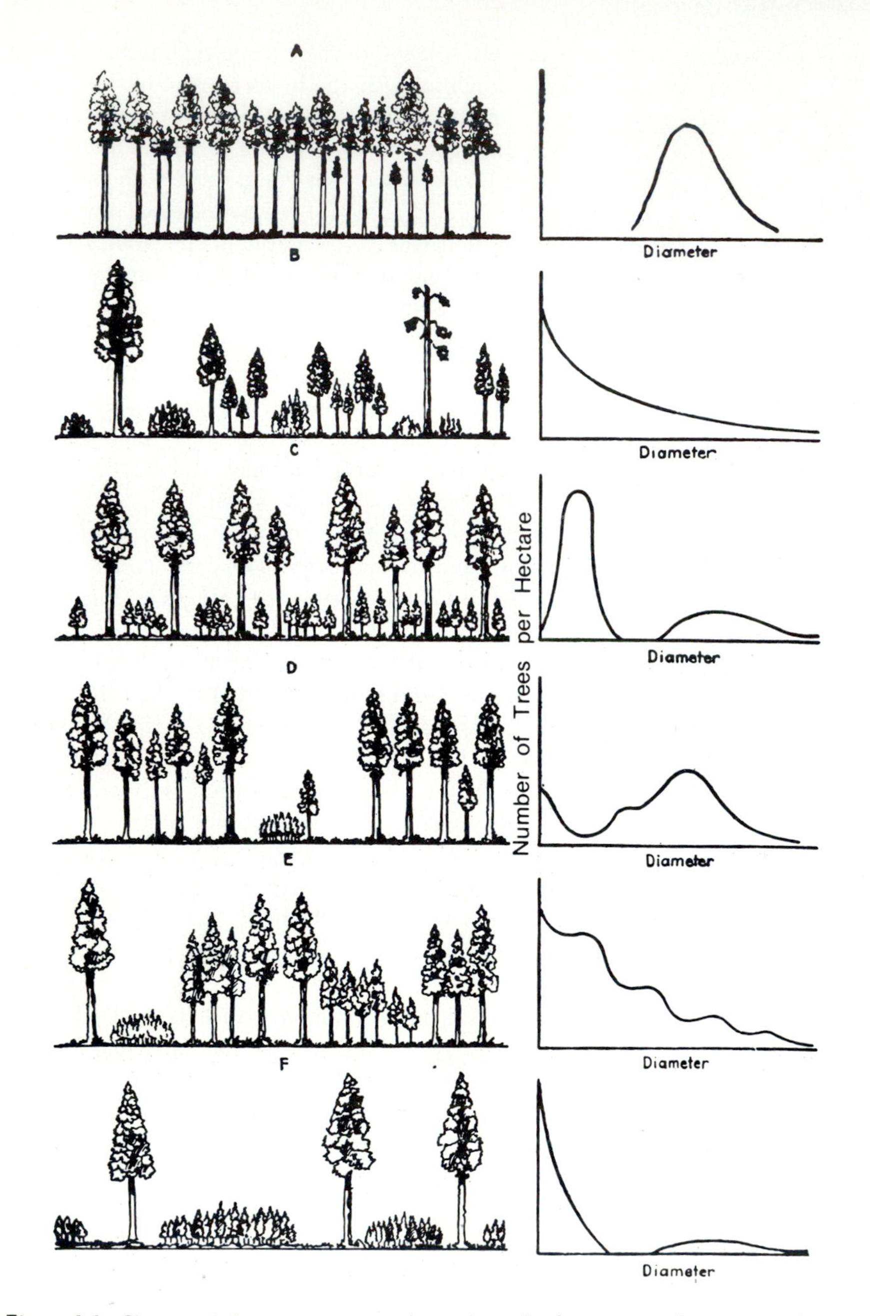

Figure 4-1 Characteristic appearances and stand graphs for even- and uneven-aged stands and some intermediate forms: A, even-aged; B, uneven-aged; C, storied; D, irregular, E, even-aged group; and F, reserve form. (*From Baker, 1950.*)

management. Here, a few trees are held on an area to produce a new crop beneath them. The same thing is often produced naturally in severe fires, or where wind and storm destroy all but a few of the stoutest individuals.

The distribution of size classes in these stands is represented by diagrams of the type shown in Fig. 4-1, where diameters are entered on the abscissa and number of trees per hectare (or other unit of area) on the ordinate. Diameter distributions in a typical even-aged stand approach a bell-shaped curve which can be skewed to the smaller diameters for tolerant species and to the larger diameters for intolerant species. In an uneven-aged stand, the frequency distribution of numbers of trees by diameter class forms an inverted-J-shaped curve which, when plotted on semilog paper, produces a straight line.

The situation with stands of different forms is somewhat analogous to the situation already considered with respect to pure and mixed stands. Even-aged stands, like pure stands, tend to have certain managerial advantages. Uneven-aged stands, like mixed stands, tend to possess certain definite biological advantages. In conifers, there is a very general correlation of even-aged form with pure stands and uneven-aged form with mixed stands. In hardwoods, most stands are mixed whether even-aged or uneven-aged—an exception is aspen in the West.

Biological Aspects

The main biological differences between even-aged and uneven-aged stands can be compared as follows.

	Even-aged	Uneven-aged
Canopy	A level, shallow canopy on slender stems	A deep, irregular canopy with sturdy boles
Wind hazard	Careful management required to prevent windthrow losses, especially for shallow-rooted species	Wind hazard very low
Small trees	Small trees are suppressed; release is unlikely.	Small trees are future-crop trees that respond to release.
Regeneration	Occurs over a short period	Distributed over whole rotation age of the crop trees
Site	Site exposed to deterioration by wind and rain erosion during period when next generation of trees is small	Site in openings always protected by surrounding trees
Control of site	Site may be lost to invading vegetation. May be the best form for controlling unwanted vegetation.	Site conditions stable, existing undesirable vegetation difficult to control
Hazards	Subject to serious fire, diseases, insect losses	Fire, diseases, and insect losses less likely to be serious
Slash	A one-time heavy accumulation to add to insect and fire hazard	Continuing light slash and low hazard

The biological disadvantages of even-aged stands can be corrected with intensive management and, therefore, are not inherent in even-aged management. However, the advantages of uneven-aged management are inherent in the method, but poor management can create serious problems.

Economic Aspects

Considerable debate has been generated, particularly in Europe, regarding the relative productivity of even- and uneven-aged stands. No differences in productivity have been demonstrated, and such differences are unlikely to exist so long as comparisons with stands of the same species are made. Since the productivity of even-aged stands of two species can differ appreciably on a particular site, a difference in productivity between even- and uneven-aged stands could be shown to go in either direction depending on how different species are paired.

The economic advantages and disadvantages of even-aged and uneven-aged stands again parallel those of pure and mixed stands. Even-aged stands have primarily economic advantages, while uneven-aged stands are associated with ecological stability. In this country, there are very large areas in uneven-aged condition, while in Europe, for various historical as well as economic reasons, the uneven-aged area has been reduced to a minor fraction of the total. Europe shifted from a concentration on monocultures to a recognition of the benefits of mixed stands, and for years considerable effort was given to reducing monocultures. However, economic conditions have forced more reliance on planting monocultures than on the time-consuming and labor-intensive process of natural regeneration. In this country, a reliance on planting for regeneration in even-aged management has become the dominant approach to silvicultural problems, and most of the plantations have been single species. The incentives have been many: (1) the successful surge of effort to develop higher-producing strains, convincing forest owners that it is practical, e.g., the Southern tree-improvement cooperatives; (2) the reclaiming of enormous acreages from brush or cull hardwoods; (3) the increasing concentration of the pulp and wood industry in the South because of the growth potential of the area and its species; (4) the burgeoning stumpage prices all over the country; and (5) the pressure of industry to get the lumber for the building boom of the 1960s at minimum cost via the clearcutting method. In the East, the conversion of uneven-aged stands to even-aged stands has been successful, except for arousing environmental concerns, because the climate favors seedlings if favorable growing conditions are created or retained. In the West, the failure to recognize the critical nature of many sites, the generally prevailing unfavorable rain patterns, and the sensitivity of climax and even pioneering species to growing under exposed conditions has left large areas of denuded forest land in spite of planting whether the original stand was even- or uneven-aged. A revision of the allowable cut on some forests to a more sustainable figure and a recognition of the error in the idea that planting can cover any silviculture sin promise that clearcutting will again become *a* method of silviculture rather than *the* method. There is even recognition in the West that uneven-aged management may be the answer to managing sensitive areas along highway-influence zones or on slopes and soils with severe erosion hazards.

The economic advantages of even-aged management can be summarized as follows: (1) the simplicity introduced into managing, inventorying, and harvesting blocks of even-aged stands; (2) the high volumes per hectare which permit the use of more efficient large equipment with corresponding lowering of logging costs; (3) the shortening of the rotation by planting and the ability to introduce and maintain genetically improved growing stock; (4) the higher-quality boles produced under even-aged conditions; (5) the more uniform growth rate during rotation and greater uniformity in the size of trees harvested; (6) the simplification in the application of any intermediate cuts, in that they can be applied uniformly over the whole stand; and (7) the lack of damage to residual stand, except as seed-tree or shelterwood methods are used in establishing the stand.

The disadvantages of even-aged management are primarily biological if management is poor and, especially, if the stand is also pure. Some of the economic disadvantages of even-aged stands are as follows: (1) a precommercial intermediate cut may on occasion be needed during stand development to meet management objectives; (2) unless commercial thinnings are practical, the stand may not return significant income until the end of the rotation (this is especially depressing on investment return where planting costs are involved since they have to be capitalized rather than treated as an operating cost); (3) there is a continuous drain on operating funds for protection; and (4) the generally lower aesthetic and recreational values associated with even-aged management practices may also be economic liabilities if considerable costs in public-relations efforts are necessary to overcome adverse social reactions.

The potential advantages of uneven-aged management are primarily biological, and these are accentuated if the stand is also of mixed species. However, uneven-aged management has its economic advantages too: (1) the growth capacity of a site can be converted to cash at short intervals; (2) small holdings can be managed for steady income; (3) less need is evidenced for special entries into the stand for noncommercial improvement work since intermediate cuts can be performed at the time of harvest at the end of each *cutting cycle* (the interval between consecutive harvests on the same area in an uneven-aged stand); (4) there is greater flexibility in harvesting since the cutting can be modified to fit changes in the market; (5) there is greater ease and certainty of natural regeneration, especially in mixed-species conditions; and (6) there is less likelihood of fire or disease destroying a stand.

The disadvantages of uneven-aged management are mostly economic: (1) there is possibly higher logging costs because of the smaller volume per hectare removed at each cutting, the greater care necessary to prevent damage to future crop trees both in destruction of young stock and in damage to older trees, and the need for smaller equipment and more labor-intensive procedures; (2) there are higher administrative costs because of individual marking of trees, the need for close supervision of the logging effort, and inventory difficulties; (3) damage to the residual stand is unavoidable, so only its extent can be changed; (4) there are higher road maintenance costs because of the frequency of access to a stand; (5) lower stem quality of bole and lower wood quality are associated with the changing pattern of growth from seedling to maturity; (6) there is danger of overcutting or undercutting with effects on maintenance of the desired diameter distribution; (7) there are compositional problems in mixed stands; (8) difficulties result from trying to introduce a genetically superior strain; (9) there is no interval in the life of a stand where damage by game or livestock may not be serious; and (10) regulating growth and yield present difficulties.

TREE CLASSIFICATIONS

While stands constitute the unit of management, the silviculturist must usually prescribe treatments on the basis of individual tree characteristics. Tree classifications are necessary if foresters are to be able to communicate their ideas and instructions at the tree-versus-tree level.

Even-Aged Stands

In order to implement all silvicultural treatments in even-aged stands, it is necessary to recognize individual trees in terms of their relative position in the canopy. Differences among trees in even-aged stands was recognized early but lacked formal systematic classification until Burckhardt in 1855 produced the classification still used in this country. His work was poorly distributed and thus failed to attain much recognition. Kraft (1884) published the scheme with modifications to give seven tree classes, and it gained wide acceptance. There are a number of more complicated classifications which incorporate a recognition of stem form and crown condition with canopy position, but they have not been adopted for use in this country (Gevorkiantz et al., 1944).

Kraft's Tree Classification Tree or crown classification in this country has five classes, although lack of sharp definition between classes tends to create other classes in the transition zones (low-codominant or high-intermediate) for some purposes. The five classes (Fig. 4-2) are as follows:

1 Dominant Trees The crowns of dominant trees rise somewhat above the general level of the canopy so that they are exposed to full light above and, to a certain degree, laterally. Sometimes in even-aged stands but far more often in imperfectly even-aged or reserve types of stands, the largest dominant trees, for lack of lateral competition from codominants, grow to be coarse, heavy-limbed, broad-crowned trees that are called *wolf trees* and are considered undesirable members of the stand.

Figure 4-2 Even-aged crown classification in a stand: D, dominant; C, codominant; I, intermediate; S, suppressed. (*From Baker, 1950.*)

2 Codominant Trees These are not quite as tall as dominants. Their crowns receive overhead light, but they may be hemmed in laterally to a certain degree by dominants. They are nearly as thrifty as dominants and with them comprise the main canopy of the stand.

3 Intermediate Trees These crowns occupy a definitely subordinate position and are subjected to sharp lateral competition from crowns of the two previous classes, although they receive some direct overhead light through holes in the canopy.

4 Suppressed Trees These are definitely overtopped members of the forest community having almost no free overhead light. They exist by virtue of sunlight that filters through the canopy or skylight that may be received through some chance opening. They are commonly weak and slow-growing.

5 Dead Trees

The proportion of each crown class found in a stand varies with species (tolerance), age and history, and density of the stand. In unmanaged stands starting with heavy reproduction, an enormous number of trees must sooner or later pass into suppression and die. How rapidly this takes place depends on tolerance and site quality. Black spruce—a tolerant species—has been studied with some care in even-aged stands (Fig. 4-3), but whether the increasing percentage of dominants and codominants illustrates a general law or merely typi-

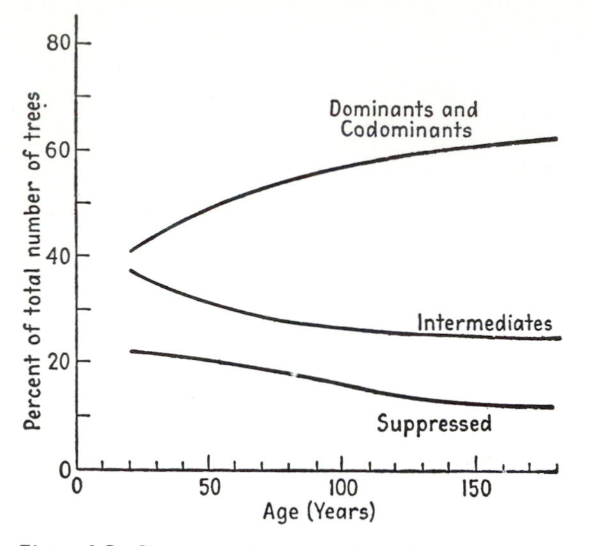

Figure 4-3 Changes in the proportions of the crown classes with age in an even-aged stand of black spruce. (*From Baker, 1950.*)

fies black spruce in the Lake states is uncertain. The question may well be academic if thinning should become as standard a management procedure as it is in Europe, where the thinning intensity controls the representation of the crown classes. It is also of small moment in plantations where the high cost of planting drastically reduces the number of trees planted.

Taylor's Tree Classification Taylor's (1939) classification for lodgepole-pine stands, which are generally even-aged, was developed from Keen's (discussed later in this chapter) because in lodgepole pine the position of a tree in the canopy is not always a good index of its growth capacity. It was designed as a guide for marking stands so as to leave a residual that would respond to release. Lodgepole pine has a tremendous capacity to respond to release if its crown condition is favorable. The method was discontinued because of the heavy windthrow losses in the severely cut stands and the upsurge of dwarf mistletoe in the released trees. However, in future managed stands, the classification may serve a useful purpose. The classification has four classes, lettered from A (high vigor) to D (low vigor), and they are based on crown length and area and apparent vigor (Fig. 4-4).

Uneven-Aged Stands

It is a measure of the prevalence of even-aged forests in Europe that in the long period during which the principles of silviculture and management were developed there was insufficient need to develop an uneven-aged tree classification system. The youth of silvicultural practices in this country is illustrated by the fact that

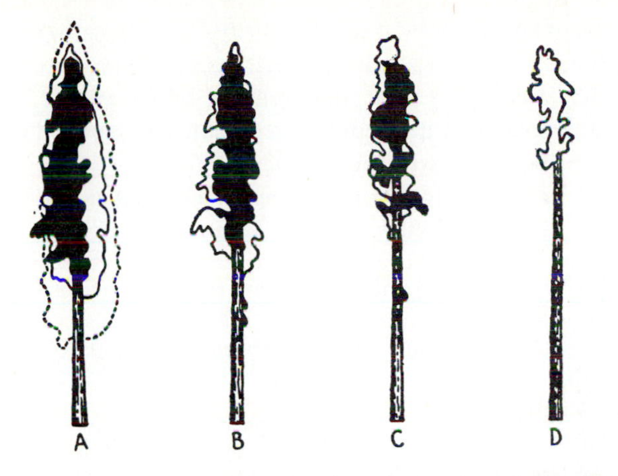

Figure 4-4 Tree classification for even-aged stand of lodgepole pine. The black area represents the minimum crown for each class, while the solid line is the maximum except for class A, which can have crowns to the dashed line. Vigor of the crown is also decisive in classifying a tree. (*From Taylor, 1939.*)

after hundreds of years of cutting a variety of uneven-aged stands, it was not until 1928 that Dunning designed a classification for the uneven-aged, mixed-conifer type in California. His principles have served as the basis for other uneven-aged-stand tree classifications. The unique element in uneven-aged stands is the fact that a sapling in an opening must be considered as dominant over its environment as a mature tree is in its space. Thus a classification for trees in uneven-aged stands must provide a guide for decisions affecting both.

Position in the canopy has been the usual criterion for vigor in a stand, but in an uneven-aged stand there is no general level of the canopy. The viewpoint of the silviculturist must be changed. In an even-aged stand, it is frequently not so much the individual tree that focuses the attention but the effect of its removal on the stand, while in an uneven-aged stand, the *focus* is the individual tree, its response to being left, and its response to the removal of surrounding trees.

Dunning's Tree Classification Dunning's primary interest was in leaving trees with the maximum growth potential. Because he was working in a mixed-conifer stand in California, the endemic insect losses were low enough to permit a heavy cut and a long cutting cycle. In addition, in leaving only high-growth-potential trees he was removing the insect-susceptible trees. This was relatively easy to do since at the time the practice was to cut 80 percent of the volume because of the need to amortize railroad construction costs.

The mixed-conifer type of the Sierra Nevada consists of an uneven-aged mixture of ponderosa pine, sugar pine, Douglas-fir, incense cedar, and white fir. Private lands were cleared of all commercial values, and any noncommercial tree that escaped the logging and fire was allowed to grow. Public lands were being marked to leave 20 to 30 percent of the original stand in thrifty trees to permit a second cut in a reasonable time. The problem was the variation among markers as to what constituted a thrifty tree that could grow well and escape insects, diseases, and windthrow. After 15 years of data collection on cutover areas and sample plots, Dunning (1928) recognized that a tree's potential for growth and survival depended on its vigor, i.e., precisely Kraft's principle. He built up his seven tree classes on the basis of the factors which influence a tree's vigor:

1 Age, with four classes—young (under 50 years), thrifty-mature (50 to 150 years), mature (150 to 300 years), and overmature (over 300 years)

2 Degree of dominance within these age groups—isolated, dominant, co-dominant, intermediate, and suppressed

3 Crown development—length, width

4 Estimate of thrift based on the preceding plus density and color of foliage, bark characteristics, and freedom from disease.

The application of these factors could be made to all the species in the type, but his classification is described in terms of its application to ponderosa pine. No effort is made to provide a guide for considering any but sound, well-formed trees as suitable to be left.

Dunning's tree classification is as follows:

Class 1. Age class, young or thrifty mature; position, isolated or dominant (rarely codominant); crown length, 65 percent or more of the total height; crown width, average or wider; form of top, pointed; vigor, good

Class 2. Age class, young or thrifty mature; position, usually codominant (rarely isolated or dominant); crown length, less than 65 percent of the total height; crown width, average or narrower; form of top, pointed; vigor, good or moderate

Class 3. Age class, mature; position, dominant; crown length, 65 percent or more; crown width, average or wider; form of top, rounded; vigor, good

Class 4. Age class, mature; position, usually codominant (rarely isolated or dominant); crown length, less than 65 percent of the total height; crown width, average or narrower; form of top, round; vigor, moderate or poor

Class 5. Age class, overmature; position, isolated or dominant (rarely co-dominant); crown, of any size; form of top, flat; vigor, poor; foliage, usually rather pale green and thin

Class 6. Age class, young or thrifty mature; position, intermediate or suppressed; crown, of any size, usually small; form of top, round or pointed; vigor, moderate or poor; some capacity for recovery from suppression

Class 7. Age class, mature or overmature; position, intermediate or suppressed; crown, any size, but usually small; form of top, flat; vigor, poor; seriously suppressed and has little chance of ever making a merchantable stem (The classes are illustrated in Fig. 4-5.)

The data accumulated on old cutover areas were summarized for the rates of growth (basal-area increase), total losses, and losses from insects and other sources

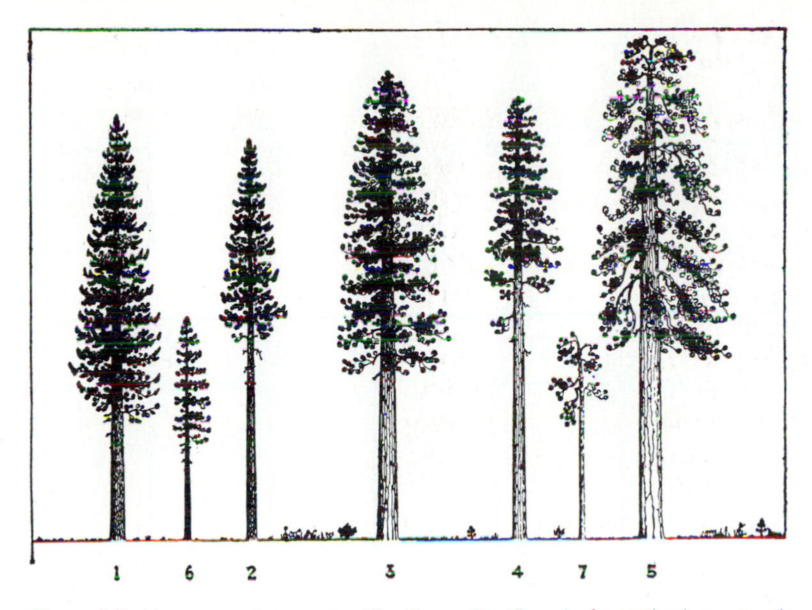

Figure 4-5 Uneven-aged tree classification using Dunning's method on ponderosa pine. (*From Dunning, 1928.*)

of mortality. Recommendations for the proportions of each class to be cut were given on the basis of these observations where a minimum residual stand was desired (approximately 20 percent). However, with the advent of trucks and tractor logging, the cut on public lands was reduced to 50 percent of the volume by increasing the number of trees left in classes 4 and 5. The original recommendation had been to cut all trees in classes 4 or 5. With a cut that left 20 percent of the volume, there was a deficiency in good seed trees, but a cut leaving 50 percent gave an adequate seed supply.

Keen's Classification On the east side of the Cascades and northern Sierra Nevada, ponderosa pine grows in pure uneven-aged stands where bark beetles (*Dendroctonus brevicomis*) are frequently epidemic. In an effort to reduce insect losses by getting over the forest quickly and removing the high-risk trees, the cut was reduced to 50 percent of the volume. This was possible because during the 1940s the use of trucks and tractors allowed a lighter cut. Dunning's classification under the 50 percent cut rule frequently failed to identify susceptible trees because there was too much difference in vigor within a class. Keen (1936) modified the method based on field observation in pure pines to create 16 smaller classes which were more homogeneous in vigor. There are four age classes, slightly different from Dunning's, because Keen developed his method to apply on poorer sites (Site IV): age group 1 = under 80 years; group 2 = 80 to 180 years; group 3 = 180 to 300 years, and group 4 = over 300 years. Each age

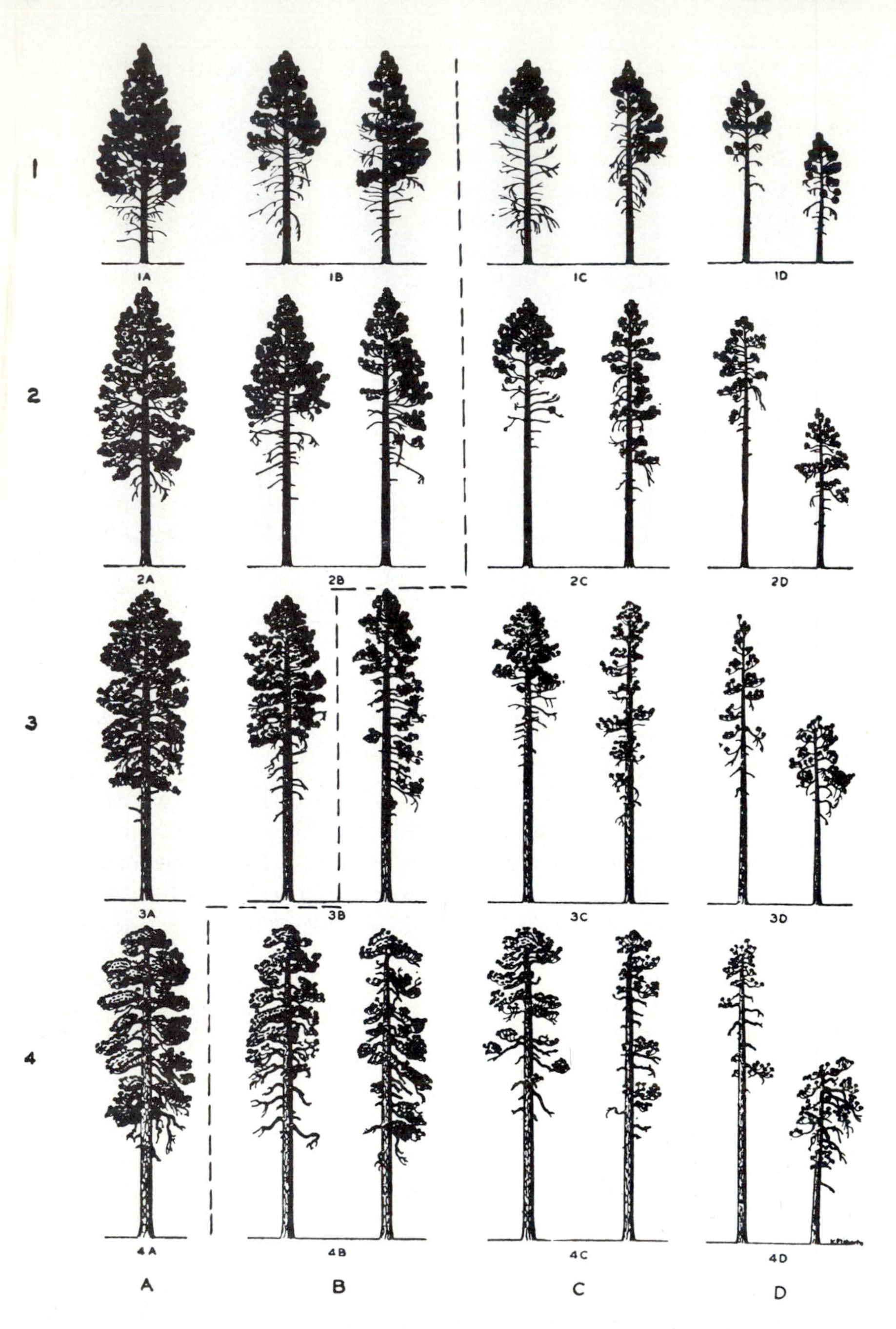

Figure 4-6 Tree classification for uneven-aged ponderosa pine using Keen's method of classification. Trees on right of dashed line approximate 50 percent of stand's volume. (*From Keen, 1936.*)

group was further subdivided in four vigor groups based on crown vigor, length, width, foliage density and condition, and position: groups A, B, C, D. Tree class 4A would be a tree that was overmature, dominant, long- and wide-crowned, and dense foliaged with long needles. Keen's method is illustrated in Fig. 4-6, in which the tree classes to the right of the dotted line are the ones that would have to be removed for a 50 percent volume cut in an average Eastside stand.

Salman and Bongberg's Classification Keen's classification, with its 50 percent removal of volume, did not meet the need for getting over the insect-infested area fast; in addition, a much larger volume would be required to actually drop the beetles to an endemic status. The epidemic was on the rise, and the presence of very mobile logging equipment made light cuts possible. In the other classifications, a definite correlation between vigor and growth or insect susceptibility was established, but the variation in actual attack, even within Keen's 16 small groups, could not pinpoint the trees to be hit first. Yet a slowing down of growth had been established as a predisposition for attack. An increment borer could establish which trees were slowing in growth, but the technique was too slow for a marking procedure. Salman and Bongberg (1942) correlated other evidence of predisposition for actual insect attack. Such characteristics as thin or sparse foliage complement (less than average number of needles per twig), short needles, the dying of twigs and branches, and the presence of active or recent top-killing infestations were evident in many or most of trees that became infested and died. There was a gradation in the way these characteristics attributed to a particular risk rating. The Salman and Bongberg classification has four risk ratings as follows:

Risk 1: Low risk—foliage of healthy appearance; needles usually long and coarse; color good dark green; foliage complement (needles per twig) normal or above normal; few twigs lacking foliage; no weakened portions of crown or dead or dying branches

Risk 2: Moderate risk—foliage mostly healthy; needle length average or better; color fair to good; foliage complement generally average; some twigs or branches dying or dead, but such injury not localized to form definite "weak" spots in crown

Risk 3: High risk—foliage, at least over a considerable portion of crown, average or shorter than average in length; foliage complement on twigs less than normal, thin, bunchy, or unhealthy; foliage color fair to poor; some to many twigs or branches lacking foliage, fading, or dead, often localized to form "weakened" portions of the crown

Risk 4: Very high risk—foliage mostly unhealthy, needles short; color poor; needle complement sparse over considerable portions of crown; some to many twigs and branches dead or dying; portions of crown definitely weakened; active top-killing or partial infestations often present

An analysis of 18,000 trees on which risk ratings were made showed that high and very high risk ratings included 11 percent of the number of trees and

15 percent of the volume. The risk ratings measured against losses over a 1- to 4-year period showed a mortality ratio (if loss in risk class 1 has a ratio of 1), of risk 1 = 1, risk 2 = 5, risk 3 = 25, and risk 4 = 136. Timber marking on the eastern side of the Sierra Nevada and other areas is presently done by risk ratings for all the species using Salman and Bongberg's risk factors. Tables of defects have been developed which assign penalty points to each risk factor on the basis of degree of susceptibility to insects or risk of dying before the next cutting cycle or stand-improvement entry. When a tree's condition accumulates nine penalty points, it is marked for cutting.

The classifications by Keen and by Salman and Bongberg were developed to identify trees which were more susceptible to attack by bark beetles. This was the major concern in marking trees for cutting in previously unmanaged old-growth stands where losses due to beetles could be enormous. The classifications are still in use but will have less applicability in managed young-growth stands.

There are other ponderosa-pine classifications for the Rocky Mountain variety of ponderosa pine which are based on an earlier maturity date (225 years instead of 300 years) and place a greater emphasis on potential growth. Beetle damage can be very destructive of Rocky Mountain ponderosa pine, but the Black Hills beetle (*Dendroctonus ponderosae*) in an epidemic attacks all merchantable trees irrespective of vigor. Therefore, risk classifications are meaningless, and the tree classifications concentrate on growth potential.

Westveld's Tree Classification If an insect epidemic can be controlled by silvicultural means in one place, there is a chance that the principles learned could be applied to another insect. Spruce budworm has been the scourge of the spruce-fir type in the Northeast, with epidemic losses of billions of board feet, and is a continuous threat to the whole type with its intermittent epidemics and recessions. Spruce-fir is recommended for handling in uneven-aged stands.

Westveld (1954) found that a tree's resistance to budworm attack depended on its vigor and that there is a considerable difference in susceptibility between balsam fir and the spruces. Balsam fir is the more susceptible species because its leaves are the preferred food source and the timing of its bud burst in the spring is correlated with the newly hatched larvae's need for a food supply of tender leaves. Spruces are much less susceptible because their bud burst is later and old leaves are less palatable than balsam fir or unpalatable, as in black spruce. Stands are vulnerable to attack depending on the amount of balsam fir in the composition, but the amount of damage depends on the vigor of the stand components. Growth rate in diameter has been shown to be correlated with a tree's capacity to survive attack. A slow-growing tree has less resistance to damage than a fast-growing tree irrespective of whether the fast-growing species is spruce or balsam fir. With the same vigor, spruce would suffer less damage than balsam fir. After establishing the relation of diameter growth to damage, Westveld designed his classification using the factors that influence growth in a particular environment: maturity, dominance, and crown length. He organized his classification using 3

Table 4-2 Westveld's Tree Classification, Illustrating the Effects of Age, Crown Class, and Crown Length on the Relative Vigor of Spruce and Balsam Fir

Young trees		Mature trees		Overmature trees	
Tree dominance and crown length	Vigor-resistance rating	Tree dominance and crown length	Vigor-resistance rating	Tree dominance and crown length	Vigor-resistance rating
Dominant:*		Dominant:*		Dominant:*	
Long crown	10	Long crown	7	Long crown	4
Medium crown	9	Medium crown	6	Medium crown	3
Short crown	8	Short crown	5	Short crown	2
Intermediate:		Intermediate:		Intermediate:	
Long crown	9	Long crown	6	Long crown	3
Medium crown	8	Medium crown	5	Medium crown	2
Short crown	7	Short crown	4	Short crown	1
Overtopped:†		Overtopped:†		Overtopped:†	
Long crown	8	Long crown	5	Long crown	2
Medium crown	7	Medium crown	4	Medium crown	1
Short crown	6	Short crown	3	Short crown	0

*Comprises both dominant and codominant tree classes.
†Comprises both overtopped and suppressed tree classes.
Source: Westveld, 1954.

age, 3 dominance, and 3 crown-length classes which provide 27 vigor-resistance classes, as indicated in Table 4-2. Vigor-resistance ratings varied between 10 (the most vigorous and resistant) and 0 (the least vigorous and resistant). In applying the classification, any balsam fir with a vigor-resistance rating below 8 would be marked for removal in order to reduce the fir content in a stand and have high-vigor trees left that were capable of surviving attack, while only spruce below a rating of 5 would need to be marked to reduce the potential for serious damage. The amount cut from a stand would depend on the degree of resistance to damage that was wanted. If a high degree of resistance was needed, then some sacrifice of productivity would be required because the fast-growing balsam fir would be reduced to trees with a maximum of 5-in diameter at breast height (dbh) at each cut.

Part Two

The Tree

Chapter 5

Ecophysiology of Tree Growth

INTRODUCTION

The silviculturist's objective is to control the density, composition, and structure of forest stands. This is done by manipulating tree growth by treatments such as thinning, improvement cuts, and final harvesting which directly control the genetic composition of plant populations and the microenvironment in which the selected plants must grow. This concept makes it apparent that the silviculturist must have a clear understanding of the particular soil-plant-environment system being managed. Each component of the system is dynamic, consequently forest manipulation must take into account the influences of phenological and climatic changes. The principles of tree growth are commonly physiological, but they must be interpreted within an ecological framework. This has given rise to the commonly used term *ecophysiology* to describe such interactions.

The growth of a particular plant in the forest depends on its capacity to express its genetic potential through physiological functioning within the environment to which it is exposed. These two sciences of forest genetics and tree physiology therefore constitute important integrated foundations upon which silvicultural principles are based. For convenience, the concepts basic to forest genetics and tree improvement are discussed in Chapter 15, where they are tied directly to their implications in tree improvement and silvicultural manipulation.

The basic physiological and environmental factors that are of concern in an ecophysiological analysis of tree growth are illustrated in Fig. 5-1. Of these, the processes that are of most concern to the silviculturist are photosynthesis and respiration, translocation, and water and mineral uptake. These processes control the net quantity of food available for plant growth, the movement of organic

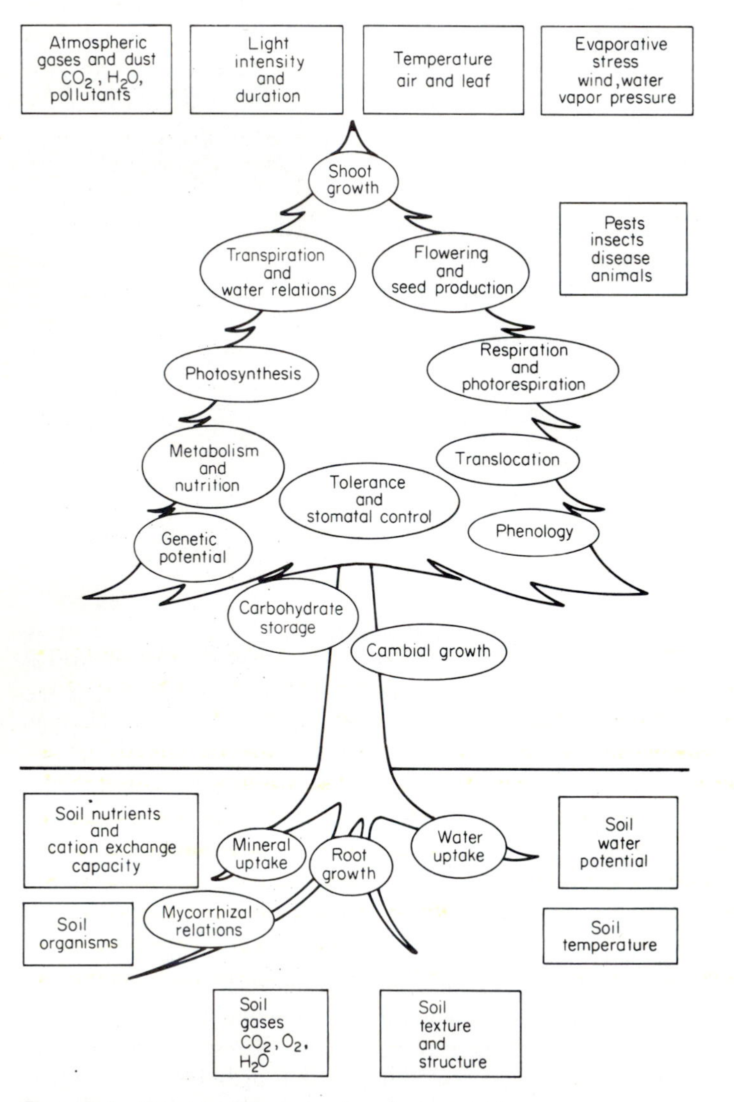

Figure 5-1 Basic physiological and environmental factors.

and inorganic materials within the plant, and the relative vigor of the plant in terms of its water and nutrient status. The rate and duration of these major physiological processes therefore essentially characterize the relative capacity of plants to develop in their environment. Since we are dealing with long-lived trees and management periods covering decades, it is essential to evaluate physiological processes on diurnal, seasonal, and annual bases.

Usually, a variety of physiological processes are proceeding concurrently within each plant. Analysis of any one process by itself is therefore difficult, and probably quite misleading since the current characteristics of one process commonly influence the rates and efficiencies of others. In addition, the current level of any process is influenced by the stage of plant development at which the process is monitored, and on the immediate past environmental conditions to which the plant has been exposed (hysteresis effects). Also, the behavior of trees under field conditions may differ markedly from the behavior of small plants under laboratory or controlled-environment conditions. Consequently, silvicultural interpretation of limited physiological analyses must be made with considerable care.

This chapter first summarizes the principal concepts underlying the major plant processes. Then, the processes of plant growth will be considered in relation to factors of the environment, with particular emphasis on diurnal and seasonal growth behavior. Finally, the ecological and silvicultural significance of the ecophysiological capacity of trees as a basis for stand manipulation and tree improvement will be evaluated.

PHOTOSYNTHESIS

The Process of Photosynthesis

Photosynthesis is the process by which carbohydrate is produced from inorganic materials by the transformation of solar energy into chemical energy. It is often described as the earth's single most important chemical process for several compelling reasons—people and all animals (heterotrophs) depend directly or indirectly on plants (autotrophs) for their food; the stability of atmospheric concentrations of oxygen and carbon dioxide is dependent on the photosynthetic process in the oceans and on land; in addition, we take advantage of the photosynthetic energy storage of past geologic ages when we use natural gas, petroleum, and coal as sources of fuel. Adding to this our utilization of wood fiber (one of the few renewable natural resources) for varied needs and we must surely recognize photosynthesis as fundamentally important to human life on this planet.

Photosynthesis is an exceedingly complex process consisting of a series of reactions which produce organic matter from inorganic substances. Carbon dioxide is taken from the air and an equal volume of oxygen is returned. Essentially, the process may be described as the absorption of light energy by chloroplasts, the splitting (photolysis) of water to provide hydrogen ions and gaseous oxygen, and the use of hydrogen ions to reduce carbon dioxide to the level of sugar.

The process basically consists of three kinds of reactions:

1 *Physical reactions:* carbon dioxide is transferred from the atmosphere into the leaf where it dissolves in water. The total resistance to this transfer is one of the most important rate-limiting factors in the process.

2 *Photochemical reactions:* 2 to 4 percent of the incoming radiation is used in photosynthesis, with wavelengths in the red and blue portion of the spectrum being most active. Energy is absorbed by chlorophyll *a* and *b* (and some ancillary pigments) and is "pumped" by large units of chlorophyll molecules to high-energy phosphate bonds in the molecule adenosine triphosphate (ATP).

3 *Chemical and enzymatic reactions:* these are a sequence of some dozen intermediate reactions from the first formed stable product, phosphoglyceric acid (PGA), to the 3- , 4- , 5- , and 6-carbon sugars.

Recently, plants have been grouped into two classes, called C_3 or C_4 plants, depending on whether they fix carbon into a 3-carbon product (as in the Calvin cycle) or whether the CO_2 is fixed into sugars via C_4-dicarboxylic acid. The two groups can be separated on the basis of photosynthetic rates, on the basis of anatomical and physiological criteria, and on the environment in which they grow. High-photosynthetic-capacity plants (C_4 type), such as maize, sorghum, sugar cane, salt bush, and some dicotyledonous plants, may have 2 to 3 times the primary production of low-photosynthetic-capacity plants (most genera including trees). The C_4 plants have high photosynthetic rates of 50 to 80 mg CO_2 dm^{-2} h^{-1}, low CO_2 compensation points[1] of 0 to 10 parts per million (ppm), and no photorespiration, they require less water, and commonly grow in stress environments such as the tropics, arid areas, mountains, and estuaries (Hatch et al., 1971; Black, 1971). Dickmann (1973) examined 14 conifers, 16 poplar clones, and 30 hardwood species and found them to be typical C_3 plants, which are characterized by having relatively low photosynthetic rates of 10 to 35 mg CO_2 dm^{-2} h^{-1}, higher CO_2 compensation points of between 30 and 70 ppm, and respiration stimulated by light.

Photosynthesis in Trees

When considering the characteristics of photosynthesis in trees and their relative capacities for carbohydrate production, we need to remember that in contrast to agricultural crops, foresters are dealing with a much more complex situation.

[1]*CO_2 compensation point* is that CO_2 concentration at which the uptake of CO_2 in photosynthesis equals the amount of CO_2 given off in respiration.

These complexities are due to the following facts:

1 We are dealing with both seedlings and tall trees.

2 Our crop is long-lived, therefore we are more concerned with seasonal characteristics, storage, and mobilization of food reserves.

3 We are often dealing with complex, irregular canopies of mixed species which have marked complexities of environment.

4 The rooting medium is usually uncultivated and heterogeneous.

5 The forest crop grows on an extreme diversity of sites, soil types, elevations, and aspects, and often under suboptimal conditions.

For these reasons, the following general interpretations of physiological data must be made with great care.

1 Stomatal Behavior *Stomata* are small pores in the epidermis of leaves through which the bulk of water and gases diffuse. They are important in that the extent to which they are open or closed determines the resistance to uptake of carbon dioxide and hence carbohydrate production, as well as the amount of water lost in transpiration. Stomatal movement therefore has considerable bearing on the relative success of plant development.

Stomata are very numerous. In hardwoods they occur only on the lower epidermis, and although they range in number from 11,000 to 100,000 cm^{-2}, they constitute only about 1 percent of the surface area of the leaf (Kramer and Kozlowski, 1960). In conifers, the stomata are arranged on all sides of the needle and may number 5000 cm^{-2} (Waggoner and Turner, 1971). Sun leaves, which grow on the exposed portions of a tree's crown, have several times more stomata per unit leaf area than do shade leaves on the same tree. The mechanism of stomatal opening is still not well understood, but CO_2 concentration, light intensity, solute potential, export of hydrogen ions, and potassium-ion flux all seem to be important (Zelitch, 1969). The rates at which stomata open and close partly depend on the tolerance of the species and the light conditions to which the trees are exposed, as indicated by the following data of Woods and Turner (1971):

Species	Opening time, min	Closing time, min
Fagus grandifolia	3	12
Acer rubrum	12	18
Quercus rubra	12	20
Liriodendron tulipifera	20	36

These data, measured at 300-footcandles (fc) light intensity, and similar findings by Davies and Kozlowski (1972) show that stomata open more quickly than they close and that the rate of movement is very slow.

Since stomata play a large part in controlling carbohydrate production and water use, it is of interest to consider the ways in which the silviculturist could influence stomatal behavior. There are two ways: first, by modifying microenvironments in the stand through choice of site preparation, treatment, or reproduction method aimed at limiting the adverse effects of temperature and evapotranspiration; or second, theoretically, by selecting those species or possibly individuals or subpopulations within the species which have stomatal behavior inherently more suited to the site conditions.

2 Within-Tree Variation in Net Photosynthesis A tree's crown is a complex structure consisting of foliage of a variety of different ages growing at different positions within the crown. These varying positions have quite different environmental characteristics, consequently expressions of photosynthetic capacity must take into account the considerable within-tree variation that occurs. Each leaf photosynthesizes at a rate which essentially reflects its own particular physiological condition and the microenvironment to which it is exposed. In considering photosynthesis of forest trees, we need therefore to determine within-tree differences, which are attributed to leaf age and position within the crown, between-tree differences, which are those between hardwoods and conifers, species, and genotype.

In discussing photosynthesis in trees, we are usually referring to *net photosynthesis*. This is defined as the difference between gross rates of photosynthesis and concurrent rates of respiration. Net photosynthesis occurs when the CO_2 taken up in photosynthesis exceeds the amount of CO_2 evolved in the concurrent process of respiration.

3 Leaf Age Photosynthetic efficiency differs among leaves of varying ages mainly because of the marked effects of different rates of respiration. The amount of assimilates used by leaves in respiration is normally 5 to 10 percent of gross photosynthate production; however, both young and aging leaves have been found to respire considerably more than this amount (Huber and Rüsch, 1961). In conifers, which usually maintain several age classes of needles concurrently, the fully expanded 1-year-old foliage is the most efficient of all age classes. As foliar age increases, net photosynthesis decreases (Fig. 5-2). This relationship has been demonstrated in many species of pine, spruce, fir, and oak.

The distribution of age classes of needles in crowns of three 50-year-old red-pine trees is shown in Fig. 5-3. On the average, the proportion of each age class was 33 percent current, 30 percent second year, and 37 percent older needles. From this diagram one can obtain an indication of the likely effects of pruning different proportions of the live crown. For example, from the diagram, removing one-third of the length of live crown would remove 21, 35, and 43 percent of the total foliage weight in trees A, B, and C, respectively.

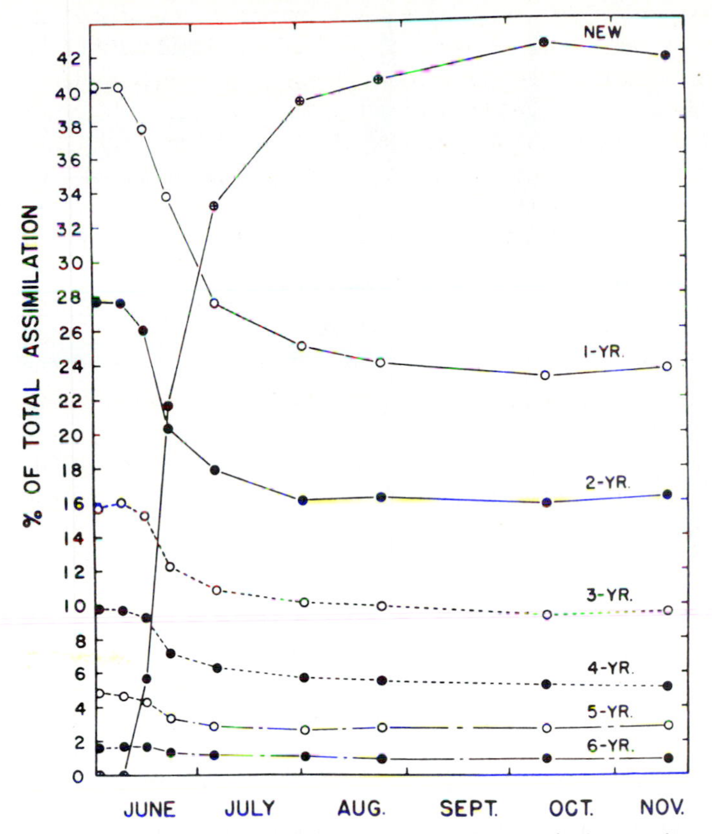

Figure 5-2 Estimated contributions to the photosynthetic capacity of a medium-sized balsam-fir tree by foliage of different ages. (*From Clark, 1961.*)

4 Position in Tree Because of characteristic differences in leaf age and environment within a forest canopy, measures of net photosynthesis of trees vary depending on the position within the tree at which photosynthesis is monitored. This is illustrated in Fig. 5-4, which describes a conifer crown as consisting of a series of overlapping cones of foliage of predominantly different ages growing in quite different environments. The innermost cone consists predominantly of relatively older, less efficient foliage, and the outermost cone consists of the youngest, most efficient foliage once the needles are fully elongated. In stand conditions, therefore, the lowest whorls, which receive relatively little light, are contributing little to net photosynthate production. This provides justification for pruning the lowest branches.

Commonly, the most productive leaves in Douglas-fir are those in the partly shaded conditions of the upper crown (Woodman, 1971). This is probably true for other moderately tolerant trees as well. This finding ties in well with reports

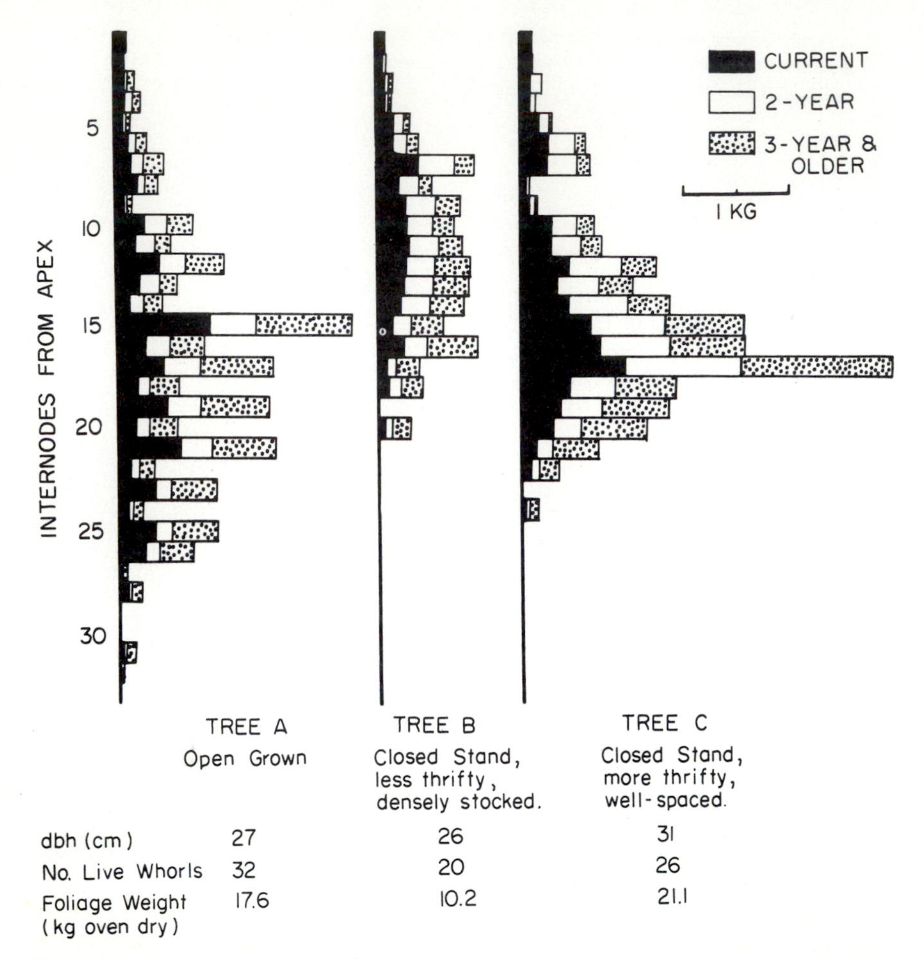

Figure 5-3 Distribution of oven-dry weight of needles by whorls and age classes in 50-year-old red pine crowns. (*From Hall, 1966.*)

of Hodges (1967), who showed that seedlings of many Pacific Northwest conifers photosynthesized best in the partially shaded boundary at the edge of forest openings. Similar superior survival and growth of naturally regenerated mixed-conifer species in California has been consistently shown to occur in partially shaded environments. This is no doubt due to the favorable temperature and evaporative stress levels that occur in these situations.

5 Differences in Net Photosynthesis among Crown Classes Differences in photosynthetic efficiency among dominant, codominant, and suppressed trees are relatively minor providing one is comparing similarly exposed foliage and expressing efficiency per unit of leaf surface. Major differences among crown classes are obtained when one evaluates the relative efficiencies of sun and shade

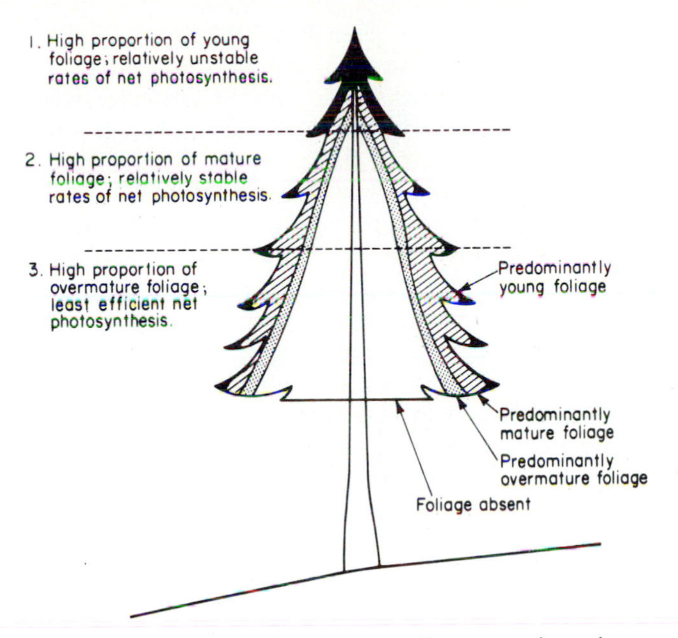

Figure 5-4 Distribution of foliage of different age classes in a conifer crown can be regarded as a series of superimposed cones.

leaves and when one appreciates the major differences in environment to which the different crown classes are normally exposed. Of particular importance are the gradients within a forest canopy of light intensity and carbon dioxide concentration. Within and beneath a dense canopy, light intensity is very much less than that experienced by dominant trees except for the transmission of sun flecks. As will be discussed more fully in a later section, the silviculturist is more concerned with photosynthetic capacity, or the total amount of carbohydrate produced by trees over a period of time, than with the relative rates or efficiencies at particular levels of environment.

A major factor causing differences in the photosynthetic capacity of trees of different crown classes and species is the enormous difference commonly found in leaf area. If we are concerned about influencing the productivity of individual trees, we are more apt to be successful if we influence leaf area. The amount of foliage in a stand is usually expressed in terms of the leaf area index (LAI), which is the amount of leaf surface area in a vegetative canopy over a given area of ground below it, expressed as a proportion of leaf surface to ground area. The ratio in forests is normally between 3 and 6. The relationship between photosynthetic capacity and leaf area is very important because the silviculturist can control leaf area of individual trees through thinning or pruning. Thus the total growth capacity of individual trees can be increased or decreased by spacing them such that they produce larger or smaller crowns.

6 Sun and Shade Adaptations At any given light intensity, rates of net photosynthesis depend on the relative tolerance of the plant in question. The same is true when comparing photosynthetic performance of sun leaves, which grow on the exposed parts of the crown, with the performance of shade leaves. Shade leaves, and also tolerant plants such as maple and white fir, commonly have higher photosynthetic efficiencies under low light conditions than do sun leaves and intolerant plants such as ponderosa pine and cottonwood. Under high light conditions, the reverse is true, intolerant plants and sun leaves have higher efficiencies per unit of leaf area. This behavior indicates that the photosynthetic mechanism of individual plants and leaves is genetically controlled and can become adapted to particular environments.

When the silviculturist drastically changes an environment in a thinning operation, the response of the released trees is partly dependent on whether their photosynthetic tissues can accommodate the increases in light intensity. Intolerant trees in the suppressed and intermediate crown classes which are released from a relatively shaded position to one that is suddenly fully exposed may show a decrease in growth or may actually die. This occurs because the shade leaves of the intolerant tree lose photosynthetic efficiency when suddenly exposed to high light intensity. The shade leaves of tolerant trees are usually better able to adapt to this change from a relatively shaded to an exposed environment.

The actual factors responsible for differing photosynthetic behaviors of intolerant and tolerant plants are not well understood but appear to involve differences in CO_2 transfer resistance through the stomata at low light intensities (Wuenscher and Kozlowski, 1971). It has also been observed that stomata of tolerant plants open and close more rapidly than those of intolerant plants, which would enable them to take better advantage of short periods of favorable light intensity under the canopy.

However, leaves from sun and shade positions, or from intolerant and tolerant plants, do vary in morphology. Exposed sun leaves are smaller, thicker, and more leathery than are shade leaves of the same age and species. Also, tolerant plants and shade leaves usually have less chlorophyll per unit of leaf area because the sun leaves are thicker and have a greater amount of palisade tissue. Because the proportion of chlorophyll on a dry-weight basis depends on the amount of other tissues present and on a leaf-area basis depends on leaf thickness, the correlation between photosynthetic activity and chlorophyll content can be poor in trees. The increase in thickness of palisade tissue has been used as an index of adaptation to sun conditions (Jackson, 1967). Figure 5-5 shows the ranking of 27 hardwoods species with respect to increasing palisade tissue in sun leaves divided by the thickness of the same tissue in shade leaves of the same species. In all species it was found that palisade ratios gradually increased, from 1.05 for tolerant dogwood to 3.0 for intolerant sweetgum. Although palisade ratios do not permit the grouping of species into clearly defined tolerance classes such as those given in Table 13-2, nevertheless they indi-

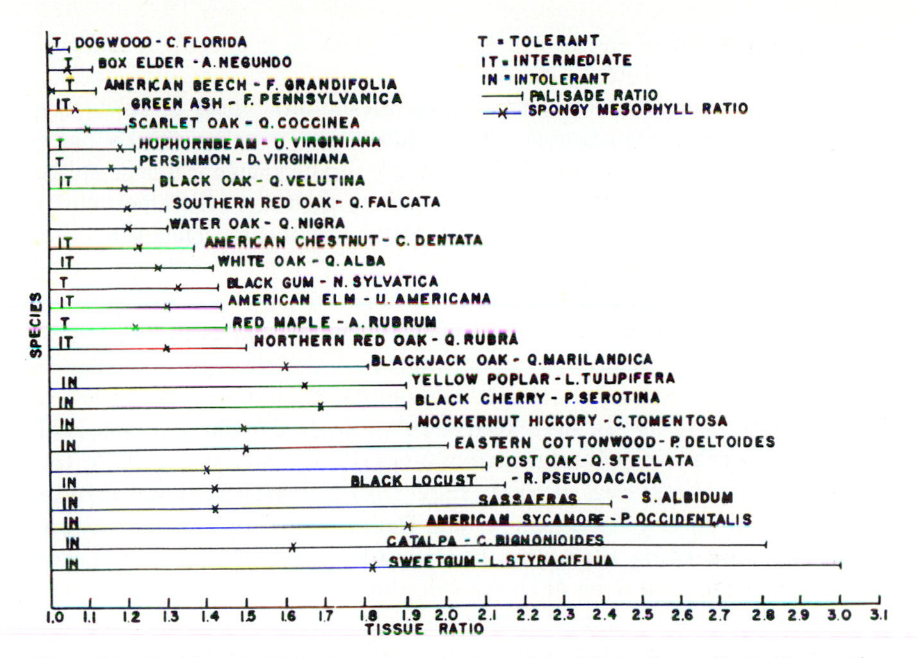

Figure 5-5 Ranking of 27 hardwood species in order of increasing palisade tissue ratios. Spongy mesophyll ratios are indicated by crosses. Baker's tolerance ratings are shown along the left-hand margin. (*From Jackson, 1967.*)

cate the extreme diversity of morphology of leaves produced in sun and shade environments.

7 Differences among Genotypes Because of the common managerial preference for fast-growing trees, it is important to establish whether or not photosynthetic efficiency is genetically controlled. If there are within-species differences in photosynthetic efficiency, and if this characteristic is heritable, planting stock with above-average growth potential can be raised. Such a selection assumes that there is a direct relationship between photosynthetic efficiency and growth, and that there is a high correlation between growth performance of seedlings and mature trees.

The question of whether different genotypes have inherently different photosynthetic capacities has been studied for some time. Results have been rather conflicting; some evidence supports a correlation between genotype and growth, other studies show no relationship, and still others show inconclusive results. This conflicting evidence results partly from the facts that photosynthesis has been measured in terms of rates per unit of leaf area and that growth has been evaluated in terms of one parameter such as height. When photosynthetic capacity of whole plants is evaluated on a seasonal basis and more elaborate evaluations of total growth are made, positive correlations have been obtained between photosynthesis and growth for specific genotypes of loblolly and jack pine (Ledig and Perry, 1969; Logan, 1971).

It seems, therefore, that inherently significant differences in productive capacity do occur among genotypes. These differences, together with the possible inherent differences in capacities to rapidly produce large leaf area and to display that foliage more efficiently, provide the silviculturist with opportunities to increase the productivity of forest crops by selecting desirable genotypes as well as by cultural treatments.

8 **Differences between Hardwoods and Conifers** Studies comparing photosynthetic efficiency of hardwoods and conifers have reported that well-exposed foliage of broad-leaved trees was 2 to 4 times more efficient than that of conifers (Polster, 1950; Jarvis and Jarvis, 1964), and that a hybrid poplar was capable of almost 10 times the rate of assimilation of pines (Huber and Polster, 1955). The situation is not entirely clear, however, since other workers have shown that the differences may be very slight (Kramer and Decker, 1944) or of similar magnitude (Kramer and Clark, 1947; Brix, 1967; and Krueger and Ruth, 1969). The lack of general agreement on differences in photosynthetic efficiency between hardwoods and conifers probably results from the fact that actual differences, if they exist, can be masked or confounded by the use of different methodology and by the problem of mutual shading of foliage which reduces the amount of radiation reaching the foliage being studied. This problem has been well illustrated by Kramer and Decker (1944), as well as by Bormann (1957), who showed that individual, well-exposed fascicles of loblolly pine reached maximum rates of photosynthesis at relatively low light intensities, whereas complete plants increased photosynthetic capacity with increasing light intensity right up to full sunlight.

It appears, then, that the photosynthetic efficiency of conifer crowns may actually be very similar to that of hardwoods. Relative photosynthetic capacity, on the other hand, may be a little more difficult to define since it depends on such factors as relative photosynthetic area, foliar arrangement and mutual shading, age classes of foliage retained, and deciduousness.

Effects of Environment on Photosynthesis

Rates of photosynthesis are influenced by both plant and environmental factors. These are illustrated in Figs. 5-6 and 5-7, which indicate that the various factors can be grouped according to whether they act directly on the process itself or indirectly by modifying the magnitude and effect of the primary factors. Silviculturists modify microenvironments through stand treatments and therefore must understand process-environment interactions and the extent to which one can control process rates and growth through stand manipulation.

In appraising the effects of an environmental factor on a particular process, silviculturists must remember that all other processes and environmental factors

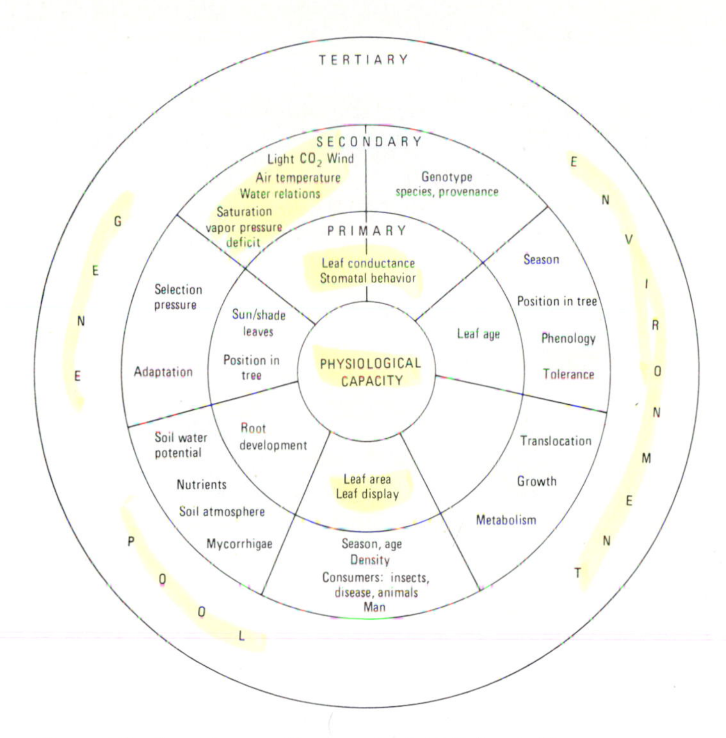

Figure 5-6 Plant factors affecting physiological capacity.

are interacting. Also, any particular growth-environment response is dependent on the condition of the plant, its stage of development, past history of treatments, season, and level of all other environmental factors. In interpreting interactions, silviculturists therefore should consider the multidimensional response surfaces that are involved rather than treating the effect of each environmental variable independently in the form of a two- or three-dimensional relationship, which, unfortunately, is all that can be reproduced on the printed page or readily conceived.

1 Light Light directly affects tree growth by its intensity, quality, and duration. Of these characteristics, light intensity is perhaps the most important to the forester because it is the most readily manipulated.

When plants are gradually exposed to increasing light intensity from darkness to full sunlight, it is commonly found that positive rates of net photosynthesis are not recorded until a certain minimum threshold of light intensity is exceeded. This *light compensation point* is the light intensity at which the amount of CO_2 taken up in photosynthesis exactly equals the amount concurrently given off in respiration (Fig. 5-8). The light compensation point for tolerant plants is typically lower than that for intolerant plants; however, it is

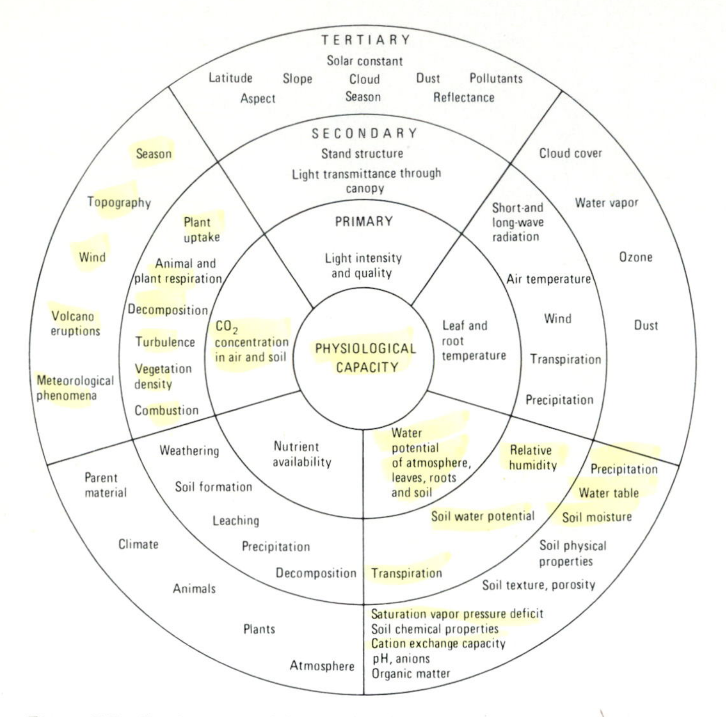

Figure 5-7 Environmental factors affecting physiological capacity.

not a fixed point for a particular plant but varies according to whether sun leaves or shade leaves are being measured; environmental conditions such as temperature, nutrient level, and water potential; and even past treatments to which the plant has been exposed. These factors modify the light compensation point by influencing concurrent rates of respiration. For example, the light compensation point of shade plants during leaf development in the spring can be considerably higher than that for the same foliage in summer (Leith and Vogt, 1959). In fact, the problem of variability in concurrent rates of respira-

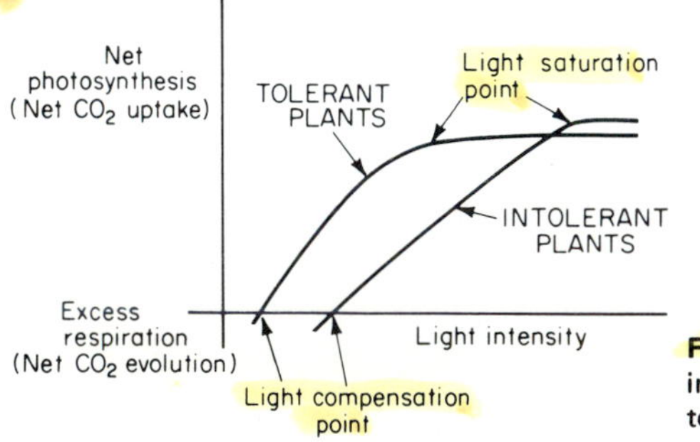

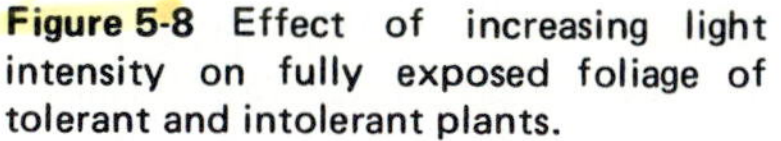
Figure 5-8 Effect of increasing light intensity on fully exposed foliage of tolerant and intolerant plants.

tion is so great that McCree and Troughton (1966) question the desirability of classifying sun and shade plants on the basis of higher or lower compensation points. They suggest that a better criterion for comparison might be the relative capability of plants to *adjust* to changes in light intensity levels.

As light intensities are increased, rates of net photosynthesis of well-exposed leaves increase. A point is reached, however, called the *light saturation point*, when increased light intensity provides no further increase in net photosynthesis. The light saturation point for tolerant plants is commonly lower than that for intolerant plants (Fig. 5-8). When light intensity is above the saturation point, fluctuations in light intensity have little influence on rates of photosynthesis. At extremely high intensities, photosynthesis can be limited by photooxidation of the chloroplasts. This effect has been observed with Englemann-spruce regeneration growing at high elevations where stunting and chlorosis have been attributed to photooxidation (Ronco, 1975). As pointed out earlier, however, crowns and canopies of both tolerant and intolerant trees would commonly not reach full productive capacity until radiation reached full sunlight because of the mutual shading of foliage.

The duration of light has considerable importance in silviculture. One aspect of light duration is *photoperiod*, which largely controls bud set and hardening-off of seedlings. Because of this phenomenon, silviculturists are able to force plants to set a bud in controlled-environment greenhouses by artificially shortening day length together with reducing the watering and nutritional regimen. This technique permits the hardening-off of planting stock at the required time before planting.

In the field, the silviculturist must appreciate the effects of light duration because the accumulated length of time that a plant is exposed to light above the compensation point largely determines the total amount of carbohydrate that is produced. This point is of little consequence for plants that are fully exposed to the sky, since these receive all the light that is available. Understory plants, however, receive light that is filtered through the overhead canopy and are subject to the passing of shadows and sun flecks. In dense stands, the length of time that light is of low intensity or actually below the compensation point may be a high proportion of the photosynthetic day. Thus the effective time that light of adequate intensity is available to understory plants can be strongly influenced by control of stand density.

The shadow patterns produced in openings of various shapes and sizes in forest stands of a given height and at a particular latitude and season of the year can be readily calculated by computer (Brown and Merritt, 1970; Halverson and Smith, 1974). This enables the silviculturist to determine for any opening the length of time that seedlings would be shaded or exposed to light intensities above a particular level.

The transmission or attenuation of light through a forest canopy is dependent on the type of canopy, whether it is composed of hardwoods or conifers, the manner in which leaves are displayed, the density of leaves, and the homogeneity of the canopy. The absorption of radiation as it passes through a forest canopy is illustrated in Fig. 5-9. The amount of light available at different levels within the forest largely influences the expression of species dominance, differentiation into crown classes, live-crown ratios, and overall crown dimensions. Therefore, if we know the light requirements of plants, we can control stand structure and productivity, the relative success of regeneration of various species, and the development of browse, cover, and vegetative screens. This control over stand structure is developed through manipulation of the available light at any height in the stand by control of species composition (the type, amount, and display of foliage) and the density of the overstory.

It can be seen, therefore, from the foregoing discussion that any analysis of photosynthetic capacity of trees in a field environment must take into account the complex problem of availability of light within tree crowns and forest canopies and the way in which light intensity and duration change diurnally and seasonally.

2 Temperature The effect of temperature on net photosynthesis is difficult to evaluate. First, net photosynthesis is the difference in concurrent rates of photosynthesis and respiration, and the temperature relations of these two processes are quite different. Second, in the field, increases in temperature are commonly associated with increases in light intensity, which confuses the issue. It is apparent, therefore, that generalizations regarding the effects of temperature on photosynthesis need to be interpreted with caution.

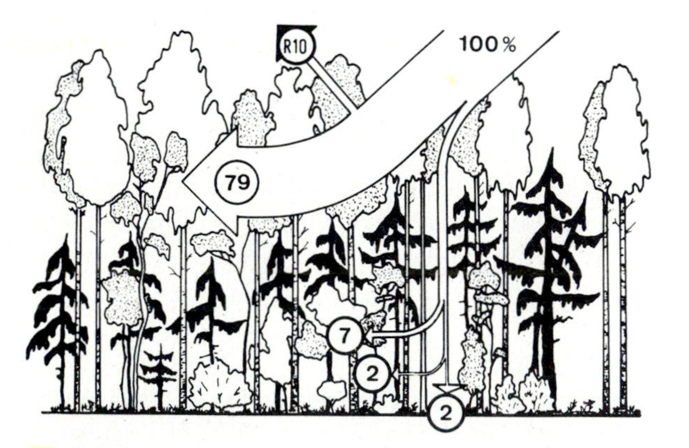

Figure 5-9 Attenuation of radiation in a boreal mixed forest. Of the incident photosynthetically active radiation, 10 percent is reflected (R) from the upper surface of the forest. Different amounts of the radiation that penetrates the stand are absorbed in each layer depending on the structure of the stand. In the forest, the greatest absorption of radiation occurs at the crowns of the trees. Only 2 to 5 percent of the incident photosynthetically active radiation reaches the ground. (*After Cernusca, 1975; Kairiukštis, 1967; from Larcher, 1975.*)

The optimum temperature range for photosynthesis varies with species and ecotype but is commonly between 18 and 25°C for temperate trees, with extreme ranges being between −5 and 40°C (Stocker, 1960; Kozlowski and Keller, 1966). The actual range of optimal temperatures for any species depends on many factors, including the age and health of the foliage and the current availability of water and light.

In broad terms, the relationship between photosynthesis and temperature is that shown in Fig. 5-10. With increasing temperature, photosynthesis increases exponentially until optimal rates of gross photosynthesis occur between 20 and 40°C. The optimum for net photosynthesis, however, is likely to be between 18 and 25°C because of the increasing impact of higher rates of respiration on net CO_2 exchange. As temperature increases, enzymatic processes are increasingly affected such that rates of photosynthesis decline. At high temperatures near 40°C, the plant begins to suffer direct heat injury resulting from coagulation of the proteins in the protoplasm. Photosynthesis ceases at the death of the protoplasm.

3 CO_2 Concentration The concentration of carbon dioxide in the earth's atmosphere above the forest canopy is approximately 0.03 percent by volume—300 ppm. Within the forest, CO_2 concentrations are commonly higher. The availability of CO_2 can commonly be a limiting factor in photosynthesis (Kramer and Kozlowski, 1960). This is most likely the case in tight canopies of tree crowns or agricultural crops during the day when there is active photosynthesis removing CO_2 from the air and very little atmospheric mixing because of stagnant air. As CO_2 concentration around a leaf decreases, a minimal level is reached, called the *CO_2 compensation concentration*, below which positive rates of net photosynthesis are not recorded. Generally, for C_3 plants, this minimum CO_2 concentration is 50 to 100 ppm; however, as mentioned in The Process of

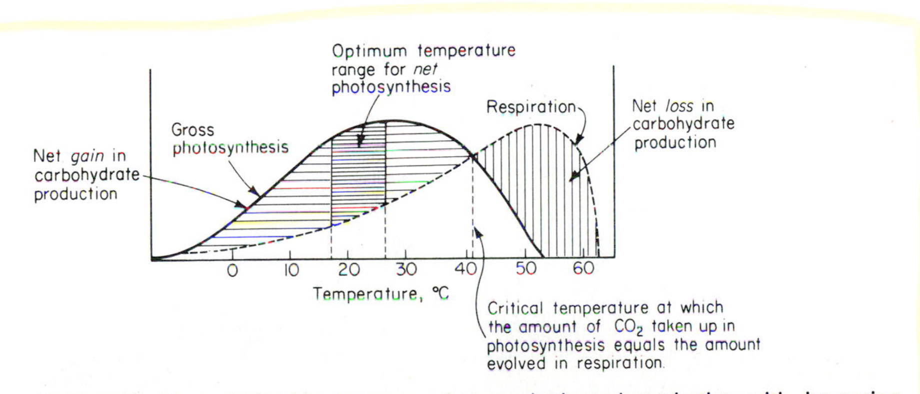

Figure 5-10 The relationship between photosynthesis and respiration with increasing temperature. Net photosynthesis, which is the difference between gross photosynthesis and respiration, is indicated by the shaded area.

Photosynthesis earlier in this chapter, there is a group of C_4 plants (not exhibiting photorespiration) which have exceptionally high photosynthetic capacity and can function at CO_2 concentrations between 0 and 10 ppm.

Atmospheric CO_2 concentration in a forest is not stable. Vertical gradients occur which fluctuate diurnally and seasonally from above the canopy down to the forest floor (Fig. 5-11). Fluctuations in concentration are mainly caused by:

1 The amount of CO_2 evolution from surface layers of the soil, which is controlled by rates of decomposition of organic matter, as well as by soil temperature, water content, and texture

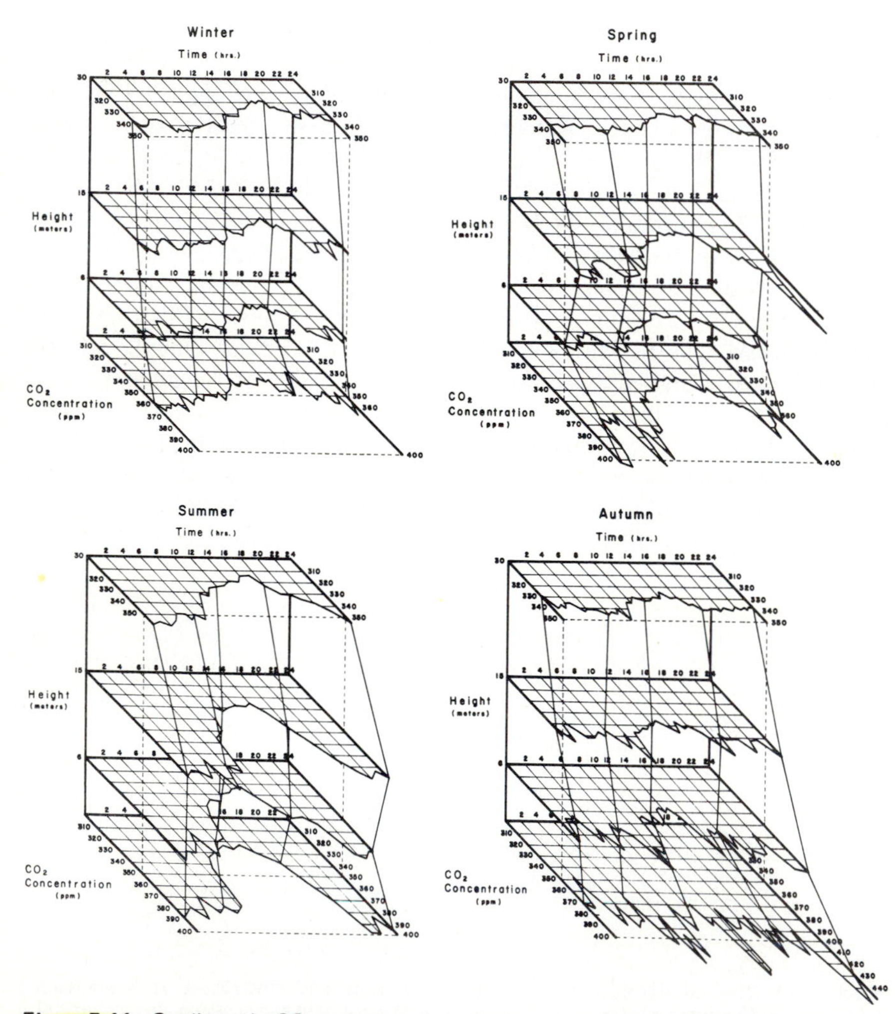

Figure 5-11 Gradients in CO_2 concentration within a mixed conifer forest in each season of the year. (*From Helms, 1976.*)

2 The degree of atmospheric mixing, which depends on the action of wind and on turbulence generated by differential heating by shafts of sunlight

3 The relative amounts of CO_2 uptake and evolution by plants

4 The transfer of large air masses by weather patterns

5 Elevation

Carbon dioxide is heavier than air and tends to sink to the forest floor when turbulence is low. Consequently, seedlings and plants with foliage near the ground probably have an ecological advantage since they are usually exposed to above-average CO_2 concentrations. This is a reasonable assumption since there are many reports in the literature describing enhanced rates of photosynthesis in CO_2-enriched atmospheres. In general, short-term rates of photosynthesis can be increased by CO_2 concentrations up to 10 times normal, and growth can be increased 20 to 50 percent. When white-pine seedlings were exposed to short-term increases in CO_2 concentration up to three times normal each day for a period of 4 months, height and diameter growth were doubled and no toxic effects were noted (Funsch et al., 1970). Sustained high levels of CO_2 may become toxic to plants, however.

4 Water Availability An extremely small portion of the total water used by a plant is consumed directly in the process of photosynthesis. Therefore, effects of water deficits on photosynthesis are due almost entirely to their indirect effects on hydration of protoplasm and stomatal closure.

Optimal conditions for photosynthesis occur when the leaves are fully turgid. This occurs when soil water is abundant and when atmospheric conditions produce low evaporative demands. As the soil dries down from field capacity and soil water potential decreases (becomes more negative), loss of turgor and stomatal closure occurs, which in turn limit CO_2 entry, causing a decline in photosynthesis. There may be differences in rates of decline depending on the drought tolerance of the species. This effect is illustrated in Fig. 5-12.

This phenomenon of decline in photosynthesis is due to decreasing water availability in the leaf, or more precisely, to decreasing leaf water potential which results in plant water stress. Figure 5-13 illustrates this decline in photosynthesis with increasing stress for Douglas-fir compared with the more intolerant ponderosa pine. The pine shows a capacity to maintain higher levels of photosynthesis as foliar stress builds up to −12 atmospheres (atm) and then appears to drop rapidly as stress increases, probably due to closure of the stomata. As stomata close, the resistance to CO_2 transfer increases, thus reducing rates of photosynthesis.

5 Nutrition The nutrition of forest trees affects photosynthesis in two ways: directly, by affecting the efficiency of the process; and indirectly, by affecting the total photosynthetic production of the tree. Experiments with 24-year-old Douglas-fir trees (Brix, 1971) have shown that the photosynthetic

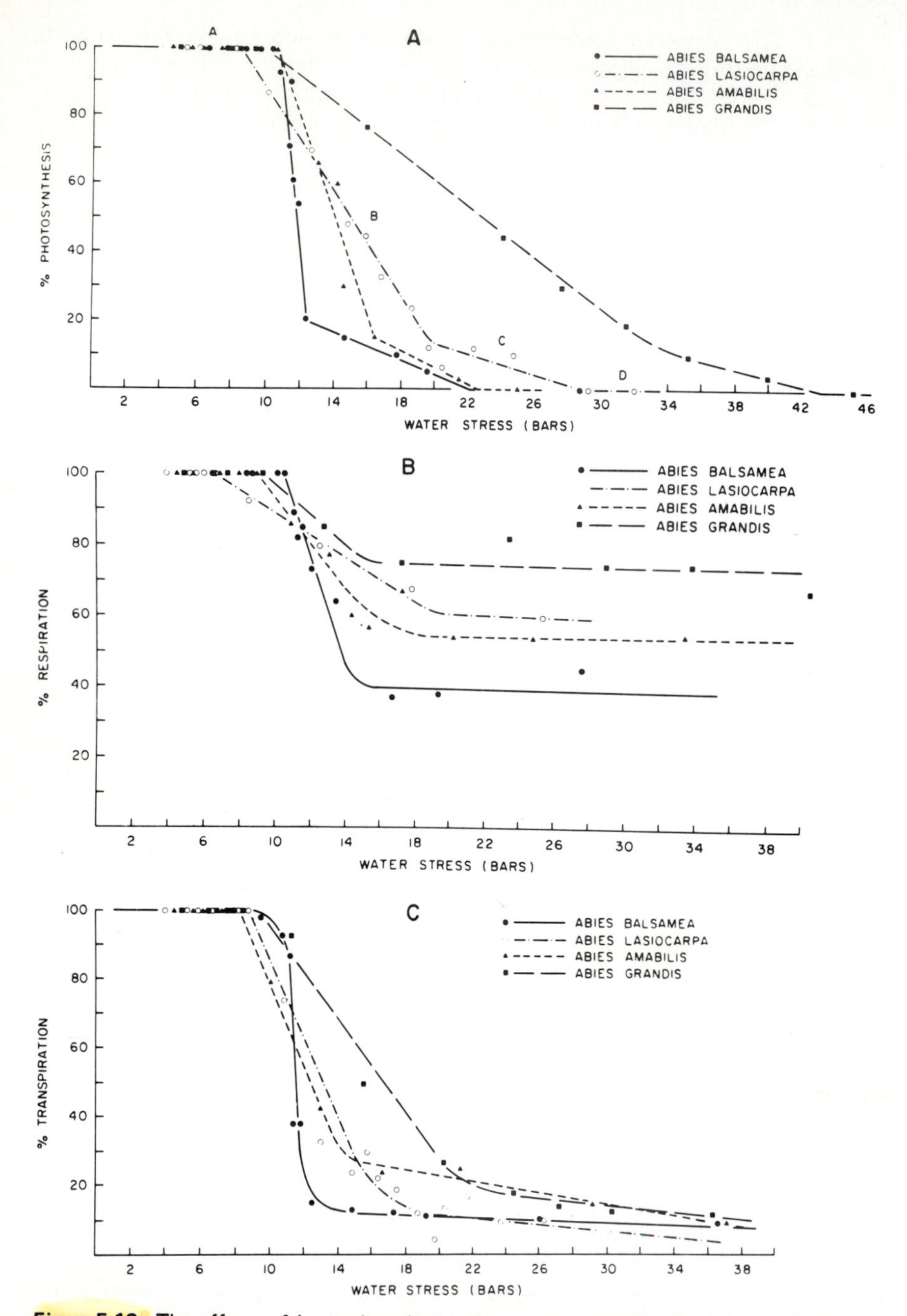

Figure 5-12 The effects of increasing plant water stress on net photosynthesis, respiration, and transpiration of several conifers. (*From Puritch, 1973. Reproduced by permission of the National Research Council of Canada from the Canadian Journal of Forest Research,* **3:** *293–298.*)

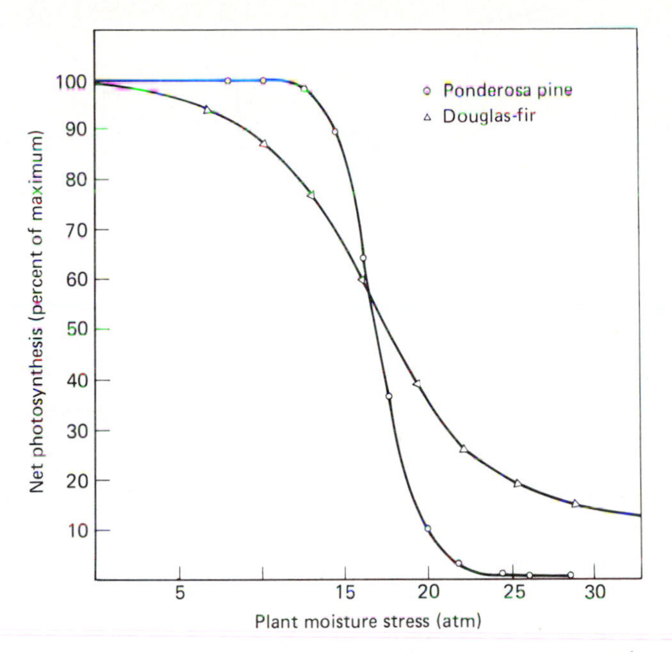

Figure 5-13 Effect of increasing plant moisture stress on net photosynthesis of Douglas-fir and ponderosa pine. (*From Cleary, 1970.*)

capacity of current shoots in the year of fertilization increased 78 percent as a result of the addition of nitrogen when the leaves were exposed to favorable temperature and water conditions and when light intensities were 5000 fc (Fig. 5-14). Photosynthetic rates were increased only when the treated foliage was exposed to light intensities greater than 2000 fc (i.e., one-fifth full sunlight).

Indirectly, the nutritional status of trees affects photosynthesis by influencing the area of individual leaves and the total size of the crown. Nutrition also influences the vigor and extent of root systems, which influence water uptake and leaf hydration.

Diurnal Patterns of Net Photosynthesis

1 General Pattern If one monitors continuously the rates of photosynthesis over a day, the type and magnitude of curve produced depend very much on plant and environmental factors. These factors include species, genotype, relative tolerance, age of leaves, position in crown, light intensity, water relations, and nutrition.

In general terms, however, the situation is as shown in Fig. 5-15. Positive rates of net photosynthesis are recorded in the morning as soon as light intensity exceeds the compensation point. In the early morning stomata are open, tem-

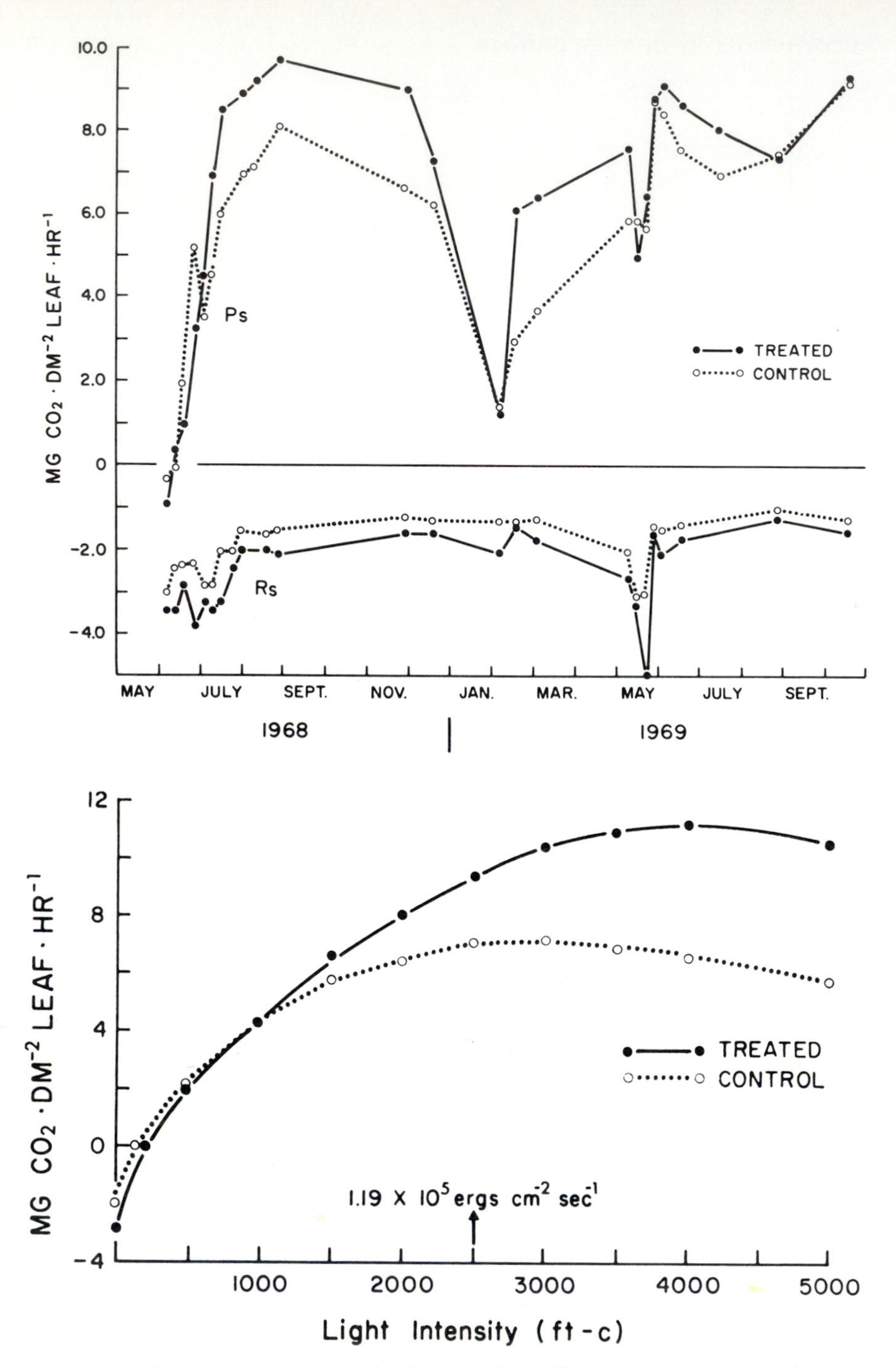

Figure 5-14 Effects of nitrogen fertilization in April 1968 on net photosynthesis (Ps) and dark respiration (Rs) of 24-year-old Douglas-fir. Gas exchange was measured on detached foliage. The diameter growth of fertilized trees increased by 51 and 54 percent in the first and second year after fertilization. (*From Brix, 1971.*)

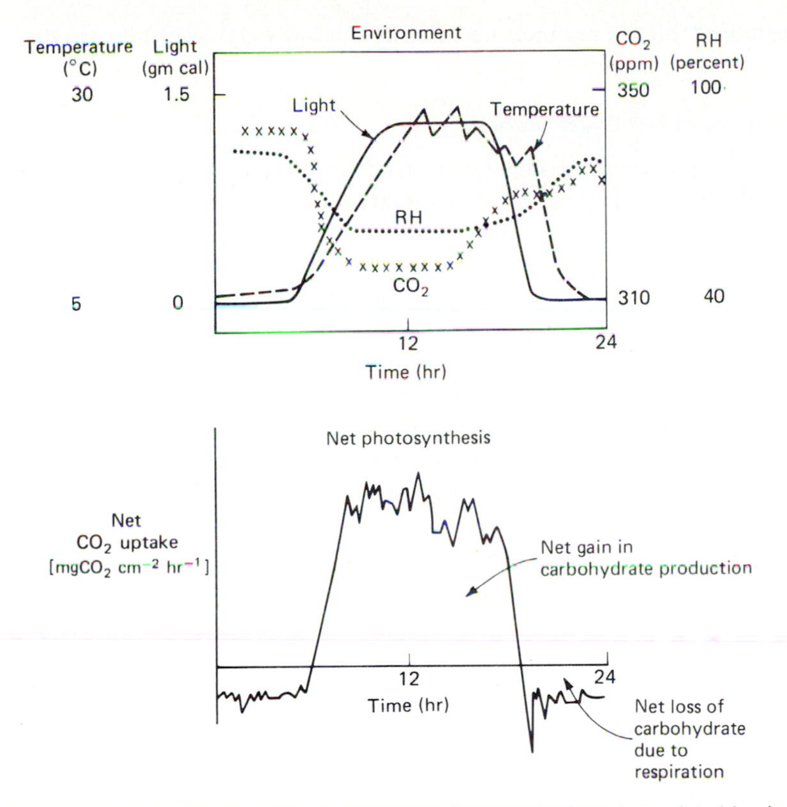

Figure 5-15 Typical diurnal pattern of net photosynthesis and associated levels of environmental conditions.

peratures are low, water is most available, tissues are turgid, and carbon dioxide content of the air is above average because winds and convection currents have not yet dispersed the carbon dioxide produced during the night by plant respiration, decomposition of organic matter, and soil respiration. Consequently, rates of photosynthesis normally increase in direct proportion to increases in light intensity. By mid to late morning, rates of photosynthesis have reached their maximum and have started to decline because temperatures are above the optimum, plant water potentials are decreasing, the stomata are partially closed, and possibly photosynthate is accumulating. Midday depressions may occur in the photosynthetic curve if environmental conditions cause leaf water stress. Later in the day, a slight increase in net photosynthesis may occur as temperature drops, and then the curve decreases with the decrease in light intensity. Of course, variations in this general pattern occur depending on plant and environmental factors.

2 Effects of Stress Photosynthetic capacity of trees is markedly influenced by the extent to which stress, caused mainly by the cumulative effects of high temperatures, high vapor-pressure deficits, and low water availability in the leaf, builds up during the course of the day.

Seasonal Patterns of Net Photosynthesis

For the silviculturist who is manipulating stands over the course of decades, it is not only important to appreciate potential rates of photosynthesis and diurnal patterns, one must also predict seasonal potentials for carbohydrate production and the likely performance over several years.

Seasonal performance depends on plant and environmental factors. In particular, patterns obtained depend on location. For example, in some locations in the Southeastern states and in coastal California, some species may essentially photosynthesize actively all year around. In northern latitudes and at high elevations, low air temperatures, low light intensities, and frozen soils may result in the cessation of photosynthetic activity. Figure 5-16 illustrates for each season of the year the net gain in photosynthate production for coastal Douglas-fir. A comparison of these seasonal patterns shows two important factors governing the amount of daily net carbohydrate production. These are (1) the length of the daily net photosynthetic period when light intensity is above the compensation point (this varies with season, as well as the position of the plant in the canopy) and (2) the effect of high ambient day and night temperatures, which increase the proportion of carbohydrate consumed by high rates of respiration.

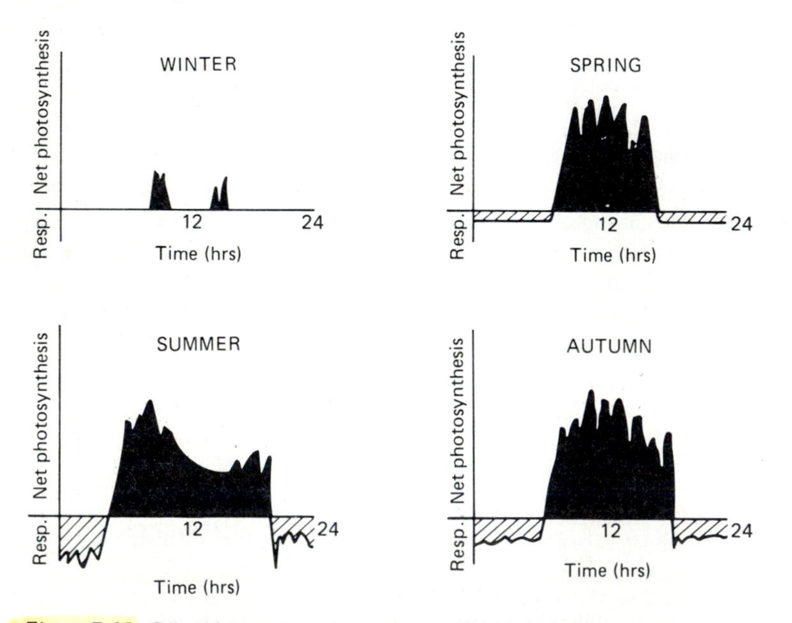

Figure 5-16 Diurnal patterns of net photosynthesis for each season of the year (Douglas-fir, Pacific Northwest).

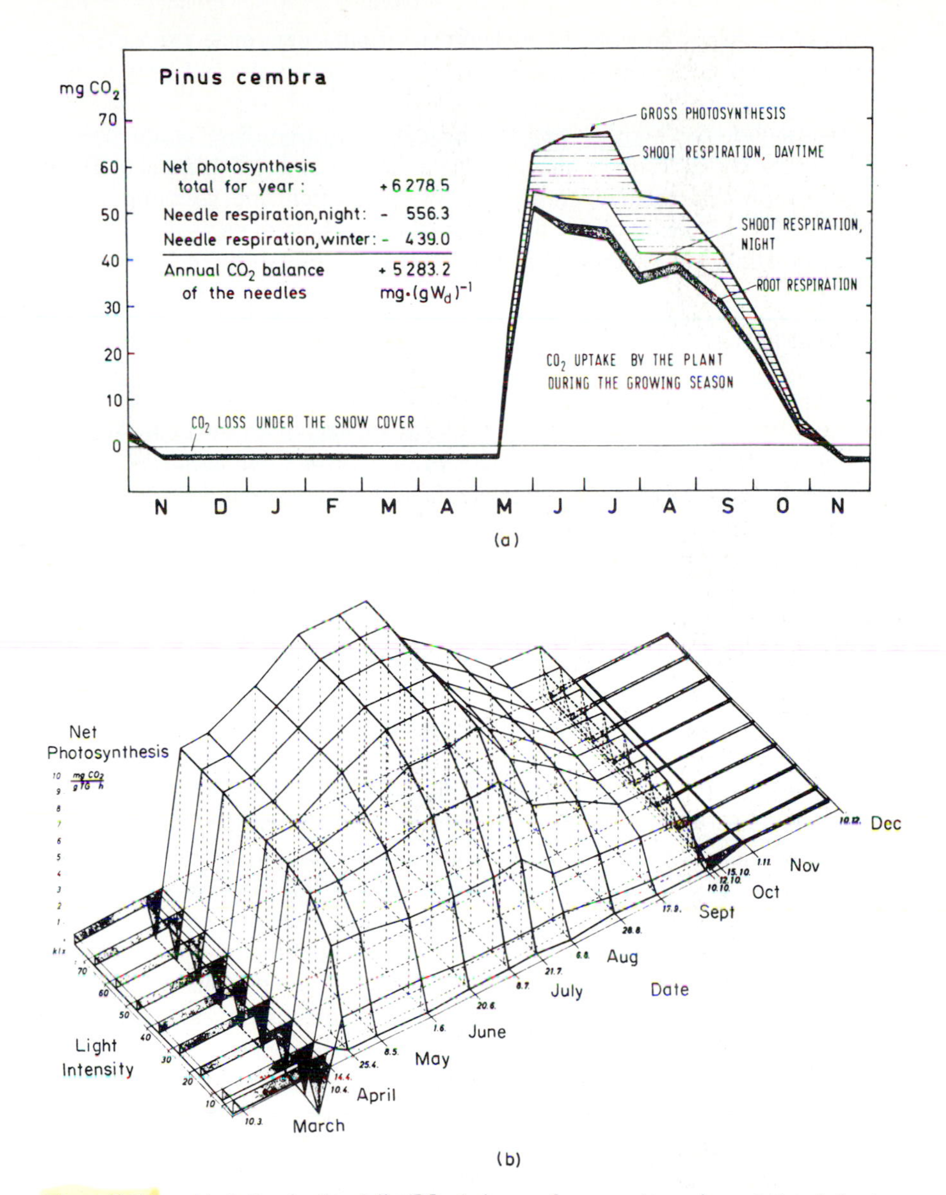

Figure 5-17 a: Variation in the daily CO_2 balance of young stone pines at the alpine tree line over a year. Part of the CO_2 gained by photosynthesis is lost the same day due to the respiration of the shoots and roots. In winter the daily balance is usually negative or at best zero: the CO_2 loss during the six winter months is subtracted from the CO_2 acquired during the growing season to obtain the annual balance. (*From Larcher, 1975.*) b: The relationship between photosynthesis and light intensity over the course of a year for beech leaves (*Fagus silvatica*) in North Germany. The shaded area indicates CO_2 release by bud respiration. (*From Schulze, 1970.*)

In controlling plant growth, the silviculturist should determine the extent to which these two factors can be modified by selection of species and treatments.

Annual Pattern Another way in which net photosynthate production of plants can be viewed is on an annual basis, as shown in Fig. 5-17*a* and *b*. These diagrams show the relative proportions of production and consumption of carbohydrate over the course of the year. Gain in carbohydrate production over the year determines the success of an individual plant in that environment.

RESPIRATION

Respiration is the use of carbohydrate and the products of photosynthesis to build and maintain all plant tissues and to produce energy for use in metabolism and nutrient uptake. Under aerobic conditions, respiration produces energy, carbon dioxide, and water. All living plants must respire—even seeds in storage. However, in unfavorable environments plants respire excessively, causing a decline in vigor and even death.

Environmental Effects

1 Light Many plants have two distinctly different kinds of respiration processes: one that occurs in the dark (and possibly also in light) and another that occurs only in light, called *photorespiration*, which has a different metabolic pathway.

The importance of photorespiration has generally been underestimated in the past and has only recently been brought to greater attention (Decker, 1970; Ludlow and Jarvis, 1971). In many studies in the past, respiration in light has been taken to be equivalent to dark respiration, and this practice has probably resulted in an underestimation of photorespiration by a factor of 3 to 4 (Zelitch, 1971). Some dicotyledonous plants and many of the tropical grasses such as sugar cane, maize, and sorghum, which are in the C_4 group of plants, appear to lack photorespiration, and this may partly account for their exceptionally high productivity (Black, 1971).

2 Temperature As temperature increases, rates of respiration commonly increase exponentially. A stage is then reached where coagulation of protein begins to occur. At this stage, rates of respiration begin to decline and then fall off rapidly as the plant material dies. This typical relationship is illustrated in Fig. 5-10, where a generalized curve for respiration is superimposed on a general curve for gross photosynthesis. Examination of this figure is instructive because it illustrates a key principle in the capacity of plants to grow in a particular environment.

In Fig. 5-10 it can be seen that as temperature rises from some limiting minimum value to some excessively high maximum level, rates of photosynthesis rise to a maximum at an optimal temperature range and then fall as high temperatures become limiting. Similarly, over the same range in temperature, respiration is increased but at an exponential rate, finally diminishing as cell organization and structure break down. An equivalence point is reached within a particular range of temperatures where the amount of carbohydrate produced in photosynthesis equals the amount consumed in respiration. If this critical temperature range is exceeded and maintained for any length of time, the plant will not survive, since respiration is consistently greater than photosynthesis. At temperatures lower than the critical level, an optimum range exists within which the net gain in carbohydrate production is a maximum. At even lower temperatures, although respiration is minimal, the capacity of the plant to gain in net food production is also much reduced.

For the silviculturist, this concept of the plant's relative capacities for photosynthesis and respiration under different temperature regimens helps explain many of the variations in plant form and community structure throughout the world's diverse environments. It also provides a basis for predicting the outcome of manipulatory treatments. For example, in a regeneration program, seedlings cannot be expected to survive temperatures that commonly exceed the range where photosynthesis equals respiration. Any site-preparation treatment should consequently take this consideration into account and should be designed to modify the microenvironment in a way that is most favorable to seedling survival and growth. Similarly, in a thinning program, where the opening up of the forest canopy can be expected to increase average air and leaf temperatures, judgments can be made as to the relative responses of understory and released tree crowns to this environmental change. Any manipulatory practice that unnecessarily exposes plants to temperatures above or below optimal levels for carbohydrate production for extended periods of time should be avoided since it will result in diminished net food production.

3 Soil Atmosphere Increases in CO_2 concentration and lack of oxygen commonly reduce rates of respiration. Oxygen in the soil atmosphere can be consumed to the point where this, plus concurrent buildup of respiratory CO_2, limits root metabolism and growth. For this reason, the silviculturist may consider treatments at the time of planting aimed at enhancing soil aeration.

4 Water The effect of increasing water stress on respiration rate depends on species. As illustrated in Fig. 5-12, for four species of true fir, respiration was not markedly affected until stress levels reached approximately −10 to −12 bars. As water stress increased, the species were differentially affected, with *Abies balsamea* being most sensitive and *Abies grandis* the least. Respiration of all species leveled off at stresses between −14 and −20 bars.

5 Nutrition As shown earlier in Fig. 5-14, fertilization can increase rates of dark respiration in shoots. Providing adequate water is available, fertilization tends to produce bigger and more succulent leaves which have greater metabolism and higher rates of respiration.

Diurnal and Seasonal Patterns of Respiration

Seasonal patterns of respiration in trees largely depend on the part of the tree being considered and the seasonal development of that component part. This is so because respiration increases with metabolic activity. When roots, buds, or leaves are actively developing, respiration tends to be high. Superimposed on this phenological trend are the effects of environment. Figure 5-16 illustrates seasonal patterns in net CO_2 evolution in leaves. Since the total respiratory losses of carbohydrate may reach 50 percent of the total produced, it becomes apparent that it is highly important that the silviculturist incorporate treatments designed to minimize these losses in respiration, particularly in seedlings. This can be done primarily with practices such as site preparation and control of canopy structure designed to reduce temperature.

TRANSLOCATION

Translocation involves the movement within a plant system of a variety of materials including gases, water, minerals, soluble carbohydrate, and hormones. This process occurs within all plant systems, including germinating seeds, but is particularly well developed in trees, which have a specialized vascular system consisting of xylem and phloem elements that enable movement of materials between the widely separated roots and leaves. The movement of soluble carbohydrate is from a point of origin (*source*) to a point of utilization (*sink*). The sources may be photosynthesizing mature leaves or centers of carbohydrate storage in leaves, stem, or roots. The sink can be any region of active metabolism, especially the cambium or developing buds, leaves, or fruit.

Prior to leaf fall, stored food materials in senescent leaves are hydrolyzed and translocated out of the leaf. Some elements also are mobilized and exported. These mobile elements include Na, Cl, S, N, P, and K, which then become available for physiological processes elsewhere in the plant, particularly in young leaves and areas of active metabolism. These mobile elements can also be leached out of the leaf in significant quantities by rain and dew. The immobile elements such as Mg, Ca, B, and Co remain in the older leaves and are returned to the soil with leaf fall. Some elements, like P, seem to be continually circulating within the plant. Mobility of Fe is dependent on P nutrition and pH. The importance of this mobilization of nutrients and its role in maintaining a nutrient balance in the forest is discussed in the section in Chap. 10 entitled Nutrient Cycling.

Pathways

Most downward movement occurs in the phloem, which consists of sieve elements, companion cells, parenchyma, and phloem fibers. All translocation occurs in living cells, and the functional life of phloem in both hardwoods and conifers is about 1 year. There are differences in the conducting elements of the two tree groups: hardwood conducting elements (vessels) have more highly specialized end walls, and conifers have less specialized elements (tracheids), with the sieve areas located primarily on the radial walls. Downward translocation of assimilates is greatly affected by the extent and respiratory activity of the root system. Pine seedlings with poorly developed roots and with low respiration rates translocate less photosynthate to the roots than do seedlings with strongly developed and active root systems. Even though it is often difficult to distinguish between cause and effect, there is evidently a relationship between rates of respiration and translocation. Possibly, this relationship provides a mechanism that explains the ecological significance of mycorrhizal associations with forest trees (Shiroya et al., 1962).

Velocity

General statements regarding velocity of movement are difficult to make because of the confusing and contradictory reports in the literature. Early studies reported high velocities, but more recent works indicate that this is not so. Perhaps the best generalization at this time is that maximum velocities in the phloem, based on mass transfer, are in the order of 40 to 70 cm h^{-1} in hardwoods and 18 to 20 cm h^{-1} in conifers. Average velocities, however, are commonly 1 to 2 cm h^{-1} in both tree groups (Shiroya et al., 1962; Canny et al., 1968; Roberts, 1964; Zimmermann and Brown, 1971).

Mechanisms

Several mechanisms have been proposed to account for the movement of sugars throughout the plant. The most commonly accepted mechanism is the pressure-flow theory first proposed by Munch in 1930. This theory proposes that movement occurs as a result of a turgor-pressure gradient which develops between a net producer cell, such as a mature leaf, and a net consumer cell, which can be a root, fruit, meristem, or any metabolizing cell.

A pressure gradient develops since the producer cell maintains high solute concentration (low potential) because of photosynthesis or active concentration of solutes and the consumer cell maintains low solute concentration (high potential) because of respiration, growth, and storage. A flow of solution occurs in response to this pressure gradient. The driving force enabling long-distance translocation is the plant's metabolism, and the process is regulated by the demand at the physiological sink and the supply at the source (Zimmermann and Brown, 1971).

Effect of Environment on Translocation

1 Light In general, increases in light intensity increase translocation to the roots by stimulating CO_2 uptake by the leaves and the production of assimilates.

This appears to be supported by the observation that plants grown under low light intensity stop translocation.

2 Temperature Translocation generally increases with temperatures up to about 30°C. As temperature increases further, translocation decreases, perhaps as a result of increases in carbohydrate consumption by respiration.

3 Water Water relations affect translocation by changing the physiological state of exporting leaves. Generally, translocation decreases with increasing water stress because of the lowering of root metabolism and decreased foliar CO_2 uptake.

WATER RELATIONS

The Importance of Water

The importance of water in the soil-plant-atmosphere system cannot be overemphasized, since availability of water is, in areas of summer drought, the most important of all factors in controlling survival and hence the distribution of vegetation. The water relations of plants are consequently a prime consideration in the development or application of all silvicultural treatments. Most water is lost in the passive process of transpiration. Physiologically, water is important as the prime constituent of protoplasm and vacuolar sap, as a solvent for gas and solutes, for transporting minerals, and to maintain turgidity. Full turgor, i.e., the maintenance of turgidity, is essential for cell elongation and growth, for maintenance of plant form, for the opening of stomata, and for plant movements, such as in leaves and flower petals.

Almost all the water used by plants is taken up by the root systems. Some can be taken up directly from the atmosphere by leaves, and this may be of importance in plants in arid areas where dew occurs (Monteith, 1963; Stone, 1963). However, the importance of atmospheric moisture appears to lie more in the reduction of evapotranspirational stress than in the direct supply of water to the plant.

The Movement of Water in Trees

Water-Potential Gradients Water moves in trees because there is a gradient in the free-energy status of water within the soil-plant-atmosphere system. By convention the free energy, or chemical water potential, of pure water is 0. The presence of solute particles lowers the water potential to some negative value, and in plant cells, increasing wall pressure increases the free-energy potential so that the tendency for water molecules to diffuse is increased. These considerations are described in an equation which expresses water potential in the following manner (Slatyer, 1967; Kramer, 1969):

$$\psi_{\text{cell}} = \psi_s + \psi_p + \psi_m$$

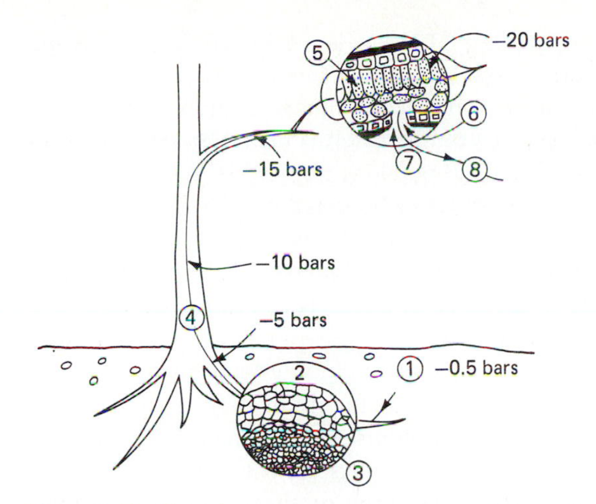

Figure 5-18 Main factors in water transport: (1) capillary size; (2) root bark; (3) endodermis; (4) vessels or fibers; (5) leaf parenchyma, mesophyll and intercellular spaces; (6) stomatal aperture; (7) boundary layer; and (8) vapor-pressure gradient from leaf to atmosphere.

where ψ_{cell} is the potential of water in the cell, and ψ_s, ψ_p, and ψ_m are contributions to ψ_{cell} by solutes in the cell, wall pressure, and matrix forces of water-binding colloids and surfaces in the cell. ψ_s and ψ_m are negative, with ψ_m commonly being very small and sometimes disregarded. ψ_p may be either positive or negative. ψ_{cell} is therefore usually negative and becomes more negative as water becomes less available. When the cell is fully turgid, ψ_{cell} approaches 0.

Water uptake occurs because the xylary fluid in the root is commonly at a lower potential (more negative value) than is the water in the soil. Due mainly to increasing solute concentrations from the root cortex to mesophyll cells, there is a gradient in water potential in the system, and water will tend to move from the roots to the leaves. This description of the mechanism of water movement based on a gradient in water potential within the soil-plant system is illustrated in Fig. 5-18. This diagram indicates that movement of water will tend to be rapid when water potential is high in the soil (i.e., close to 0, when soils have abundant available water) and low in the leaf (i.e., very negative, when temperatures and wind are high and atmospheric water vapor is low, resulting in high evapotranspiration rates).

When water moves from the mesophyll cells into the substomatal cavity, it undergoes a change in state from liquid to vapor, and movement into the atmosphere is by diffusion. Thus water in the plant moves in response to a vapor-pressure gradient.

Pathway of Water Movement Water moves upward in trees along the path of least resistance. In conifers, this is along the axial direction of the xylem tracheids (i.e., grain), and in hardwoods, along the direction of the xylem vessels. These water-conducting elements of the xylem are dead. In conifers and diffuse-porous hardwoods, several growth rings are involved in conduction. In ring-porous hardwoods, water movement occurs almost entirely through the large earlywood vessels of the current year (Zimmermann and Brown, 1971). In conifers, water moves from one 0.5- to 3.0-mm-long tracheid to another by passing through thousands of cross walls containing bordered pits on the radial walls. Hardwoods have a more complex system, which, in addition to bordered pits, develops a xylem vessel in which end walls disappear. This provides vessel elements which in diffuse-porous trees may be 1 m long and in some ring-porous trees may extend from the roots to the leaves.

The xylem of ring-porous trees ceases to function effectively after 1 or 2 years, whereas the xylem of diffuse-porous trees functions for many years. Consequently, in diffuse-porous trees and in conifers, many annual rings are functional in water uptake, although the most active rings may be near the cambium.

Most conifers have spiral grain and the pitch of the spiral often changes from year to year, providing a stem in which successive growth rings have tracheids aligned at somewhat different angles. In extreme cases, this produces trees with interlocked grain. Spiral grain results in a spiral pattern of water uptake which has been well illustrated by Rudinsky and Vité (1959) and by Werner (1971). The patterns are not entirely consistent within a species or genus, and although trends have been demonstrated (Fig. 5-19), there is considerable variation among trees in pitch and direction of spiral. The interesting silvicultural point raised by Rudinsky and Vité is that the angle at which tracheids are laid down may be increased by pruning, which reduces the transpiration surface. They also showed that the release of white-fir trees increased the transpiration surface, reversed the direction of tracheids, and decreased their degree of spirality.

It has also been shown that *Pinus contorta, P. ponderosa, P. resinosa,* and the five southern pines spiral to the right, and *Pinus lambertiana, P. monticola, P. banksiana, P. strobus,* and *P. sylvestris* spiral to the left (Kozlowski and Winget, 1963; Werner, 1971).

In a somewhat similar manner to conifers, the vessels of ring-porous hardwoods may "interweave" to provide a broader distribution of water from any one point of uptake. These complex patterns of water uptake in both conifers and hardwoods provide an effective mechanism which probably safeguards growth and development against damage to the stem or loss of roots.

Resistance to Water Movement The major resistance to water uptake is across the suberized endodermis of the root. This resistance gives rise to the almost daily development of an absorption lag. Water is transpired from the system faster than it is absorbed through the roots. This gives rise to midday deficits and a tendency to wilt which is readily observed in herbaceous plants. In trees, because of their coriaceous leaves with cuticular covering, the foliage may not appear to wilt but may nevertheless be under considerable water

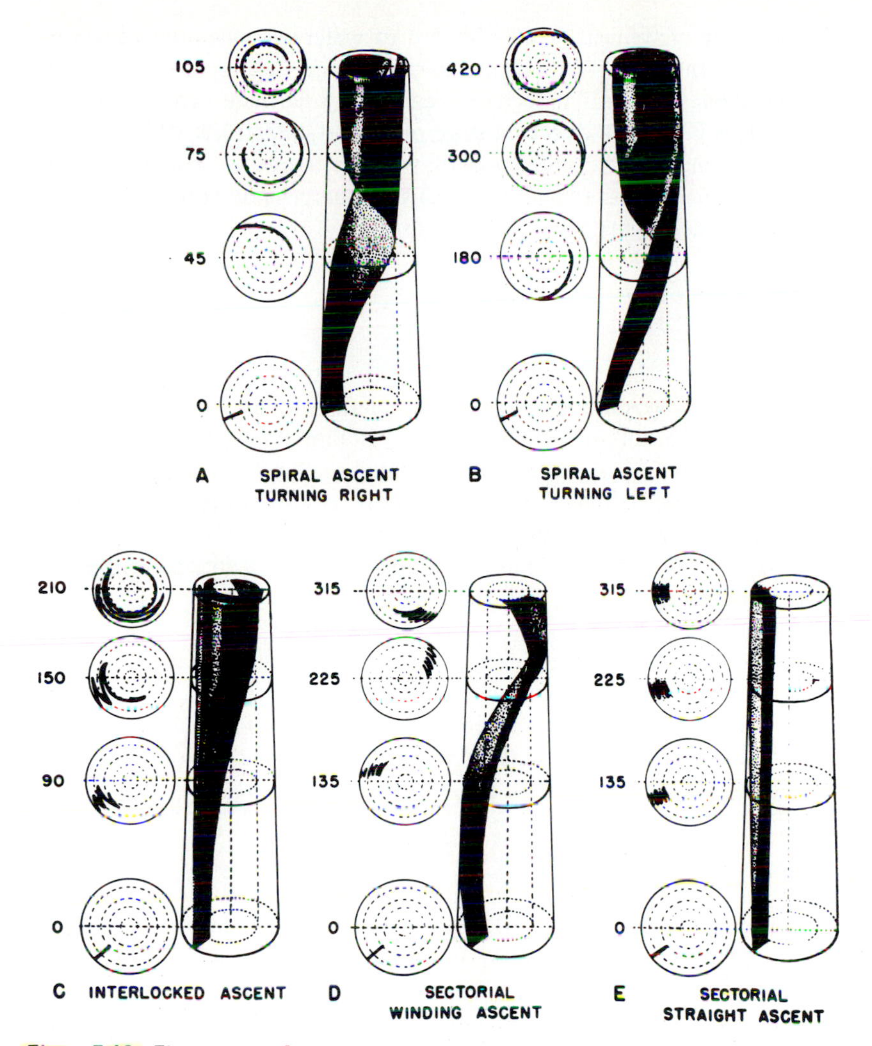

Figure 5-19 Five types of water-conducting systems in various conifers as shown by the tracheidal channels that were dyed by trunk injection. The numbers give the height in centimeters of the transverse section above injection. a: Spiral ascent, turning right: *Abies, Picea, Larix,* and *Pinus* (Rehder's section III; *Taeda*). b: Spiral ascent, turning left: *Pinus* (section I: *Cembra*). c: Interlocked ascent: *Sequoia*, *Libocedrus*, and *Juniperus.* d: Sectorial winding ascent: *Tsuga* and *Pseudotsuga.* e: Sectorial, straight ascent: *Thuja* and *Chamaecyparis.* (*Courtesy of Boyce Thompson Institute for Plant Research. From Rudinsky and Vité, 1959.*)

stress. An even greater resistance to water flow is in the soil itself once it is below field capacity. This fact indicates the importance of a high degree of root-soil contact and the importance of root hairs and mycorrhizae. In both hardwoods and conifers, stem resistance is relatively low (Zimmermann and Brown, 1971). Blocked vessels or tracheids seem to have no effect on movement because of efficient lateral movement. The resistance to water movement in the leaf can vary because of the capacity of plants to control stomatal aperture. Measures of leaf conductance (or leaf resistance) can readily be made with a diffusion porometer. This provides a direct measure of stomatal aperture, which is probably the single most useful indication of whether a tree or seedling is being stressed under particular site conditions. This is true because under any kind of environmental stress, stomata will tend to close.

TRANSPIRATION

Mechanism

Transpiration is the evaporation of water from plants and involves the movement of water through the soil-plant-atmosphere continuum. As water is lost from the leaves through evaporation, more water is pulled up the stem and through the roots in a continuous column. The stem and roots are more or less passive in this process, for which reason it is often termed *passive uptake.* The roots, of course, do not act simply as wicks; they must constantly grow to maintain permeability and exploit moist soil.

In rare cases when transpiration is very low, especially when the soil is moist and atmospheric water levels are high, water may be pushed through plants by root pressure. As salts are transported into the xylem, they increase the xylem solute concentration, resulting in osmotically induced water movement. Root pressure is rarely found in conifers; it may lead to exudation of water from special leaf cells called *hydathodes.* The process is called *guttation* and, incidentally, is *not* the cause of maple sap flow.

Velocity and Quantities

The velocity of water transport up the xylem of trees is influenced by the steepness of the gradient in water potential from the atmosphere to the soil solution. Velocities also vary with species, but commonly quoted values are: conifers, 1 to 2 m h^{-1}; diffuse porous hardwoods, 1 to 6 m h^{-1}; and ring-porous hardwoods, 20 to 40 m h^{-1}. The quantity of water transpired has been estimated to be approximately 430 to 560 mm yr^{-1} (17 to 22 in yr^{-1}) for hardwoods in South Carolina (Hoover, 1944) and about 100 to 250 mm yr^{-1} (4 to 10 in yr^{-1}) for pine in Europe (Ivanov et al., 1951; Isakov, 1974). This amounts to about one-third of precipitation. The possibility of errors resulting from differing procedures and conditions is so great that these velocities and quantities must be regarded as very rough estimates. Because the amount of water transpired is, among other factors, strongly influenced by the amount of water available, there is no good generalization regarding the relative amounts of water transpired by hardwoods and conifers.

Attempts at decreasing water use of trees and seedlings by the application of antitranspirants generally have not been useful in forestry. Incentives for the use of antitranspirants come from the desire to increase yields from forest watersheds or to enhance survival of seedlings when outplanting. Work done in red-pine forests (Waggoner and Turner, 1971) indicate that evapotranspiration can be reduced by up to 30 percent immediately after spraying with phenylmercuric acetate. However, after spraying three times in each growing season for three successive years, the overall reduction was 3 to 10 percent. The treatment causes partial stomatal closure, and since this also reduces photosynthesis, some growth loss might occur.

Effect of Environment on Transpiration

1 Light Transpiration is markedly influenced by light intensity because of the direct influence of light on stomatal aperture.

2 Temperature The temperatures of the soil, leaf, and atmosphere also affect the rate of water uptake. Cold soils reduce uptake because they reduce root permeability as well as water movement and slow down root growth and metabolism. Leaf temperature is of special interest since it influences leaf metabolism, photosynthesis, respiration, and transpiration directly. During the day, leaves exposed to radiation from the sun may be 1 to 10°C higher than ambient air temperature, whereas leaves in the shade may have temperatures about the same as the atmosphere. At night, leaf temperatures may be 2 to 3°C lower than ambient air temperature due to reradiation into the atmosphere. Transpiration has the effect of cooling leaves. This reduction may result in transpiring leaves being up to 10°C lower than ambient air temperature, particularly when the heat load is large at air temperatures greater than 30°C (Gates, 1968).

3 Vapor-Pressure Deficit This term describes the difference between the water-vapor content of the air surrounding the leaf and the water-vapor content of the stomatal cavity. The greater the difference, or deficit, the greater the tendency for the plant to lose water or transpire. Vapor-pressure deficit therefore is a major factor controlling transpiration. It is strongly influenced by temperature, wind, and relative humidity.

4 Water Availability Transpiration depends on available soil water, and rates of transpiration are increased by supplying more water to the plant. This effect can be seen in the high rates of transpiration of riparian vegetation and irrigated trees. However, even under conditions of high soil water availability, when the air is warm, windy, and at low relative humidity, transpiration can exceed water uptake, and wilting and stomatal closure can occur. This, in turn, generally reduces photosynthesis and growth.

When a tree is exposed to conditions of decreasing water availability, the first process to be retarded is transpiration, followed by photosynthesis, and then respiration. The sensitivity of trees to increasing stress conditions is largely influenced by the relative tolerance of the plant in question. In general, it has been found that as soil water potential becomes more negative (i.e., soil water is less available), the more intolerant trees are the first to show reductions in transpiration rates as a result of their capacity to close stomata early. The more tolerant species maintain high rates of transpiration with increasing stress conditions because they maintain their stomata in an open condition at lower water potentials than do the intolerant trees. For example, in California on a warm summer's day, the leaf water potential of mature ponderosa-pine trees was -18 bars, whereas the tolerant white-fir trees had developed stress levels of -23 bars.

Conservation of water through genetic and adaptive means of stomatal control is therefore of paramount importance in influencing the internal water balance of plants growing in areas with dry summers. Stomata are the mechanisms which are largely responsible for determining the ecological success various plants have in satisfactorily developing in relatively dry environments.

MATCHING ENVIRONMENTS WITH PHYSIOLOGICAL REQUIREMENTS

The most important unifying concept that must be understood—one that underlies all of silviculture—is the recognition that the silviculturist's prime objective is to control the growth of trees, and hence the structure and composition of stands, by controlling gene pools and environments. Stands are dynamic, therefore treatments must accommodate change. Thus, the ultimate success of any treatment is dependent on the silviculturist's ability to predict the behavior of vegetation following natural or manipulated changes in environment.

The only way these predictions can be made is with an appreciation of the way in which genetics, physiology, and environments interact in the control of tree growth.

To grow successfully, a tree must

1 Produce more food by photosynthesis than it needs to sustain basic metabolism and compensate for respiration.

2 Have sufficient control over its internal water relations such that water is conserved, cells are maintained at full turgor, and stomata are open for an adequate period during the day for carbohydrate production.

The silviculturist must select genotypes and maintain environments such that these two growth requirements can be met. However, conditions favorable for one physiological process are not necessarily favorable for another, and we are commonly confronted with selecting a compromise environment which is, for the most part, favorable for the plant as a whole. For example, conditions which tend to maximize gross photosynthesis (such as high light intensity, open stomata, and large, well-displayed leaf area) also tend to result in high rates of respiration and transpiration. Clearly, potential conflicts can exist that can only be resolved if we can predict which processes are likely to be limiting in any given situation. In many parts of California, Utah, and Idaho, for example, where summer precipitation is infrequent, most mortality of seedlings is due to adverse water relations, excessive respiration, or external biotic factors such as insects, disease, or browsing. Capacity to photosynthesize usually does not limit survival, consequently the site-preparation treatment is more likely to be successful if the microenvironment is manipulated to conserve water, lower radiation and temperature loads, and limit the amount of exposed leaf area, despite the fact that this tends to reduce photosynthesis. This relative capacity of a plant to produce adequate carbohydrate and maintain adequate water relations in a given environment is sometimes expressed in terms of a net photosynthesis/transpiration ratio.

Success in manipulating plant growth therefore depends on the ability to predict the environmental requirements of species and genotypes and to determine the extent to which one can create the desired environment through treatments. The key concept involved here is that a plant may be considered as having a multidimensional response surface described primarily in terms of net photosynthesis, transpiration, and leaf conductance as functions of the interactions among environmental factors such as light intensity, temperature, and vapor-pressure deficit. Conceptualization of such interactions enables the silviculturist to predict the likely capacity of a particular plant to develop within a particular microenvironment. This provides a rational ecophysiological basis for the selection of particular genotypes or development of particular microenvironments through silvicultural treatment. For example, light intensity can be modified by manipulating crown cover, soil water availability enhanced by reducing plant competition, or temperature levels changed by control of shade.

Thinning, fertilizing, spraying, mulching, or bedding are essentially aimed at enhancing the operational environment of selected plants. The precise definition of each treatment in terms of level or intensity can best be made if one is aware of the general physiological requirements of the selected trees and the levels of microclimatic factors that are needed for satisfactory growth. The same kind of analysis is useful when considering the choice of the most appropriate reproduction method for ensuring the regeneration of particular species.

Chapter 6

Stem Cambium Development

INITIATION OF CAMBIUM DEVELOPMENT AND GROWTH RINGS

Early in each growing season, the 2- to 3-cell-wide layer of cambial cells around the stem is reactivated by becoming hydrated, the cells expand radially, radial walls become thinner, and the cytoplasm becomes less dense. This cambial reactivation is associated with the well-known phenomenon of bark slippage in the spring, when bark is easily peeled from a log. Some 2 to 4 weeks later, cambial divisions start to occur, slowly at first, then more rapidly during the grand period of growth when the cambium is 10 to 15 cells wide. Finally, in late summer, the width of the cambium and the rate of cell division declines. The maximum rate of cambial division varies between species and locations. *Thuja occidentalis*, near Toronto, Canada, produces up to one division every 4 to 6 days (Bannan, 1962), whereas red pine in the Lake states produces 4 to 5 divisions per week (Zahner, 1963). More rapidly growing Douglas-fir, southern pines, and tropical species may undergo more rapid cambial division.

It is now well accepted, in accordance with the original observations made by Hartig in 1853, that the initiation of cell division is strongly associated with the translocation of hormones from expanding buds and leaves (Larson, 1962; Brown, 1970). Thus cambial development starts near the base of the buds and moves basipetally down the tree. The time delay between the initiation of cambial activity in the crown and when it occurs at the base of the tree is too short to be readily detected in ring-porous hardwoods but may be 3 weeks to a month in large diffuse-porous trees. The corresponding time delay in conifers is approximately 1 week (Wilcox, 1962).

The cambial cells divide to produce thick-walled, lignified wood cells (xylem) on one side and a layer of thin-walled cells (phloem) on the other in a proportion of about 6:1. Each new layer of phloem in the inner bark crushes the previous year's phloem and makes it nonfunctional. Also, successive cork cambia are formed in the older portion of the inner bark to produce outer layers of suberized cork cells which eventually die, are sloughed off, and give trees their characteristic bark patterns (Brown, 1970).

Conifer Growth Rings

In conifers, the cambial cells divide to produce relatively few cell types—mostly tracheids. Consequently, the wood produced is very uniform. Figure 6-1 shows a schematic drawing of typical southern-pine wood where the bulk of the woody matter consists of tracheids. The earlywood tracheids have large diameters and relatively thin walls. Part way through the growing season there is a transition from the production of earlywood to latewood tracheids which are characteristically more lignified, thicker-walled, and have smaller lumens. Resin canals are commonly visible on transverse sections. This transition from earlywood to latewood occurs first near the base of the tree, moves acropetally up the tree, and is associated with shorter day conditions, reduction in auxin synthesis accompanying cessation of height growth and needle elongation (Larson, 1960), and the development of internal water stress when daily transpiration consistently exceeds the rate of water uptake.

This transition from earlywood-type to latewood-type tracheids may be gradual or abrupt depending on the rate with which diffusible auxin concentrations are reduced, whether the species has a sharp threshold sensitivity for auxin stimulation (Wort, 1962), and the suddenness and severity of drought conditions.

The cessation of cambial division seems to be associated with a reduction in availability of both auxin and carbohydrate. This occurs in late summer when environmental conditions of shorter photoperiod, low temperature, and low water availability lead to a reduction in physiological activity within the crown. Cessation of cambial activity occurs first within the crown and upper bole and progresses basipetally toward the base of the tree.

The tracheids mature rapidly, lose their cytoplasm, and become functional in water conduction within a few days to a few weeks (Zahner, 1963). Several to many growth rings are functional in water transfer, but most movement is through the newly differentiated earlywood tracheids which have thin walls and

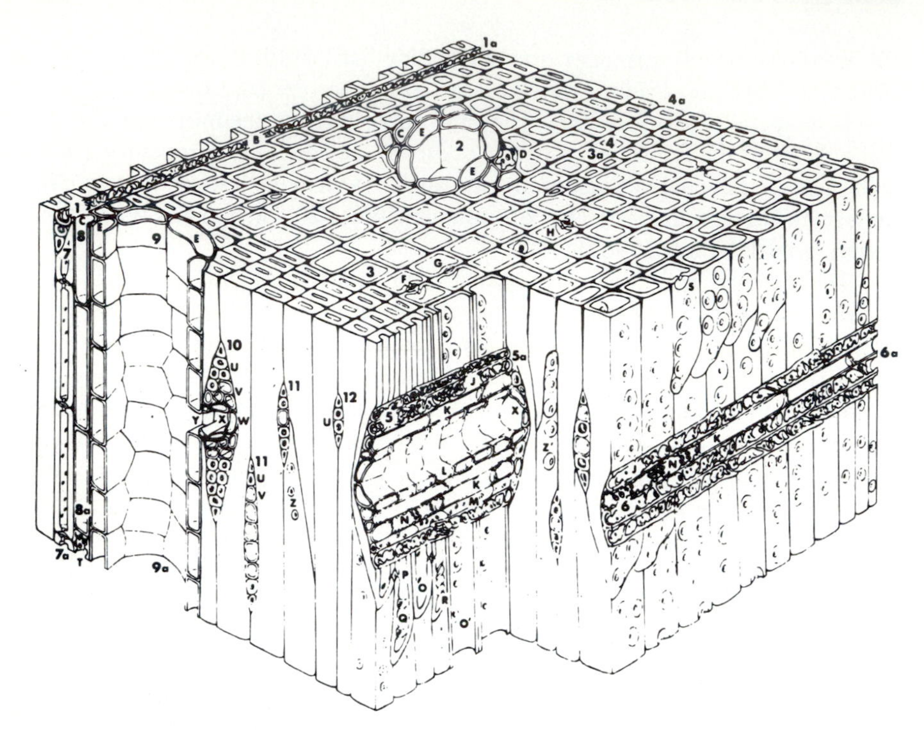

Figure 6-1 Schematic drawing of typical southern pine wood. *Transverse view:* 1–1a, ray; B, dentate ray tracheid; 2, resin canal; C, thin-walled longitudinal parenchyma; D, thick-walled longitudinal parenchyma; E, epithelial cells; 3–3a, earlywood longitudinal tracheids; F, radial bordered pit pair cut through torus and pit apertures; G, pit pair cut below pit apertures; H, tangential pit pair; 4–4a, latewood longitudinal tracheids. *Radial view:* 5–5a, sectioned fusiform ray; J, dentate ray tracheid; K, thin-walled parenchyma; L, epithelial cells; M, unsectioned ray tracheid; N, thick-walled parenchyma; O, latewood radial pit (inner aperture); O′, earlywood radial pit (inner aperture); P, tangential bordered pit; Q, callitroid-like thickenings; R, spiral thickening; S, radial bordered pits (the compound middle lamella has been stripped away, removing crassulae and tori); 6–6a, sectioned uniseriate heterogeneous ray. *Tangential view:* 7–7a, strand tracheids; 8–8a, longitudinal parenchyma (thin-walled); T, thick-walled parenchyma; 9–9a, longitudinal resin canal; 10, fusiform ray; U, ray tracheids; V, ray parenchyma; W, horizontal epithelial cells; X, horizontal resin canal; Y, opening between horizontal and vertical resin canals; 11, uniseriate heterogeneous rays; 12, uniseriate homogeneous ray; Z, small tangential pits in latewood; Z′, large tangential pits in earlywood. (*After Howard and Manwiller, 1969; from Koch, 1972.*)

many bordered pits. These earlywood tracheids thus offer lower transfer resistance than the more lignified and thicker-walled latewood cells. Tracheids are vertically aligned and may spiral or become interwoven.

Storage tissue in most softwoods is confined to parenchyma ray cells which are small boxlike cells running horizontally across the growth rings in radial bands. In softwoods, these ray cells are always very fine and occasionally have a central cavity which allows the transverse flow of resin.

The difference in density between species of softwood is largely determined by the relative proportions of thin-walled earlywood tissue and thick-walled latewood tissue.

Hardwood Growth Rings

In hardwoods, the dividing cambium usually produces three kinds of cells: vessels, fibers, and parenchyma. *Vessels* are vertically aligned and form a pipelike system for transporting water up the tree. In general, hardwood vessels do not develop in a strictly linear manner but interweave, cross over, and join together within the growth ring (Zimmermann and Brown, 1971). *Fibers* are vertically aligned, fusiform (needlelike) in shape, and constitute the matrix through which the vessels pass. The fibers vary in cell thickness and are largely responsible for giving hardwoods their strength. The density of hardwood depends primarily on the cell-wall thickness of the fibers and the proportion of vessels present. *Parenchyma* cells in hardwoods are very similar to those in softwoods. These radial bands, called *rays,* are living, thin-walled cells adapted for food storage. The radial arrangement of rays permits the transfer of food material across annual rings. Rays are commonly small, as in aspen, but in oak and beech they are very broad and produce the characteristic "figure" on radially cut surfaces.

Within genetic constraints it appears that it is primarily the hormone concentration and, to a lesser extent, the carbohydrate nutrition and water relations of the cambium which control the type of cell differentiating from the cambial mother cell (Zahner, 1970). Once developed, it is the relative numbers, dimensions, and sequences of vessels, fibers, and parenchyma which give each hardwood species its characteristic appearance and properties. In ring-porous species (oaks, ash, hickories, black locust, teak), annual growth rings are very distinct, since a band of very large diameter earlywood vessels is contrasted with the numerous small vessels found in the latewood zone. In diffuse-porous species (walnut, cherry, beech, aspen, poplar, mahogany, eucalyptus), growth rings are less distinct, since the vessels are smaller in diameter and are more uniformly distributed throughout the growth ring.

Most hardwoods of tropical origin grow continuously and show no readily distinguishable growth rings.

MULTIPLE RINGS AND MISSING RINGS

Growth rings in trees are not necessarily produced annually. Some species, particularly those in the subtropics, can produce several rings in one year in response to a temporary cessation followed by a resumption in growth. False rings, which need not extend completely around the stem, may occur in conifers, particularly in the upper parts of the stem. These are due to adverse hormone conditions or water and carbohydrate relations which affect only a portion of the stem (Larson, 1963).

In extremely adverse situations, particularly in overstocked stands where individual trees have very short crowns of low vigor, it is even possible for annual growth rings to be entirely missing at the base of the tree. In a particular growing season a small ring may be produced within the crown region of the stem, but there may be insufficient production of hormones and carbohydrate for the continued development of the ring near the base of the tree.

Since the development of the crown has such control over initiation, duration, and cessation of cambial activity, it would appear that the most important silvicultural decisions affecting growth-ring characteristics would be initial spacing, thinning, fertilizing, and pruning. All these treatments directly influence crown size and the width of the ring produced.

EFFECT OF ENVIRONMENT ON CAMBIAL DEVELOPMENT

In both hardwoods and conifers, the initiation, duration, and cessation of cambial activity is highly dependent on conditions of the crown. After bud break, rapid shoot growth and continued leaf development result in high auxin production and the production of earlywood. Earlywood is thus produced from photosynthate not required for shoot or root development (Larson and Gordon, 1969). The transition to latewood varies with the age and developmental state of the new shoot. However, the initiation of latewood commonly occurs shortly after bud set and usually coincides with new needle maturity. Transition to latewood is encouraged by low temperatures, drought, and short photoperiod, all of which adversely affect shoot extension and leaf development and lower the level of diffusible auxin (Zimmermann and Brown, 1971).

It can be seen, therefore, that there are distinct opportunities for silvicultural control over the characteristics of the growth ring. The timing of initiation of cambial activity is relatively difficult to influence since it appears to be primarily controlled by heredity, photoperiod, and temperature. It may, however, be possible in some areas to lengthen the period of cambial activity by keeping the soil above threshold levels for root activity through manipulation of crown cover and spacing. Both the transition from earlywood to latewood production and the timing of cessation of cambial activity offer more opportunity for manipulation since they are both controlled by hormonal relations, carbohydrate supply, and water relations. These factors can be directly influenced by controlling the vigor and duration of crown development through treatments which manipulate spacing, thus influencing temperature, water-potential gradients, and transpirational flux. One of the best examples of the control over earlywood-latewood transition is that of Zahner et al. (1964), who showed that flattened latewood cells began to develop in late July at breast height in red pines subjected to drought, whereas latewood cells were not initiated until 9 weeks later in late September in irrigated trees (Fig. 6-2). In this study, the

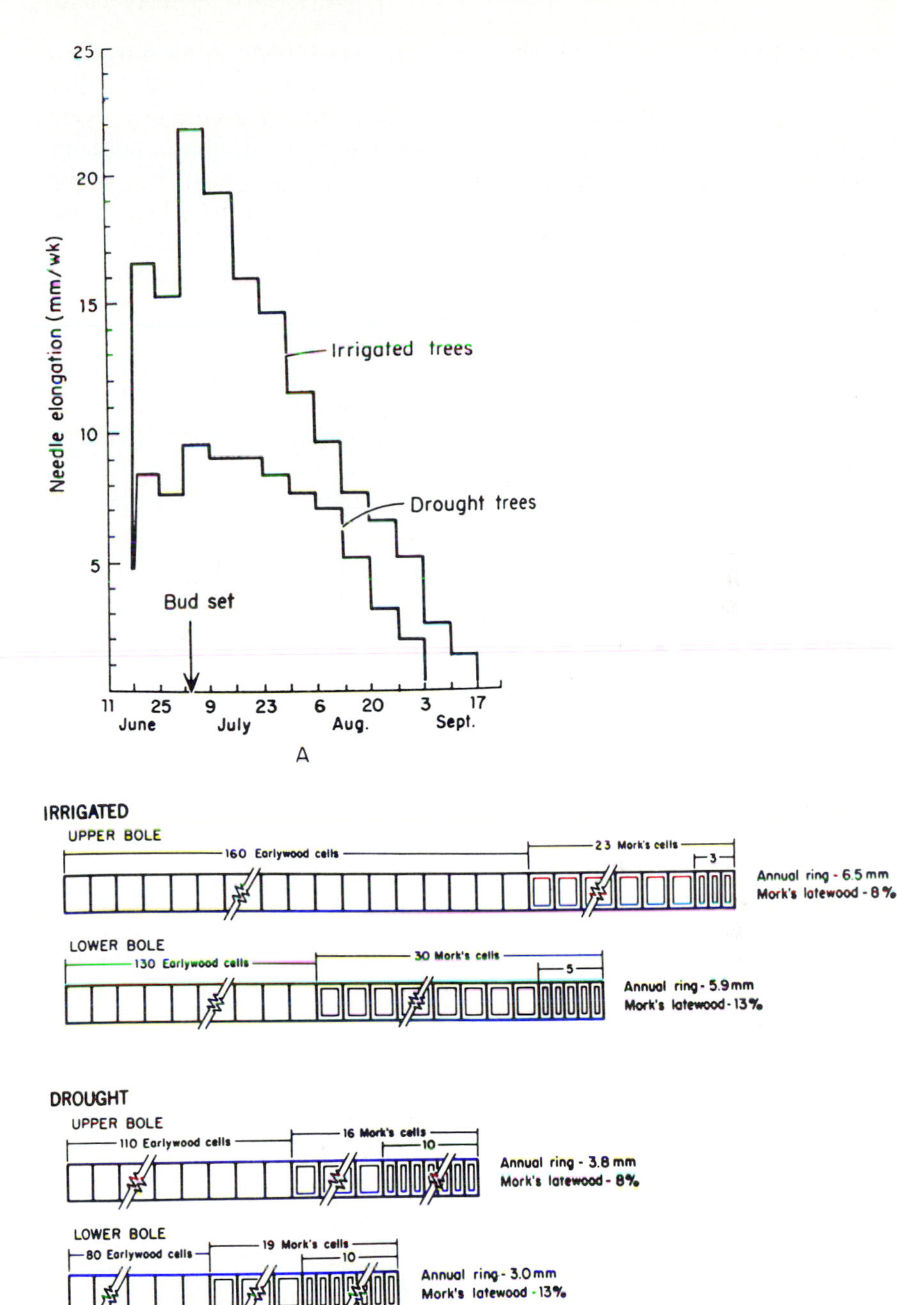

Figure 6-2 Effects of drought and irrigation treatments on: (A) weekly needle elongation and (B) earlywood-latewood formation in red pine. (*From Zahner et al., 1964.*)

changeover from earlywood to latewood cells at breast height in the droughted trees was not well correlated to cessation of needle elongation since, at the time of the changeover, needle elongation was still continuing. As Zahner et al. point out, however, the changover may still be associated with a decline in hormone availability since the low rate of needle activity in the droughted trees simply may have resulted in low concentrations of auxin reaching the lower bole. In the irrigated trees, it appears that the greater rates of needle activity produced sufficient auxin to maintain the formation of large earlywood-type tracheids throughout the entire bole, and that the changeover occurred when crown activity ceased. It can be seen from Fig. 6-2 that although the treatments had a marked influence on date of transition from earlywood to latewood, there was essentially no change in percent latewood and hence no change in ring density.

Thinning forest stands can also have an effect on the initiation of flattened summerwood tracheids in conifers. The changeover from earlywood- to latewood-type tracheids has been shown to occur 2 weeks later in thinned plots of red and jack pine (Zahner and Oliver, 1962). This delay has been attributed to a delay in the depletion of soil moisture, which in turn was responsible for the maintenance of auxin synthesis at higher levels and for a longer period in the thinned stands. It is debatable whether the effect of a thinning treatment would continue beyond the first growing season or two because once the root systems of the remaining trees have expanded to fully exploit the available soil mass, transpirational use of water by individual trees will tend to increase and the loss of water per unit area will tend to return to what it was prior to thinning. This occurs because rates of transpiration tend to be a function of the water available.

SPIRAL GRAIN

Spiral grain is of nearly universal occurrence in both hardwoods and conifers. Extensive studies of Western species report spiral grain in excess of 8° occurring in 25 to 99 percent of trees sampled (Northcott, 1957; Brazier, 1965; Lowery, 1965). In conifer stems, the grain commonly spirals to the left when the tree is young, passes through a period of straightness at an age of about 10 years, and then at an age of about 25 years spirals to the right. In pines, the twist is more intense when spiraling is to the left. In hardwoods, the pattern of spiraling is generally reversed, with the first few rings having the grain oriented to the right and then subsequently changing to the left.

The causes of spiral grain have been summarized by Naskoweak (1963), who attributes the phenomenon to the orientation and mode of division of cambial initials, which determine the axis of inclination of new initials.

Spiraling is a major cause of lumber degrade, particularly in young-growth softwoods because it is at the younger ages that the degree of spiral is greatest. The angle of the spiral is commonly 4 to 6° but may get as high as 40°. Boards cut from wood which has a 7° longitudinal spiral grain suffer a loss in strength of about 6 percent. Boards with greater than 30° spiral have virtually no bending strength whatsoever. The problem becomes alarming when poles with spiral grain have been used for telephone poles. They can twist and untwist with changes in atmospheric conditions, and the effect on the ends of crossarms can result in the breakage of the telephone lines (Banks, 1969).

In managing young stands, therefore, the silviculturist can prevent considerable economic loss in degrade by taking positive steps to limit the occurrence of spiral grain. This can be achieved primarily by the selection of seed source, since the degree of spiraling is under strong genetic control.

REACTION WOOD

Reaction wood is a name given to wood of distinctive anatomical characteristics which is formed in leaning trees. It is of particular concern to silviculturists since reaction wood is one of the principle causes of degrade in small stems (together with stem straightness and knot size) and is therefore of considerable importance in young-growth management. There are several good reviews of this subject, and these can be found in such literature as Westing (1965), Hughes (1965), Low (1964), Zimmermann and Brown (1971), Panshin and de Zeeuw (1970).

Reaction wood occurs in both leaning hardwoods and conifers and is formed under the influence of gravity. Auxins produced by actively elongating shoots move basipetally from the crown to the base of the tree. In a vertical tree, the distribution of auxin around the stem is usually even and the cambium divides at an approximately even rate all around the stem. In leaning stems (and branches) the flow of auxin is laterally displaced by gravity and accumulates in higher concentrations on the lower sides of the stem. It is an interesting physiological phenomenon that these slightly higher concentrations on lower surfaces produce opposite effects in gymnosperms and angiosperms. In conifers, higher concentrations of auxin on lower surfaces of leaning stems cause localized increased cambial activity and the formation of compression wood, eccentric stem cross section, and the tendency for the tree to resume its vertical position. In hardwoods, it is the lower concentrations of auxins on the upper surfaces of leaning stems which lead to the formation of tension wood on these upper surfaces and eccentric radial growth which again tends to restore the tree to its original vertical position.

Compression wood appears in stem cross sections of gymnosperms as crescent-shaped darker zones within the annual ring, rather like unusually wide bands of latewood. It is characterized by having shorter and rounder tracheids with numerous intercellular spaces, high lignification, higher specific gravity, and lower percent cellulose.

These characteristics combine to produce wood which, most important, has exceptionally high longitudinal shrinkage. In addition, compression wood has a lower strength/weight ratio, a greater density and is therefore more difficult to nail, and has lower pulpwood yields. Compression wood can be of very common occurrence in conifers, and according to Low (1964), it accounts for 20 to 50 percent of the standing volume of Scotch pine growing in windy areas. Thus it can be the most serious defect with regard to utilization.

Tension wood is difficult to distinguish macroscopically. The most important anatomical characteristics are thick, gelatinous fibers, low proportion of lignin, and high proportion of cellulose. It produces wood which is difficult to saw, resulting in a rough woolly surface, has high longitudinal splitting, and produces severe cellular collapse on drying. Tension wood does, however, produce higher pulp yields due to the higher proportion of cellulose, but the strength of the pulp is lower.

Silvicultural Control of Reaction Wood

Since reaction wood is one of the most important causes of degrade in young growth, it is important to determine the extent to which it can be reduced. The most important treatment is to remove trees which have been caused to lean by snow damage, injury, or forked tops. With hardwoods which have a lower degree of apical dominance it is important to control uneven spacing. It may, however, be economically disadvantageous to remove small, leaning trees before they reach merchantable size since the value of small logs with bends and sweeps will be relatively low due to problems of handling, transportation, and utilization. As log diameter increases, the effect of bends and sweeps on log value rapidly decreases. Thus the silviculturist may prefer to delay the removal of a bent stem until it reaches a diameter that is valuable. In these kinds of decisions, a compromise must be made between the increasing value of the log as it becomes larger and the potentially decreasing values resulting from producing logs with a high proportion of compression wood.

JUVENILE WOOD

The first 10 to 15 annual rings of wood laid down by conifers next to the pith in any internode have distinctive anatomical characteristics which differ from the subsequently laid down mature wood. A core of juvenile wood (or corewood) therefore exists throughout the tree, and the characteristics of this wood are strictly related to age. In a 10- to 15-year-old tree, all wood is of the juvenile type. The quantity of juvenile wood produced is consequently dependent on rate of growth, which in turn is largely a function of the size and vigor of the live crown. After age 10 to 15, the annual sheath of wood produced has juvenile characteristics in the upper bole and mature characteristics in the lower bole. Juvenile wood can therefore be described as wood formed under the direct influence of the crown.

Juvenile wood is characterized as having low specific gravity due to a low percent of latewood, low cellulose content, shorter tracheids, higher spiral grain, and commonly, compression wood. These characteristics result in wood which has low strength, low pulp yields, and high shrinkage. It is therefore generally undesirable for many uses. When lumber is produced from old trees, the proportion of juvenile wood is low. However, in short-rotation management, this proportion is quite significant. The best approach to minimizing the problem is to capitalize on the large degree of natural variation in juvenile-wood characteristics that occurs in natural stands and to increase the overall desirability of juvenile wood through genetic selection (Zobel et al., 1972). The alternative suggestion of restricting the amount of juvenile wood produced by growing stands with close spacing until age 10 to 15 and then releasing them is impractical. Greater value can be obtained by growing trees as rapidly as possible. This maximizes the amount of juvenile wood, but the mature wood is laid down on a larger-diameter stem. The greatly enhanced values derived from maximizing total wood volume vastly outweigh any disadvantages inherent in juvenile-wood characteristics. Maximizing growth rates, however, does delay the transition from juvenile to mature characteristics (Bengtson, 1968; Larson, 1969).

WOOD QUALITY

Wood is used for a tremendous array of purposes including fuel, furniture, pulp, and construction. For each product there is a type of wood that has the highest quality for that use. Thus definition of the term *wood quality* varies depending on end use. The actual properties of wood which make it more or less useful for a product depend primarily on its specific gravity, fiber length, cell-wall thickness, and cellulose and lignin content. The silviculturist is often asked the extent to which wood quality can be controlled, and to answer the question some knowledge is required of the nature of wood-quality characteristics and the extent to which they can be influenced by genetics, manipulation of the environment, and rate of growth.

Specific Gravity

Of all wood-quality criteria, specific gravity is the most important since it is directly related to the strength of lumber and to kraft pulp yields. Table 6-1 compares the density of some common hardwoods and softwoods. Strength of boards is becoming of increasing importance because of the growing trend toward stress-grading of lumber. Also, from the standpoint of a pulpwood operation, if the average specific gravity of wood is raised 10 percent, then the yield of kraft pulp is raised 10 percent. This is equivalent to the harvesting of 10 percent more volume or adding 10 percent to the size of an industry's land base.

Considerable within-species variation in specific gravity occurs in conifers, and variation as high as 50 percent has been reported (Mitchell and Wheeler, 1959). This means that an individual tree can have twice the mass of other trees of the same size. The opportunities for selection therefore can be quite large. In

Table 6-1 Specific Gravity (Green) of Some Common North American Trees

Hardwoods					
Ring porous		**Diffuse porous**		**Conifers**	
Oak, live	0.81	Maple, sugar	0.56	Pine, slash	0.56
Locust, black	0.66	Birch, yellow	0.55	Pine, longleaf	0.54
Honeylocust	0.66	Maple, red	0.49	Larch, western	0.48
Hickory, shagbark	0.64	Sweetgum	0.46	Pine, loblolly	0.47
Oak, white	0.59	Sycamore	0.46	Douglas-fir	0.45
Oak, red	0.57	Poplar, yellow	0.40	Pine, jack	0.40
Chestnut	0.40	Alder, red	0.37	Pine, ponderosa	0.38
		Aspen, quaking	0.35	Spruce, Engelmann	0.36
		Willow, black	0.34	Pine, eastern white	0.34
		Basswood, American	0.32	Fir, white	0.35
				Fir, balsam	0.34
				White cedar, northern	0.29

Source: Panshin and De Zeeuw, 1970.

selecting trees, the breeder must recognize that there is considerable within-tree variation in specific gravity. First, it varies within each growth ring–in the southern pines, latewood with a density of about 0.70 is 2 to 3 times that of earlywood (Koch, 1972). More uniform textured woods such as white fir and Monterey pine are easier to nail when dry and easier to season and finish, whereas species such as Douglas-fir, with considerable differences in specific gravity across the ring are more difficult to nail when dry but have greater strength properties. Second, specific gravity varies with height in the tree (Bethel, 1941; Echols, 1972); within any one sheath of wood there is about a 10 percent drop in specific gravity from the base to the top. This is due to the differences between mature- and juvenile-wood characteristics and is thus influenced by the size of the crown and the position of the base of the live crown. Third, specific gravity varies from pith to bark. Again, this is due to the transition from juvenile- to mature-wood characteristics.

Effect of Growth Rate on Specific Gravity The effect of growth rate on wood specific gravity has been the subject of considerable attention. Reports in the literature are often conflicting. However, it appears to be commonly agreed that increased growth rate has usually no marked effects on specific gravity (Spurr and Hsiung, 1954; Zobel et al., 1969; Choong et al., 1970). Fertilizer and thinning treatments vary in their effect on specific gravity. If the treatment increases the relative proportion of earlywood in a ring, specific gravity is lowered; if the percent of latewood is increased, specific gravity will be increased (Erickson and Lambert, 1958; Nicholls, 1971*b*; Megraw and Nearn, 1972). In hardwoods, the general response to fertilization and thinning appears to be a slight increase in specific gravity in ring-porous species and no effect at all in diffuse-porous species (Mitchell, 1972; Saucier and Ike, 1972).

The effect of irrigation depends on whether water deficits are causing cessation of growth. For example, irrigating ponderosa pine in eastern Washington and Monterey pine in South Australia resulted in the production of wood with 20 to 30 percent more latewood in the lower bole (Howe, 1970; Nicholls, 1971*a*), whereas irrigating red pine in the Lake states increased total growth, but the proportion of latewood cells, and therefore specific gravity, remained unchanged (Zahner et al., 1964).

Even if increased growth rate does cause some small reduction in specific gravity, the effect is insignificant compared with the increased wood volume obtained from thinning and fertilizer treatments. The following example from Scandinavia illustrates the effect (Table 6-2). In other words, fertilization decreased specific gravity by 5 percent but increased productivity by 35 percent. This, of course, is attractive from the standpoint of pulp production. From the standpoint of lumber production, it must be recognized that the 5 percent lowering of specific gravity would result in a 10 to 20 percent reduction in bending strength, compression parallel to the grain, and toughness.

Geographic Variations in Specific Gravity Specific gravity varies in different localities due to site differences and broad changes across a species range. Higher site quality has the same effects as described for high growth rates. Generally, young dominant trees with large crowns produce a higher proportion of earlywood than latewood and thus have lower specific gravities. In contrast, trees in dense stands with smaller crowns, including suppressed trees, produce a higher proportion of latewood and have higher specific gravities (Pillow, 1954; Larson, 1964; Elliott, 1970). In general, higher latewood percent is associated with areas having low elevation, low precipitation and relative humidity, and dry, warm conditions (Edlin, 1965). For example, in Douglas-fir at low elevation, specific gravity was 0.55 g cm^{-3} due to a latewood percent of 47 percent, whereas at high elevation the levels were 0.44 g cm^{-3} and 29 percent, respectively (Lassen and Okkonen, 1969). Similar results have been shown for ponderosa pine (Echols and Conkle, 1971).

Across the range of southern pines specific gravity increases from north (0.47) to south (0.51) and also from west to east. The difference appears small, but it represents approximately 68 kg (150 lb) dry weight of pulp per cord

Table 6-2 Effect of Commercial Fertilizer Treatment on Volume Production, Specific Gravity, and Dry Weight Production of Norway Spruce

	Volume		Specific gravity	Dry weight	
	m^3/ha/yr	ft^3/ac/yr		kg/ha	tons/ac
Before fertilizing	7	125	0.475	3325	6.6
After fertilizing	10	175	0.451	4510	9.0

Source: Klem, 1968.

(Mitchell, 1965). Douglas-fir shows a much smaller increase from north to south due to greater variation in sites, climate, and elevation. As might be expected, specific gravity of Norway spruce in Germany increases from north to south (Elliott, 1970). Thus there seems to be a typical increase in specific gravity toward the tropics. Although these systematic changes are well substantiated, the magnitude of the differences between stands is no greater than that between trees within a stand. However, the potential exists for selecting trees in breeding programs for either high or low density should the objective be desirable. This is especially true since specific gravity is one of the most highly heritable wood-quality characteristics, with a broad-sense heritability of between 0.5 and 0.8 depending on species (Dadswell et al., 1961; Einspahr et al., 1964; Elliott, 1970).

Tracheid Length

Tracheid length is not as important a criterion of wood quality as it was several years ago when in order to make high quality papers it was necessary to have a pulp with a high proportion of long fibers. Today this requirement for long fibers has been somewhat reduced by advances in paper-making technology. Long fibers are still important, however, since they impart strength to wood and give paper high tearing strength and folding resistance. Within a single ring of loblolly pine juvenile wood, earlywood tracheids range in length from 2.86 to 3.33 mm and latewood tracheids range from 3.15 to 3.87 mm. In mature loblolly pine, earlywood tracheids range in length from 3.69 to 4.70 mm and latewood tracheids range from 3.87 to 4.89 mm (Koch, 1972). It has also been shown that taller trees have significantly longer tracheids than shorter trees, thus possibly providing a basis for predicting height-growth potential in seedlings (Echols, 1958). Within any one sheath of wood, tracheids have a maximum size near the base of the live crown, the decline in length in the upper portion of the tree being associated with the inherently shorter tracheids in juvenile wood. Generally, for the same reason, tracheid length increases from the pith toward the bark except where compression wood is present. Tension wood in hardwoods produces no change in fiber length (Bisset and Dadswell, 1950).

Increases in growth rate following thinning and fertilization may be accompanied in conifers by decreases in length of tracheids and reductions in cell-wall thickness (Posey, 1964; Cown, 1972), although some studies have reported conflicting evidence. Fielding (1967) points out, however, that differences in tracheid length in response to growth rate are small relative to the within-tree and between-tree variations. This may result in faster-grown trees in a stand having tracheids which are longer than those in slower-grown trees. In hardwoods, increased growth rate following irrigation or fertilization treatments is commonly accompanied by increases in fiber length (Saucier and Ike, 1972; Einspahr et al., 1972).

Geographically, tracheids and fibers generally decrease in length from south to north (Echols, 1958; Winstead, 1972). Since the broad-sense heritability of tracheid length is approximately 0.7 (Einspahr et al., 1964; Fielding, 1967), it is also highly suited to selection in tree-breeding programs.

Silvicultural Control of Wood Quality

The three major causes of degrade in young-growth lumber are knot size and frequency, reaction wood, and spiral grain. Anything the silviculturist can do to minimize the effects of these factors will substantially increase value production from forest stands. Silviculturists can influence wood quality by treatments that affect spacing, live-crown ratio, and growth rate, and also through tree breeding.

Spacing Control of spacing permits the silviculturist to exert some direct influence over three important criteria of quality: knot size, the size of juvenile wood, and stem straightness in hardwoods. Of all important wood-quality criteria, branch size has the lowest heritability–broad-sense heritability of 0.3 (Fielding, 1967). This means that size of branches is largely controlled by environment, or the space they have for development. Branch size, particularly of intolerant trees, can therefore be minimized by close spacing, but of course this would probably be achieved at the expense of loss of wood production and so in most cases would not be desirable. Close spacing can also minimize the extent of juvenile wood produced but, again, only with slowing of growth and loss in volume. Spacing control has a particularly favorable effect on the straightness of hardwood stems, which have a tendency to develop sweeps and bends if spacing is wide or irregular. Not only are crooked stems difficult and more costly to harvest, transport, and utilize, they also contain a considerable proportion of reaction wood which lowers their quality. Once again, however, a compromise must commonly be sought because closer spacing to control stem straightness also results in a loss of individual tree growth. Total wood production per unit area, however, may not be affected.

Live-Crown Ratio The live crown of the tree defines the position where auxins and carbohydrate are produced, and the presence and relative abundance of these materials have strong influences on the extent of juvenile wood, the proportion of earlywood to latewood, and the position in the stem of maximum ring width and hence stem taper. Live-crown ratio can be controlled directly by pruning, or with intolerant trees by close spacing. Close spacing will cause lower branches to die even in tolerant conifers, but unless rotations are long, these dead lower branches usually stay attached to the tree. Furthermore, the close spacing necessary to cause lower branches to die commonly results in a considerable loss in volume.

Growth Rate Growth rate can be increased by species selection, genetic selection, spacing, fertilization, and irrigation if this is feasible. The effects of increased growth rate generally include a decrease in tracheid and fiber length, percent cellulose, and perhaps percent latewood and specific gravity; increases may occur in percent lignin, ring width, volume, and perhaps percent earlywood and hence lower specific gravity. These are trends only, and in some circumstances the reverse or no change will occur. Second, the amount of change is quite small—usually smaller than the within-tree and between-tree variations. And third, increased volume production due to increased growth rate commonly more than offsets any possible undesirable change in wood-quality characteristics. The environmental conditions under which trees grow certainly have predictable effects on the quality of the wood produced. The effects are small, and the greatest gains can generally be made by tree-breeding programs which select for highly heritable traits, such as specific gravity, tracheid length, branch angle and spiral grain, and fast-growing trees. From industry's standpoint, efficiencies can be obtained if the silviculturist can increase the uniformity of wood and reduce the variation that normally occurs in wood-quality characteristics.

Chapter 7

Shoot and Crown Development

SHOOT GROWTH

Silviculturists are usually concerned with the relative rate at which individual trees or species increase in height growth. This is important at the seedling stage since individuals with more rapid growth will develop more rapidly beyond the influence of weed species and pests. In later stages of development, height growth becomes a major determinant of volume increment. Consequently, tree breeders commonly use rapid height growth as an important criterion in selection. For these reasons it is important to understand how trees grow in height, to recognize patterns of growth behavior, and to determine the extent to which these growth characteristics can be influenced by genetic or environmental manipulation.

Shoots increase in length as a result of the development of primordia which usually have been held dormant for a period within the bud. Differences in timing when these primordia actually elongate result in three basically different kinds of shoot growth: free growth, fixed growth, and fixed and free growth.

Free Growth

Free, or indeterminate, growth is typified by the growth of the epicotyls in first-year seedlings of both hardwoods and conifers, and in the growth of such genera as *Juniperus*, *Thuja*, and *Chamaecyparis* which set no buds. Here, elongation of the shoot results from the *concurrent* initiation and elongation of new stem units. [A *stem unit*, described by Lanner (1969; 1976) from a term introduced by Doak (1935), defines that portion of the shoot bearing a primary scale and its axillary structure if any. Stem-unit number can be determined by counting the needles or leaves, or on pines by counting the scales—cataphylls—along the stem.]

Free growth also accounts for the development of "foxtails" in some pines, particularly some southern pines and in Monterey pine in low latitudes. Here, a favorable environment results in continuous growth without the development of lateral buds.

Fixed Growth

Fixed, or determinate, growth is typified by the development of shoots of north temperature conifers and hardwoods such as *Carya*, *Juglans*, *Quercus*, and *Aesculus*, where there is a distinct dormant period between the initiation and elongation of stem units. In this familiar pattern, shoot growth is a 2-year process involving the formation of a bud during the first year, its over-wintering, and its extension into a shoot during the second year. In the case of red pine, for example, during any year, stem-unit primordia for the following year's shoots begin to form during the flush of growth in the spring. The maximum rate of primordium formation of about nine stem units per day coincides with the maximum rate of elongation of the needles on subtending shoots (Sucoff, 1971). Timing and rate of stem-unit formation vary with species, ecotype, and site. During the summer, apical meristematic activity can be retarded or halted by adverse water or nutrient relations or by adverse temperature, and is rarely continuous (Cannell et al., 1976). Elongation appears to be dependent on buds having had their chilling requirement followed by exposure to a threshold-level heat sum.

Fixed growth implies that under normal conditions the potential length of a shoot of some northern coniferous and hardwood species is governed by (1) the number of anatomical stem units which were laid down in the bud in the previous year and whose numbers depend on the conditions of that year, and (2) the climatic conditions in the current year which influence the extent to which the stem units elongate. Since the number of stem units is reflected in the size of the bud (Kozlowski et al., 1973), silviculturists can predict relative final shoot length from relative bud size during the dormant period prior to elongation. Thus, since final shoot size and the number of stem units are largely predetermined, the *duration* of growth must be a function of growth *rate* (Lanner, 1976). It also becomes apparent that the nature and extent of silvicultural manipulation of a tree's growing environment in any season will directly affect the magnitude of height growth in the following year. There are many examples of this, one of which is a late-summer irrigation of red pine during the period of bud development which resulted in an almost doubling of shoot growth in the following year (Clements, 1971).

It follows, therefore, that there are two ways in which the silviculturist can directly influence height growth. First, genotypes must be selected or environments must be manipulated such that the number of stem units within the annual bud is maximized. This can be done only by increasing the *rate* of stem-unit initiation or by increasing the length of the stem-unit initiation period (Lanner, 1976). The extent to which these are under genetic and/or environmental control is not yet known. Second, growing conditions can be influenced by controlling spacing, shade, or water relations to enhance stem-unit initiation and subsequent elongation.

The behavior of a species may vary geographically. For example, in black spruce, *Picea mariana* (Pollard and Logan, 1974), growth was found to be correlated with needle complement, which in turn was controlled by a combination of both fixed and free growth. The relatively faster growth of the southern provenances was interpreted as a result of less stringent requirements for free growth. This provides the southern provenances with initial gains in height during the seedling stage, which probably results in lasting advantages in later height development.

Most shoots resulting from fixed growth form only one whorl of branches per year. Such shoots are said to be *monocyclic* or *uninodal*. Some pines (jack pine, lodgepole pine), however, have several whorls of branches laid down in their dormant buds, and when these buds elongate, the shoot will be *polycyclic* or *multinodal*.

Fixed and Free Growth

A combination of fixed and free growth is characteristic of the southern pines, *Pinus radiata* and such hardwoods as *Populus*, *Betula*, *Liriodendron*, some *Acer*, and perhaps some *Eucalyptus*. Increase in height is achieved by the elongation of the spring shoot (fixed growth) followed by the flushing of one to several summer shoots (free growth). Each shoot produces lateral branches, which makes it difficult to determine the end of a year's growth (Fig. 7-1). The proportion of an annual shoot developed due to fixed or free growth and the number of successive summer shoots may vary from year to year and probably depends on the climate of the current growing season. Each summer shoot begins activity before the previous shoot has completed elongating. This results in a rhythmic overlapping pattern of growth throughout the growing season with no single,

Figure 7-1 A year's growth of a shortleaf pine. The upper two internodes of this growth are summer shoots and the branch whorls at their bases are similar to those at the bottom of spring shoots. (*From Tepper, 1963.*)

well-defined peak in growth (Tepper, 1963; Griffing and Elam, 1971). This pattern of height growth is illustrated in Fig. 7-2.

Pinus radiata has a slightly different height-growth behavior in that it has a capacity for virtually continuous development when grown in Southern Hemisphere plantations, particularly when juvenile (Pawsey, 1964; Fielding, 1966; Doran, 1974). Flushes of growth occur in a similar manner to that previously described for the southern pines. However, the subordinate shoots of Monterey pine, i.e., all shoots other than the leader, exhibit monocyclic development

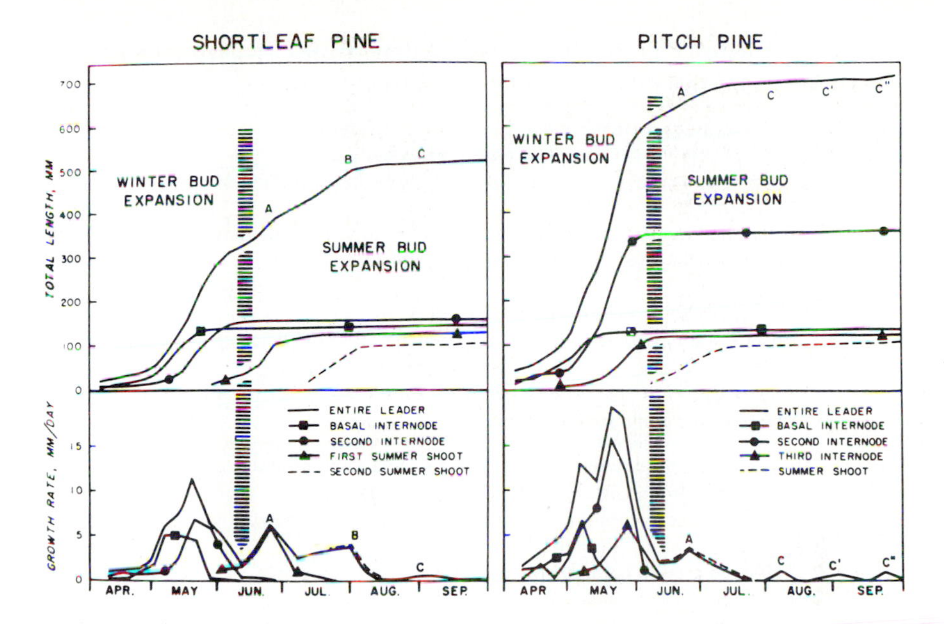

Figure 7-2 Total length and growth rate of the leader and internodes comprising the leaders of individual pine seedlings in New Jersey. Letters indicate peaks in growth, of the first summer shoot (A), second summer shoot (B), and winter bud formation (C, C′, C″). (*From Tepper, 1963.*)

with one grand period of growth in the spring, which is similar to the patterns of leading and subordinate shoots of many conifers in the north temperate zone (Cremer, 1973). This pattern becomes obscured, however, if the subordinate shoots are bearing "flowers."

The development of shoot growth in pines is richly diverse and may conform to any of the three modes previously described. These modes have been tentatively divided into 10 distinct patterns by Lanner (1976) (Table 7-1), who emphasizes that the patterns do not necessarily categorize species throughout their life cycles or under all conditions of growth. He suggests that these patterns reflect adaptive strategies imposed by natural selection acting through climatic constraints. Thus the occurrence of free growth in pines beyond the seedling stage is related to mildness of climate.

In trees with both fixed and free height-growth development, there are gradual increases throughout the growing season in accumulation of dry weight in all segments of the tree. This is accomplished by an overall increase in nitrogen content depending on whether nitrogen is transferred into or out of that component. This general situation is illustrated in Fig. 7-3, which shows that throughout the year the greatest accumulation of dry weight is by the stem and current foliage. The foliage mass doubles even though the oldest foliage is

Table 7-1 Tentative Classification Scheme Differentiating between Currently Known Developmental Patterns of Shoot Growth in *Pinus*.

Category	Description	Pattern
Mode I	The annual shoot is elaborated entirely by free growth.	
A	The annual shoot originates as an embryonic plumule.	
1	There is a single cycle of free growth followed by formation of a resting bud.	*Seedling pattern 1*
2	Additional cycles of free growth occur by formation of temporary nonresting buds that elongate while stem-unit initiation is progressing.	*Seedling pattern 2*
B	The annual shoot does not originate as a plumule.	
1	There is a single cycle of free growth which often continues for several years.	*Foxtail pattern*
2	Several cycles of free growth occur by formation of temporary nonresting buds that elongate while stem-unit initiation is progressing.	*Caribaea pattern*
Mode II	The annual shoot is elaborated by both free and fixed growth.	
A	Free-growth components (summer shoots) constitute a significant part of the annual shoot, sometimes exceeding the fixed-growth component (spring shoot) in length.	
1	The fixed-growth component (spring shoot) is monocyclic.	*Elliottii pattern*
2	The fixed-growth component (spring shoot) is polycyclic.	*Echinata pattern*
B	Free-growth components (summer shoots) usually constitute only a minor part of the annual shoot.	*Piñon pattern*
Mode III	The annual shoot is elaborated entirely by fixed growth.	
A	The annual shoot is monocyclic.	*Resinosa pattern*
B	The annual shoot is polycyclic.	
1	All cycles of the shoot are initiated in the bud consecutively.	*Contorta pattern*
2	The final shoot cycle is initiated during the winter, long after earlier cycles.	*Pinaster pattern*

Source: Lanner, 1976.

shed during the year. It can also be seen that current foliage accumulates nitrogen continuously. These data of Smith et al. (1971) show that the net accumulation of each tree during its fifth year of development was 7.6 g of N and 2.1 kg dry matter. The plantation as a whole accumulated 34.1 kg N and 9400 kg dry matter per hectare (ha).

The important point to recognize here is that particularly in trees with both fixed and free shoot growth, current leaf development constitutes a considerable proportion of dry-matter and nitrogen accumulation. Smith et al. estimate that 80 percent of the total nitrogen accumulation by loblolly pine may occur during the first 30 years of plantation development at a mean annual rate of 7.1 kg ha^{-1}. During the initial 4 years, the nitrogen accumulation rate was 9.5 kg ha^{-1}. In the early years of stand development, therefore, the nutrient-rich crown is developing at an accelerating rate. If the silviculturist wishes to enhance the potential for initiating larger numbers of stem units and subsequent height growth, it is therefore important to evaluate nutrient availability prior to planta-

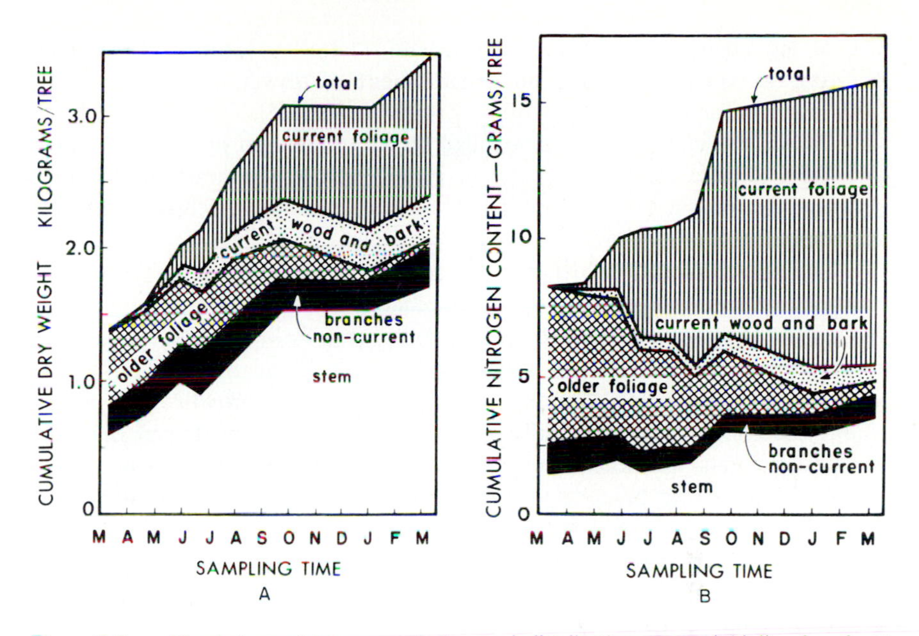

Figure 7-3 a: Total dry weight accumulation and distribution among loblolly pine shoot system components during the fifth year. b: Total *N* accumulation and its distribution among tree parts. (*From Smith et al., 1971.*)

tion establishment and to fertilize, if desirable, at that time and possibly also at a later date during the *first* decade of the rotation. Fertilizing at this time, of course, increases growth of competitive plants. However, if the objective is to maximize growth rates of crop plants, other treatments might be necessary to control competition.

A second important point raised is that since current foliage can potentially accumulate a high proportion of the nitrogen available to the tree, the application of fertilizer is likely to result in enhanced canopy development which may mask responses in the more commonly used growth-evaluation criteria of diameter, height, or stem volume. Consequently, some index of crown development should be used as one criterion of response of stands to fertilization.

CROWN SHAPE

Trees commonly have distinctive shapes which are well recognized, particularly by urban foresters, horticulturists, and landscape architects who frequently choose trees for planting on the basis of ultimate size and deciduousness, as well as whether they are columnar, conical, spreading, or weeping. Tree shape is also a diagnostic feature, and species can sometimes be recognized on the basis of their characteristic silhouettes. In forestry, potential size and shape of a tree's crown have important bearings on such factors as initial spacing, control of log quality, necessity for intermediate treatments, control of screenings or vistas in recreation areas, and enhancement of flowering and fruiting for both regeneration needs and provision of food for wildlife.

Crown size and shape are controlled by a combination of inherent genetic and environmental factors. A detailed review of these interactions has been provided by Brown (1971), and much of the following discussion has been summarized from this source.

The genetic factors influencing the shape of a tree's crown are those which control leaf arrangement and bud development. The alternate and opposite arrangement of leaves and buds result in the repeated forking or false dichotomy of maple and ash crowns. Many conifers, on the other hand, have a characteristic whorled branching habit. The expression of a particular growth habit is commonly associated with the relative growth of the leader and lateral branches. In most conifers, the leader outgrows the lower laterals, resulting in a cone-shaped or *excurrent* branching habit with a clearly defined central bole. In most dicotyledonous trees, the laterals grow as fast as, or more rapidly than, the leading shoot, resulting in the *decurrent* or *deliquescent* growth habit in which the central stem commonly disappears from repeated forking to form a large, spreading crown.

As Brown (1971) points out, most foresters assume that the excurrent branching habit of conifers is an expression of strong apical dominance, and that hardwoods, with a decurrent, profusely branching habit, have weak apical dominance. Brown states that the situation is much more complex in trees than is the case in herbaceous plants, in which the phenomenon of apical dominance was first described. In conifers, only partial inhibition of lower bud development occurs, whereas in hardwoods almost all the lateral buds on current-year shoots are completely inhibited. In decurrent forms, during the second growth season after a period of dormancy, one or more lateral buds will elongate as rapidly as, or more rapidly than, the terminal bud. This gives rise to the characteristic repeatedly branched stems. This same branching pattern can occasionally arise during the current season if Lammas shoots are formed (Brown, 1971). These complex patterns of bud inhibition on individual tree branches, including complications due to the time sequence involved in the formation and release of lateral buds, make use of the term *apical dominance* misleading when applied to tree crowns (Brown, 1971). As Brown states:

> If by definition, the term apical dominance connotes bud inhibition by an active apex on currently elongating shoots, then form in trees must be explained in a manner that will be consistent with the original meaning of the term. Another term, *apical control*, seems better suited for describing the physiological condition governing the excurrent or decurrent pattern of growth. One could then explain the excurrent pattern of growth in terms of strong *apical control* made possible by the initial expression of weak apical dominance or incomplete bud inhibition, so that the terminal leader always maintains complete control over the partially suppressed branches below.

> Only by this pattern can truly conical shaped crowns arise and be indefinitely maintained. Conversely, strong apical dominance resulting in the complete inhibition of lateral buds on the current year's shoots gives rise to the decurrent habit of growth. The release and rapid growth of two or more uppermost lateral buds the following spring tends to suppress the terminal leader so that apical control is lost and repeated forking occurs in the crown. Although this terminology, viz., strong *apical control* versus weak *apical dominance* and vice versa, may at first seem contradictory, it is consistent with our observations and the general pattern of growth and form in woody plants. (p. 134)

The branching habit and resulting crown shape of trees is also influenced by age and factors of the environment. All forest trees initially possess excurrent growth habit for some period in early development, otherwise none would exist with an unbranched bole for any period of time (Brown, 1971). The timing of the shift from excurrent to decurrent varies with species; it is early in the case of *Rhus* and *Aesculus* and gradual with declining vigor of uppermost competing lateral buds in young cottonwoods (Brown, 1971). Also, the shift is earlier on poorer, drier sites. In conifers, it is common for crowns to become flat-topped and decurrent with age, and again, this is more obvious on poorer sites.

Another characteristic of tree crowns, especially conifers, is the change in branch angle from the top to the bottom of the live crown. Commonly, branch angle near the top of the tree is acute, in the midcrown region branches are horizontal, and toward the base of the live crown branches tend to droop. Thus, as the tree develops, a given branch increases in length, changes its relative position within the crown, and changes its branch angle.

The position of branches is also controlled by gravity and light. Gravity controls the characteristic turning up of branches by controlling the unilateral displacement of auxin on the lower side of branches, which causes rapid differential growth response. Response to small changes in auxin concentration varies considerably because of differences in sensitivity of cells in different tissues to the action of auxin (Brown, 1971). The positioning of branches (and roots) in space is called *plagiotropism*, and this results in certain species having a characteristic shape or pattern of branches. *Phototropism*, on the other hand, is the response of branches, particularly in hardwoods, or the tendency of plant growth in general, to develop in response to light direction. Thus tree crowns may become one-sided or leaning in response to light coming primarily from the side as a result of openings in the canopy. An additional response to light is the stimulation of dormant buds to produce epicormic shoots on the stems and branches of some species after a sudden change in crown exposure resulting from thinning or the removal of substantial portions of the original crown by fires, frost, insects, or other damage. These epicormic branches can be so numerous that the stem and branches can be covered in a dense mass of young foliage. This may be temporary, and after some years the crown may recover and the tree may return to its usual, characteristic appearance.

It can be seen, therefore, that the inherent shape of tree crowns is modified to some extent by age, site, and environmental conditions. As trees develop and form a canopy, the development of individual crowns is influenced by competition and spacing, and this forms the basis of classification into crown classes (dominants, codominants, intermediates, and suppressed). Trees in a dominant position may also develop into coarse, heavily branched crowns and are then commonly called "wolf trees." Silviculturists have considerable control over crown size and shape (as well as root development) through control of spacing. It is through exercising this control that silviculturists can manipulate individual tree growth and the characteristics of stand growth to meet management objectives.

CROWN DEVELOPMENT AND STEM FORM

The stem of a tree has a very complex shape depending on species and crown characteristics (Larson, 1963). This is largely because the annual sheath of wood laid down by a tree in any one growing season is not uniformly thick over the length of the stem. In conifers there is usually a region near the base of the live crown where ring width tends to be maximum. This means that stand-grown trees with relatively short live-crown ratios will have more cylindrical stems than open-grown trees which have deep crowns resulting in more stem taper. This is largely a function of availability of carbohydrate supply, since there is probably more food available toward the lower part of the live crown where the bulk of the foliage occurs. Even when food supplies are adequate, ring widths usually diminish toward the base of a tree because stem diameter is larger. Ring area or volume, however, may increase toward the base of the live crown. Stem form is consequently influenced by stand condition or by silvicultural treatments which modify live-crown ratio.

Thinning results in the leave trees tending to have more taper because the treatment increases live-crown ratios. Pruning, however, tends to produce stems with less taper because the treatment decreases live-crown ratio. Thinning and pruning have compensating effects, but the degree of potential change in stem form is dependent on the intensity of the treatment and the extent to which growth rate is increased or decreased by the treatment. Severe thinning, for example, can result in "thinning shock" or decreased growth rate, and excessive pruning can remove such a proportion of the functional crown that height growth may slow down and rings may fail to form at the base of the stem.

PATTERNS OF GROWTH

Growth Interrelationships

The growth of any living organism conforms to the familiar sigmoidal relationship illustrated in Fig. 7-4. The characteristic shape of this relationship varies

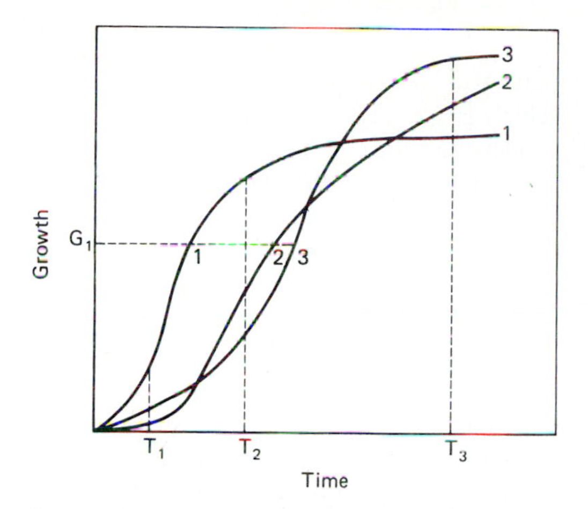

Figure 7-4 Theoretical patterns of growth representing three different genotypes or the impact of three different kinds or levels of treatment.

for each situation of genotype and environment. The growth curves illustrated could represent the growth performance of three species, three genotypes of one species, or the behavior of one genotype under three different kinds of silvicultural treatment. For each case, the silviculturist needs to predict, in broad conceptual terms, the likely extent of the juvenile period of growth, the rate of growth (slope of the curve) during the grand period of growth, the time when rate of growth declines, and the extent to which these characteristics are controlled by genetics and environment.

The length of the juvenile period of slow growth is important in situations where seedlings need to out-grow grass competition or pest damage, and the rate of growth during the grand period of growth has important bearing on the time taken for the tree to reach a given stem diameter. In terms of progeny selection, if the curves in Fig. 7-4 represent genotypes, the ranking of individuals at time T_1, T_2, and T_3 would be genotypes 1, 3, 2; 1, 2, 3; and 3, 2, 1, respectively. In the long run, genotype 3 is the best producer, but if weeds and pests cannot be effectively controlled, genotype 3 may not escape competition and the better selection would be genotype 1. This decision would also depend very strongly on the time period taken for the tree to reach merchantable size. If G is the threshold size between precommercial and commercial thinning, then it can be seen that there is considerable difference in the time it takes for each genotype to reach this size. This can be of decisive importance if the manager is carrying the compounding cost of regeneration at a high interest rate.

In a similar manner, if Fig. 7-4 illustrates the effect of three kinds or levels of site preparation on the growth performance of a single genotype, the silviculturist can evaluate the biological and economic costs and benefits of the alternative treatments.

Seasonal Growth Patterns

There is a periodicity in the development of roots, stems, branches, foliage, and leader growth of individual trees which is a function of genotype and growth environment. It is important to understand this periodicity, not only of a particular tree but also of each potentially competing species of vegetation, because the effectiveness of silvicultural treatments might well vary depending on the timing of application. Examples of this are the timing of nursery practices, outplanting, herbicide application, and treatments to enhance seed production.

If one follows the phenological development of a particular genotype in a particular environment, it is apparent that shoots, stems, and roots have characteristically different patterns of growth. Seasonal patterns of growth depend on species, age, and environment. Two examples are shown in Fig. 7-5. The

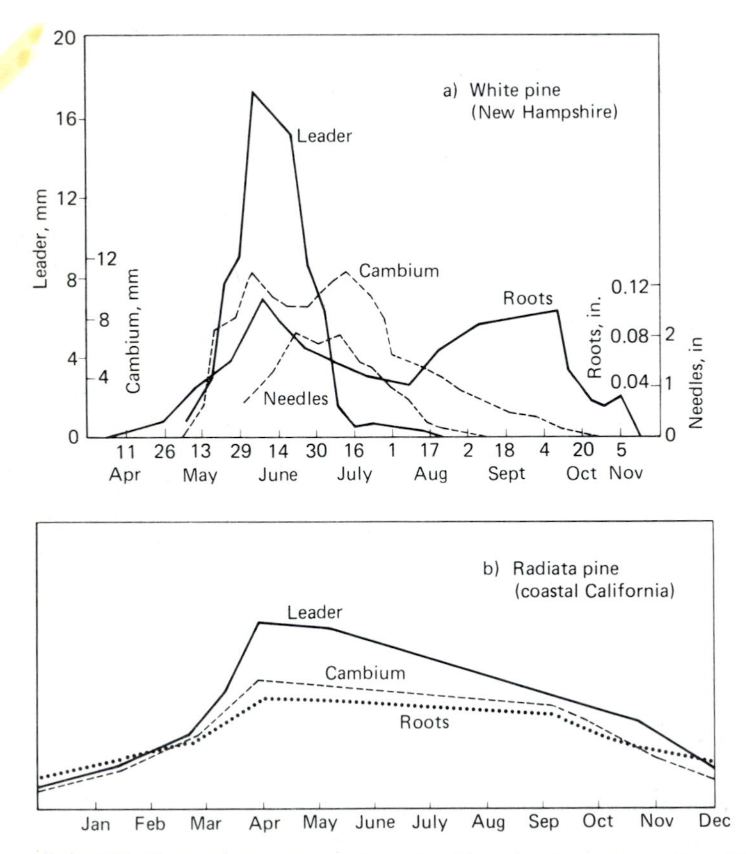

Figure 7-5 Seasonal growth patterns of white pine in southern New Hampshire and of a hypothetical well-watered *Pinus radiata* in coastal California. Different growth patterns are exhibited depending on species, age, and environment. (*From Kramer and Kozlowski, 1960.*)

upper diagram shows the following growth characteristics for moderately intolerant white pine growing in an environment of cold winters and relatively short growing seasons:

1 Height growth—about 90 percent of all height growth occurs over a period of 6 to 9 weeks. Much of this rapid growth is dependent on the number of stem units present in the bud and the extent of stored reserves of carbohydrate. The extent of these two factors is dependent on the *previous* year's environmental conditions. The current year's summer and fall conditions will affect carbohydrate storage, the extent of primordia initiation, and consequently *next* year's height performance.

2 Stem growth—most stem growth occurs over a 15- to 20-week period, extending well into the summer and therefore strongly influenced by the current year's environmental conditions and the current year's production of photosynthate.

3 Root growth—root growth ceases in the winter when soils are cold or frozen and is depressed in summer due mainly to adverse effects of temperature and water relations. There is also some evidence that in red pine, summer root-growth depression occurs regardless of the environment and may represent a change in growth balance within the tree (Merritt, 1968).

In contrast, Fig. 7-5*b* illustrates the growth patterns of Monterey pine, which has "continuous" growth in the favorable coastal environment. Figure 7-5*a* and *b* clearly illustrates the differences in seasonal patterns of growth which depend on species and climate. Silvicultural treatments may need to recognize differences in phenology of plants within a particular plant aggregation.

The important silvicultural question is to what extent the initiation, duration, and time of cessation of tree growth can be influenced. If we could answer this question, we could put an importance value on proposed treatments in terms of costs and the benefits to be obtained from influencing growth periodicity. Initiation of height growth and stem growth can probably not be significantly manipulated. However, the cessation of stem growth is strongly influenced by adverse temperature and water relations in late summer. This can be influenced to some extent by control of spacing. Zahner and Oliver (1962) have demonstrated, for example, that cambial stem development in thinned red-pine stands continues for 3 weeks longer than in unthinned stands. Continuing development of root systems is needed to sustain tree growth. Manipulation of crown and litter cover to limit soil cooling in winter and soil heating in summer can enhance the continued uptake of water and nutrients. Each silvicultural treatment must be evaluated in terms of the conditions it will create and its likely impact on seasonal growth patterns.

Other approaches to growth-pattern evaluation have been developed, notably one by Duff and Nolan (1953) and one by Forward and Nolan (1964). Through detailed stem analyses of growth rings in each internode of 25- to 30-year-old red pine trees they were able to characterize three types of growth

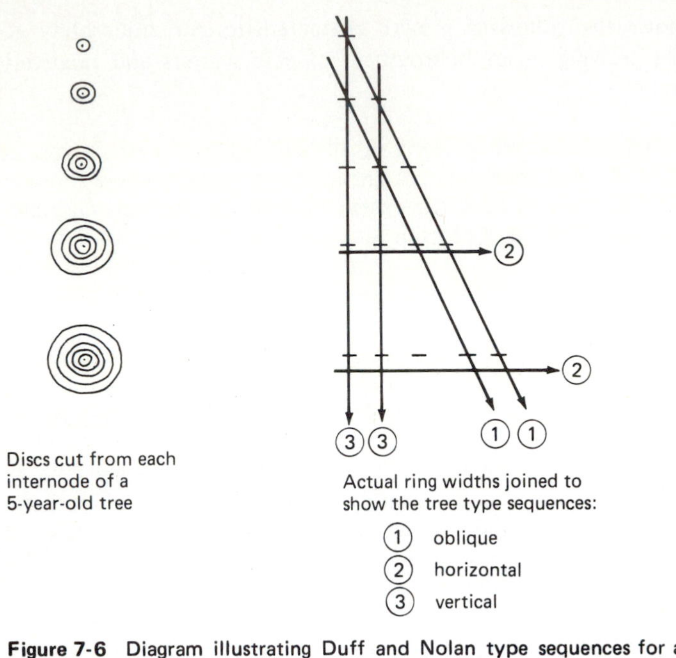

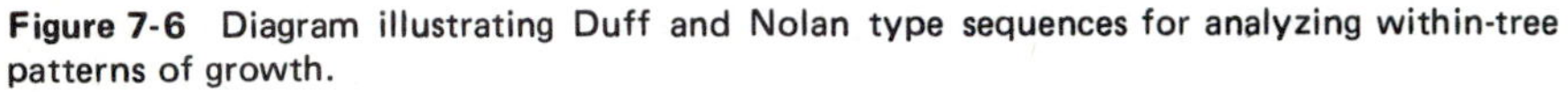

Figure 7-6 Diagram illustrating Duff and Nolan type sequences for analyzing within-tree patterns of growth.

patterns: (1) an oblique sequence where the widths of rings within each annual sheath of wood are compared at each internode; (2) a horizontal sequence where ring widths are a function of ring number from the pith (as observed in an increment boring); and (3) a vertical sequence where ring widths at a fixed number of rings away from the pit are compared from each internode (Fig. 7-6). A comparison of the observed patterns enabled Duff and Nolan to distinguish between systematic (genetic) and random (environmental) factors that influence growth throughout the life of a tree. In this manner, they identified the vertical sequence of patterns as being the best for determining the impact of fluctuating weather conditions on tree growth. This detailed stem-analysis approach has been effectively used to study competitive relationships among trees (Walters and Soos, 1962; McDougall, 1963).

Chapter 8

Flowering, Fruiting, and Seed Production

In those silvicultural systems relying on natural seeding, we must necessarily concern ourselves with predicting the likelihood that the overstory or neighboring stand will provide an adequate amount of quality seed from desirable genotypes—at the time when we require it. Reproduction methods requiring direct seeding or planting necessitate the advance collection and processing of sufficient quantities of quality seed for reforestation needs. So regardless of whether stands are to be artificially or naturally regenerated, we find that we need considerable information on factors that influence the timing and amount of seed production.

Considerable information on all aspects of seed, seed production, and germination is provided in the USDA Handbook No. 450, "Seeds of Woody Plants in the United States."

GENERAL CHARACTERISTICS

Flowering and seed production in forest trees differ from other plants in the following three major respects:

1 There is the major difference in method of reproduction between *angiosperms*, which possess true flowers with stamens and carpels and in which the ovule is always enclosed in the carpel, and *gymnosperms*, which do not form an ovary around the ovules (i.e., the ovules are naked) and in which the flowers consist of male (staminate) and female (ovulate) cones or strobili.

2 The age at which trees become sexually mature is later than most other plants, and there is considerable difference in the age of maturation between species.

3 There is commonly a very extended developmental period between flower initiation and seed maturity, covering, in the case of some pines, 3 to 4 years.

In addition, similar to some other plants, it is often difficult to distinguish initially between male, female, and vegetative buds, and there may be large time differences between the development of male and female flowers. Further, flowers can be unisexual or bisexual, and may be borne on the same tree or on separate trees (Table 8-1). *Monoecious* trees are those in which the male and female flowers are on the same tree; and *dioecious* trees have male and female flowers on separate trees.

Table 8-1 Flowering Characteristics of Common Tree Species

Monoecious Male and female flowers on same tree	**Dioecious** Male and female flowers on separate trees	**Polygamous** Bisexual flowers with both stamens and pistils	**Variable**
Angiosperms	**Angiosperms**	**Angiosperms**	**Polygamomonoecious**
Alnus	*Ilex*	*Cornus*	*Celtis*
Betula	*Populus*	*Liriodendron*	*Aesculus*
Carya	*Salix*	*Robinia*	**Polygamodioecious**
Fagus	**Gymnosperms**	*Ulmus*	*Fraxinus*
Juglans	*Juniperus* (mostly)		**Monoecious-dioecious**
Liquidambar	*Gingko*		*Juniperus*
Platanus			**Monoecious, dioecious, polygamodioecious, or polygamous**
Quercus			*Acer*
Gymnosperms			
Abies			
Chamaecyparis			
Larix			
Libocedrus			
Picea			
Pinus			
Pseudotsuga			
Sequoia			
Thuja			
Tsuga			

For these reasons it is difficult to make valid general statements characterizing flowering and seed production in woody plants.

SEXUAL MATURITY IN TREES

There is a juvenile or vegetative period during which young trees do not normally develop reproductive structures. There are examples of precocious early flowering, which is probably an inherent characteristic, but this does not contribute significantly to total seed production. Generally, trees develop the capacity for reproduction only after an early period of rapid height growth. However, the actual attainment of physical size is probably not in itself a determinant of the end of juvenility. There is evidence that the capacity to flower is dependent on a change in condition of the apical meristem brought about after a necessary number of cell divisions (Puritch, 1972).

Table 8-2 Age at which Forest Trees Begin Producing Abundant Seed Crops

	Seed production		
	Early, 10–20 years	Intermediate, 20–40 years	Late, 40–60 years
Conifers	Pines: Jack, pitch, lodgepole, knobcone, Virginian, sand, Monterey, bishop, slash, loblolly Other: Tamarack, black spruce, northern white cedar, Port Orford cedar, southern white cedar, cupressus	Pines: Red, eastern white, short-leaf Other: Red and white spruce, balsam fir, Douglas-fir	Pines: Sugar, western white, ponderosa, limber, white bark Spruce, true firs
Hardwoods	Willow, cottonwood, aspen, alder, gray birch, paperbirch, pin cherry, red maple, big leaf maple, box elder, scrub oak, and other fast-growing short-lived trees that produce small seeds	Hickories, maple, basswood, ash, elm, sycamore, chestnut, buckeye.	Beech, oak

Source: Adapted from Baker, 1950.

Table 8-3 Periodicity in Production of Good Seed Crops in Common Tree Species

	Periodicity in production of good seed crops (some seed commonly produced each year)			
	Annual or 1–2 years	Generally 2–3 years	Generally 3–5 years	Generally 3–10 years
Conifers	*Chamaecyparis thyoides* *Juniperus occidentalis* *Pinus contorta, P. clausa* *Sequoia, Sequoiadendron*	*Abies amabalis,* *lasiocarpa* *Juniperus virginiana* *Pinus palustris,* *edulis, rigida,* *elliottii* *Picea engelmanii* (2–6 years) *Taxodium* *Tsuga canadensis,* *mertensiana,* *heterophylla* (3–4 years)	*Abies balsamea,* *grandis,* *procera,* *concolor* *Chamaecyparis* *nootkatensis,* *lawsoniana* *Larix laricina,* *occidentalis* *Libocedrus decurrens* *Picea mariana,* *sitchensis,* *glauca* (2–6 years), *rubens* (3–8 years) *Pinus strobus,* *banksiana,* *virginiana,* *monticola* *Thuja*	*Pinus jeffreyi,* *ponderosa,* *resinosa,* *echinata,* *lambertiana* *Pseudotsuga*

Hardwoods	*Acer, Betula, Celtis,* *Carya glabra, Platanus* *Populus balsamifera* *trichocarpa, deltoides* *heterophylla* *Quercus falcata,* *phellos* *Robinia, Salix, Sassafras* *Tilia*	*Carya tomentosa,* *ovata,* *myristicaeformis,* *laciniosa* *Cornus, Diospyros* *Fagus, Juglans* *Liquidambar, Nyssa* *Lithocarpus* *Quercus velutina,* *macrocarpa* *garryana, stellata* *coccinea, shumardii*	*Alnus* *Carya cordiformis* *Populus grandidentata* *tremuloides* *Prunus serotina* *Quercus lyrata,* *michauxii, prinus* (2–7 years), *rubra* (2–5 years) *Ulmus*	*Quercus alba*

Some species of tree are notable for forming abundant flower buds at a consistently early age, whereas others tend to flower only at an intermediate or late stage in life. As the tree matures, the abundance of flowers in any one year gradually increases. A listing of trees which generally can be described as falling into one of these three groups is provided in Table 8-2.

PERIODICITY AND FRUITFULNESS

It has long been observed that trees commonly do not produce an abundant crop of seeds every year. There is commonly a cyclic periodicity of seed production in which a year of excellent production is followed by a varying period in which production is low. This cyclic periodicity ranges all the way from a remarkable Chinese bamboo which flowers at intervals of about 33 years to trees such as the willows which have little or no natural annual fluctuation. Another interesting example is the yucca-like century plant (*Agave*) which flowers once at an age of 5 to 60 or 100 years, and after the fruit is ripe the plant dies.

Forest trees can be broadly grouped into four categories depending on whether good crops of seed are produced annually, every 2 to 3 years, 3 to 5 years, or 3 to 10 years (Table 8-3). In each of these categories, in the "off" years, some seed is commonly produced. At best, this categorization can be only a very rough guide to periodicity since there is so much variability within a single species due largely to site and climatic differences. In addition, it can readily be seen that it is common for a single genus to be represented in more than one category. In very general terms, pioneering-type hardwoods tend to be the only trees which are likely to produce successive crops of abundant seed annually.

There are several well-documented illustrations of the periodicity of seed production in forest trees. One is the 35-year record obtained on the Stanislaus National Forest in California (Fig. 8-1), which shows that for ponderosa pine, sugar pine, and white fir, heavy seed crops occurred at intervals ranging from 2 to 9 years, with an average periodicity of 4 to 5 years. During the same 35-year period, good seed-collection years occurred a little more frequently, at intervals ranging from 1 to 7 years, with an average interval of 2 to 3 years (Schubert and Adams, 1971).

A 48-year record of Douglas-fir cone crops in Washington and Oregon has been compiled by Lowry (1966) which shows that the time interval between abundant cone crops varied between 2 and 11 years, with an average interval of about 5 years.

The periodicity, or cycles, of cone-crop production was given wide attention in Germany, and the conclusions of the studies were summarized by Schwappach (1895). The essential conclusions can be stated briefly as follows:

1 A good seed year for one species is not necessarily a good one for another species.

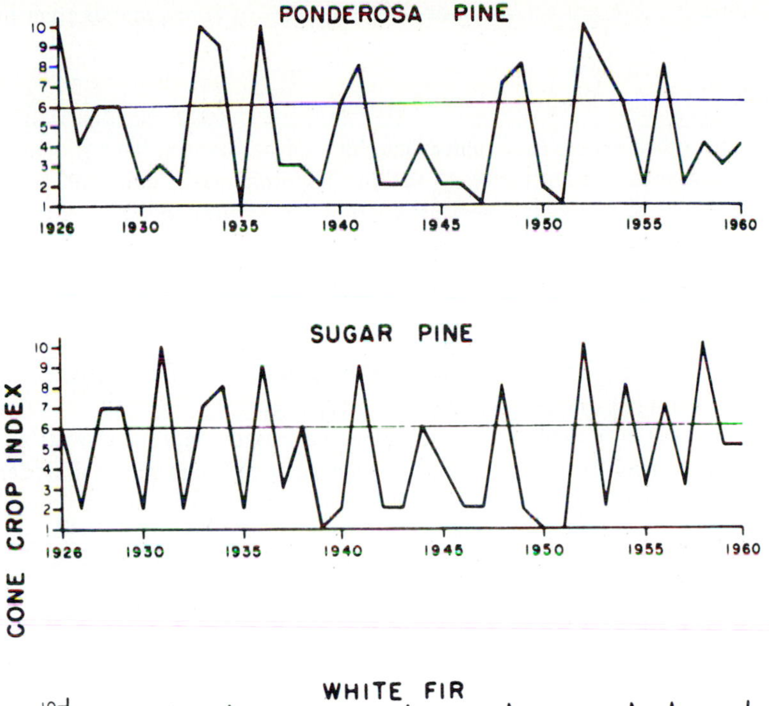

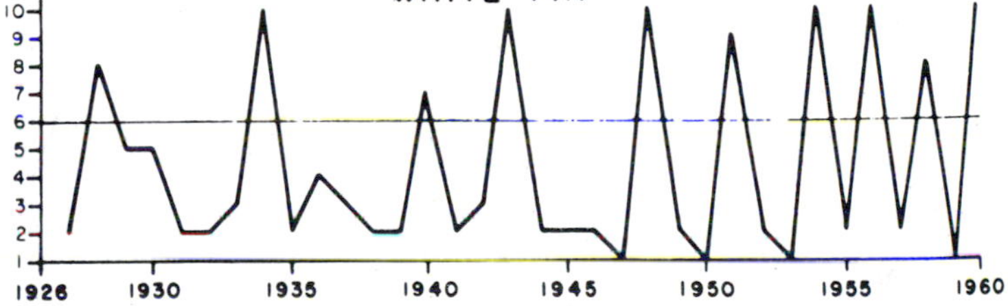

Figure 8-1 Average cone crop index for dominant trees larger than 19.5-in dbh, Stanislaus National Forest, 1926–1960. Number of cones: 1, none; 2, very light; 4, light; 6, medium; 8, heavy; 10, very heavy. (*From Schubert and Adams, 1971.*)

2 The variation in annual crops of seed is wide in some species (beech) and narrow in others (pine).

3 Some species have short cycles like spruce and fir, while others have long cycles like beech.

4 Total crop failures are rare.

5 Uniformly good crops over large areas (size of Montana or New York plus New England states) are rare.

6 In some years good crops are found over very large areas on a local basis.

7 In the best seed years some trees of seed-bearing size will be barren (about 10 percent), and conversely, in years of crop failure some trees will bear seeds.

8 There is more annual variation in seed crops of forest stands than in isolated trees.

9 A good seed year cannot be foreseen in advance of the formation of flower buds.

These and other examples of periodicity in seed production have provided a basis for evaluating the influence of various environmental factors on fruitfulness and the extent to which seed production can be influenced by silvicultural treatments.

TYPES OF TREES BEARING CONES

Cone production of a particular genotype is influenced by (1) crown development and vigor, and (2) tree class or canopy position.

Dominant trees with well-developed and well-exposed, vigorous crowns are notoriously heavy seed producers. Even less vigorous trees, particularly if under environmental stress, produce significant crops of cones as long as the crowns are well exposed. The size of the tree, or crown, at which prolific seed bearing occurs varies between species—partly in relation to the tolerance of the species and partly to the position within the crown of seed development. In the mixed-conifer forest of California, the more tolerant white fir bears significant cone crops when at relatively small stem diameter, probably because at that diameter it has a relatively large crown compared with other more intolerant species. Another factor might be that true firs bear cones only in the upper few whorls of branches (Fig. 8-2) which are well exposed. Ponderosa pine more commonly produces heavy cone crops when the trees have attained diameters of 76 to 127 cm (30 to 50 in). At this size, crowns are well developed, which is important since cones are born within a large portion of the upper crown. Sugar pine, however, does not become a prolific cone producer until it has obtained the much larger size of 127 to 178 cm (50 to 70 in) dbh (Fig. 8-3). Crowns of sugar pine need to be much larger to produce heavy crops of cones since the cones are only born at the ends of the branches.

In terms of tree class or canopy position, dominant trees are usually by far the best producers of seed, while suppressed trees in the stand produce virtually no seed at all. This trend is shown in Table 8-4, which shows that dominant California pines clearly produce virtually all the seed in the stand, whereas in the other species, dominants contribute a little more than half. Again, the variation is probably associated with the relative tolerance of the species, the position of seed production within the crown, and the relative exposure of the seed-bearing portion of the crown by the various crown classes.

A strong correlation has also been shown between Dunning's 1928 tree classes (see Dunning's Tree Classification in Chap. 4) and cone production. For pines, classes 1, 3, and 5 are the most productive, whereas for fir, the most productive was class 1 (Schubert and Adams, 1971).

Figure 8-2 Cone production in white fir is restricted to the upper parts of the crown.

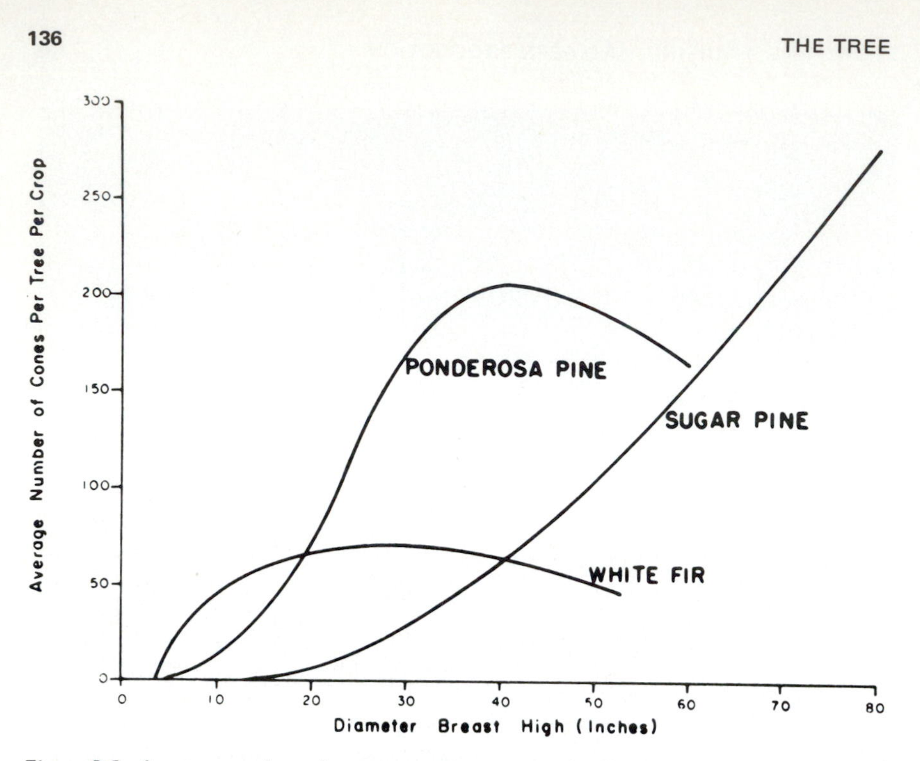

Figure 8-3 Average number of cones per tree per crop by diameter classes for ponderosa pine, sugar pine, and white fir on the Stanislaus National Forest, California. (*From Schubert and Adams, 1971.*)

FLOWER INITIATION AND CONE DEVELOPMENT IN CONIFERS

In many tree species, flowers, initials or primordia, are formed in the growing season preceding flowering. The actual timing of initiation depends on species, and varies from year to year according to weather conditions, site, genetics, and management practices (Kozlowski, 1971; Krugman et al., 1974). The sequence in steps between the laying down of bud primordia and the shedding of mature

Table 8-4 Cone Production by Species and Crown Class

	Cone production, %				
Species	Dominant	Codominant	Intermediate	Suppressed	Source
Norway spruce	44.4	39.1	16.3	0.2	Baker, 1950
Western white pine	54.3	44.5	1.2	0	
Lodgepole pine	55.0	29.3	10.5	5.3	
Ponderosa pine	99.03	0.92	0.05		Schubert and Adams, 1971
Sugar pine	98.46	1.51	0.03		
White fir	87.56	11.81	0.63		

seed varies considerably among species. Again, species can be listed as to whether this sequence occurs over a period of 1, 2, or 3 years (Table 8-5).

To better appreciate the variation in the sequences of developmental steps from primordia to seed, it is useful to compare the calendar of events in several conifers. Table 8-6 compares cone development in pines, true fir, and Douglas-fir.

In pines, primordia are laid down 8 to 10 months prior to "flowering." The duration of flowering can be quite short. It has been shown for *Pinus monticola*, for example, that the period of pollen dissemination averaged $8\frac{1}{2}$ days and the period of ovulate receptively was $9\frac{1}{2}$ days (Bingham and Squillace, 1957). After pollination, the pollen tube starts growing, develops during the summer, but ceases growth prior to winter. In the following spring, the pollen tube resumes growth and fertilization takes place. The female cone then rapidly develops and mature seeds are formed, usually at the end of the second summer after flowering.

In Douglas-fir, true fir, and spruce, bud primordia are initiated in spring in the axils of some leaves. Initially, vegetative and reproductive initials are indistinguishable. Some fascinating research has been done to clarify the manner in which buds develop (Owens, 1969; Ebbel, 1971; Allen and Owens, 1972; Puritch, 1972). It appears that throughout April the lateral bud primordia enlarge, but

Table 8-5 Typical Developmental Period Required to Produce Mature Seed in Common Tree Species

Developmental period for seed		
1 year	**2 years**	**3 years**
Conifers:	Conifers:	*Pinus:*
Pseudotsuga	Most *Pinus**	*leiophylla*
Abies	*Sequoiadendron giganteum*	*torreyana*
Sequoia sempervirens	*Juniperus*	*pinea*
Libocedrus	*Chamaecyparis*	*canariensis*
Larix		
Picea		
Almost all hardwoods:	Hardwoods:	
Quercus	*Lithocarpus*	
macrocarpa	*Quercus*	
prinus	*velutina*	
muehlenbergii	*falcata*	
virginiana	*laurifolia*	
garryana	*rubra*	
lyrata	*nutalli*	
stellata	*palustris*	
michauxii	*coccinea*	
bicolor	*shumardii*	
alba	*nigra*	

*Some tropical pines such as *P. teocote* and *P. oocarpa* produce mature seed in less than 2 years.

Table 8-6 Phenology of Seed Production in Common Conifer Species

					Year 1				Year 2			
	Winter	Spring	Summer	Autumn	Winter	Spring	Summer	Autumn	Winter	Spring	Summer	Autumn
Most Pines			Differen- tiation of primordia			↑ Cones visible ⟷ Scales open (flowering) ⟷ Pollination (Scales close, pollen grains germinate but pollen tube stops growing)				↑ Cones and pollen tube commence growing ⟷ Fertilization ⟷	Cone reaches final length ⟷	Seeds ripen and fall
Douglas-fir true fir, spruce			Differentiation of Primordia			↑ Cones visible ⟷ Scales open (flowering) ⟷ Pollination ⟷	Fertilization ⟷ Cones reach maximum size ⟷	Seeds ripen and fall				

in some cases certain buds stop dividing and cease to develop, i.e., they "abort." Other buds develop further and form bud scales, but some of these also cease growing and are called "latent" buds because they can be stimulated later to resume activity and continue development. Sometime during April-May, a transition occurs in that certain undetermined primordia become reproductive buds. By the end of May, vegetative and reproductive buds can be distinguished by cytochemical tests. By early July, they can be distinguished anatomically.

The significance of these observations has been well summarized by Puritch (1972). The cause of the change of primordia from undetermined to either vegetative or reproductive is still obscure. However, the transformation is perhaps one of the main physiological processes controlling cone production, and it is during this time that silvicultural techniques to induce increased cone production should be applied. Also, it is believed that the alternative manner in which buds may either develop, abort, or become latent is the major cause of periodicity in cone production.

About 9 months after differentiation of primordia in Douglas-fir, in the following spring, flowering and pollination take place. Fertilization of the ovule occurs about a week or two later, cones develop and mature, and seeds are released in the fall of that same year.

FLOWERING IN DECIDUOUS TREES

In deciduous trees there is considerable variation in the initiation of flowering. A general review has been provided by Krugman et al. (1974). The female flower structure of sweetgum (*Liquidambar styraciflua*) is morphologically complete by early summer and overwinters that way (Flint, 1959), whereas the female initials of white oak (*Quercus alba*), formed during late summer, are still morphologically immature at the time buds break the following May (Turkel et al., 1955). In many species which produce unisexual flowers, the female flowers start earlier and differentiate more rapidly.

Also, not all trees flower in the spring. Flowers may appear before leaves (as in *Populus*, *Fraxinus*, *Ulmus*), before or concurrent with the unfolding of leaves (as in *Carya*, *Celtis*, *Alnus*, *Betula*, *Quercus*, and *Salix*), or after leaves (as in *Fagus*, *Aesculus*, *Acer nigrum*, *Tilia*, *Nyssa*, and *Robinia*) (Kozlowski, 1971; Krugman et al., 1974). Temperate zone trees flower once annually, but tropical species may flower several times a year.

TREE-TO-TREE VARIATION IN FLOWERING

The age at which trees flower, and the calendar date of initiation of flowering, vary considerably between individual trees of the same species in the same stand. Also, it has commonly been observed that some individual trees are consistently early or late in flowering, or consistently produce flowers more or less abundantly. If one examines the time period over which individual trees within a stand come into flower, one finds that pines commonly produce male flowers over a period of a few days to 10 weeks, averaging about $2\frac{1}{2}$ weeks. Female flowering in pines and Douglas-fir also extends over about 3 weeks, but the range in time is considerably shorter—only 2 to 5 weeks (Kiss and Sziklai, 1965). If one notes the onset of flowering of individual trees over a period of successive years, one finds that these dates may cover a period of from 4 to 75 days (Bingham and Squillace, 1957).

This variability may be due partly to microsite conditions, but it is more likely to be largely the result of inherent mechanisms. The extent to which this tendency to flower is determined genetically is expressed in terms of broad-sense heritability, which, for slash pine, is quite high and has been estimated to be approximately 0.50, with a range between 0.4 and 0.7 (Varnell et al., 1967). In other words, about half the variation in a tree's tendency to flower can be explained in genetic terms. This high level of genetic control has enabled tree breeders to reduce the usual age of flowering in species (Krugman et al., 1974). This becomes critically important in seed-orchard programs where early seed production makes the total program much more practical and economically attractive.

STAND-TO-STAND VARIATION

Initiation of flowering also varies between different stands, and definite trends have been noted with changes in latitude and elevation. As latitude becomes more northerly, bud initiation occurs later in the year to the extent of about 5 days for each degree of latitude northward. As elevation increases, the timing of bud initiation becomes progressively later, with the delay being about 5 days for each added 300 m (1000 feet) in elevation (Bingham and Squillace, 1957; Silen, 1967; Puritch, 1972).

FACTORS INFLUENCING PERIODICITY AND FRUITFULNESS

The phenomenon of periodicity in seed production is intriguing and the cause of much debate and speculation. It has, of course, considerable economic importance in agriculture, and much of our knowledge of fruitfulness in woody plants has come from studies of fruit trees. In forestry in North America, since we have commonly been involved in extensive forms of management, periodicity in seed production has largely been regarded as an interesting but mildly inconvenient phenomenon. As we progress into more intensive forms of management, however, where we require prompt regeneration from the better genotypes and much of our seed requirements will be produced from seed orchards, it becomes increasingly apparent that there is an urgent need to understand the factors influencing flowering and periodicity.

It now appears that the occurrence of a good seed year in conifers is largely determined by the proportion of buds in any one year which either abort, become latent, or develop into reproductive rather than vegetative buds. This is undoubtedly influenced by tree metabolism and climatic factors. Once reproductive buds are formed, their development into mature, viable seed is influenced by site, climate, animals, insects, and diseases.

Site

Seed production begins earlier and is heavier and more consistent in the southern range of a species. Also, the tendency to early seed production is enhanced on warm, dry south exposures. This tendency may be partly attributable to the likelihood that on those poorer, more exposed sites there is an earlier differentiation into crown classes and that crowns are more exposed. There seems, however, to be a direct environmental effect on seed production, as has been shown by growing ramets of a single clone on both moist and dry sites. Ramets grown vigorously with ample moisture produced fewer cones than slower-growing ramets on drier sites (Schmidtling, 1974).

On poor sites, whether resulting from low nutrient status or low moisture availability, some limited amounts of seed are produced at an early age, even though the trees are often small and stunted. These are the so-called stress crops, and are analogous to the stimulation of seed crops by trees subjected to injury.

Climate

It is natural to expect that the cycling of weather patterns may play some role in determining the periodicity of seed crops. Several analyses have been made, including a 48-year study of Douglas-fir seed-production records, to determine the extent to which seed production is correlated with weather patterns (Lowry, 1966; Owens, 1969; Van Vredenburch and La Bastide, 1969; and Eis, 1973). These independent studies of Douglas-fir in both North America and the Netherlands indicate that good cone crops are associated with the sequence of climatic conditions, as shown in Table 8-7.

With regard to pines, the relationship between climate and seed production varies according to species, location, and time of year (Wenger, 1957; Shoulders, 1973; Dewars and Moehring, 1970; Matthews, 1963; and Puritch, 1972). In general, it appears that above-average cone crops are associated with unusually warm, dry conditions at the time of initiation of primordia—27 months before cone maturation—and unusually high rainfall at the time of flowering—15 to 18 months before cone maturation. Lester (1967), for example, has shown that for red pine, both April mean temperature and July-September mean temperature 2 years before cone maturation were positively correlated with cone counts.

Table 8-7 Relationship between Climatic Conditions and Cone Production

Season	Summer	Autumn	Winter	Spring	Summer	Autumn	Winter	Spring	Summer	Autumn
Douglas-fir cone development					Differentiation of flower primordia			Flowering, pollination, fertilization		Seed fall
Climatic conditions favoring heavy cone production	Cool cloudy summer	Cold sunny Dec., Jan., Feb.		Wet Mar., Apr.	Warm dry June		Warm January	Warm June		

Source: From information from Lowry, 1966; Owens, 1969; Van Vredenburch and La Bastide, 1969; and Eis, 1973.

However, strong relationships are not clear, particularly with respect to the effects of water stress occurring at varying times prior to cone maturation.

Physiological Causes

Physiology of flowering in trees is complex, and it is possible that a simple phytochrome-linked florigen-mediated control system, such as is traditionally described for some annuals, may not be satisfactory for woody plants. Romberger and Gregory (1974) argue that it is more likely that numerous biochemical or physiological systems are involved in the control of flowering in every bud on the tree. And for floral development to be initiated, all systems must be "permissive." If any system is "nonpermissive," then vegetative growth prevails.

Traditionally, flowering has been linked to carbohydrate status, nitrogen status, and hormonal activity.

Carbohydrate Status It has been thought for some years, following the original theory proposed by Klebs in 1910, that the stimulation and periodicity of flowering is due to the relative proportion of carbon (C) to nitrogen (N) in the plant. This C/N theory was based on the common observation that highly fertilized plants, particularly fruit trees, go "all to tops" and fail to produce heavy crops of fruit in spite of their vigorous vegetative growth.

The C/N theory of initiation and periodicity in flowering proposes that when either the available nitrogen or the amount of carbohydrate is very low, the plant will be unfruitful because it is starved. If, however, the levels of both nitrogen and carbohydrate are reasonably high, a high ratio of carbohydrate to nitrogen leads to abundant fruitfulness; a low ratio leads to strong vegetative growth coupled with poor fruit production. The theory has been attractive because it has permitted, better than any other general theory, a "reasonable" explanation of many of the commonly observed characteristics of seed production in trees, for example:

Seed production begins earlier and is heavier and more consistent in locations where photosynthesis and carbohydrate accumulation are favored.

Seed production is commonly observed in trees growing on poor sites where it has been assumed that uptake of nitrogen may be limited.

Heavy seed crops consume large amounts of carbohydrate, and a period is required to build up depleted food reserves before another seed crop can be sustained.

Attractive as the theory has been in the past, there is a reason to doubt its validity. Criticism comes from its intangibleness and from the observation that stimulation of flowering has been achieved without a corresponding increase in carbohydrate reserves. It has also been difficult to explain why application of nitrogen fertilizer to forest trees with well-exposed crowns commonly increases seed production. To maintain the C/N theory, one must argue that nitrogen fertilization increases the C/N ratio despite concurrently increasing the nitrogen content of the foliage.

Without doubt, heavy seed production consumes large amounts of carbohydrate, resulting in annual rings of markedly reduced width and reduced foliage production. This can readily be seen in tolerant conifers such as the true firs, which may have as many as 12 age classes of foliage on a branch. Years in which heavy seed crops were produced can be easily determined by observing the presence of a particular age class of needles which is substantially shorter than the average (Tappeiner, 1969). In Monterey pine, Fielding (1960) has estimated that the carbohydrate consumed in seed production is 16 percent of the annual production, which is equivalent to a loss of 2.1 m^3 ha^{-1} (30 ft^3 per acre) over a 40-year rotation. This consumption of food supplies by seed crops is of potential concern in selection of individuals for seed orchards. The breeder wants seed orchards to bear heavy seed crops, but ideally the progeny from seed orchards should not produce any seed at all if one is involved in maximizing wood production in even-aged silviculture incorporating artificial reforestation. The dilemma could be solved if we could artificially stimulate normally nonprecocious, rapidly growing trees to produce abundant seed only while they are in the seed orchard.

Another way in which carbohydrate consumption induces periodicity in seed production has been suggested by Puritch (1972). Consumption of food reserves results in less foliage, fewer bud loci, less substrate, and therefore more latency and abortion in buds. This causes subsequent cone crops to be relatively small until carbohydrate supplies again build up to support large crown development.

It can be seen from this discussion that the relationship between carbohydrate supply and flowering is not clearly understood, with some authors attributing direct relationships and others discounting its importance (Jackson and Sweet, 1972). For many years, the C/N theory has been commonly regarded as a plausible concept largely due to the absence of a more acceptable alternative. However, it has yet to be demonstrated whether changes in carbohydrate supply are really part of the cause or simply one of the effects of flowering in trees.

Nitrogen Status While some workers have considered changes in carbohydrate supply to be of prime importance in influencing flowering, others have considered relative nitrogen status to be dominant. Schmidtling (1974) reviews this general topic and concludes that nearly every response to treatments which have induced flowering can be better explained in terms of increased nitrogen availability at the critical time of primordia formation than in terms of carbohydrate supply. He proposes that nitrogen fertilization increases flowering in conifers because it results in physiological or biochemical conditions, possibly associated with amino acid production and especially free arginine, comparable to those induced by stress.

Biota

Insects Insects are the most important biotic agent reducing fruit and seed production. The damage is due mostly to six orders which, listed roughly in order of destructiveness, are moths (*Lepidoptera*), flies (*Diptera*), beetles (*Coleoptera*), wasps (*Hymenoptera*), tree bugs (*Hemiptera*), and thrips (*Thysanoptera*) (Stark, 1960). Relatively few insects attack the reproductive bud and immature female cone of conifers, but numerous insects damage or destroy the cones and fruits. The most common type of seed injury results from feeding by larvae of moths and chalcids within the cones, fruit, or seeds. Some insects, such as midges, do not attack or destroy the seed directly, but their activity prevents the seed from developing (Krugman et al., 1974).

The impact of insects in reducing seed crops is most noticeable in light seed years. This is so because insect populations build up in good seed years because of the abundance of food, causing extensive seed losses in the following year and a subsequent decline in the insect population as a result of reduced food availability. These dramatic changes in insect populations probably play important roles in influencing periodicity in seed production and probably tend to preclude the occurrence of two successive good seed years. The magnitude of the impact can be very large. Records of losses in seed production in ponderosa and Jeffrey pine, true fir, and Douglas-fir in western forests have shown that insects have, on occasion, destroyed essentially all the potentially available seed (Schubert and Adams, 1971).

In species with extremely small seed, such as in *Eucalyptus*, foraging by ants after seed dispersal is a major cause of loss.

Diseases The most important diseases attack the vegetative portion of the crown and hence influence seed production through causing a general reduction in vigor. Cone crops are never completely destroyed by diseases such as cone rusts, but the problem can be severe in high-value seed-producing areas.

Birds and Mammals Losses due to birds can be a severe local problem, for example, longleaf pine seed falling on a freshly burned area, but generally these losses do not constitute a serious problem. Mammals can be a problem, and although squirrels can harvest large quantities of cones from the tree, the major impact is after seed dispersal as a result of the foraging activities of ground squirrels, chipmunks, rodents, and other mammals.

TREATMENTS TO INDUCE SEED PRODUCTION

Stress

It has long been known that trees subjected to stress are stimulated into flowering. The stress involved may arise from a variety of sources such as drought, flooding, insect and disease damage, logging damage, induced moisture stress, and incompatible grafts. The capacity of trees to become more reproductive under conditions of stress has obvious ecological advantages and no doubt has been under positive selection pressure as plants have evolved.

Girdling

Of all treatments to induce flowering, girdling or strangulation has been most commonly studied. This practice arose out of the C/N theory of flower induction. Theoretically, if one was to either enhance carbohydrate accumulation in the crown or impede the uptake of nutrients by the roots, then flowering should be stimulated. This situation is created by girdling, which restricts downward transport of carbohydrate in the phloem and the uptake of water and nutrients in the xylem. This treatment has commonly been successful in stimulating flowering, with the response time varying from the second year after treatment of Douglas-fir (Ebbel, 1971) to the third year after girdling shortleaf pine (Bower and Smith, 1961). Of course, a major disadvantage of the treatment is its damaging nature, which may completely destroy the tree's value for wood production, as a component of a seed orchard, or for other purposes.

It is unclear as to how girdling stimulates flowering. In several studies, for example, it has been shown that girdling did not affect carbohydrate reserves (Bilan, 1960) or only weakly influenced food supply (Ebbel, 1971). In addition, altered carbohydrate distributions in the tree may be the result, rather than the cause, of seed production. An interesting study was conducted by Ebell (1971) in which he girdled one stem of two double-stemmed 20-year-old Douglas-fir trees. The girdled stem produced 7.4 times as many cones as the ungirdled stem in the year after treatment, falling to 1.6 to 2.3 times in the subsequent 3 years.

The timing of the girdling treatment is critically important. The optimum time in Douglas-fir is 1 month before vegetative bud break, which coincides with the beginning of differentiation of the floral primordia for the following year. Variable response is obtained when girdling is done close to the time of vegetative bud break, and an adverse effect is obtained if the treatment is applied later than 1 week after vegetative bud break (Stephens, 1964; Ebell 1971).

Strangulation is not as effective as girdling because it is more damaging, gives slower response, and provides less control over timing the desired alteration of physiological processes. Similarly, the pruning of roots and branches can also be used to influence carbohydrate and nutrient relationships, but the results of these treatments are also variable.

Fertilization

A considerable number of investigations have been made on the relationship between nutrition and fruiting. Generally, fertilizer applications increase seed production (Matthews, 1963; Steinbrenner et al., 1960; Day et al., 1972) and can also induce precocious flowering (Schmidtling, 1971). The reasons for increased cone production with fertilization are not entirely clear. Certainly, the general nutrition of the plant is enhanced and the nitrogen content of the foliage increased. In Douglas-fir it has been shown that the foliage of the 10 best cone producers contained an N content of 1.24 percent, whereas the N content in the foliage of the 10 worst cone producers was a significantly lower 1.13 percent (Kozak et al., 1963). But it appears that responses are not due primarily to improved mineral nutrition alone, but are associated with specific chemical stimulation from critically timed changes in the type of nitrogen metabolism (Ebell, 1972*a* and *b*).

There is also strong evidence that the form in which nitrogen is applied is important. The nitrate form, rather than the ammonium form, enhances seed production. The nitrate form seems to increase the concentration of amino acids, particularly free arginine, which is thought to regulate the development of reproductive tissues in conifers (Barnes and Bengtson, 1968; Ebell and McMullen, 1970; Puritch, 1972; Schmidtling, 1974). Also, it is argued that natural periodicity of seed production in conifers is caused by fluctuations in amino acid levels. This is substantiated by the finding of unusually high concentrations of arginine following cone-inducing treatments.

Viewed from a slightly different standpoint, it appears that the enhanced nitrogen status of fertilized trees does not have any direct influence on the total number of buds per shoot, which is similar on control and fertilized trees. Rather, fertilizing with nitrate nitrogen at the time of bud differentiation reduces bud latency and abortion and increases the proportion of buds that complete development and produce mature seed.

The timing of the fertilizer application is important, with the greatest stimulation of flowering occurring when the treatment coincides with the period of differentiation of the floral primordia (see Table 8-6). Fertilizing earlier or later than this period is ineffective (Stoate et al., 1961; Ebell, 1972*b*; Schmidtling, 1974). The effect of differences in timing of fertilizer applications on flowering in loblolly pine is shown in Fig. 8-4.

Not all genotypes react in the same way to fertilizer treatments, and since the tendency to flower is highly heritable, genetic effects could easily mask treatment effects if not taken into account. In a well-controlled study of 32 ramets from each of seven clones of loblolly pine treated with either N, P, or N + P, 33 percent of the variation was attributable to clone effects, whereas the fertilizer treatment accounted for only 30 percent of the variation (Schmidtling, 1974). This general situation is illustrated in Fig. 8-5, which shows the different responses of the seven clones to fertilizer treatment. It can be seen readily that in general the inherently more fruitful clones responded best to fertilization.

It is no small wonder, then, that the literature contains considerable variability in reports on the effects of fertilization on flowering. This variability is due to differences in genotypic responses, nutrient availability in the soil, the fact that applications are made at different times of the year and at different

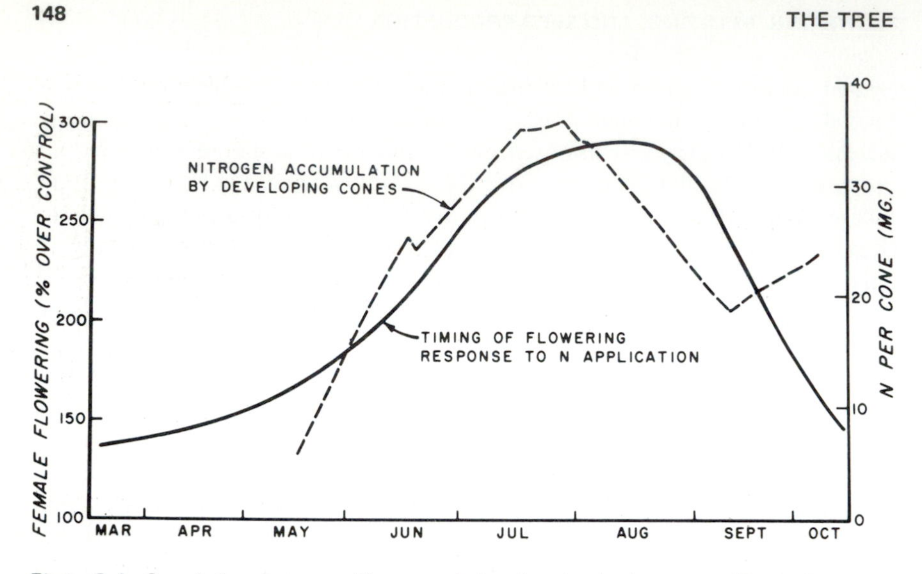

Figure 8-4 Correlation between *N* accumulation by developing cones (*From Dickmann and Kozlowski, 1969*) and timing of flowering response to *N* fertilization (*From Schmidtling, 1974.*)

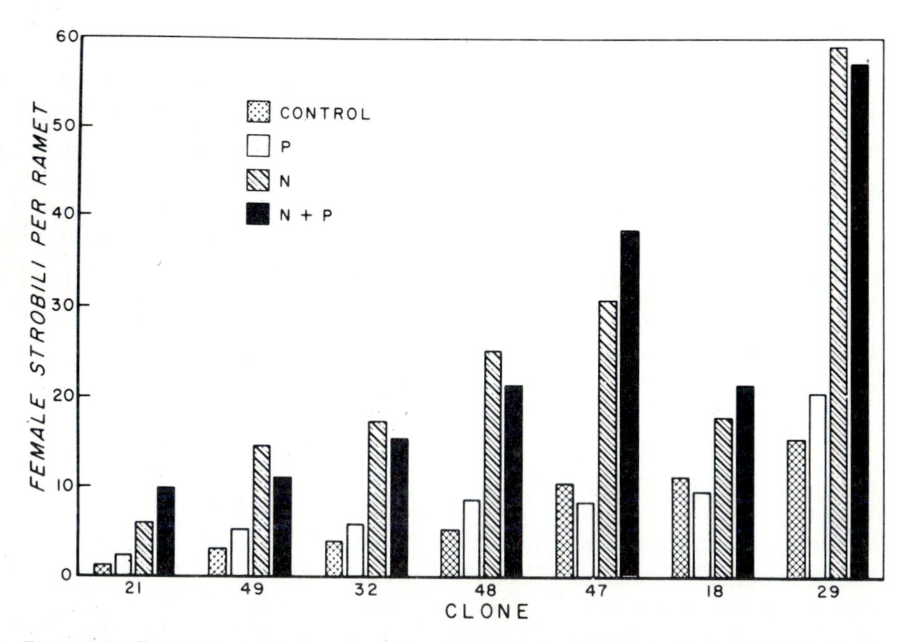

Figure 8-5 Flowering response of various loblolly pine clones to *N* fertilization. (*From Schmidtling, 1974.*)

times relative to flower-bud differentiation, differences in the form of the fertilizer used, and the beneficial or detrimental effects of rainfall and general weather conditions before and after application.

Hormones

Auxins are known to inhibit flowering in herbaceous plants, and there is some evidence that they also have some effect on forest trees. The only consistently successful treatment for cone induction through the use of natural plant hormones has been the application of gibberellic acid, but this treatment has been effective only on species in the *Taxodiceae* and *Cupressaceae* (Japanese cedar, Arizona cypress, western red cedar, coast redwood, and giant redwood), with no significant effects on the *Pinaceae* (pines and Douglas-fir) (Pharis and Owens, 1966; Pharis and Morf, 1968; Pharis, 1970). More recent work using very large dosages of less polar gibberellic acid, however, has produced precocious flowering in Douglas-fir and lodgepole pine (Pharis et al., 1974). It is possible that further successes in inducing flowering by hormonal substances will be attained in the future despite disappointing results in the past.

Spacing

It has been consistently shown that well-spaced trees produce more seed than those which are growing closely together. Consequently, thinning has often been used, in conjunction with fertilization, to increase seed production in many species (Florence and McWilliam, 1954; Steinbrenner et al., 1960; Reukema, 1961; Goodman, 1962; Stiell, 1971; Brender and McNab, 1972).

The effect of thinning is most pronounced in the more intolerant species, which bear their cones in the free-growing portion of the crown above the level of crown contact by adjacent trees. In this situation, increasing spacing provides trees with a higher proportion of the crown which is fully exposed and therefore in a more likely condition to bear flowers. It is commonly accepted that trees under stress from limited availability of water tend to produce more seed. Thinning, however, at least initially, tends to make more soil water available per tree, which would seem to diminish the likelihood of enhancing seed production. This apparent contradiction can be explained by the fact that thinned trees, because they have greater crown exposure and more water available to them, have increased rates of transpiration often leading to leaf water potentials that are more negative than those in the crowns of unthinned trees (Sucoff and Hong, 1974).

The amount of crown exposure required to maximize seed production becomes important in the design of spacing in seed orchards. Similarly, predictions of the amount of seed likely to be produced by fully exposed crowns is important in determining the number of seed trees needed to provide adequate and prompt production of seed for natural regeneration. For loblolly pine, the recommended stocking for seed trees is a basal area of 6.7 to 6.9 m^2 ha^{-1} (25 to

30 ft^2 per acre), which is equivalent to about 11 to 13 trees 51 cm (20 in) dbh or 7 to 9 trees 63 cm (25 in) dbh (Brender and McNab, 1972).

From this discussion of initiation and periodicity in flowering of forest trees it can be seen that much has been learned since 1910 when the C/N theory was proposed as the controlling mechanism. There is much still to be learned, and the urgency to attain this knowledge increases as we require greater control over the promptness, adequacy, and quality of regeneration. Initiation and periodicity in flowering are known to be associated primarily with genetic, carbohydrate, and nitrogen relationships—all of which can be influenced by treatments.

The timing of these treatments is critically important. In the past it was thought that the destiny of primordia as vegetative or reproductive buds was fixed at the time of formation, and that the number of cones capable of being produced was a function of the number of reproductive buds initiated. As a result, treatments to induce cone production in conifers were timed to coincide with the time of primordia initiation. It is now known, using Douglas-fir as an example, that in sexually mature trees the number of cones produced is not entirely dependent on the number of primordia initiated but is determined by the proportion of primordia that develop as reproductive buds and do not become latent or abort. In both good and poor seed years, the total number of buds per shoot is about the same. In good seed years, a higher proportion of these buds develop into reproductive buds. In applying treatments, therefore, it must be recognized that there is a crucial period just after initiation during which bud primordia are differentiating when treatments must be applied in order to affect the proportion of buds that become reproductive rather than vegetative (Puritch, 1972). The second objective of treatments is to minimize the tendency for a proportion of these reproductive buds to become latent or to abort.

Chapter 9

Root Growth

ROOTING CHARACTERISTICS

The roots of trees are obviously of fundamental importance in tree growth and development. Not only do they provide mechanical fastening to maintain the tree's upright structure, but they are essential for water uptake and mineral absorption. In fact, the health and vigor of root systems are so fundamental to the health and vigor of the tree as a whole that, ideally, silvicultural treatments should be equally based on root as well as crown characteristics. Unfortunately, relatively little is known regarding root systems because of the inherent difficulties in studying roots without at the same time altering conditions for their growth.

Rooting characteristics of trees vary greatly among species, among individuals of the same species, and even among different roots of an individual tree. The lateral spread of root systems is commonly 2 to 5 times the radius of the crown, being greater on poorer sites and under more xeric (drier) conditions. The extensive lateral development of tree root systems means that commonly the highest concentration of fine feeder roots is at some distance from the trunk. This fact has obvious bearings on the placement of fertilizer.

The depth of roots is to some extent influenced by genetics, with some species tending to be shallow-rooted and others having a tendency to produce a tap root. In general, it seems that species with large seeds rich in reserve foods such as *Quercus*, *Carya*, *Castanea*, *Juglans*, and *Pinus* quickly develop a long tap root which rapidly reaches greater soil depths and thus contributes significantly to survival in drought-prone areas. This rapid juvenile development of a deeply penetrating root is commonly achieved at the expense of shoot growth, which initially may be rather slow. When the tree is more mature, the inherent juvenile characteristics of root development commonly become completely dominated by conditions of the site such that continued characterization of genera and species as being either shallow-rooted or deep-rooted may be quite unwarranted (Sutton, 1969). It is common, for example, for the character of root systems of mature trees to be dominated by such factors as soil type, nutritional status, drainage characteristics, and the presence or absence of peat, gleys, hardpans, and organic matter.

DISTRIBUTION AND BIOMASS

The distribution of roots, then, is a function of species, age of tree, environment, and unit of measure. In terms of environment, the presence in the upper soil layers of high humus content and high aeration, nutrient, and moisture content leads to the greatest development of fine roots in these upper horizons and consequently gives a tendency for a shallow root system for most tree species. With regard to units of measure, the distribution of roots depends markedly on whether one is evaluating root biomass, root surface area, or the length of absorbing roots. For example, most of the root system, in terms of length, is in fine roots which commonly represent 85 to 99 percent of total root length but only 14 to 60 percent of total root weight (Lyr and Hoffmann, 1967). When the extent of total absorbing root surface is considered, the proportion contributed by fine roots is naturally higher still. The quantity of fine roots (those less than 2 to 3 mm in diameter) in conifer stands seems to be approximately 2 to 3 kg per tree, which corresponds to about 7 metric tons per hectare (2.8 tons per acre). Quantities of this magnitude have been recorded for plantations of white spruce in Canada (Safford and Bell, 1972) and for plantations of Monterey pine in Australia (Moir and Bachelard, 1969). Most of these fine roots are near the surface of the soil where most of the soil humus and nutrients occur. Similarly, in 12-year-old slash-pine stands it has been shown that 50 percent of the total root surface occurs in the top 0.3 m of soil, and that the proportion found below a depth of 1.35 m was 6 to 24 percent, depending on nearness to individual trees (Schultz, 1972).

The linear extent of root systems in forest stands is considerable. In a 28-m-tall mixed conifer stand in northern Idaho, for example, the total length of roots having diameters greater than 2 mm within the total rooting depth of 3 m amounted to 850 m/m^2 of soil surface (Leaphart and Grismer, 1974). If one were to consider the additional length of fine roots and root hairs, the total length of the root system in forest stands would be astronomical.

From these remarks it can be seen that root systems constitute a significant amount of biomass, commonly 20 to 25 percent of total dry-weight production (Ovington, 1961; Young and Carpenter, 1967; Karizumi, 1974), and this must be taken into account when considering the total productivity of forests.

GROWTH RATES AND PERIODICITY

From a silvicultural point of view, the dynamics of root development are more important than the actual static size of the root system. This is so because capillary water conduction in most soils is very low, and once the soil is below field capacity, the uptake of water is primarily dependent on the capacity of the root to grow and tap new water and nutrient sources held in adjacent soil volumes. High rates of root increment therefore indicate that a plant is capable of quickly penetrating large volumes of soil.

During the process of germination, the primary root appears and elongates rapidly using the reserves of energy and nutrients in the seed. In 10 days, the primary roots of white spruce and balsam fir are 3 to 5 cm long. Rates of further elongation are somewhat slower and dependent on soil conditions. The initial tap-root form is frequently transitory, and lateral development, which commences soon after primary root elongation, rapidly surpasses the length of the main root. Maximum daily elongation of roots is commonly 5 to 16 mm for conifers, 56 mm being reported for black locust (Sutton, 1969; Kozlowski, 1971). Growth of any one root is not necessarily continuous but may be cyclic in nature (Wilcox, 1962).

Within a week or so of formation, a general browning of the root occurs which is caused by a suberized layer commonly thought to retard water uptake and loss. The root tips which emerge from the older, brown-colored root are generally white in color but may also be brilliant red (*Larix* and western red cedar) to yellowish or dark colored. The color is not always a species characteristic, since it can depend on the seedling nutrient status (Hacskaylo, 1962).

The seasonal periodicity of root growth is determined by both crown activity and soil conditions. If conditions are favorable, root systems can develop all year around. Therefore, roots, unlike the stems of many plants, apparently have no inherent dormant period (Kramer and Kozlowski, 1960). The two factors commonly limiting root growth are soil moisture stress in summer and low soil temperatures in winter. Consequently, the seasonal pattern of root growth is generally bimodal, with minimum activity in midwinter and midsummer. The rhythm of root growth differs between species and environmental conditions, but maximum growth in terms of number of growing roots and total growth in length in most tree species occurs in early summer. Seedlings with early termination of shoot growth (*Quercus* type) often exhibit strong root growth in midsummer. Roots of deciduous trees usually develop maximum growth rates in early summer and can continue growth after leaf abscission. Conifers tend to have a more uniform growth throughout the vegetative period (Lyr and Hoffman, 1967).

A young root has four zones of differing anatomical characteristics: (1) the root cap, which is a protective layer of thin-walled mucilaginous parenchyma cells; (2) the subapical meristem, which is the growing point that gives rise to cells of all other primary root tissues; (3) the zone of elongation; and (4) the zone of maturation, which develops into the permanent root tissues of the epidermis, cortex, and stele (Kuntz, 1973). Root hairs are formed continually behind the root tip, especially in nonmycorrhizal roots of conifers, and these are of considerable importance in the absorption of water. Thousands of root hairs are developed on seedlings only a few weeks old, and these are continually being sloughed off as the root system develops (Karizumi, 1974). Within the root, the impregnated and thickened cells of the endodermis provide the greatest resistance to the uptake of water and solutes. The annual production of xylem and phloem results in annual rings in roots, but these are not as pronounced as those in stems. Long-lived roots may develop a heartwood surrounded by a sapwood.

ENVIRONMENTAL EFFECTS ON DEVELOPMENT

The development of roots depends to a large extent on the soil environment.

Temperature

Low temperatures retard growth, metabolism, and maturation of roots. In addition, uptake of water and nutrients decreases, and is perhaps inadequate to meet the needs of the shoot. At low temperatures, water becomes more viscous and tissues become less permeable. At high temperatures, increased rates of respiration decrease root growth. This general relationship has been demonstrated by Barney (1951), who showed that daily rates of root growth in loblolly pine increased rather uniformly from 0.36 ± 0.16 mm at 5°C to a maximum of 3.41 ± 0.24 mm at 20°C, and then decreased to 0.24 ± 0.34 mm at 35°C. There are considerable differences between species, however. The root systems of true firs, for example, operate much more efficiently at low temperature than do those of pines.

Inefficient root activity as a consequence of low soil temperature can commonly be a cause of failure of a species to become established after spring planting. It may also be a reason for the natural tree line in alpine and northern regions (Lyr and Hoffmann, 1967). In both situations, the plant suffers from dessication and water stress due to high rates of transpiration and limited water uptake.

Oxygen

The degree of soil aeration is important because the roots of most trees begin to respire anaerobically as oxygen concentrations become minimal, resulting in the production of toxins such as carbon dioxide and ethanol, the cessation of root growth, and root death. Notable exceptions to this are species such as swamp cypress (*Taxodium distichum*), river red gum (*Eucalyptus camaldulensis*), and to some extent redwood (*Sequoia sempervirens*), which are capable of continued growth under anaerobic conditions of flooded soils for periods of up to at least several weeks.

Fertility

The development of roots is closely associated with the fertility of the soil. The more fertile the soil, the greater is the root development in terms of mass per unit volume, and also the greater the root penetration. The impact of nutrition on root development is seen in the optimal development of roots in the upper, most fertile layers of the soil, and also in the enhanced root development around the placement of fertilizer within the soil profile. Fertilizer applications tend to encourage the development of shallow roots, and this is often accompanied by a decrease in deep rooting.

Mechanical Impedance

Root growth is impeded by compacted soils because of the direct physical restrictions as well as lower aeration and less water percolation. Hardpans also restrict root growth by preventing root penetration, with the result that masses of roots develop above the barrier. Water can also accumulate above a hardpan, and this may directly limit root growth. In most cases, impediments to root growth cause shallow root systems with consequent potential for windthrow.

Soils are usually not uniformly compact, and root penetration and development occur in less dense areas and where there is less resistance due to the presence of old root systems and structural fractures.

ROOT GRAFTS

It is becoming increasingly apparent that individual trees in a forest stand do not necessarily function as separate entitites because of the occurrence of natural grafting which links the root systems of several trees into essentially one functional unit or community. There are three types of root grafting: self-grafting, between roots of the same tree, which is quite common; intraspecific grafting, between different trees of the same species; and interspecific grafting, which is quite rare, with most apparent unions being nonfunctional (Fig. 9-1). Intraspecific grafting is quite common and has been reported in many kinds of stands, including at least 150 different conifers and hardwoods (Graham and Bormann,

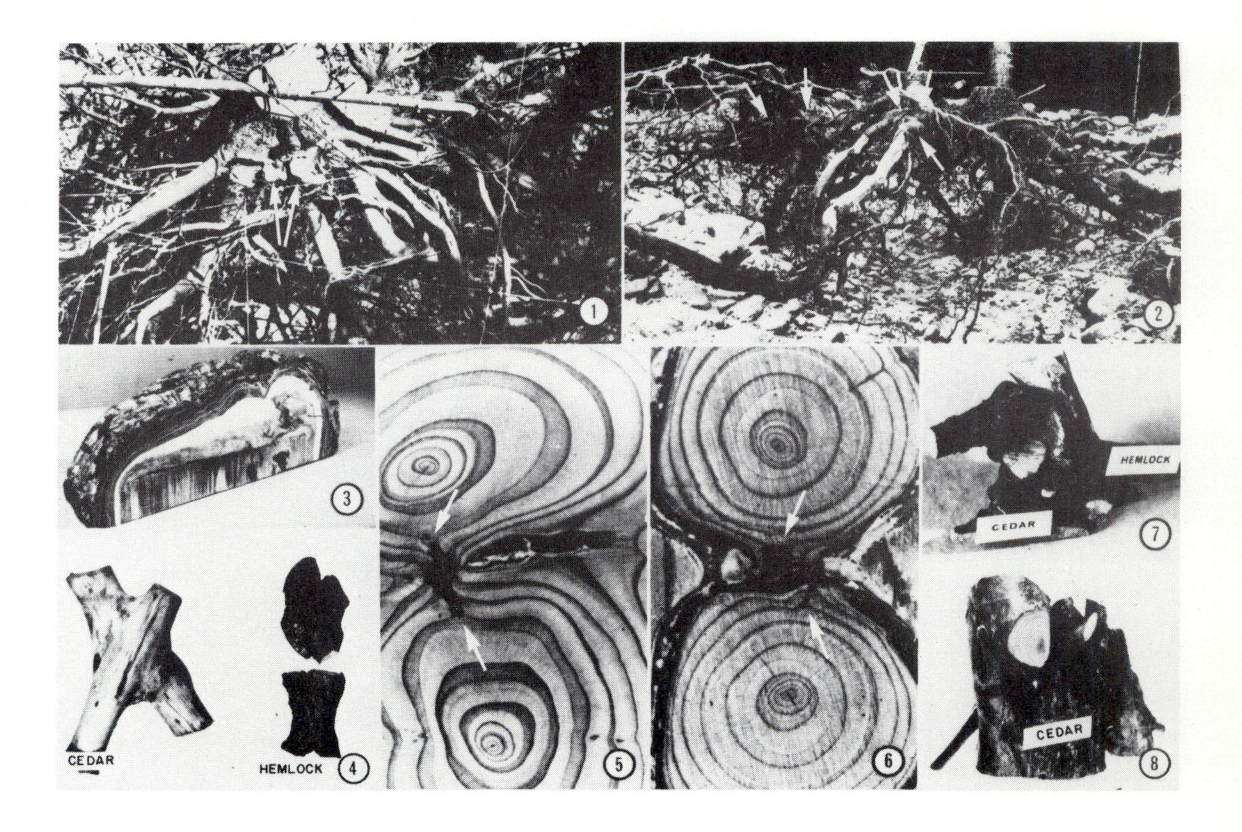
1
2
3
CEDAR
HEMLOCK
4
5
6
HEMLOCK
CEDAR
7
CEDAR
8

1966). In red pine, grafting was found to be common between 10 and 35 cm below the surface in plantations older than 15 years (Armson and van den Driessche, 1959). Similarly, in Monterey pine plantations in Australia, root grafting was found to begin when stands were 10 years old and increased in frequency until stands were at least 25 years old (Wood and Bachelard, 1970). In 16 plantations and 4 natural stands of slash pine which were lightly thinned at 10 to 35 years of age, investigations of intraspecific root grafting were made 2 to 10 years after thinning (Schultz, 1972). The frequency of grafting at a spacing of 1.8 m was double that at a spacing of 3 m within the rows. The proportion of stumps grafted to potential crop trees was 7 to 15 percent in plantations and 15 to 25 percent in natural stands, and it was estimated that about 70 percent of all grafts would die by the fifth year after thinning. In 27- to 83-year-old loblolly pine plantations only 7.6 percent of 2054 stumps (4 trees per acre) were found to be grafted to living trees (Schultz and Woods, 1967). In Douglas-fir, 45 percent of selectively cut trees were found to be root grafted, and half of these were still actively growing 22 years after harvesting (Eis, 1972). Details of many other examples of root grafting are provided by Kozlowski (1971). If functioning grafts occur, the cut stumps show evidence of growth and callous-tissue development. The callous tissue exhibits annual rings, thus it is possible to determine the time during which these living stumps have been maintained by the graft. Lanner (1961) examined living stumps of ponderosa pine, sugar pine, Douglas-fir, incense-cedar, white fir, and red fir on the western slopes of the Sierra Nevada which had followed logging in stands 35 to 200 years old. The oldest living stump was 87 years old, and Lanner quotes Newin (1916) as having found a living stump 200 years old. Keeping old stumps alive through root grafts must constitute a significant drain on the growth of the parent tree.

The occurrence of natural root grafting means that the grafted group of trees can function as a single physiological unit in response to environment. For example, it has been observed in red pine stands that in years of drought trees have died in groups (Stone et al., 1973). If the trees in some stands are grafted together, the concept of competition and the relationship between dominant, codominant, and suppressed trees may have to be reexamined. The

Figure 9-1 1: Excavation of root systems of a dominant (78 years) and two suppressed (36 years) hemlock (arrows). Despite frequent contact with large roots of the dominant tree, grafting took place only among small roots. 2: Excavation of a group fo three Douglas-fir, 39 years old. Note massive grafts in the foreground (arrows). 3: Vigorously growing stump of a codominant Douglas-fir, 32 years after logging. 4: Orientation of wood cells and bark on intraspecific grafts, indicating translocation between grafted roots. 5: Intraspecific graft on red cedar. Callus tissues, growing opposite each other, united probably during the third year and were overgrown by xylem during the following year. 6: On two hemlock roots with heavy bark, the stimulation did not occur opposite each other. Callus tissue flattened against the bark and the graft did not materialize. 7: Interspecific graftlike union showing intact bark on the area of contact. 8: Nonfunctional interspecific graftlike union of cedar and hemlock. (*From Eis, 1972; reproduced by permission of National Research Council of Canada, from the Canadian Journal of Forest Research,* **2**: *111–120.*)

exact role and importance of root grafting in the uptake and transfer of water and minerals are not well understood. Grafted suppressed trees may obtain a portion, if not all, of the carbohydrate necessary for growth of roots and lower boles from the dominant trees to which they are grafted (Eis, 1972), and thus the photosynthate produced from the small, suppressed crowns may be needed only to sustain growth and development of the crown and upper bole. This kind of relationship between trees may cast into doubt the common generalization that smaller trees with light crowns are more efficient and produce larger stems per unit of crown area than larger trees.

The occurrence of natural root grafting has important implications regarding the transfer between trees of diseases such as Dutch elm disease, oak wilt, elm phloem necrosis, *Fomes*, and *Armillaria*. In the South and parts of the West, borax is sometimes applied to cut stumps in thinning operations to reduce the risk of infection. Root grafts also can have importance in stability and windthrow, girdling, and between-tree competition (Graham and Bormann, 1966; Sutton, 1969; Kozlowski, 1971; Kuntz, 1973). The possibilities of root grafting also requires one to question the use of herbicides where there is a possibility of the injected poison being translocated to a neighboring tree–a phenomenon known as "backflash." There is also the question of whether the thinning of root-grafted stands stimulates the growth of the remaining trees, which are suddenly provided with extra root systems to which they are grafted. Most evidence indicates that these live stumps do not promote better diameter growth in the crop trees to which they are grafted (Graham and Bormann, 1966).

The silvicultural significance of root grafting is difficult to determine and must depend on the extent of grafting and the proportion of carbohydrate needs that potentially can be donated by one tree to another. Generally, no significant differences in growth response have been found after thinning grafted and nongrafted trees. In some cases, however, understory trees have died following release; this may have been due to the fact that a neighboring thinned tree was actually a host tree connected to the suppressed tree by a natural root graft. Alternatively, the released suppressed tree might not have been able to adjust to the changed environment. The translocation of any material undoubtedly follows a source-sink relationship, and therefore the significance of grafting depends on the relative size of the sink in the recipient tree compared to the needs for carbohydrate in the potential donor tree.

ROOT-GROWTH CAPACITY

Factors Affecting Seedling Survival

In regenerating forests by planting, survival of seedlings commonly depends on three factors: soil temperature, availability of water at planting depth, and the capacity of the seedlings to produce roots.

Soil Temperature Regardless of seedling quality and physiological condition, a seedling planted in the spring will not produce new roots unless soil temperatures are above a certain minimum. In California, this minimum level is about 10°C for ponderosa pine and somewhat less for true fir species. Consequently, success in planting is dependent on knowledge of when soil temperatures are high enough for the initiation of new roots.

Availability of Water at Planting Depth There must be adequate water available within the root zone, otherwise the seedling will become dessicated. The combination of soil temperature and water availability requirements is illustrated in Fig. 9-2. Planting time must be related to these two prime site factors.

Capacity to Produce Roots Even if site conditions are such that soil temperature and soil water availability are both adequate, the seedling will survive only if it has the physiological capacity to produce new roots. This capacity varies considerably according to the genetic characteristics of the seedling and the environmental conditions under which it is grown.

In many forest areas at the time of planting, the site may have adequate moisture and optimal temperature. Under these conditions there is little evaporative stress on the plant, and the necessity to rapidly produce new roots immediately after planting is not acute. However, in climates where the spring planting season is restricted to a few weeks between snow melt and the onset of a dry summer, it is imperative that the seedling have the capacity to initiate new roots immediately after planting.

The capacity of seedlings to produce new roots depends on the climatic condition of the nursery, the genetic characteristics of the particular species, the

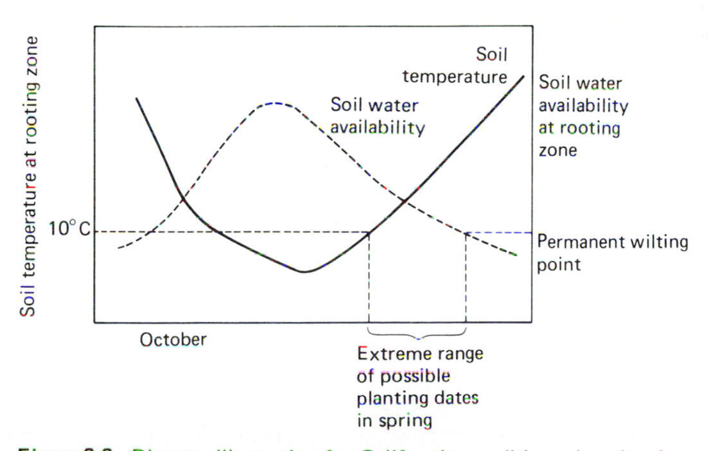

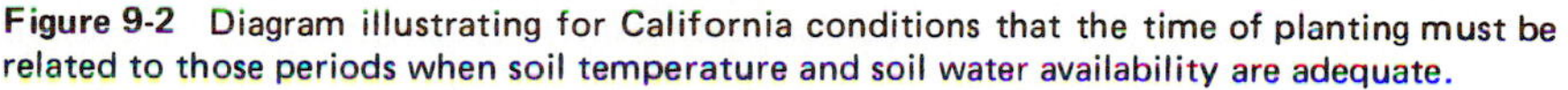
Figure 9-2 Diagram illustrating for California conditions that the time of planting must be related to those periods when soil temperature and soil water availability are adequate.

treatments in the nursery prior to lifting, the conditions of seedling storage and transportation, the care in planting, and the environment of the planting site. Poor survival can result, particularly in difficult planting sites, if any one of these factors is unfavorable. Procedures for appraising root-growth capacity (or root-regeneration potential) have been developed (Stone and coworkers 1959; 1963; 1971), and these have led to the design of nursery treatments, particularly with regard to time of lifting seedlings from the nursery bed, that ensures the production of stock with a higher potential for outplanting survival. Studies on ponderosa pine have shown that root-growth capacity increases through the fall and then decreases in the spring. This change is not accompanied by any change in the appearance of the shoot. The controlling factor appears to be the accumulated hours of cold to which the seedling is exposed prior to lifting. This depends on current climatic conditions, and thus the most appropriate time for lifting will vary from year to year as well as between nurseries. Figure 9-3 shows that

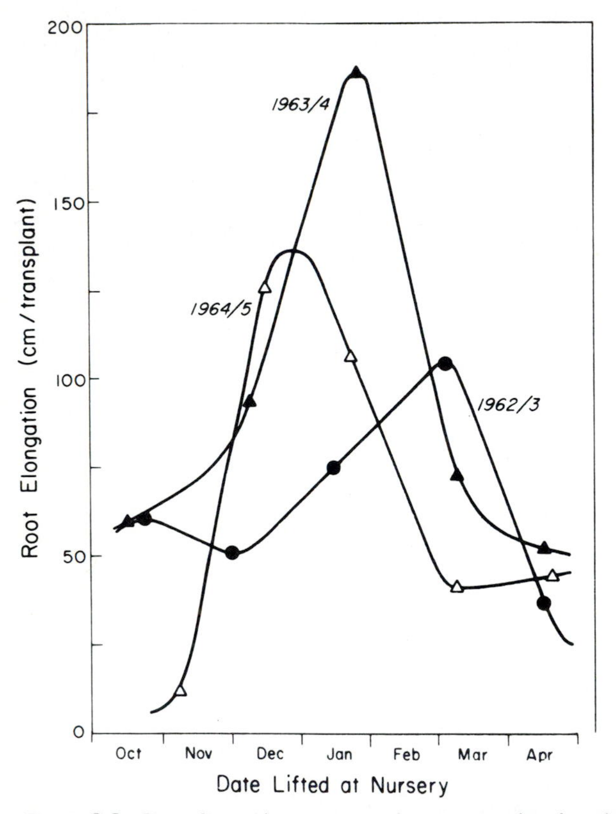

Figure 9-3 Data from three consecutive seasons showing the number of hours that the seedlings had experienced temperatures ≤10°C at the time that maximum root-growth capacity was measured. The date of maximum root-growth capacity is earlier if the season is cold. *(From Stone and Jenkinson, 1971.)*

the faster the accumulation of cold conditions in the nursery, the earlier does root-growth capacity reach a maximum.

The development and maintenance of high levels of root-growth capacity by seedlings in the fall vary markedly with the geographic source of the species (Fig. 9-4). The effect of increasing elevation is to increase the time taken to reach maximum root-growth capacity in the nursery; the lower the elevation is, the longer the period in which maximum root-growth capacity is maintained. Similarly with respect to latitude, northernmost sources of ponderosa pine reach maximum root-growth capacity later, and hold that capacity for a shorter period, than do southern sources.

Stone and Jenkinson (1971) have also demonstrated that root-growth capacity is influenced by cold storage. Fresh ponderosa-pine seedlings lifted from October to March and tested immediately were shown to have maximum root-growth capacity when lifted in February. This same stock, when stored 3 months at -1°C, decreased root-growth capacity to the lowest level (Fig. 9-5). For stored stock, highest root-growth capacity was obtained from stock lifted in late December. In other words, cold storage may increase or decrease root-growth capacity depending on time of lifting.

The silvicultural significance of these findings is that the time of lifting seedlings in the nursery will have a marked effect on their potential for surviving outplanting. This may be less important in mild climates where survival is usually high. But in difficult environments, seedlings must have the capacity to immediately produce new roots because they immediately begin to transpire after planting.

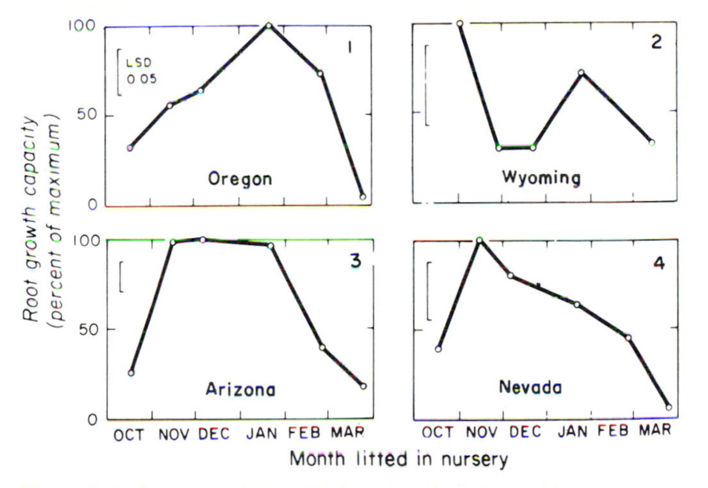

Figure 9-4 Patterns of root-growth capacity in ponderosa pine throughout the lifting period: (1) one peak occuring late; (2) two peaks, one early and one late; (3) one peak, but capacity for maximum root growth is maintained over a long period; (4) one peak which occurs early, followed by a rapid decline. (*From Jenkinson, 1975.*)

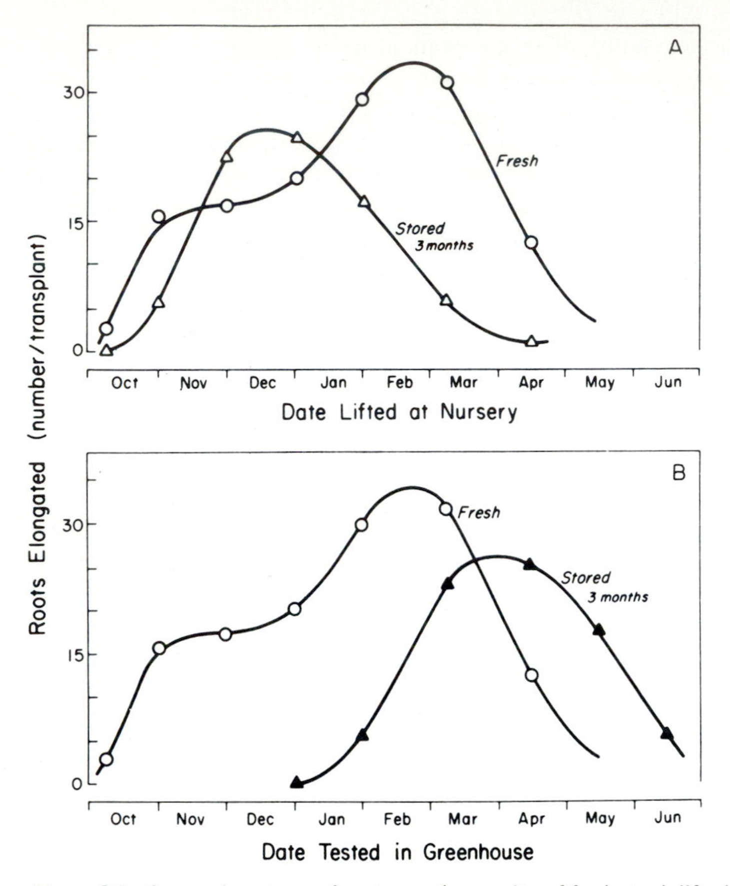

Figure 9-5 Seasonal patterns of root-growth capacity of fresh stock lifted at different times of the year and of samples of the same stock tested after storage for 3 months at −1°C. a: Curves for fresh and stored stock are superimposed to show that storage *increases* root-growth capacity of stock lifted during the period of mid-November to mid-January. Stock lifted at all other times has its root-growth capacity decreased by 3 months of cold storage. b: Cold storage extends the time seedlings with high root-growth capacities are available for planting. (*From Stone and Jenkinson, 1971.*)

ROOT PATHOGENS

Soil-borne pathogens are of universal occurrence and commonly cause root diseases in forest trees. Their significance often becomes important only in situations of more intensive management such as in campgrounds, plantations, or forest nurseries. However, in the more extensively managed areas of native forests, root pathogens can cause major losses.

The most important root-disease fungi in natural stands include *Fomes annosus*, which is very widespread and attacks both hardwoods and conifers; *Poria weirii*, which is an important root disease in Douglas-fir; and *Armillaria mellea*, the shoestring fungus which is associated with oak and conifers. Other root diseases of importance are *Verticicladiella wagnerii*, which aggressively attacks ponderosa pine in the cooler sites in Western United States, and *Phytophthora cinnamomi*, which is causing extensive die-back in pine forests of the Southeast and in Australian eucalypt forests. *Polyporus schweinitzii*, a brown cubicle rot, is a root disease attacking both pine and fir in the West; however, it is relatively unimportant. All these diseases have in common the capacity to spread from tree to tree via the roots, causing enlarging pockets of dead and dying trees. Disease centers can originate by invasion of stump surfaces, basal wounds, fire scars, or by direct invasion of roots. Once established, they may persist for many years in old root systems and may attack regeneration of any susceptible species within the disease center (Parmeter, 1973).

Root diseases of trees commonly cause the rotting of roots, leading to death, as with *Verticicladiella*, or more commonly resulting in windthrow. Infection is commonly by wind or rain washing of spores which gain entrance through fresh stumps, basal wounds, or by local root contact or root grafting. Symptoms of root diseases are very similar to those caused by nutrient deficiencies in that needles become short, tufted, and chlorotic, with fewer age classes present. In addition, the tree shows a general decline in vigor. Resinosus may occur, as with *Armillaria*, or conks may develop, as with *Polyporus*. Also, the development of root disease may predispose the tree to subsequent attacks by insects, which then become the immediate cause of death. An example of this interaction is the relationship between *Verticicladiella wagnerii* and the western pine beetle *Dendroctonus brevicomis*, in an attack on ponderosa pine.

In order to control the spread of root diseases, strong arguments can be made in favor of maintaining mixed stands, although in some cases the problem of control may be best managed in vigorous plantations of one species. Infection centers are probably best handled by clearcutting and scarification, severe burning, and perhaps soil fumigation. These practices could not be applied in uneven-aged or shelterwood systems of management, however. Where infection is through freshly cut stumps, such as with *Fomes*, the problem can be minimized by shorter rotations, which are likely to cause less basal damage to residual trees because of smaller logs and smaller, more maneuverable equipment and which leave smaller stumps that dry out more quickly. Disease problems can be more severe on poorer sites, as is the case with littleleaf disease (*Phytophthora cinnamomi*) on shortleaf pine.

Despite these managerial problems, the net effect of root rots in the ecosystem is certainly positive in that they break down organic matter in the soil and play a vital role in the cycling of nutrients.

CONTROL OF ROOT DEVELOPMENT IN NURSERIES

Undercutting is the pruning of seedling roots in the nursery bed to limit the growth of long, unbranched main roots. *Root wrenching* uses the same equipment employed in undercutting, but the horizontally mounted blade is tilted in order to "lift" or wrench the seedlings in the nursery bed. *Root pruning* is the pruning of root systems to a standard length after the seedlings have been lifted and graded prior to storing or planting.

In growing nursery stock suitable for outplanting, it is desirable to lift the seedlings at a time during the late fall or winter when they are dormant and the

Figure 9-6 Untreated seedlings (left) and seedlings conditioned by undercutting and wrenching (right), (*N.Z. Forest Service Photo by H. G. Hemming. From Van Dorsser and Rook, 1972.*)

buds have hardened into their resting stage. During this stage, seedlings can be readily transplanted, or placed in storage, since they are not actively growing. It is commonly found, however, that without nursery treatments the seedlings develop inadequate root systems, which decreases their potential survival when outplanted. This problem can be solved by transplanting in the nursery; however, this is very costly and requires excessively large nurseries. Alternatively, the seedlings may be undercut in the beds, which tends to slow down shoot growth and promote the development of a more fibrous root system (Hastings, 1923; Stoeckler and Jones, 1957; Shoulders, 1963). In favorable climates where nursery stock does not become truly dormant but continues to grow and remain succulent over winter, seedlings develop large shoot/root ratios and are prone to mortality on outplanting. This has led to a practice called *wrenching* (Rook, 1971), which may be one or a series of treatments. This treatment partially lifts the seedlings, breaking off newly developing roots. Successive wrenching treatments in the fall causes a virtual cessation of shoot growth and stimulates a vigorous development of fibrous roots (Fig. 9-6). Wrenched seedlings develop a more balanced shoot/root ratio which increases outplanting survival. The procedure has become standard practice for Monterey pine in New Zealand (Van Dorsser and Rook, 1972), slash and longleaf pine (Shoulders, 1963), red pine (Stoeckler and Jones, 1957), and Douglas-fir (Tanaka et al., 1974).

SILVICULTURAL MANIPULATION OF ROOT GROWTH

Tree growth is largely dependent on the continuing capacity of root systems to metabolize and take up water and nutrients. Root growth tends to be low in winter, due primarily to cold soil temperatures, and can also be low in areas of summer drought, due to higher soil temperature and lower soil water availability. The silviculturist should therefore consider the extent to which these potentially adverse effects can be ameliorated by manipulation of crown cover and depth of litter or slash on sites of different slope and aspect. Since root respiration rises exponentially with temperature, a change in soil temperature of only 1 or 2°C in winter or summer could make a substantial difference in tree growth, especially in seedlings. Similarly, one needs to predict the extent to which conditions of water availability, soil atmosphere, and soil nutrition are influenced by proposed treatments, such as site preparation, fertilization, prescribed burning, harvesting, ripping, ploughing, and bedding, which have the potential for changing the physical, chemical, and biotic environments of roots.

Increasing mechanization of ground systems for harvesting and thinning may potentially reduce growth of residual stands directly by damaging root systems or indirectly through soil compaction. The potential for soil or root damage appears to depend largely on the following conditions of site, stand, and operational factors (Froehlich, 1976). *Site factors* include topography, soil type and depth, soil moisture at time of operation, and degree of soil protection from litter or slash. *Stand factors* include tree species, age or size, stand density before and after operation, and rooting characteristics. And *operational factors* involve ground pressure and vibration of equipment, operator skill and attitude, load size being removed, pattern of extraction routes, and frequency of entry. The interaction of these factors is illustrated in Fig. 9-7. In the Pacific Northwest, the species regarded as the most resistant to root damage or soil compaction are ponderosa and lodgepole pine, followed by Douglas-fir, Sitka spruce, and western hemlock toward the more sensitive end of the scale. The degree of compaction depends largely on soil type and wetness, type of equipment used, and frequency of entry. Small crawler tractors have bearing pressures of 20 to 28 kPa (3 to 4 psi) compared with 172 to 207 kPa (25 to 30 psi) for rubber-tired skidders. Clay-loam soils are highly susceptible to compaction, and in one test area in Oregon had a soil density in undisturbed areas of 1.04 g cc^{-1}. Use of a small tractor for thinning 35-year-old Douglas-fir causes soil densities to increase to a maximum of 1.64 g cc^{-1}, depending on whether the skid trails are lightly

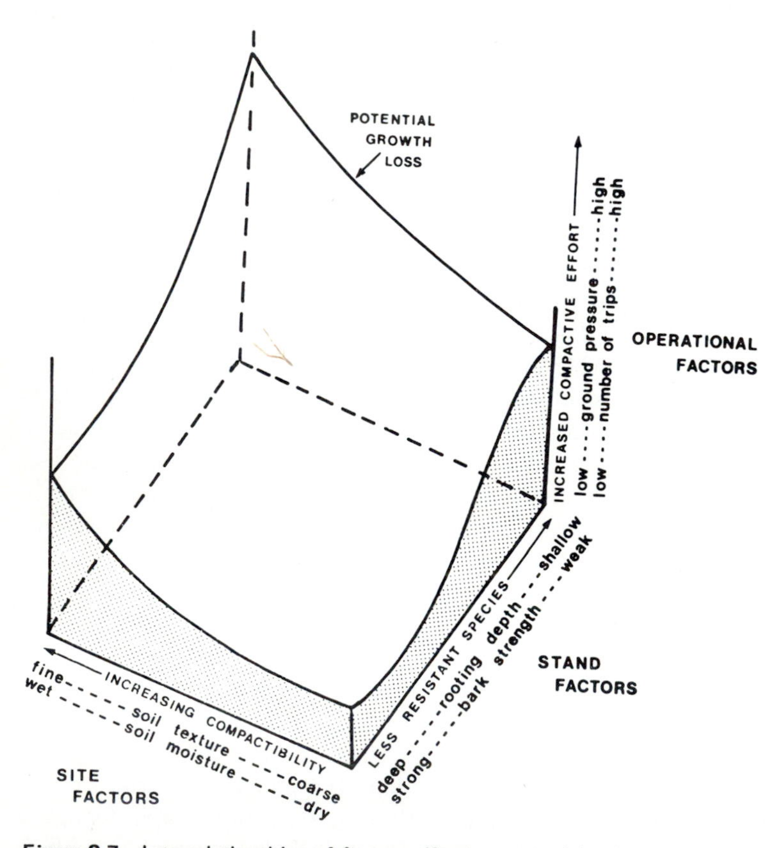

Figure 9-7 Interrelationships of factors affecting potential reduction in tree growth following thinning by machine. (*From Froelich, 1976.*)

used, secondary, or major trails (Froehlich, 1976). Highest soil compaction occurs when the soils are wet.

In the extensive flatwood country of the southeastern Coastal Plains, where soils are poorly drained sands or sandy loams, and also on some upland soils, bedding is an increasingly common site-preparation treatment. This practice improves survival by draining the soil and thus increasing the effective soil volume available for root development. Studies on plantation establishment of slash pine in Florida have shown that fertilization increased fine-root biomass by 145 percent and total root biomass by 58 percent, and that maintenance of the water table at depths of 46 and 92 cm increased root biomass by 69 and 53 percent, respectively (White et al., 1971).

An effectively functioning root system is a necessary prerequisite for diameter growth responses to silvicultural treatments. We commonly evaluate growth responses by measuring changes in diameter growth at breast height. But increased growth at an arbitrary 137 cm (4 feet 6 inches) above the ground is dependent on current source-sink relationships in the tree. Many so-called failures of silvicultural treatments, or delays in responses, are due to the fact that root systems, live-crown ratios, or general tree vigor is inadequate. Before growth responses can be evident at breast height, the tree must be generally in balance. Consequently, any particular treatment may indeed have dramatic effects on growth of root systems and other parts of the tree in addition to possible effects on the diameter of the stem at breast height.

Part Three

Forest Site

Chapter 10

Soils, Water Relations, and Nutrition

Soil is the top part of the regolith which has been exposed to the intense weathering action of climate and the biochemical action of plants. Soil is usually distinguishable from other unconsolidated material by the presence of organic matter and by the horizons developed in the profile as it matures. Soil has been the most essential natural resource in human development, and its severe abuse has impoverished many cultures. This has been especially important in the lands surrounding the Mediterranean Sea, where the destruction of the forests and the soils that supported them has drastically impaired productivity and supply.

A knowledge of soil science is basic to the silvicultural management of a forest because the quality of the soil is one of the chief constraints on silvicultural practices. As soils become deeper, less stony, and better supplied with nutrients and water, the silvicultural options increase in number and variety. Silvicultural considerations such as choice of species, determination of site productivity, prediction of seedling survival and growth, and determination of stand growth and levels of stocking are greatly influenced by soil factors, such as soil depth, water-holding capacity, development of iron pans in spodic soils or calcium carbonate pans in aridic soils, nutrient deficiency, and chemical toxicity problems. Thus silviculturists are concerned with how soils form, with the characteristics they develop in terms of physical, chemical, and biological properties, with their dynamics in terms of water and nutrient cycling, with the nutritional status of stands of trees in relation to soil nutrient storage, and, finally, with the silvicultural practices that are needed to control the soil space of the forest to achieve desired forest-management objectives.

SOIL FORMATION

Soils in a forest area will be as variable as the major factors which influenced their formation. The independent factors in soil formation are parent material, climate, topography, living organisms, and the time involved in soil formation (Jenny, 1941). Thus, in any area, the silviculturist may be dealing with a mosaic of soils, depending on local variations in these factors. There may be local changes in parent material related to changes in geologic materials, to variations in form of deposition, such as glacial moraines, alluvium, and bog deposits, or to periodic deposits such as sand in dune areas. The age of these parent-material surfaces will determine the degree of weathering and soil profile formation. The *manner* in which these soil profiles form, such as under extreme leaching or with accumulation of soluble materials, will depend on climate, defined mainly in terms of precipitation. The *rates* of these soil-forming processes will depend mainly on temperature.

Local variations in topography resulting from changes in slope, aspect, or depth to water table may bring associated soil changes. For example, steep mountain slopes will generally have younger soils in equilibrium with higher erosion rates, whereas the oldest soils will be on the less steep slopes. Areas with a high water table will have soils formed under conditions of poor aeration with typical accumulations of organic matter. With a uniform or a variable areal distribution of soil-forming factors in a forest, the forester will be dealing with either a simple or complex mosaic of local soils.

The soil-forming factors create the site conditions with which the silviculturist must work, and the soil profile represents the integration of these factors over time. Inasmuch as the history of the site is coded into the soil profile, an examination of the soil should give evidence of that history and aid the forester in making sound silvicultural prescriptions.

SOIL PROFILE

Soils develop over time through the gradual weathering of minerals in the original parent materials which forms secondary minerals, such as clays, which with increasing time migrate down the soil profile to deposit in accumulation zones. Any vegetation adds material to the soil surface, enriching it in organic matter and its inorganic constituents. The result is a series of layers in the soil, called *horizons*, which are diagnostic in classifying soils. A generalized profile is represented in Fig. 10-1, along with an actual soil profile. In the generalized case, an accumulation of organic material on the surface produces a 0 horizon which may be divided into 01 (A_0), in which the original plant or animal structural components are still recognizable to the eye, and 02 (A_{00}), in which the organic material has been reduced to an amorphous form. An A horizon develops from the accumulation of organic materials and the leaching of soluble and dispersible materials from the topmost mineral layers, while a B horizon develops

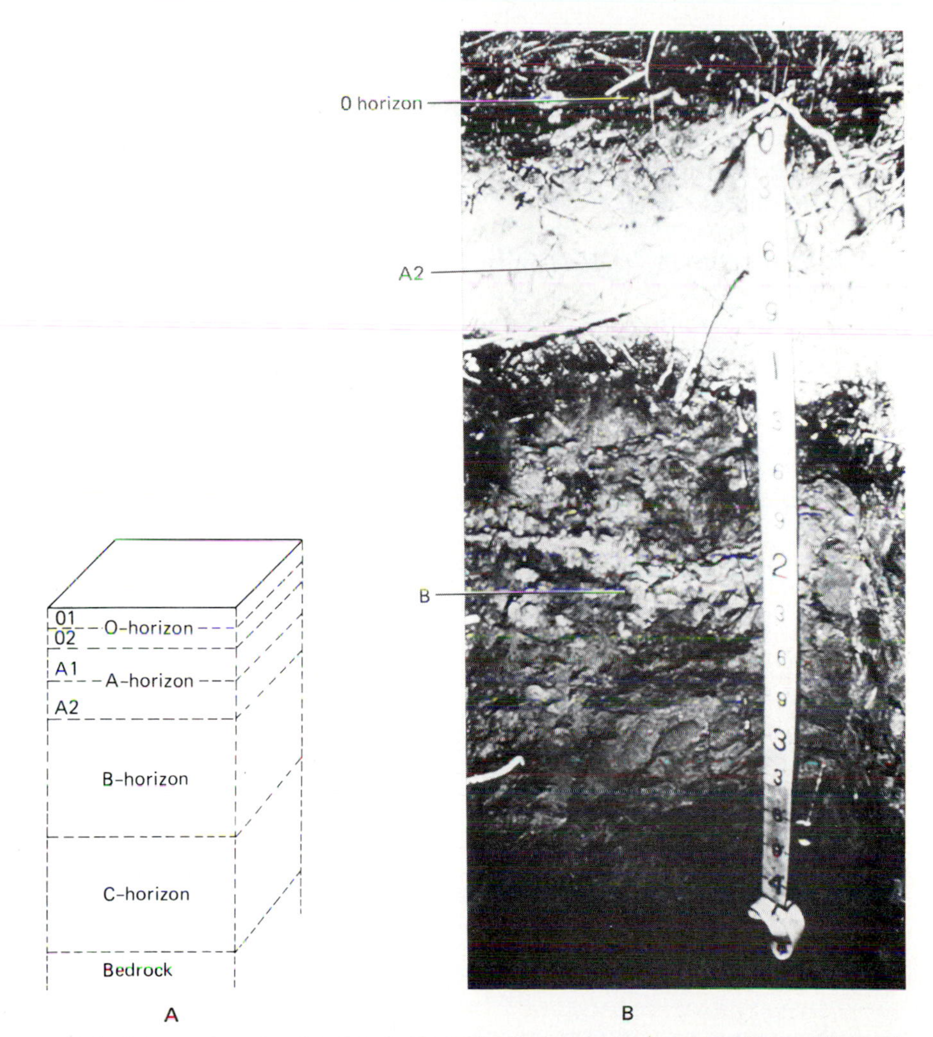

Figure 10-1 A: A pedon that has developed in place. A pedon is the smallest sampling unit of a soil that is large enough for the nature of its horizons to be studied, usually one square meter. B: Profile of a Spodosol showing 0, A2, and B horizons from upper New York. (*From R. W. Miller.*)

as a result of an accumulation of materials migrating from the 0 and A horizons. A C horizon consists of unconsolidated material, in which weathering action and biological activity are usually minor, and may gradate into bedrock. The A1 horizon is dark from the incorporation of humus material through the activity of biota, while the A2 horizon is lighter colored and characterized by the eluviation of clay particles, some humus colloids, and iron and aluminum oxides.

The soil profile gives a ready estimate of the progress of the soil-forming agencies as influenced by time (Fig. 10-2). A California study of soil development on various parent-rock types shows that a sequence of soil profiles of increasing degree of development is related either to climatic differences or topographic variation and that analogous sequences occur on each of the parent-rock types, with variations reflecting differences in the physical and chemical properties of the rocks (Zinke and Colwell, 1963). Changes in soil properties in each profile-development sequence for the various rock types showed a fairly consistent relationship with development:

1 Color generally showed progressive changes from grayish brown to reddish brown with increasing soil development.
2 Coarse-fragment content generally decreased with increasing development of forest soils.
3 Clay content of the soil profile increased with the degree of development on any of the parent rocks.

The changes in properties of the soil profile with developmental maturity may be related to forest-management problems. In immature soils that have a

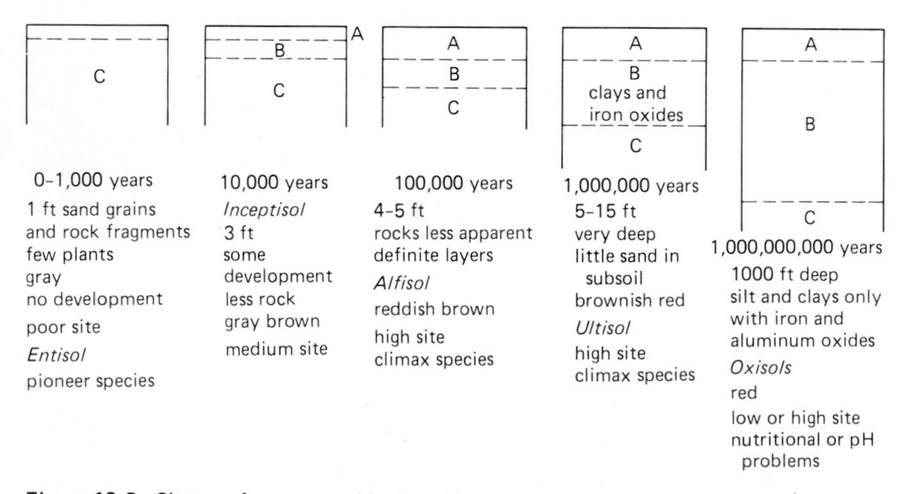

Figure 10-2 Changes for some residual soils in profile characteristics and soil order on low slope sites with time that are almost independent of parent material. Oxisols are lacking in United States but highly developed in Amazon River drainage. (*Personal communication, Paul Zinke, Berkeley, California.*)

high amount of stoniness, the lower water-storage capacity presents a more droughty situation than that on better developed soils. Mature residual soils, with their high clay content on every parent material, present a building problem for unpaved roads because of poor traction when the soil is wet which lowers the allowable gradient and increases the amount of surfacing material required to obtain stabilization.

SOIL CLASSIFICATION

Forest-management guidelines have general limitations in that they are commonly derived from studies on limited areas that are applied over extensive areas. In view of the diversity of sites on which a particular species may grow and the probability that the results of a study were greatly influenced by the character of the soils associated with it, the application of broad management guidelines must be tempered by a recognition of soil differences and their effects. Soil classifications based on the character and properties of a profile have been used to translate the results from one site for prediction of the results at another site. These classifications have been ineffective and most unwieldy, as illustrated by this statement from Wilde (1953):

> The true genetic nature of many soils was obscured by a smoke screen of such terms as "weakly podsolized soils," "grey-brown podsolic soils," "crypto- or phanero-podzolic soils," "latent podsols," "podsolized brown earths," "podzolized lateritic soils," "yellow podzolic soils," "red podsolic soils," and "brown podsolic soils." Some of these soils cannot be classed among podzolic soils even by a long stretch of the imagination, since they are formed under essentially different climatic and floristic conditions and show distinctly different morphology.

In view of the need for better communication among soil scientists and for the transfer of information about soils to other disciplines, a new classification was recently developed in the United States (Soil Survey Staff, 1975) that is totally divorced from the old terminology, thereby eliminating any confusion of terms. Another advantage of the classification is that the names of the categories are derived from Latin and Greek roots expressing the properties of the soils. This new system is the official system in the United States at the moment, but may well be a case of reverse metrification, if accepted worldwide. The classification is divided into six categories, order, suborder, great group, subgroup, family, and series, which are determined by quantitatively measurable properties of the soils as they exist in the field. The soil orders and suborders of the United States under the new classification and some of their diagnostic characteristics are given in Fig. 10-3 and Table 10-1. The principal forest soils in this country are the Alfisols, Inceptisols, Mollisols, Spodosols, and Ultisols.

While the surface appearance of a soil and the plant community which it

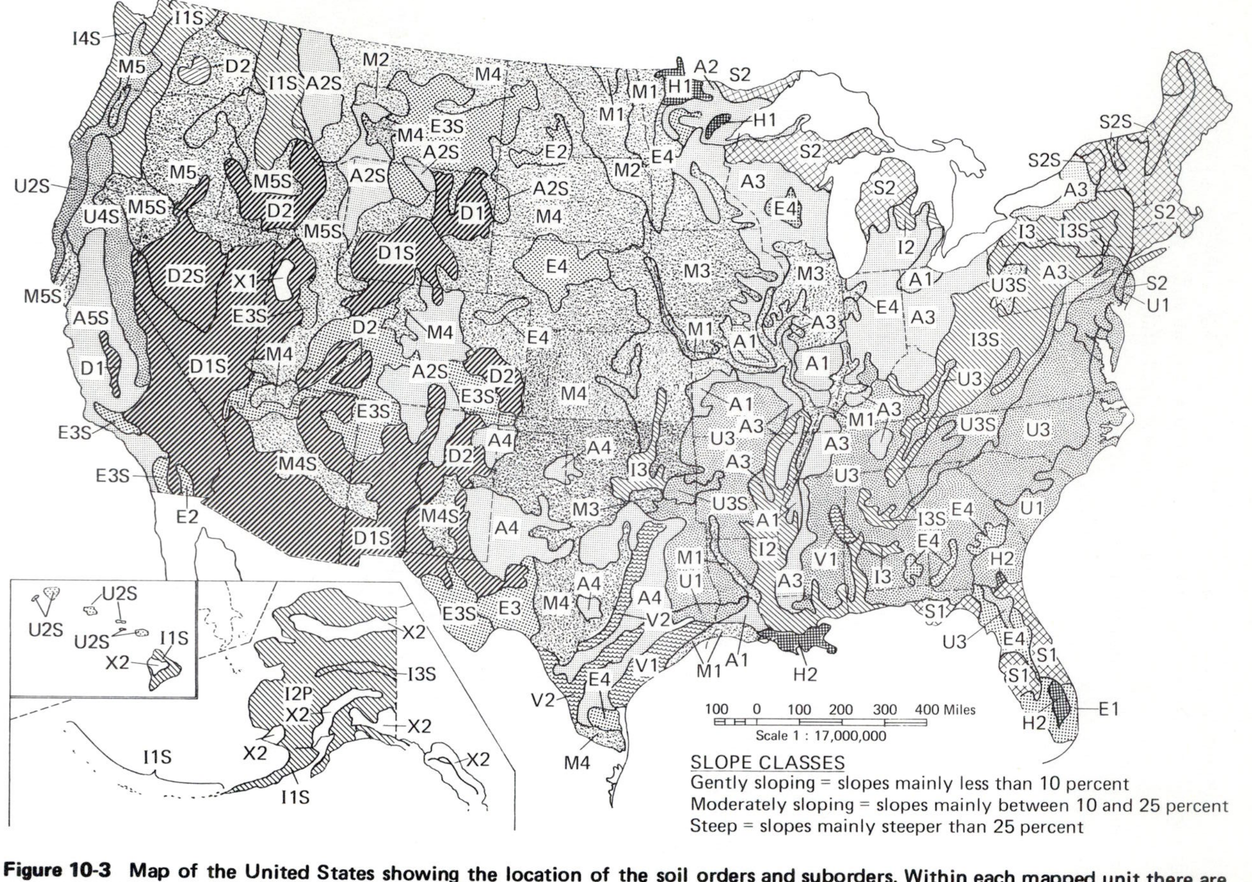

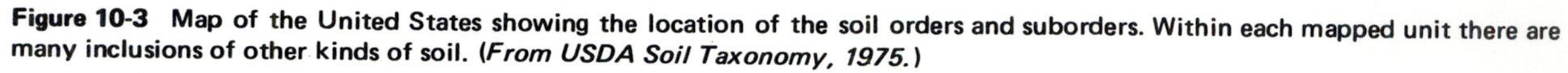

Figure 10-3 Map of the United States showing the location of the soil orders and suborders. Within each mapped unit there are many inclusions of other kinds of soil. (*From USDA Soil Taxonomy, 1975.*)

Table 10-1 Legend Providing Brief Descriptions of the Orders and Suborders Given in Fig. 10-3.

Legend

Only the dominant orders and suborders are shown. Each delineation has many inclusions of other kinds of soil. General definitions for the orders and suborders follow. For complete definitions, see Soil Survey Staff, "Soil Taxonomy," Soil Conservation Service, U.S. Department of Agriculture, 1975. (Approximate equivalents in the modified 1938 soil classification system are indicated for each suborder.)

ALFISOLS: Soils with gray to brown surface horizons, medium to high base supply, and subsurface horizons of clay accumulation; usually moist but may be dry during warm season

A1 AQUALFS (seasonally saturated with water) gently sloping; general crops if drained, pasture and woodland if undrained (Some Low-Humic Gley soils and Planosols)

A2 BORALFS (cool or cold) gently sloping; mostly woodland, pasture, and some small grain (Gray Wooded soils)

A2S BORALFS steep; mostly woodland

A3 UDALFS (temperate or warm, and moist) gently or moderately sloping; mostly farmed, corn, soybeans, small grain, and pasture (Gray-Brown Podzolic soils)

A4 USTALFS (warm and intermittently dry for long periods) gently or moderately sloping; range, small grain, and irrigated crops (Some Reddish Chestnut and Red-Yellow Podzolic soils)

A5S XERALFS (warm and continuously dry in summer for long periods, moist in winter) gently sloping to steep; mostly range, small grain, and irrigated crops (Noncalcic Brown soils)

ARIDISOLS: Soils with pedogenic horizons, low in organic matter, and dry more than 6 months of the year in all horizons

D1 ARGIDS (with horizon of clay accumulation) gently or moderately sloping; mostly range, some irrigated crops (Some Desert, Reddish Desert, Reddish-Brown, and Brown soils and associated Solonetz soils)

D1S ARGIDS gently sloping to steep

D2 ORTHIDS (without horizon of clay accumulation) gently or moderately sloping; mostly range and some irrigated crops (Some Desert, Reddish Desert, Sierozem, and Brown soils, and some Calcisols and Solonchak soils)

D2S ORTHIDS gently sloping to steep

ENTISOLS: Soils without pedogenic horizons

E1 AQUENTS (seasonally saturated with water) gently sloping; some grazing

Table 10-1 Legend Providing Brief Descriptions of the Orders and Suborders Given in Fig. 10-3 (*Continued*)

E2	ORTHENTS (loamy or clayey textures) deep to hard rock; gently to moderately sloping; range or irrigated farming (Regosols)
E3	ORTHENTS shallow to hard rock; gently to moderately sloping; mostly range (Lithosols)
E3S	ORTHENTS shallow to hard rock; steep; mostly range
E4	PSAMMENTS (sand or loamy sand textures) gently to moderately sloping; mostly range in dry climates, woodland or cropland in humid climates (Regosols)
	HISTOSOLS: Organic soils
H1	FIBRISTS (fibrous or woody peats, largely undecomposed) mostly wooded or idle (Peats)
H2	SAPRISTS (decomposed mucks) truck crops if drained, idle if undrained (Mucks)
	INCEPTISOLS: Soils that are usually moist, with pedogenic horizons of alteration of parent materials but not of accumulation
I1S	ANDEPTS (with amorphous clay or vitric volcanic ash and pumice) gently sloping to steep; mostly woodland; in Hawaii mostly sugar cane, pineapple, and range (Ando soils, some Tundra soils)
I2	AQUEPTS (seasonally saturated with water) gently sloping; if drained, mostly row crops, corn, soybeans, and cotton; if undrained, mostly woodland or pasture (Some Low-Humic Gley soils and Alluvial soils)
I2P	AQUEPTS (with continuous or sporadic permafrost) gently sloping to steep; woodland or idle (Tundra soils)
I3	OCHREPTS (with thin or light-colored surface horizons and little organic matter) gently to moderately sloping; mostly pasture, small grain, and hay (Sols Bruns Acides and some Alluvial soils)
I3S	OCHREPTS gently sloping to steep; woodland, pasture, small grains
I4S	UMBREPTS (with thick dark-colored surface horizons rich in organic matter) moderately sloping to steep; mostly woodland (Some Regosols)
	MOLLISOLS: Soils with nearly black, organic-rich surface horizons and high base supply
M1	AQUOLLS (seasonally saturated with water) gently sloping; mostly drained and farmed (Humic Gley soils)
M2	BOROLLS (cool or cold) gently or moderately sloping, some steep slopes in Utah; mostly small grain in North Central States, range and woodland in Western States (Some Chernozems)

M3 UDOLLS (temperate or warm, and moist) gently or moderately sloping; mostly corn, soybeans, and small grains (Some Brunizems)

M4 USTOLLS (intermittently dry for long periods during summer) gently to moderately sloping; mostly wheat and range in western part, wheat and corn or sorghum in eastern part, some irrigated crops (Chestnut soils and some Chernozems and Brown soils)

M4S USTOLLS moderately sloping to steep; mostly range or woodland

M5 XEROLLS (continuously dry in summer for long periods, moist in winter) gently to moderately sloping; mostly wheat, range, and irrigated crops (Some Brunizems, Chestnut, and Brown soils)

M5S XEROLLS moderately sloping to steep; mostly range

SPODOSOLS: Soils with accumulations of amorphous materials in subsurface horizons

S1 AQUODS (seasonally saturated with water) gently sloping; mostly range or woodland; where drained in Florida, citrus and special crops (Ground-Water Podzols)

S2 ORTHODS (with subsurface accumulations of iron, aluminum, and organic matter) gently to moderately sloping; woodland, pasture, small grains, special crops (Podzols, Brown Podzolic soils)

S2S ORTHODS steep; mostly woodland

ULTISOLS: Soils that are usually moist with horizon of clay accumulation and a low base supply

U1 AQUULTS (seasonally saturated with water) gently sloping, woodland and pasture if undrained, feed and truck crops if drained (Some Low-Humic Gley soils)

U2S HUMULTS (with high or very high organic-matter content) moderately sloping to steep, woodland and pasture if steep, sugar cane and pineapple in Hawaii, truck and seed crops in Western States (Some Reddish-Brown Lateritic soils)

U3 UDULTS (with low organic-matter content; temperate or warm, and moist) gently to moderately sloping; woodland, pasture, feed crops, tobacco, and cotton (Red-Yellow Podzolic soils, some Reddish-Brown Lateritic soils)

U3S UDULTS moderately sloping to steep; woodland, pasture

U4S XERULTS (with low to moderate organic-matter content, continuously dry for long periods in summer) range and woodland (Some Reddish-Brown Lateritic soils)

VERTISOLS: Soils with high content of swelling clays and wide deep cracks at some season

V1 UDERTS (cracks open for only short periods, less than 3 months in a year) gently sloping; cotton, corn, pasture, and some rice (Some Grumusols)

V2 USTERTS (cracks open and close twice a year and remain open more than 3 months); general crops, range, and some irrigated crops (Some Grumusols)

Table 10-1 Legend Providing Brief Descriptions of the Orders and Suborders Given in Fig. 10-3 (*Continued*)

☐ AREAS with little soil.

X1	Salt flats
X2	Rockland, ice fields

Nomenclature

The nomenclature is systematic. Names of soil orders end in *sol* (L. *solum*, soil), e.g., ALFISOL, and contain a formative element used as the final syllable in names of taxa in suborders, great groups, and sub-groups.

Names of suborders consist of two syllables, e.g., AQUALF. Formative elements in the legend for this map and their connotations are as follows

and	Modified from Ando soils; soils from vitreous parent materials
aqu	L. *aqua*, water; soils that are wet for long periods
arg	Modified from L. *argilla*, clay, soils with a horizon of clay accumulation
bor	Gr. *boreas*, northern; cool
fibr	L. *fibra*, fiber; least decomposed
hum	L. *humus*, earth; presence of organic matter
ochr	Gr. base of *ochros*, pale; soils with little organic matter
orth	Gr. *orthos*, true; the common or typical
psamm	Gr. *psammos*, sand; sandy soils
sapr	Gr. *sapros*, rotten; most decomposed
ud	L. *udus*, humid; of humid climates
umbr	L. *umbra*, shade; dark colors reflecting much organic matter
ust	L. *ustus*, burnt, of dry climates with summer rains
xer	Gr. *xeros*, dry; of dry climates with winter rains

Source: USDA Soil Taxonomy, 1975.

supports may supply appreciable information, an examination of the profile is necessary to confirm its character and to classify it. In defining the pedon characteristics, no reference is made to climate, vegetation, geology, or soil-forming factors. Even though soil genesis appears to be neglected, the various orders have characteristics, such as certain horizons, which are associated with climate and with the effects of vegetation and geology, as Fig. 10-3 illustrates. A brief description of the principal soil orders is given to highlight the interactions among the soil-forming factors that created them.

Soil Orders

Alfisols Profiles of Alfisols show an ochric epipedon (light-colored A1 horizon) and a very thin 0 horizon; the A1 is less than 10 cm, while the A2 may vary from 0 to 25 cm; the argillic (clayey) B horizon is a diagnostic feature. There is no accentuated accumulation of humus or iron in the B horizon. With the short, dry periods and the medium to high base saturation that are characteristic of Alfisols, an Alfisol is not to be confused with a Spodosol but may be with a Mollisol.

Inceptisols These soils develop on volcanic material and other unconsolidated deposits like glacial till and have some horizon development but no appreciable zone of accumulation. In volcanic material, the profile may have a mollic (dark with more than 50 percent base saturation) epipedon, making it difficult to separate morphologically from a Mollisol because the soil starts off dark. It is distinguishable, however, by the presence of volcanic glass particles, low bulk density, and high exchange capacity. In other recent deposits, the Inceptisols have either an ochric or umbric (dark with less than 50 percent base saturation) epipedon, while the B horizon lacks sufficient clay to be argillic. With further development, an Inceptisol would become an Alfisol in the Midwest and West, while in a sandy textured soil in the Northeast, an Inceptisol would probably become a Spodosol.

Mollisols A salient feature of a Mollisol is the dark mollic epipedon, a dark A1 horizon which is 18 to 25 cm deep and which develops as the result of the incorporation of organic matter into the A1 horizon. This incorporation results from the activity of bacteria and insects in decomposing the organic matter and of burrowing animals in mixing the organic material into the soil creating a friable, porous soil. The mollic epipedon must also have a base saturation greater than 50 percent. Mollic epipedons occur in other soil orders too, particularly in Vertisols and volcanic-ash Inceptisols. The B horizon may vary widely in clay content, since an argillic horizon is not a diagnostic feature of Mollisols. For example, cold soils at high elevations may not have a strong B-horizon development. A "mull soil" is the common designation under the old classification for the surface horizon of forest soils developed under deciduous hardwoods, and the term corresponds to a mollic or umbric epipedon in that the horizon is dark, deep, friable, and porous as a consequence of biotic activity. In "mull soils," however, the base saturation may be above or below 50 percent. Mollisols occur under aspen and oak.

Spodosols These soils develop primarily under the influence of a cold, wet climate on coarse textured soils, though Spodosols do occur in warm climates. The diagnostic horizon of a Spodosol is the dark B horizon that results from an accumulation of organic matter and amorphous aluminum oxide with or without the presence of iron oxides. What little clay may be present accumulates below the diagnostic horizon and may form a fragipan. In Spodosols, the 0 horizon may be quite deep, especially under a conifer stand, and its decomposition is dominated by fungi with little activity of other biota. This is due to the development of acid (low pH) conditions. There is commonly no A1 horizon, and the A2 horizon forms an eluviated albic layer that may be thin or thick. Humic acids and the eluviated aluminum and iron oxides are carried through and out of the A2 horizon and accumulate in the B horizon as the soil solution reaches the lower level of the profile. This precipitation may be caused by pH increase, chelation, adsorption, lodging, or decomposition changes between storms. With the continual elutriation of the profile by acidified water, the percentage base saturation of a Spodosol is quite low.

Under the old soil classification, a profile developed under typical spodic conditions would be classed as a "podzol" and the surface horizon of thick organic matter with a low pH would be called a "mor layer." Wherever a layer of sclerophyllous organic material accumulates in cooler climes, as under many conifer stands, fungal decomposition tends to dominate, which lowers the pH and induces acidified water elutriation of the profile—a process recognized as "podzolic" or spodic without any implication that a Spodosol would be developed. The process which is simply a combination of chelation, leaching, and translocations exists so widely in soil formation that the array of terms, which Wilde and most soil professionals consider confusing, has a germ of truth; yet Spodosols are eighth in area among the soil orders and occupy only about 4.3 percent of the world land area.

Ultisols An Ultisol has a profile very similar to an Alfisol in that the A1 horizon is usually less than 10 cm thick and there is an argillic B horizon. The distinguishing features are the lower base saturation (less than 35 percent at 1.8 m below the surface) and the usually warm climatic regime under which it develops.

Whereas the soil orders and other categories of the classification hierarchy provide information needed for soil surveys and for grouping similar soils so that their relationships are evident, the lowest category—the soil series—and its properties are the most important in influencing silvicultural decisions and land management.

SOIL PROPERTIES

The importance of soil in silviculture stems from its three functions in tree growth: mineral nutrition, moisture supply, and physical support. Hydroponics offers a substitute for soil in minor instances (such as in some specialized kinds of agriculture), but the role of soil in providing humanity with its food and fiber grows more important and more limiting as the world population mushrooms. Productivity of soil is a function of soil properties and climate. Soil properties are divided into physical and chemical categories. It is generally considered that physical properties are the most important in forest management, and this may commonly be true because the nutrient level of a soil is strongly correlated with physical properties. However, physical and chemical properties can independently dominate growing conditions, such as in severely compacted soils, soils in which there is either an excess or a deficiency of nutrients, or soils with toxic pH levels.

Physical Properties

Texture, depth, and topographic situation are physical attributes of soil which cannot be modified except in the case of erosion. On the other hand, soil structure, porosity, and organic-matter content of soil can be markedly influenced by forest-management practices.

Texture Texture depends on the proportion of the various sizes of *mineral* particles in a particular soil. The proportions of particle sizes are used to name a soil or describe the texture of a particular horizon (Fig. 10-4). Thus a Hoquiam clay loam is the name of a series in western Washington where the word *Hoquiam* describes the profile characteristics of the soil and the *clay loam* refers to the texture of the surface soil. Soil particles are divided into six sizes:[1] clay (less than 0.002 mm), silt (between 0.002 mm and 0.05 mm), sand (0.05 mm to 2.0 mm), gravel (2.0 mm to 7.6 cm), cobbles (7.6 cm to 25.4 cm), and stones (over 25.4 cm). Sand is further divided into very fine, fine, medium, coarse, and very coarse. In general terms, soils are described as coarse textured when sand is the dominant fraction (sandy soils) and fine textured when there is a predominance of clay and silt (clayey soils). An accurate textural measurement requires a laboratory determination of the amount of each particle size, but reasonable estimates can be made in the field after some practice with known soil textures by moistening a sample and rubbing it between thumb and forefinger.

Soil Structure Analyzing the texture of a soil to determine its distribution of particle sizes usually requires considerable effort to ensure that the soil's aggregates are reduced to their component particles. Soil structure refers to the

[1]There are different systems of sizing particles, so the sizes that are given here are those used by the U.S. Department of Agriculture.

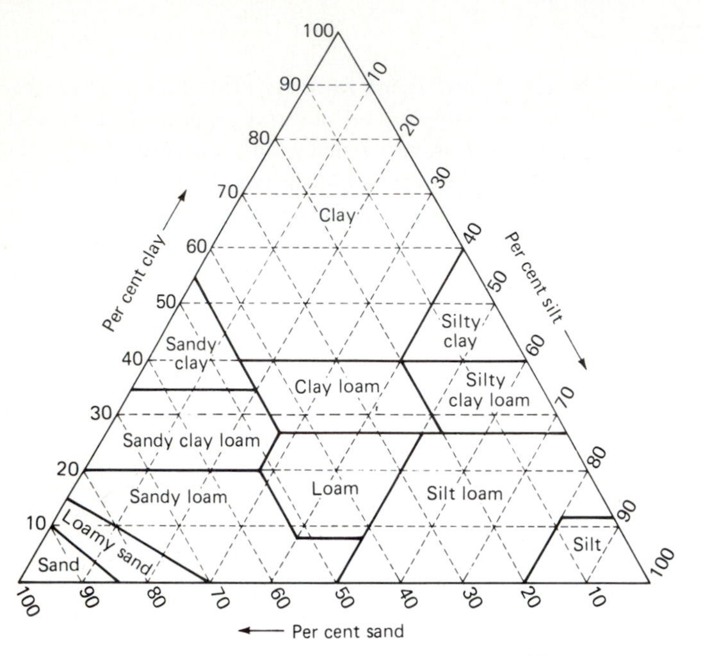

Figure 10-4 Guide for textural classification of soils. (*From Soil Survey Staff, 1960*)

kind and amount of aggregation of particles. The coarser-textured soils usually have less structure than the fine-textured soils. Sandy soils lack cohesive properties, while the clay and silt of fine-textured soils tend to form aggregates. In surface horizons, granular and crumb structures are most common, while prismatic and blocky structures occur in fine-textured subsoils. Granular structures may develop in surface soils as a reaction to a number of factors such as high base saturation (especially with calcium), humus content (especially in the form of colloidic organic matter), the excretions and activity of various biota, root activity, and perhaps the mechanical processes of freezing and thawing and wetting and drying.

Porosity Aeration and infiltration rate are functions of soil porosity. The pore space of a soil is the percent of the total volume occupied by air and water. A sandy soil has less pore space than a clayey soil, yet a sandy soil has better aeration and infiltration rates than a clayey soil unless the clayey soil is strongly aggregated. In sandy soils the large capillaries permit more rapid flow of air and water, while in a clayey soil the capillaries are very small and impede the movement of air and water. Compaction or puddling of soil reduces its porosity, while root cavities, animal activity, and aggregation of particles increases it. Pore space usually decreases from the top of the A horizon through the B horizon. In tight B horizons, as in argillics, and in calcic or spodic layers with their low aeration and slow water permeability, a barrier to root penetration develops.

Calculation of the porosity of a soil depends on a knowledge of the average specific gravity of the soil particles and the bulk density of the soil. An average figure for the specific gravity of soil particles is usually accepted as 2.65. The bulk density of a soil is the oven-dry weight (g) of the soil divided by the volume (cm^3) of the sample in place and may vary between 0.6 and 2.0. Pore space would equal the inplace volume of the sample minus the volume of solid material (oven-dry weight of sample in grams divided by 2.65).

The impact of aeration on the seedling growth of four bottomland hardwood species under four watering regimens is illustrated by the data in Table 10-2 (Dickson et al., 1965). The adaptation of tupelo to growing under continuously flooded conditions is shown by its superior growth in saturated soil; green ash showed little difference in height growth under saturated conditions and soil moisture maintained at the moisture equivalent; sycamore grew best at the moisture equivalent; while pin oak grew relatively poorly compared to the other species but similarly grew best at the moisture equivalent. Since all these species are adapted to flooding, they survived the experiment under all conditions. If black walnut and longleaf and ponderosa pine had been tested too, they would have died under saturation conditions and would probably have grown best in height at the 50 percent moisture equivalent regimen.

Depth of Soil Soil depth is a significant factor in the evaluation of the site quality of an area since an increase in soil depth tends to be associated with a larger nutrient supply, greater water-holding capacity, and less windthrow

Table 10-2 The Effects of Four Water Regimens on the Growth of Four Bottomland Tree Species

(The trees were established in pots before treatment was initiated and the experimental period ran 84 days)

Moisture condition in pot	Species (Average height growth in cm as a mean of 12 seedlings)			
	Tupelo	Green ash	Sycamore	Pin oak
Saturated (constant)	67.33	41.04	27.63	2.71 c
Moisture equivalent (ME) (daily)	39.21	36.75	49.04	11.33
50 percent of ME, then back to ME	14.54 a	23.92	34.83	10.04 a c
Wilting point, then back to ME	8.87	7.21 b	16.50	4.75 b

Figures not connected by the same line are significantly different at the 5 percent level. Lines with the same letters are not significantly different.

Source: Dickson et al., 1965.

hazard. Profile depth and soil depth may not be equivalent terms for comparing sites because the amount of rock in a profile reduces the effective depth in terms of available nutrients and moisture. Both deep profiles and deep soils reduce windthrow hazard. As has been indicated, the depth to an impervious layer may determine the functional depth of a soil.

Organic Matter A forest is continuously replenishing the supply of organic matter on the forest floor by shedding leaves, twigs, and bark fragments. Depending on the climate, organic matter accumulates in appreciable quantities (Spodosols) or is more rapidly decomposed by the biota, with considerable amounts being incorporated into the mineral soil profile (particularly Mollisols and Alfisols and, to a lesser extent, in most of the other orders). The incorporated organic materials are concentrated in the surface horizons, but root growth adds organic matter to the profile as far down as 150 cm or more in orders like the Aridisols and Entisols. Organic content of the A horizon is about 1.0 to 5.0 percent, with Aridisols down to 0.2 percent. This quantity seems like a minor amount until it is realized that it is a percentage of the dry weight of soil. Total organic matter in a 1-m profile ranges from 200 to 600 metric tons per hectare, with a high value in the Adirondacks of 806 metric tons per hectare (Romell, 1932).

Topographic Situation

Aspect, steepness, and position on the slope influence the physical properties of a soil because profile development reflects the microclimatic differences associated with each change. Soils are younger on upper slopes because of natural erosion processes, and the steeper the slope, the greater the erosion rate will be, with a correspondingly slower profile development. On the usual south slopes in the north temperate zone, the more severe climate with its higher evapotranspiration rate reduces the effective precipitation and slows profile development too. The geology in high relief areas or in glaciated areas is more complex, so differing parent materials may be reflected in a soil's properties and profile. Topography also influences soil properties, e.g., rainfall generally increases with increasing elevation along with other changes in the total macroclimate under which a soil develops.

Chemical Properties

The two essential contributions of soil to the growth of trees are physical support and supply of the minerals and water necessary for the photosynthetic process. The whole development of a soil is actually a chemical process dependent on the amount of rain, temperature, the existence of a gravitational force to pull the water with its accumulated solutes and particles through the profile, and time. Erosion and freezing and thawing are physical agents in soil formation, but for a particular soil series they only impede or accelerate the chemical process. Thus the chemical processes create a profile that stores moisture and nutrients and permits the root development necessary for tapping that moisture and those nutrients.

Parent Material Chemical properties of a soil are a reflection of the parent material on which it developed. In many instances, the chemical influence of the parent materials is strongly reflected in the discrimination among species as to which species of a community will persist on a particular soil. There are many communities endemic to California that are peculiar to the chemical nature of their soils, such as incense-cedar-dominated cover at lower elevations on the western slope of the Sierra Nevada on the magnesium rich soils of the serpentine outcrops. In the Wasatch Mountains of Utah, Douglas-fir seems to restrict itself primarily to limestone soils, while Engelmann spruce seems limited to the Wasatch conglomerate in the same area. In the Ozark Mountains, where immature soils have developed on single rock formations, distinct communities form according to whether the parent material is chert, limestone, or sandstone. Eastern red cedar is limited to limestone-based soils along with shagbark hickory, white ash, black walnut, and sugar maple, while mockernut hickory, black gum, and red maple are predominant on chert or sandy soils but do not occur on St. Joe limestone (Read, 1952).

The influence of parent material on the mineralogy of a soil developing on it is particularly pronounced in an immature soil. Soils developing from the same parent material in two localities will be entirely different morphologically and chemically if the climates are entirely different (Lutz and Chandler, 1946). On the other hand, soils from different parent materials under the same climate will resemble each other when weathered enough to have lost their rock.

Profile Mineralogy The textural fractions of the soil differ tremendously in their chemical activity because soil chemistry is primarily a phenomenon of interactions on the surface of soil particles. Sand and silt particles are predominantly quartz, with relatively little surface for their mass (Brady, 1974). Among the other minerals may be iron and aluminum oxides, feldspar, micas, and various secondary silicate minerals. Clay-sized particles may be of any of the minerals, but as clays become finer they are increasingly composed of aluminum silicates that are predominantly of two crystalline structures: kaolinite tends to dominate the fines under moist climates, with iron and aluminum oxides increasing with soil maturity, and montmorillonite tends to dominate the fines in soils developing under drier climates. Kaolinite is a tightly balanced structure with no substitution of cations within its structure and little shrinking or swelling on drying or wetting. Montmorillonite's structure is such that cation exchange can take place in spite of an apparent balancing of charges on the clay particles, and its shrinking and swelling under varying moisture conditions is great (Fig. 10-5).

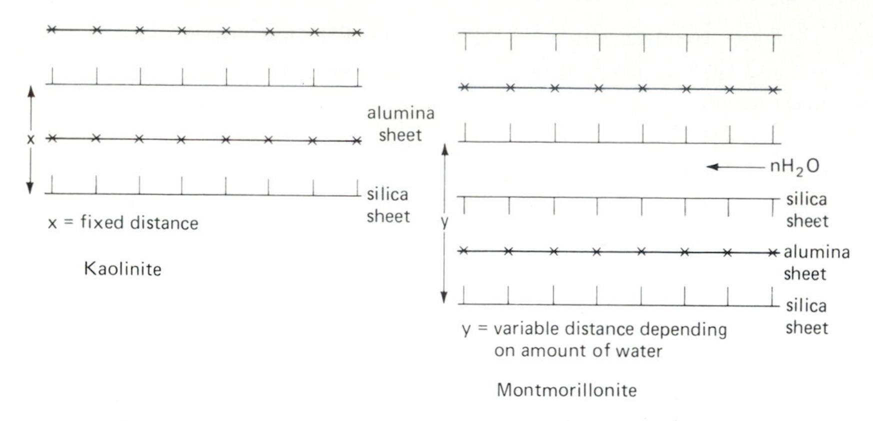

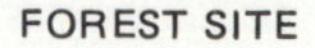

Figure 10-5 Diagrammatic structure of two principal clays. In kaolinite some of the oxygen ions of the gibbsite layer are shared with a single silica layer, while in montmorillonite the gibbsite oxygen ions are shared by the silica layers on both sides of it because the top silica layer is inverted.

Cation Exchange Capacity Cation exchange capacity (CEC) is a measure of the total exchangeable cations available in a soil, and it is expressed as the number of milliequivalents (meq) in 100 g of soil. (An *equivalent* is equal to the gram-atomic weight of a cation divided by its valence). The amount of cation exchange capacity depends on the presence of negative charges on the soil particles, and it is highly correlated with the amount of particle surface area—primarily that in the colloidal clays and organic matter (Table 10-3).

It is evident from Table 10-3 that a large particle has very few cations on its surface that are free to exchange with cations in the soil solution, and that the cation exchange capacity does not increase in direct proportion to surface area. It is noteworthy that the cation exchange capacity of the soil texture classes is almost directly proportional to the amount of clay in each soil. Also, in evaluating the impact of the early seral stages on succession, the contribution of a small increase in organic matter content of a soil to the cation exchange capacity and its significance for site improvement should be kept in mind.

High site quality is associated with a high cation exchange capacity, but the reverse is not necessarily true. In a soil with a 0 horizon, the rain is acidified when decomposition is slow and the hydrogen ions displace other cations like calcium, magnesium, potassium, sodium, and ammonium ions which are gradually elutriated to lower levels of the profile. The extent to which hydrogen ions dominate the cation exchange capacity is indicated by the percentage base saturation value: if the percentage base saturation is 60 percent, then 40 percent of the CEC sites are occupied by hydrogen ions. Thus, although the A horizon maintains its cation exchange capacity, it provides little of the nutrient cations needed for growth, and the nutrient anions (phosphate, sulfate, and nitrate ions) are moved out or tied up in an unavailable form. Also, when the B horizon has such poor structure as to be almost impermeable to roots, then its high exchange capacity, even though high in basic ions, does little for tree growth.

Table 10-3 An Example of the Relationship of Surface Area to Cation Exchange Capacity for Pure Fractions and Representative Soil Texture Classes

Particle class	Average surface area, $cm^2\ g^{-1}$	Average meq $100g^{-1}$	Soil texture class		Surface area, $cm^2\ g^{-1}$	Cation exchange capacity, meq $100g^{-1}$
Sand*	136	0	Sandy loam	60% sand 30% silt 10% clay	28,300	6
Silt†	4000	5	Loam	40% sand 40% silt 20% clay	55,600	11
Clay†	270,000	46	Clay	20% sand 20% silt 60% clay	163,000	29
Colloidal soil organic matter‡	—	284		1 g organic matter	—	2.8

*Average value derived from Baver et al., 1972, p. 17.
†Silt and clay averages derived from Whitt and Baver, 1930.
‡Derived from Pratt, 1957.

In fertilizing with a calcium source, the objective is to render the infiltrating rain more neutral in pH and to displace hydrogen ions on the exchange sites with calcium ions. In the order of their strength of adsorption, the various cations can be arranged as $H > Ca > Mg > NH_4 > K > Na$. In other words, calcium can displace hydrogen easier than potassium can, but hydrogen can readily displace any of the other cations.

It is evident, then, why such physical properties of soil as structure and depth become of major importance, since it is the total supply of cations and anions available to the trees' roots in the soil profile that determines growth rate. It will be demonstrated in the discussion of site quality that the rapid expansion of the root system to provide access to the total nutrient supply in a coarse, low-productivity soil may give a false impression of higher site quality compared with slower root penetration in a fine soil that actually has higher potential productivity. This problem tends to confuse the determination of site index in young stands.

WATER SUPPLY

Some consideration of the dynamic nature of soil and the sources of that dynamism have been given in the discussion of soil formation, profile development, and the influences of infiltrating water. However, these characteristics only partly describe the opportunities and constraints controlling the development of root systems in a soil for a tree community growing on it. Of course, the macroclimate and seral stages in community development have determined many of the characteristics of the particular soil, so the circularity of the interactions of soil, plant community, and macroclimate with the activities and stresses in that soil (such as deep percolation, evaporation, and moisture stress) becomes apparent. The supply of water is a source of constraint on growth and productivity within a particular temperature regimen. Since water is only available to a tree through its root system, the role soil plays in supplying it is of fundamental importance.

From Holdridge's system of climatic zonation it is evident that water supply is not directly related to the amount of precipitation because the variation between two environments in evaporative stress (due largely to temperature differences) and in the seasonal distribution of precipitation combine to change the effectiveness of a given amount of precipitation. Water supply in terms of its effect on productivity of a site is defined as the amount of water available to a tree during the growing season. In a forest, the forester cannot manipulate the amount or distribution of rainfall or the water-storage capacity of its soil, but the use of available water can be modified by stand treatment. Inasmuch as the moisture supply in the surface soil is particularly important during the regeneration period, it is fortunate that moisture can be conserved by proper manipulation of stand density and ground cover. An understanding of the water cycle in a forest is therefore a basic necessity to sound stand manipulation.

Water Cycle in a Forest

The cyclic nature of water in the life of a stand is illustrated in Fig. 10-6. It shows the six ways in which precipitation ultimately returns to the atmosphere, with interception, transpiration, and soil evaporation having the shortest turnaround time, with runoff and percolation taking a longer time, and with the water utilized in photosynthesis or stored in cells eventually being released. The amount of precipitation following each pathway depends on a multitude of factors, some of which will be discussed under the various elements of the cycle.

Interception An average value for interception is 20 percent of total precipitation and seems to vary little with changes in species. Among the variables affecting the seasonal and perhaps the actual amount of interception for any stand are whether its composition consists of conifers, hardwoods, or a mixture, whether the precipitation occurs as snow or rain, whether the precipitation occurs in frequent small storms or in a few large storms, whether the hardwoods have their leaves or have shed them, whether the canopy is closed or open, and whether the bark is smooth or rough.

A dense stand of red fir or sugar maple has a greater surface to wet before no further interception loss is experienced in a particular storm than a dense

Figure 10-6 The water cycle in the forest. Under varying circumstances values may depart rather widely from the averages, which are indicated by the width of the pathways.

stand of jack pine or aspen. If the interval between storms is long enough, the interception will be lost to evaporation and the stand will remove a similar amount from the next storm. When a series of small storms occurs, all the precipitation may be lost to interception and subsequent evaporation, while in a sustained rain the interception loss may be a small fraction of the total. While in most conifers interception capacity is reasonably constant throughout the year, deciduous hardwoods have much lower interception losses in winter. With an open stand interception is lower, but a stand open enough to reduce interception would show a loss in production due to understocking. Consequently, attempts to enhance growth by conserving moisture through opening up a stand would tend to be self-defeating. An exception to this might be where the objective is to increase deep seepage for more stream flow. In addition, in more open stands, soil evaporation loss would be greater but probably less than the reduction in interception. Smooth-barked species have appreciably less interception for a given canopy closure than rough-barked species because smooth bark is more conducive to stem flow. Stem flow also depends on the angle the branches have to the trunk. Stem flow is estimated as varying between 0.1 and 21 percent of precipitation depending on the intensity of the precipitation, crown area, and stem basal area (Kittredge, 1948).

Snow interception losses depend on stand conditions, crown form, and type of snow. Stands with uneven-aged, irregular canopies have low interception losses because most of the snow is dumped from the branches. If stands are fairly open, there may be some net gain in water supply because the openings accumulate snow beyond that found in the open (Kittredge, 1948). Pole stands with even spacing may have considerable interception losses, especially where freezing and thawing tie the snow to the branches creating a roof of snow which then sublimates without contributing to soil moisture.

Interception losses are difficult to estimate because the measurement of the actual precipitation in the open lacks precision and the measurements of the through-fall and stem flow have considerable variation. Resolution of such problems requires considerable sophistication in instrumentation and sampling design.

Another form of interception is the condensation of fog moisture in a canopy, with the condensate dripping to the ground. An appreciable amount of soil moisture is provided in this manner along parts of the West Coast where the upwelling of cold water from the depths of the Pacific creates an almost continuous fog bank during the summer. The west edge of Peru where no rainfall may fall in some years has tree growth on the lee side of coastal ridges supported only by moisture from condensed fog. Kittredge (1948) concludes his analysis of fog drip with the statement that in regions of frequent fogs or fine misty rains, fog drip may increase the precipitation reaching the ground by amounts up to 2 to 3 times the precipitation in the open.

Evaporation from Soil Evaporation of soil moisture is estimated at 25 percent of precipitation entering the soil in Eastern United States, where growing-season rains predominate, but would be less in Western United States, where there are long, dry summers. Moisture evaporation is controlled by a complex interaction of many factors, and fortunately, some important factors in this complex can be controlled by the forester. The factors can be divided into the climatic factors, which influence the supply of moisture and the capacity of the atmosphere to take up moisture, and the edaphic factors, which affect the supply of moisture at the surface of the soil.

Climatic factors that control evaporative losses are (1) air temperature, which determines the capacity of the air to hold water, as illustrated by the changes in vapor pressure with temperature and humidity in Fig. 10-7; (2) air humidity, which determines the saturation deficit—the higher the saturation deficit, the greater the capacity of the air to pick up moisture (Fig. 10-7); (3) wind movement, which is essential for rapid evaporation—the higher the wind velocity, the higher the evaporative stress because wind sweeps the moisture away from the evaporating surface, while in still air, only the slow diffusion process is operating; and (4) the rainfall pattern, which determines in large measure the moisture available at the surface of the soil, in that frequent rains repeatedly wet the surface soil.

Macroclimate factors such as air temperature, humidity, and rainfall pattern cannot be influenced. However, the forester has substantial control over canopy cover and stand density which directly influences microclimatic conditions of soil temperature, within-stand humidity, wind movement, and evaporative

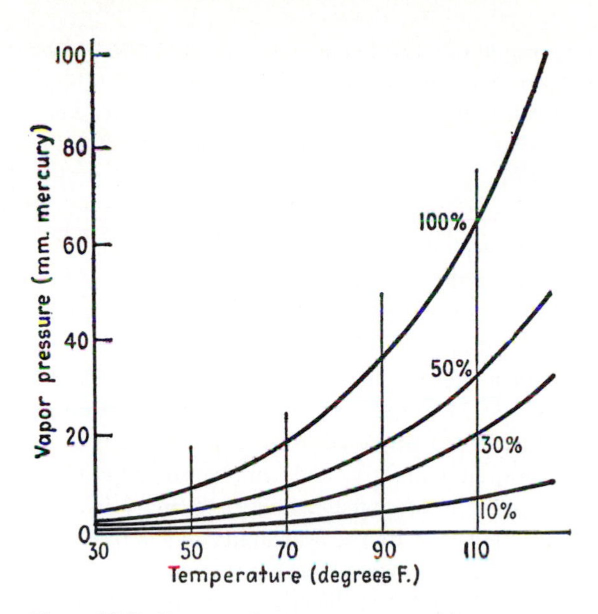

Figure 10-7 Changes of vapor pressure with temperature at saturation (100 percent) and at relative humidities of 10, 30, and 50 percent.

stress. The protective value of slash in reducing wind movement and temperature at the soil surface is too frequently ignored in analyzing the success or failure of plantings or natural regeneration.

Edaphic factors that control evaporative losses are (1) the amount of moisture in the surface layer of soil, which depends on the interval between rains and the texture of the soil as well as the mulch value of any 0 horizon; (2) temperature of the soil surface; and (3) the rate of movement of moisture or water vapor from lower in the profile to the surface, which is influenced by texture, compaction, and the associated capillary movement. As the surface particles of a soil surface dry, the moisture retreats into the interstices between the particles, and the moisture is pulled from below by this lowering of the water potential, which in turn reduces the evaporation rate. In coarse and medium sand, the pore spaces are too large for good capillary replacement of surface losses, so there is an initial rapid drying of the surface layer. In clay soil, with its fine capillaries, the capillary action brings water to the surface from some depth; however, the depth of drying will be slower and shallower than for coarser soils. Mulch on the surface dries out rapidly but is a poor conduit for bringing moisture from the underlying soil to the surface, so mulch reduces surface evaporation from the soil. However, mulch has a high absorption rate for rain, so light rains would be unlikely to add any moisture to the soil below the mulch. Of critical importance is the fact that soil evaporation only removes the moisture from the upper few centimeters of soil; the depth is greater in coarse soils than in fine soils, but evaporation is minimal below about 20 to 30 cm in very coarse soils after months of exposure. However, in natural regeneration, the top 5 cm of soil depth are critical and must remain moist until the seedling roots can penetrate below the drying layer. This can be accomplished often by stand and site condition manipulation.

Transpiration The details of transpiration and its function in the life of a tree are covered in Chap. 6. In terms of the water cycle of a forest, transpiration accounts for all the water loss from the soil other than gravitational water, which goes into deep seepage, soil evaporation, and the minor amount used up in tissue growth. Actual transpirational losses in terms of annual precipitation have wide limits, with estimates from 25 to over 50 percent where species tap ground water for their use. In addition to the usual macroclimate controls on evaporative stress, i.e., temperature, air humidity, and wind movement, transpiration is controlled by the response of stomatal openings to internal, morphological, and microclimatic conditions and by the availability of soil moisture to the root system. Trees differ in their responses to rapid transpiration losses when adequate soil moisture is available, but they differ much more in response to soil drought conditions. Xerophytic trees transpire like mesophytes when soil moisture is plentiful, but they can close down their water loss to negligible quantities in soil droughts and survive while mesophytic trees cannot.

Percolation and Runoff The purpose of discussing percolation and runoff together is to recognize that it is the failure to maintain an adequate infiltration rate that causes runoff. There are exceptions, such as when long periods of rain completely saturate the soil profile and further precipitation causes surface runoff. Maintenance of a suitable infiltration rate depends on protection of the usually highly permeable forest soils from poor management practices. Infiltration is greatly reduced by compaction by heavy equipment when the soil is too moist. It is also impaired when bare mineral soil is exposed to the physical impact of rain, which causes the soil pores to be plugged by fine soil particles. The conservation of normal infiltration rates can be attained through care to reduce the severity of surface disturbance and the judicious use of slash to break up areas of exposed soil. Of course, revegetation of an area with grass or herbs would retain the infiltration rate, but this greatly increases the difficulty of reforesting the site.

Percolation of water below the root zone feeds rivers and springs. The actual amount varies greatly from almost none in some areas of the piñon-juniper type to most of the rainfall in the Oxbow country of the Olympic peninsula with its 3500 mm (140 in) of rain. On a national basis the water cycle is apportioned as 70 percent (530 mm) (21 in) lost to evapotranspiration and 30 percent (230 mm) (10 in) to stream flow and ground water (Hewlett and Nutter, 1969).

Regional Water Cycles

In a country as diverse as the United States, regional water cycles should provide an opportunity to visualize climatic patterns and their impact on the distribution of precipitation among the several elements of the water cycle. Open juniper forests develop with rainfall as low as 250 mm (10 in) on coarse soils, while the lower limit for ponderosa pine is about 400 mm (16 in). About 1270 mm (50 in) approaches the optimum for growth, since additional precipitation does not add much more to forest productivity because the forest does not use it. By using two charts drawn up by Meyer (1928) which allow a monthly estimate of interception and evaporation from an average loam soil and of transpiration if mean monthly temperatures and rainfall are known, Baker (1950) was able to draw up water-cycle balance sheets for nine parts of the country (Fig. 10-8). These approximations are based on some simplifying assumptions: transpiration is from a full canopy, evaporation includes interception, the soil is a deep loam, and there is a soil storage capacity of 200 mm (8 in) of available water within reach of the roots. In addition, the profiles at the beginning of the season on October 1 were assumed to have no available water within reach of the roots.

The general trend of increased soil moisture storage during winter when evaporation is low and transpiration is negligible and fluctuations in the rate of depletion (or accumulation) with climatic differences in summer is shown by Fig. 10-8.

In the cool, humid climate of Maine, with its short growing season, the normal rainfall for every month except July and August is sufficient to supply the needs of transpiration and offset the losses by evaporation, with the considerable excess draining into the streams. In eastern Virginia, the warmer climate produces greater transpiration and evaporation, so, in spite of a higher rainfall than Maine, the soil reservoir is slower in getting replenished and during the summer (May to October) the transpiration and evaporation losses exceed the rainfall but not enough to empty the soil reservoir.

In central Georgia, the long growing season, warm climate, and heavy rains promote high water losses. In Fig. 10-8, the losses exceed rainfall, except from November to February, and although the monthly differences are not very large, the year ends with the soil reservoir totally depleted. This water regimen does not agree with the high productivity of Southern forests, but by assuming a sandy soil, which has one-third less evaporation than a loam soil, the water balance shifts to one in which rainfall exceeds losses in all months except October, April, and May. It is a more realistic pattern and emphasizes the importance of soil texture–moisture relationships. In Missouri, the pattern is similar to the one in Virginia. However, although the dry winters delay the accumulation of stored soil moisture, the wet summers have only June, July, and August where the losses exceed the rainfall.

In Colorado, at high elevations in the eastern portion of the spruce-fir type, the climate is cool, the growing season is short, and the rainfall is princi-

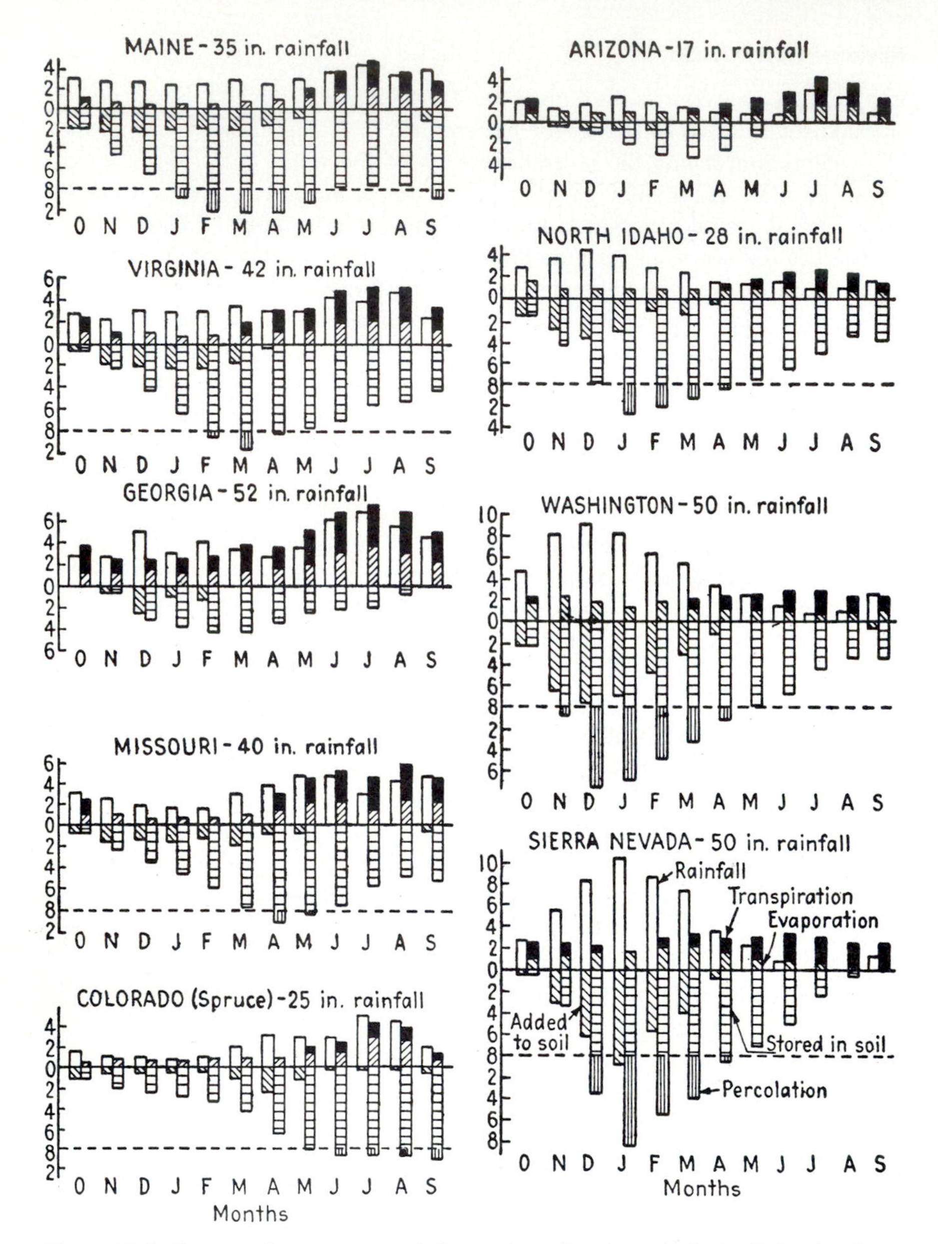

Figure 10-8 Pattern of water accumulation and use in a hypothetical soil, bearing forest and having 8 in of available water capacity within reach of tree roots. The season starts October 1 with soil depleted to the permanent wilting point. [*Computations for graph from Mayer's charts.*] (Meyer, 1928)

pally in the spring and summer. Although the annual rainfall is only 635 mm (25 in), there is an excess in rainfall over the evapotranspiration losses for every month of the year. This generalization is consistent with research findings which show that over 90 percent of the stream flow originating in the middle and southern Rocky Mountains comes from the high-elevation grassland and wooded areas, with the lower-elevation grasslands, sage brush, and piñon-juniper contributing very little.

In Arizona, in the low-altitude piñon-juniper, in spite of the bimodal precipitation pattern (winter and summer peaks), precipitation only exceeds evaporation and transpiration losses in late fall and winter and never fills more than half the profile to field capacity. During the summer, the diagram shows no water available to balance the excess losses over rainfall. One reason for this may be that the assumption of a full canopy is in error since the moisture is only sufficient to support an open woodland.

In northern Idaho, in the western-white-pine area, the high winter and low summer rainfall pattern provides a rapid renewal of stored soil moisture in the winter which is adequate to handle the excess of losses over rainfall in the summer. In Washington, in the Douglas-fir region, at about the same latitude as northern Idaho, the rainfall pattern is the same, but there is an abundance of moisture in the winter with a large contribution to stream flow and losses exceed rainfall only in June, July, and August. In California, in the mixed-conifer type of the Sierra Nevada, at about 1500-m (5000-ft) elevation, the extreme form of the Pacific rain pattern occurs, with high winter precipitation and no rain in July and August. In the most productive zone in the Sierra Nevada, the soil reservoir is filled in the winters, with a large contribution to percolation, but in the warm, dry summers it is completely exhausted.

Where the September data show a reserve of moisture in the soil, the contribution to stream flow is greater than shown in the figure. In estimating the probable effects of a drought in any one of the areas, it should be emphasized that the first precipitation of the season usually contributes to storage. Stream flow occurs only after the profile's water storage capacity is filled. An unusual drought, a warmer than usual summer, a combination of the two, or a sequence of droughts is an event which stands that are dependent on soil moisture stored within the limited range of their root systems must have the capacity to handle. It may be the one event in a hundred years that dictates which species survive on a particular site.

Soil Moisture

Water uptake by roots is dependent on the supply of moisture in the soil. The storage capacity of a soil depends on its texture, depth, and structure. The availability of soil moisture depends on water potential, root distribution, and temperature. Soil moisture is available to the root in two ways: the root grows to it or the moisture moves to the root. Root activity is not well understood because most of the information is buried in the soil and there have been very few efforts to unearth it except to measure lengths, depths, and volumes of soil occupied. There were 21 m of roots in 0.028 m^3 (1 ft^3) of A-horizon soil under a 20-year-old loblolly-pine stand (Roberts, 1948), while blue grass had 9.1 m (30 ft) in 16.4 cm^3 (1 in^3) of soil (Dittmer, 1938), which indicates why grass gives trees such severe competition. For trees, there seem to be two general periods of root growth–in the spring and fall, with some growth in summer and least in the winter. If root growth is slow in summer when the demand for moisture is greatest, the movement of water to the roots has to be a major factor in supplying the tree's needs.

Texture of Soil When water enters the soil, the infiltration rate is a function of the moisture content of the soil and soil texture. If it is approaching saturated flow, which would be when the descending front was moving through a soil with greater than -0.1 bar water potential, then movement would be greater in sandy soils than in fine soils. The reason is that in saturated flow the rate is proportional to the fourth power of the radius of the pores, so a pore with a 1.0 mm radius would have a flow equivalent to that in 10,000 pores having 0.1 mm radius. This would not be true if the adhesive forces between the water molecules and the walls of the capillary could be reduced to zero. In forests, saturated flow would occur with heavy rains; a high organic matter content in a fine soil would facilitate flow by maintaining larger pores.

Unsaturated flow is the common condition for moisture movement in field soils when the water potential is less than - 0.3 bar. It is the movement which brings water to a root tip when the available water has been adsorbed in its immediate vicinity. In this case, flow is slower in a sandy soil than in a fine soil because the capillaries in sandy soil are large and the moisture has a more discontinuous distribution. Capillary flow in a fine soil is greater because the fine capillaries are most likely to have continuous water connections as well as lower vapor pressures above the menisci.

Moisture in the soil can be divided into a number of categories: on a physical basis into gravitational, capillary, and hygroscopic water and on a plant-use basis into available and unavailable water. Gravitational water has the maximum free energy (greater than -0.3 bar water potential) and moves through the profile under the influence of gravity. Capillary water is held in pores small enough in radius so that the capillary forces are equal to or greater than the force of gravity. Hygroscopic water is bound to the soil particles at water potentials of - 31 to - 10,000 bars. The only water in the soil with less free energy than hygroscopic water is that within the crystalline structure of the soil minerals.

The term *available water* has been accepted as meaning water readily available for use by trees for growth. *Unavailable water* is not so clearly defined since part of this water is useful to some trees as an aid in surviving drought. If a quantity of water is allowed to infiltrate into a soil profile and if the profile is assumed to be of uniform texture, then the water fills the capillary capacity of the soil as it moves down until there is no further gravitational movement. The water held by the soil against the pull of gravity is called the *field capacity* and is measured by determining the amount of water in a sample divided by the oven-dry weight of the sample, expressed as a percentage. After a silt-loam soil has been flooded, it is assumed that the field capacity has been attained after draining for 48 h; with finer-textured soils, it may require a longer drainage period. Soil physicists object to field capacity being considered as a soil characteristic because it is not a stable value. The lack of stability is caused by the continued very slow downward movement of the moist front into drier soil under the pull of the lower water potential at that front. However, from a practical field point of view, field capacity represents the capacity of the soil to hold water against gravity flow and ensures that moisture is available in a wetted soil for use in dry periods. The water potential of a soil at field capacity is - 0.3 bar, and field capacity is determined for a soil sample in a laboratory by maintaining a -0.3 bar pressure on a porous plate. The amount of moisture in a soil at field capacity depends on the texture of the soil; coarse soils have low field capacities and fine soils have high field capacities (sandy, 6 percent; sandy loam, 14 percent; loam, 21 percent; silt loam, 26 percent; clay loam, 28 percent).

Evaporation can reduce the water content of the surface soil, but only transpiration can reduce the moisture in the rest of the profile. If the roots continue to draw moisture from soil with no replacement by rain or irrigation, a soil moisture condition is reached at which the roots will no longer be able to supply the water necessary to maintain the turgor of the leaves, and the plant wilts. If the plant will not recover when placed in a saturated atmosphere for 24 h, then the soil moisture has reached the *permanent wilting percentage* (PWP, wilting coefficient) and the plant will not recover unless the soil is watered. The water potential at the PWP has been arbitrarily set at -15 bars by using sunflowers as test plants. The PWP can be obtained in the laboratory on the porous plate by applying -15 bars pressure. The PWP is the weight of water in a sample divided by the oven-dry weight of the sample, expressed as a percentage. Wilting may take place at any time the transpirational loss exceeds the amount of water taken up to replace it. On any bright, hot, windy day in late summer, the sunflowers along the highways in Utah may wilt in the afternoon and be fresh and turgid the next morning. Thus the wilting is not always an indication of inadequate soil moisture, only of evaporational loss greater than root adsorption. The amount of water in a soil between field capacity and the PWP is called *available water*. The energy expenditure of a root in absorbing moisture just above the PWP is greater than the energy needed close to field capacity, but the availability of soil moisture between these two levels is approximately the same. However, moisture movement to the root is slowed as PWP is approached, and this may cause a reduction in growth rate, some lowering of transpiration at midday, and a tendency toward temporary wilting. The PWP is also a function of soil texture (sandy, 3 percent; sandy loam, 7.5 percent; loam, 11 percent; silt loam, 13.5 percent; clay loam, 16 percent), but it appears to be independent of the species of plant. Any variation between species where wilting has been recognizable has been slight and may be due to the variation in fibrousness of the root systems. Conifers and other sclerophyllous-leaved species are difficult to assess since the leathery foliage does not wilt, as it commonly does in herbaceous plants.

The *hygroscopic coefficient* (HC) is the amount of water in a soil (expressed as a percentage of the oven-dry weight) when it is in equilibrium with an almost saturated atmosphere. At the HC the water potential is -31 bars. Water in the soil between the PWP and the HC is not available for growth of any tree, but species can withdraw part of the water in this range for survival, with the amount depending on how xerophytic the species is.

Depth and Structure of Soil Storage capacity of a soil for water or nutrients is strongly influenced by the effective depth of the profile and its structure, since it is the amount of soil available to the roots, not the morphological depth of the profile that counts. The effective depth may be reduced by a high percentage of rock, by an impervious horizon in the profile, such as a strongly cemented B horizon as may develop in a Spodosol or a thick calcic layer in an Aridisol, or by a fluctuating water table. Structure influences infiltration and evaporation, as is the case when a fine soil is dispersed or puddled by bad management. This puddling impedes infiltration, and the resulting increased capillarity increases evaporation losses. The same soil with a crumb structure would have a greater infiltration rate and less evaporation because the pores between the aggregates would be too large and broken up for good capillary action. Textural changes within a profile can also change the field capacity of a soil, since fine-textured upper horizons may have the free movement of water downward impeded by a coarse-textured horizon. Profile structural conditions may even create a parched water table where an impervious lower horizon exists.

Soil Temperature The importance of soil temperature in silviculture is best illustrated by its effect on seedling survival at the time of planting (see Chap. 16). Changes in soil temperature lag behind changes in air temperature, and the greater the depth in the soil, the greater is the lag. At a depth of 20 cm, diurnal changes are small, and at greater depths there may even be little seasonal differences. In the arctic there is permafrost at relatively shallow depths, and at lower latitudes the temperature at depth is a function of the mean annual temperature. Because of the lag, the soil is warmer than air temperature in winter and spring and colder in summer and fall.

Higher temperatures increase the rate of evaporation from the soil and alter the rate of decomposition of the 0 horizon either by increasing the growth of soil organisms or by drying out the forest floor. A low surface temperature reduces evaporation but will also reduce the activity of the decomposing organisms and may create a selective factor for the kind of organisms. Thus the adjustment of canopy density becomes a silvicultural tool for creating a particular microenvironment.

Surface tension and viscosity of water decrease with increasing temperature while vapor pressure increases. These physical properties influence the availability of water in the soil. In a cold soil, the movement of water is slowed, so the PWP would be reached at a higher soil moisture percentage in a cold soil than in a warm one. In silviculture some poor decisions can be made in the selection of species for a site if a species' requirement for a favorable soil temperature is not met. Ponderosa pine planted on a cool western-white-pine site or at high elevation has done very poorly.

Temperature influences the solubility of various nutrient elements in the soil solution, so the decreased oxygen dissolved in a colder soil may be an influencing factor for some species.

Drought Resistance Trees and other plants demonstrate marked differences in their drought resistance by the ecological nature of the sites they can occupy. Drought resistance, like dormancy, has been a fertile field for hypotheses, and many mechanisms proposed for the ability of various plants to evade or endure droughts are not appropriate to our temperate zone trees. Trees show several adaptations to the occurrence of drought: some fail to perpetuate in areas where their drought resistance is insufficient for survival to seed-bearing age; others develop deep root systems that tap deep soil moisture for use during drought periods; others drop their leaves so summer mesophytes become winter xerophytes with low moisture losses; and xerophytic evergreen conifers shut down their stomates to restrict water loss to minimal levels. The three key conditions of water availability are from field capacity to near the PWP, near the PWP, and below the PWP. In the range from field capacity to PWP, xerophytes were once thought to be able to conserve moisture because of such adaptations as thick cuticle, hairy surfaces, deep-set stomata, waxy coating, and small leaves. However, unnecessary conservation of moisture by closing stomata would be self-defeating from an ecological point of view because it would tend to slow growth and allow competing species with less stomatal control to gain in stand position. Under truly xeric conditions, however, a relatively greater capacity to close stomates would have definite survival value. This concept is illustrated in mixed-conifer stands in the West. Transpiration losses by mature trees of ponderosa pine, incense-cedar, and white fir show that the xerophytic ponderosa pine at moderate evaporative stresses loses almost as much water as the more mesophytic species. When the evaporative stress is severe, the water loss of ponderosa pine is much lower than that of the incense-cedar and white fir. Even though under these conditions stomata of ponderosa pine are relatively more closed than those of the other two species, carbohydrate production is apparently adequate to retain the pine's position in the stand. In contrast with trees, small seedlings growing in water culture show ponderosa pine at high evaporative stresses losing appreciably more water than more mesophytic conifers.

It has already been indicated that at the PWP, leaves of a plant wilt; however, with hardwood trees, the symptoms are most likely to start at the top of the crown, with the edges of the leaves turning brown and drying inward toward the main veins. In a severe drought, deciduous trees lose part or all of their leaves. Since the root systems of trees tap a large volume of soil, the dramatic wilting experienced by a potted plant is not possible, even though the top foot or two of soil may be reduced to the PWP, because the tree has deeper roots in more moist soil. In potted materials, a drastic reduction in transpiration occurs and growth ceases at PWP.

At moisture levels below the PWP, potted mesophytic plants are morphologically unable to reduce transpiration below a rather high minimum, which continues to exceed the intake of the roots. Consequently, these plants die shortly after the soil reaches PWP. On the contrary, more xerophytic plants have a morphology that cuts off transpiration almost completely when the stomata are closed, so water loss is reduced below the level of intake by the roots. Below the PWP, roots cannot extend themselves, so water uptake comes from a low rate of capillary movement in fine soils plus moisture condensation from the vapor phase. Soil humidity, even at the HC, is over 90 percent, and is almost 100 percent at PWP. However, the sharper-curved menisci in the vicinity of absorbing roots have a low vapor pressure. Therefore, water vapor condenses on the surface with the lowest vapor pressure. As the xerophyte continues to lose moisture, a point is reached where the loss exceeds the rate of condensation at the root tip, and the plant dies. In an experiment with potted ponderosa-pine seedlings, the pine lived for 3 weeks after the PWP determined by sunflowers had been reached (Fowells and Kirk, 1945). At any point before they showed signs of being dead, the pine could have been revived by adding water. In the field, on droughty sites or in drought years, the root systems of even mesophytic trees can withdraw moisture from the soil almost to the HC after dropping leaves, but deeper roots are probably contributing to the survival (Batchelor and Reed, 1923).

Control of Soil Moisture through Silviculture

There are a number of points within the water cycle that a forester can manipulate by silvicultural practices. In the West, where summer droughts cause problems in securing natural regeneration and plantation survival, moisture-conserving practices merit consideration. In addition, the high value of water because of its scarcity provides an alternative objective to maximizing timber production, especially on public lands in high mountainous areas where the mean annual growth loss would be small and the potential for manipulating stream flow large.

Water-cycle management might involve a number of practices of varying effectiveness in conserving moisture, such as:

1 Maintaining a 0 horizon to facilitate the infiltration of precipitation into the soil, which reduces the likelihood of erosion. The 0 horizon as a mulch reduces soil evaporation but in an area of light rains may increase the total evaporation.

2 Maintaining enough shade to conserve moisture in the surface layer of soil to favor seed germination and seedling establishment.

3 Clearcutting stands increases the amount of precipitation that reaches the soil surface, but it also tends to increase runoff and soil evaporation. The duration of increased stream flow depends on the rate and species of plant invasion. The timing of the increased flow could well be adverse to conserving moisture by adding to flood peaks unless the size and locations of stands are carefully selected. A sequence of narrow clearcuts on north to east slopes would help to flatten peak runoff and increase stream flow.

4 Partial cuttings reduce interception losses and increase evaporation from the soil but offer an opportunity to manage the flow peak, especially as opposed to clearcutting exposed sites. Stream flow would be increased for a short period following each cut, and the duration of the increase would depend on how fast surrounding root systems expanded into the open area and how fast newly established regeneration and understory species exploited the area.

5 The presence of ground cover increases the rate of soil water depletion. Whether the ground cover's contributions in terms of aesthetics, game production, nitrogen fixation, and soil amelioration through its contribution of leaves and organic matter to the 0 horizon may be desirable depends on whether advantages of ground cover compensate for the water loss, particularly when the costs of manipulating the ground cover are considered.

FOREST NUTRITION

In this chapter we have already considered two major site components: soil development and soil water status. The third major component is soil and plant nutrient status. These three components have the common characteristic of being dynamic, in that for any location their status changes with time, vegetative cover, and stage of plant development. While the water supply is a prime factor constraining growth of vegetation, the supply of nutrients is of prime importance in determining site quality or potential productivity. Nutritional considerations in forest management involve an appreciation of the nutrient capital of the site, nutrient cycling, the role of mycorrhizae, tree requirements for nutrients, and an analysis of the effects of silvicultural treatments on the nutrient capital of the system.

Nutrient Capital of the Site

The nutrient availability of a site is of particular importance in determining its potential productivity. In general, the higher the nutrient capital, the more flexi-

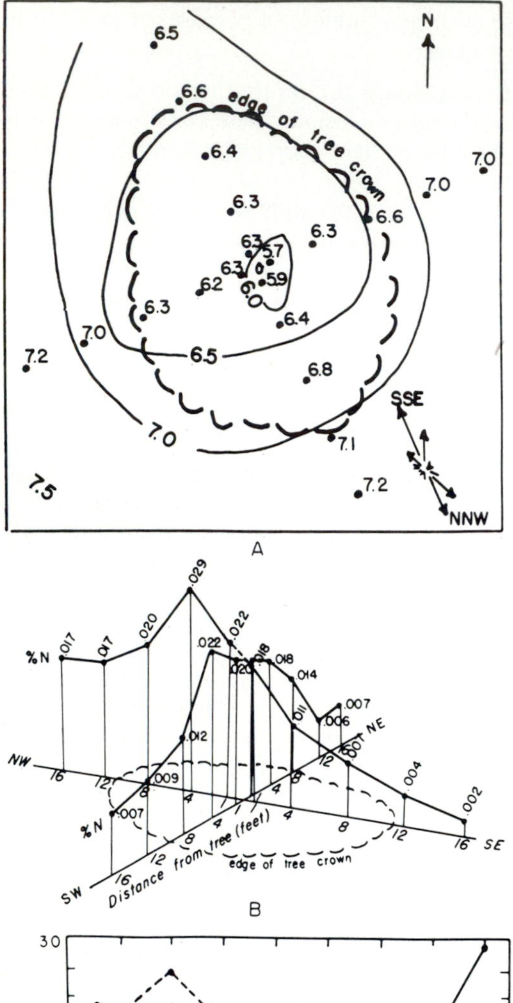
N
6.5
6.6
edge of tree crown
6.4
7.0
7.0
6.3
6.6
6.3
6.3
5.7
6.3
5.9
6.2
6.0
6.3
6.4
7.0
6.5
7.2
6.8
SSE
7.0
7.1
7.5
7.2
NNW
A

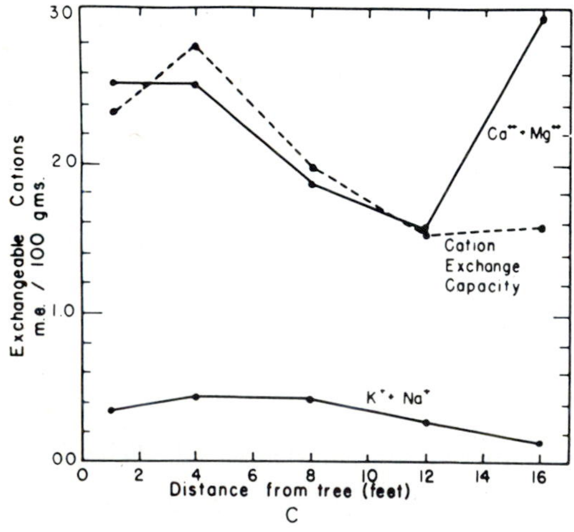
%N
.017
.017
.020
.029
.022
.022
.020
.018
.018
.014
.011
.007
.006
.012
.009
.007
.004
.002
NW
NE
SE
SW
Distance from tree (feet)
edge of tree crown
B

Exchangeable Cations m.e. / 100 gms.
Ca⁺⁺ + Mg⁺⁺
Cation Exchange Capacity
K⁺ + Na⁺
Distance from tree (feet)
C

bility one has in considering alternative silvicultural treatments. Forest ecosystems in nature are usually in a state of balance, where nutrient inputs approximately equal nutrient outflow. Silvicultural practices interrupt the balance and cause either additional inputs or, more commonly, increased outputs. If the nutrient capital of a site is high, these silvicultural disturbances will cause insignificant effects to the total nutrient balance. On low-fertility sites such as sands, however, the same silvicultural practice could be extremely unwise and result in loss of site productivity because of the high proportion of nutrients incorporated in the vegetation as opposed to the nutrient capital of the soil.

Soil fertility depends largely on the age, depth, cation exchange capacity of the soil and on climate. The growth of vegetation modifies the physical and chemical properties of soil, and it can be shown that successional development results in a buildup of nutrient supply which supports the next seral stage. This modification of soils by plants is difficult to conceive, but significant changes can occur in a relatively short time. This has been well illustrated by Zinke (1962), who analyzed the extent to which a single 45-year-old shore pine with a crown width of 3.6 m modified a sand dune on which it was growing near Crescent City, California. By intensively sampling to the 6.4-cm depth along a number of 4.8-m transects radiating out from the tree stem, Zinke showed that there was (1) a decrease of over 1 pH unit from the base of the stem to the end of the 4.8-m transects (Fig. 10-9*a*); (2) an increase of four times the amount of nitrogen in the sand within the influence of the tree's crown compared to that in the sand dune away from the tree (Fig. 10-9*b*); and (3) within the influence of the tree there was an appreciable increase in cation exchange capacity and either an increase or decrease in CA^{2+}, Mg^{2+}, K^{+}, and Na^{+} concentrations, as shown in Fig. 10-9*c*.

Nutrient Cycling

A forest system depends on a continued recycling of minerals for continued growth. Additions to the system come primarily from precipitation and dust, and from physical and chemical weathering of mineral soil and subsoil. Losses to the system occur primarily from leaching, burning, forest harvesting, and erosion. These considerations are discussed more fully later in this chapter. A more detailed discussion of nutrient cycling is provided by Armson (1977).

The Nitrogen Cycle In most forest systems, nitrogen is the most important element because of the quantities required by trees, its mobility in the system, and the commonly small proportion of total nitrogen available in soil for plant

Figure 10-9 A: The pH of surface soil under a tree of *Pinus contorta.* Isolines and recorded values of pH are indicated. Wind rose (arrow flying with the wind) indicates relative proportions of wind directions in the area. B: The nitrogen contents (percent by weight) in the surface mineral soils. C: Exchangeable cation composition of surface mineral soil in the transect extending to the southeast. (*From Zinke, 1962.*)

use. In many countries of the world, nitrogen is the element most commonly used in forest-fertilization programs. For these reasons, the cycling of nitrogen is discussed here in some detail.

In the forest, the nitrogen cycle is a complex interaction between the many nitrogen users within the system and the conditions under which each operates. In order to simplify the discussion of the nitrogen cycle, it can be broken into two parts: the primary cycle and the secondary cycle. In the field it is not possible to make such a division because the two cycles are operating simultaneously, with the dominance of one over the other depending on specific conditions.

The Primary Nitrogen Cycle The annual fall of leaves and twigs enriches the forest floor with its compliment of minerals, protein, and carbohydrate. All the elements present in a green leaf are not returned to the soil at the time of litter-fall since a proportion of the mobile elements N, P, and K have usually been previously leached from the leaves to the soil by precipitation or translocated out from the leaf during senescence and transported to other growing points in the tree.

Depending on the activity of the litter decomposers, the most recent leaf fall may constitute a high percentage of the 0 horizon, where material decomposes rapidly, as with hardwoods in warm, moist climates, or a low percentage if decomposition is slow, as under conifers in cold, moist climates. The primary cycle begins with the breakdown of this fresh material, but it is also functioning in the other organic matter in the 0 horizon and in the soil horizon. The fauna and flora of the forest floor work on the litter and use the carbohydrate for their energy to break proteins into amino acids (the process of *aminization*, Fig. 10-10), then the amino acids are reduced to ammonia (the process of *ammonification*). There are many organisms capable of carrying the cycle to this point, e.g., worms, insects, fungi, and bacteria. Once produced, the ammonia can be picked up by the roots and used by the trees. Under anaerobic conditions, which eliminate many of the decomposers except fungi and bacteria, this is as far as the primary cycle goes. The slow breakdown creates acid conditions, with the usual consequence of humic acids filtering through the profile.

In an aerobic situation, the ammonia is oxidized to nitrite by the soil bacteria *Nitrosomonas* and *Nitrosoccocus*, which is a fast process, then further oxidized to nitrate by *Nitrobacter*, which is a very fast process. This process is called *nitrification*. The formation of the nitrate completes the cycle, and this material is taken up by the root system.

The Secondary Nitrogen Cycle In the primary cycle it is assumed that the trees take up available ammonium and nitrate ions, but in reality there is an intense competition between the tree roots and the soil microorganisms (Fig. 10-10). The soil microorganisms are favored when there is a high carbon/nitrogen ratio (above about 20), while the roots have more nitrogen available at low C/N ratios. Freshly fallen leaves have a high C/N ratio (up to 40), so while this ma-

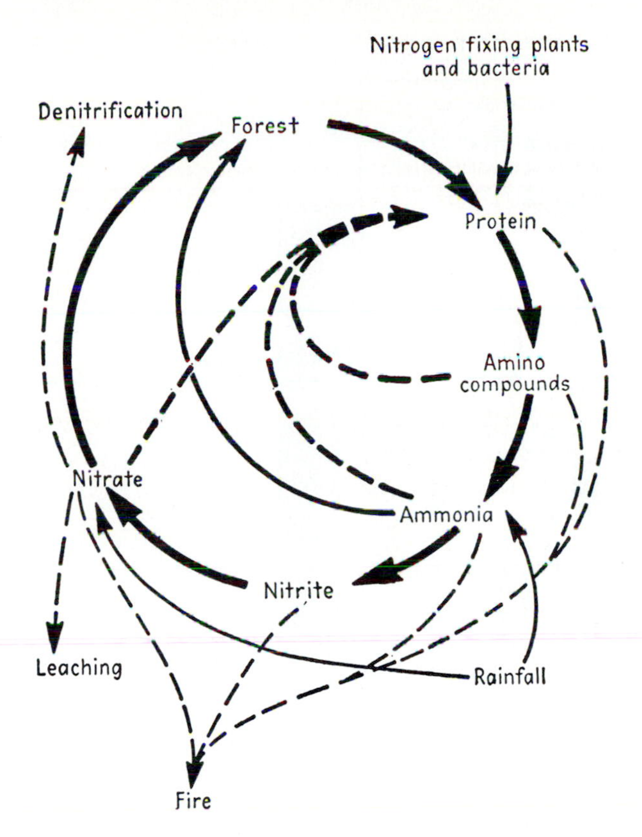

Figure 10-10 The nitrogen cycle. The main cycle is shown by the heavy full line. The secondary cycle in the soil is shown by heavy broken lines. Additions to the circulating fund are marked by light solid lines, losses by light broken lines.

terial is being oxidized, soil organisms multiply tremendously. The trees must wait for the release of the nitrogen by these organisms when their numbers decrease as the C/N ratio is lowered.

A conceptual diagram illustrating the nitrogen cycle in loblolly pine is given in Fig. 10-11. Similar studies have been made in Douglas-fir, and the amounts of nitrogen in different parts of the system are shown in Fig. 10-12.

The amount of nitrogen present in soils varies markedly with soil type, location, and climate. There is a marked decrease in the amount of total nitrogen in the soil, for example, with an increase in the mean annual temperature, as is the case from Canada to the coastal plains of the South. This is probably associated with the lowered incorporation of organic matter in the soil, since most soil nitrogen is in an organic form. At the same mean annual temperature there is more soil nitrogen where the effective precipitation is the highest (a lower evaporation over precipitation ratio, Holdridge's measure of the effectiveness of

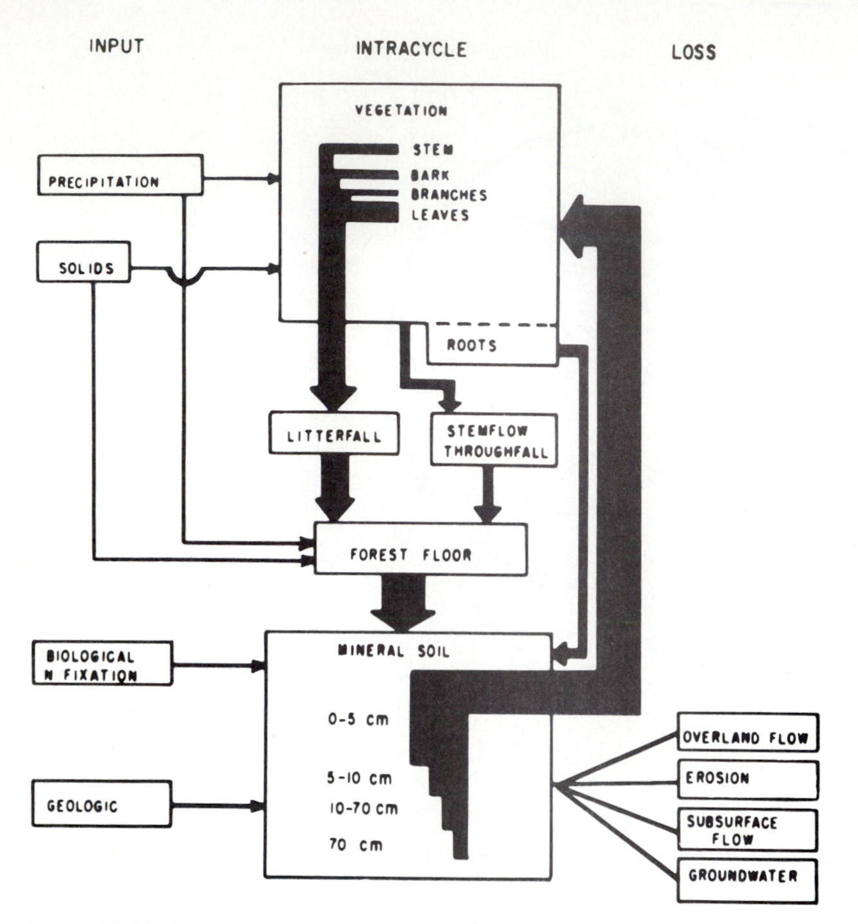

Figure 10-11 The nutrient cycle in a loblolly-pine stand. (*From Jorgensen et al., 1975.*)

precipitation). In general, quantities of nitrogen in the soil have been reported as shown in Table 10-4.

Nitrogen Accumulation Studies in Douglas-fir stands of ages varying from 9 to 95 years have shown that nitrogen accumulates rapidly and reaches a steady state at about the time when the stand has a maximum foliar biomass (Fig. 10-13*a*). Nitrogen is relatively mobile in plants and is retranslocated from older tissues to younger tissues. Thus, after foliar biomass is maximized, the trees increasingly satisfy their demands for nitrogen by using nitrogen available from internal remobilization and translocation rather than taking up nitrogen from the soil (Cole et al., 1975).

Understory vegetation increases in nutrient content until canopy closure, when its above-ground mass declines (Fig. 10-13*b*).

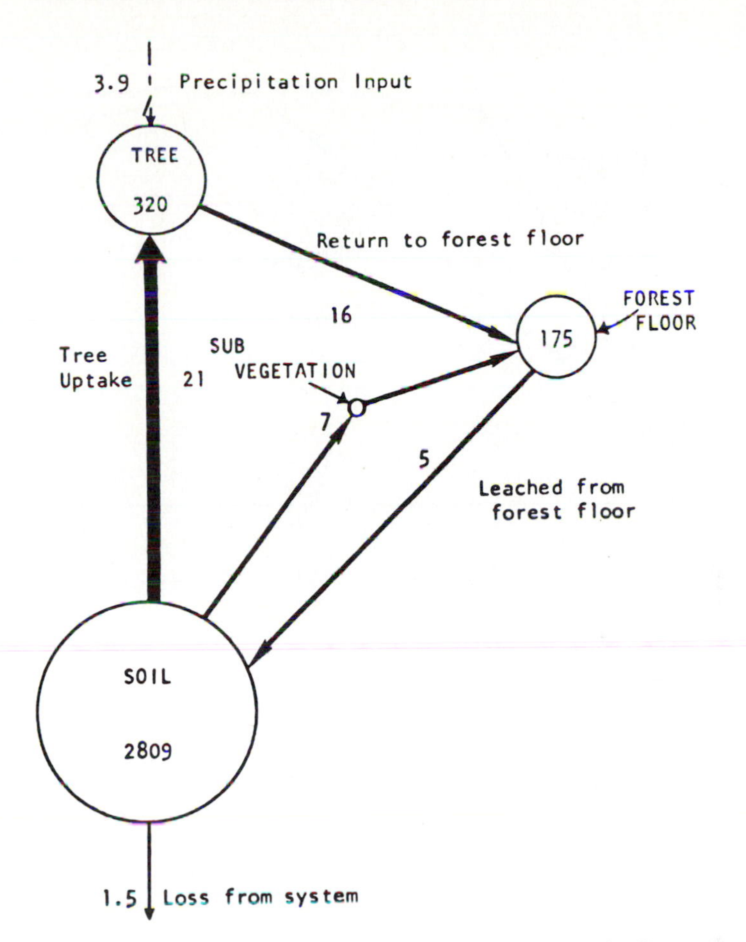

Figure 10-12 Nitrogen cycle in a 36-year-old Site IV Douglas-fir ecosystem. Numbers are kilograms per hectare. (*From Cole et al., 1968.*)

Table 10-4 Total Nitrogen Present in Various Forest Soils

Forest cover	Total nitrogen in soil, kg ha^{-1}	Reference
Black spruce (Quebec)	1500	Weetman and Webber, 1971
Loblolly pine	1753	Jorgensen, 1971
Douglas-fir, low site (Washington)	2809	Cole, 1968
Lodgepole pine (California)	6360	Zinke, 1950
Douglas-fir (Oregon)	5600	Brown et al., 1973
	10600	
	20000	

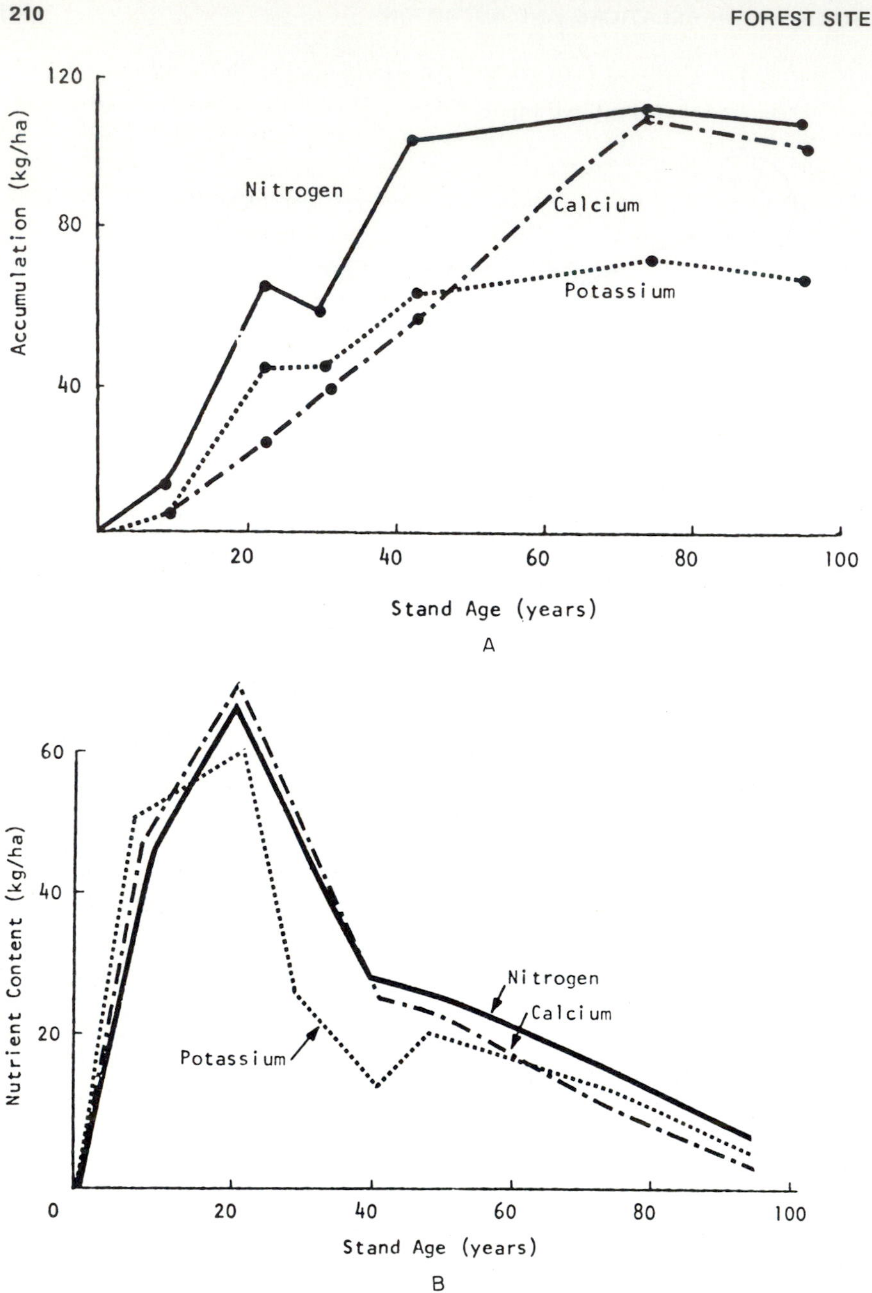

Figure 10-13 A: Accumulation of nitrogen, potassium, and calcium in the foliage of various aged stands of Douglas-fir (Site IV, Everett series soil). B: Accumulation of nitrogen, potassium, and calcium in the understory vegetation of various aged stands of Douglas-fir (Site IV, Everett series soil). (*From Cole et al., 1975.*)

Additions of Nitrogen to the System While most of the inorganic nutrients accumulate from the dissolving of minerals in the profile, nitrogen is added to the system mostly from precipitation or by nitrogen-fixing organisms. The amount of nitrogen brought in by precipitation and dry fallout varies with location but may be from 2 to 13 kg ha^{-1} $year^{-1}$. The proportion of this input which is in the ammonia or nitrate form varies depending on location. This is a substantial quantity and, over a rotation, may exceed the losses resulting from harvesting.

Nitrogen gas (N_2) makes up 80 percent of the atmosphere and is, surprisingly, useless for direct plant use. Some plants such as legumes, alder, *Ceanothus*, and scotch broom have the capacity to take nitrogen from the soil and, by means of symbiotic bacteria in root nodules, convert it to a usable form. Natural stands of alder may fix from 15 to 156 kg ha^{-1} $year^{-1}$ of nitrogen (Van Cleve et al., 1971; Armson, 1977). A mixture of alder with Douglas-fir can add 40 kg ha^{-1} $year^{-1}$ of N to the system (Tarrant and Miller, 1963), whereas alfalfa on good sites can produce up to 700 kg ha^{-1} $year^{-1}$. Bacterial fixation of nitrogen gas in well-aerated forest soils has been commonly believed to take place in the presence of organisms such as *Azotobacter.* A more widespread nitrogen-fixing bacterium is *Clostridium*, which functions in anaerobic conditions but is probably not as efficient as *Azotobacter*. These processes are less active in coniferous soils because of their higher acidity; however, fixation still occurs.

Losses of Nitrogen from the Cycle Since nitrate and ammonium ions are very soluble in water, there is a tendency for them to be lost by leaching. This can cause a problem of lowering of nutritional levels in the forest or lead to undesirable enrichment of aquatic systems. However, since in forest soils these ions may constitute less than 1 percent of the total nitrogen in the soil, it can be seen that the leaching of nitrogen is normally not a serious problem in undisturbed systems. The slower the decomposition rate, as in northern forests, the less likely there will be much leaching of nitrogen, while losses may be greater in southern forests.

In untreated forests, leaching losses are enhanced by changes in environmental conditions which increase the availability of CO_2 through increased rates of decomposition of organic matter or increase in the activity of soil organisms. Increased CO_2 raises the number of bicarbonate ions in the soil solution, as well as equivalent hydrogen ions, which are capable of displacing cations from the exchange complex. These cations can then potentially be leached from the soil. Increased activity of soil organisms causes nitrification, which produces both hydrogen ions and highly mobile nitrate ions and thus increases the potential for leaching.

Denitrification occurs when the soil organisms in an anaerobic situation use the nitrate as a source of oxygen and reduce it to nitrogen gas. In addition to the very low concentrations of nitrate at any time in forest soils, under anaerobic conditions the decomposition of the protein in the organic material of the soil stops with the ammonification process, so there is little or no nitrate to denitrify.

Significant losses of nitrogen from the system can also accompany harvesting, burning, and erosion. These losses are discussed in a later section of this chapter.

Other Minerals in the Nutrient Supply The number of essential elements known to be required for plant growth continues to expand as purification of chemicals and reduction of contamination in growth media permit a more precise examination and isolation of the influences of various minor elements. The essential nutrient elements, other than the carbon, hydrogen, and oxygen derived from the air and water, required in large quantities are nitrogen, phosphorus, potassium, calcium, magnesium, and sulfur. The essential elements that are required only in small quantities are iron, manganese, boron, molybdenum, copper, zinc, chlorine, and cobalt. In the future, areas that are presently quite productive may show a decline in production or the development of deficiency symptoms resulting from the removal of too much of a particular minor element. For example, a disease of orange trees in Orange County, California developed after the trees had been producing well, and it was corrected by shooting a small wedge of zinc into the tree trunk with a glazer's tool.

Calcium, in addition to being a major nutrient element, is of great importance in soil fertility through the effect it has on the physical and chemical properties of the soil. It is a constituent of many primary minerals and also is supplied commonly at the rate of 2 to 20 kg ha^{-1} $year^{-1}$ in precipitation and dry fallout. Kimmins (1977) states that inputs of calcium from the atmosphere may be as high as 52 kg ha^{-1} $year^{-1}$. Calcium is usually in adequate supply in the soil. However, it becomes deficient under conditions of high acidity, although low pH does not necessarily mean that calcium is deficient. Calcium is quite subject to leaching, a tendency which is accentuated where excessive hydrogen ions displace it in cation exchanges, and it needs to be recirculated by being brought to the surface in leaf and twig fall. Hardwoods recycle more calcium than conifers because of their higher general requirements for this element. Where site amelioration requires addition of calcium, then sources which have a strong acid anion such as calcium sulphate should be avoided since, after the cations have been exchanged in the soil, the released hydrogen ions produce a strong acid. Preferably, some form of lime fertilizer should be added.

Calcium is a relatively immobile element in the plant because it is tightly bound as a cell-wall constituent or as a precipitate, such as calcium oxalate, and it continues to accumulate in the tree (Fig. 10-13*a* and *b*). Quantitative estimates of the amounts of calcium cycled in hardwood and conifer stands are illustrated in Fig. 10-14.

Phosphorus is the most likely element among the important nutrients to become a problem. It is present in low amounts in most forest soils because it is a constituent of only one secondary mineral, apatite. Inputs of phosphorus from

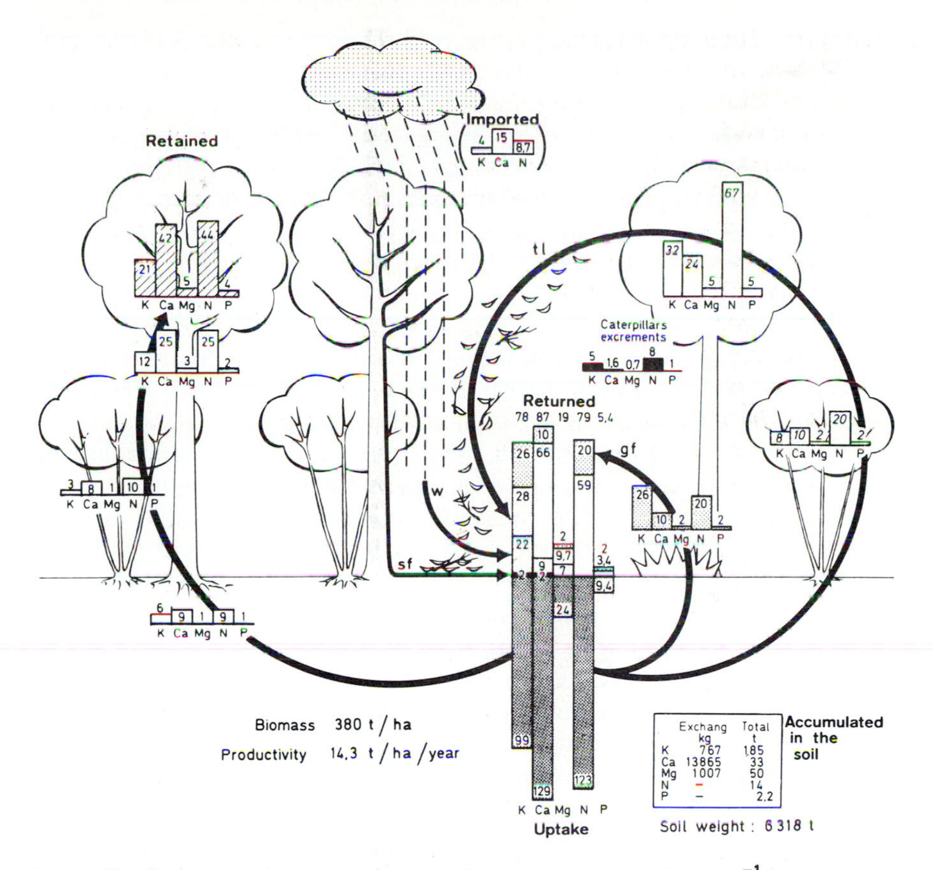

Figure 10-14 Annual mineral cycling of K, Ca, Mg, and N and P (in kg ha^{-1}) in a *Quercus robur–Fraximus excelsior* forest with coppice of *Corylus avellana* and *Carpinus betulus* at Wavreille-Wève, Belgium. Retained: in the annual wood and bark increment of roots, 1-year-old twigs, and the above-ground wood and bark increment. Returned: by tree litter (*tl*), ground flora (*gf*), washing and leaching of canopy (*W*), and stem flow (*sf*). Imported: by incident rainfall (not included). Absorbed (uptake): the sum of quantities retained and returned. Macronutrients contained in the crown leaves when fully grown (July) are shown on the right-hand side of the figure in italics; these amounts are higher (except for Ca) than those returned by leaf litter, due to reabsorption by trees and leaf-leaching. Exchangeable and total element content in the soil are expressed on air-dry soil weights of particles smaller than 2 mm. (*From Duvigneaud and Denaeyer-De Smet, 1970.*)

dust and precipitation are in the order of 0.1 to 0.9 kg ha^{-1} year^{-1} (Kimmins, 1977). The available phosphorus on a site tends to be concentrated in the 0 horizon (Table 10-4). Thus slash-disposal practices that bunch the slash with bulldozers taking most of the 0 horizon into the slash piles are likely to produce a phosphorus-deficiency problem. In addition to being present in small quantities, phosphorus is very sensitive to environmental conditions. Phosphorus becomes insoluble above pH 6.5 by combining with calcium, and below pH 4.0 by combining with iron or aluminum. If there is an excess or a deficiency of calcium, then phosphorus availability is also reduced. Root exudates, however, produce organic chelates which form organic complexes with phosphorus. These complexes may then be available for uptake. Thus the plant is not necessarily dependent on direct absorption of phosphate ions from the soil solution.

Potassium is taken up in much larger quantities than is phosphorus, and a great deal more is taken up than is required for normal development. This "luxury uptake" may not occur in trees to the extent that it does in fast-growing crop plants. Soils are usually well supplied with potassium, and it is added to an ecosystem by precipitation at the rate of about 0.1 to 7.7 kg ha^{-1} $year^{-1}$ (Kimmins, 1977). It exists in the structure of certain clays and is released as a reaction to the dilute acid created by water and carbon dioxide. Sandy soils have potassium in proportion to their silt and clay content. Pure quartz sand would be deficient in potassium. It may be leached quite readily once in the soil solution but can be held in the soil, especially if the soil contains montmorillonitic clays. In some crops, the presence of an abundance of potassium is associated with resistance to root rots, so it would be interesting to know if root-rot-susceptible species have a resistance deficiency of potassium especially in highly leached soils.

Potassium is relatively mobile and is readily leached from foliage by rainfall. Hardwood forests may supply about 16 kg ha^{-1} $year^{-1}$ of potassium to the forest floor from this leaching action (Fig. 10-14). Similar to nitrogen, potassium accumulates in the forest canopy until a steady-state condition occurs at the time of maximization of canopy biomass (Fig. 10-13*a* and *b*).

Magnesium is present in a number of primary minerals, so it is generally adequate in most soils. Inputs from the atmosphere may be up to 11 kg ha^{-1} $year^{-1}$ (Armson, 1977). Similar to potassium, much of the magnesium in a soil is associated with the clays, especially the montmorillonites, and is released by weathering. High proportions of magnesium to calcium ions in a soil, as in soils developed from serpentine rocks in California, limit establishment to the few species of trees, such as incense-cedar and Jeffrey pine, which are capable of tolerating high Mg content.

Sulfur, from minerals such as pyrites and gypsum, generally occurs in soils to about the same extent as phosphorus, but it is much more readily available. It is oxidized to sulfate in much the same manner as nitrogen to nitrate by a number of specialized bacteria and in anaerobic conditions may be reduced to hydrogen sulfide which is volatilized. Similar to nitrogen, about 5 to 15 kg ha^{-1} $year^{-1}$ of sulfur are added to a site by precipitation. In highly industrialized areas, such as Northern Europe and the Northeastern United States, especially when high-sulfur coal was burned, the amount of sulfur added has contributed to the phenomenon of "acid rain," which may have a pH as low as 3. On the West Coast, near the large metropolitan centers, pH values of precipitation are commonly 4 to 5. This form of pollution has potentially serious ecological consequences but, on the positive side, has reduced the need for sulfur fertilizers. With the emphasis on maintaining pollution-free air, sulfur could become limited under intensive timber management because it is very easily leached and the available quantities in forest soils are small.

Nutrient Uptake

While water moves across the live cells of the roots, epidermis, cortex, and pericycle to the xylem in response to a gradient in water potential from the leaves to the soil solution, the absorption of nutrient ions requires an expenditure of energy because the concentration of ions within the cells is usually greater than their concentration in the soil solution. Root uptake of ions from the soil solution is selective, and plants frequently build up high concentrations of some ions like potassium. Research using isotopes has shown, however, that the movement of ions is in both directions, since some of the desired ions move out of a root while the same ions are being taken up. An osmotic gradient probably accounts for the movement of ions from the pericycle to the xylem, after which mass movement carries them to the protoplasm of the leaves and any living cells in between.

For many plants, most nutrient uptake occurs from a few millimeters behind the root tip to a point where the roots become suberized. The movement of nutrients through a soil to the plant root surface is accomplished in three major ways (Barber, 1962): (1) by mass flow of the soil solution, in which movement of ions is associated with the movement of water resulting from plant water uptake; (2) by diffusion of individual ions, which becomes a dominant factor when mass flow is limited and where a concentration gradient exists due to nutrient uptake; and (3) by the movement of ions directly from the exchange site on a soil particle to the exchange site on the plant root without associating with an anion in the soil solution as an intermediary step (Jenny and Overstreet, 1939).

Nutrient availability is influenced by the rate at which nutrients move through the soil to the root surface and the rate at which root growth occurs. Since mass flow is a dominant cause of nutrient movement in moist soils, nutrient supply is strongly influenced by rates of transpiration. In coarse-textured soils (therefore generally poor nutritionally), capillary movement may stop while a considerable part of the available water has not been used. At that point, the mass movement of nutrient ions stops too. Water will move in the vapor phase to the point of absorption, but nutrient supplies can be provided only by growth of the root or by diffusion. In fine-textured soils, roots have good physical contact with the finer fractions, which enhances absorption, whereas in coarse-textured soils, this physical contact is poor.

Mycorrhizae

Uptake of nutrients by plants is considerably enhanced by the presence of a symbiotic relationship between the smallest order of secondary plant roots and particular fungi. These associations are called *mycorrhizae*, in which the fungus invades the root cortex and leaves the apical meristem and vascular cylinder fungus-free (Hacskaylo, 1972). Mycorrhizae are exceedingly common in the plant kingdom, occurring in all plants except those in the *Cruciferae*, *Chenopodiaceae*, *Cyperaceae* (sedges), and aquatics (Hacskaylo, 1972). Their occurrence is essential for adequate plant development. Some associations between host plants and mycorrhizal fungi appear to be quite specific, while other host plants have been shown to be associated with many different fungi, even on the same root. Trappe (1977) estimates that some 2000 species of fungi are potential mycorrhizal associates of Douglas-fir.

Considerable knowledge has been gained regarding mycorrhizae, and an extensive bibliography has been prepared by Hacskaylo and Tompkins (1973). Their taxonomy is still in a state of flux, but mycorrhizae are classified into three groups:

1 *Endomycorrhizae* have intracellular hyphal infections. They are by far the most common group, occurring in numerous herbaceous and woody plant families, including most angiosperms. They are found, however, on relatively few tree species, including *Liriodendron*, *Acer*, *Sequoia*, *Sequoiadendron*, *Cupressus*, *Thuja*, *Taxodium*, and *Juniperus*.

2 *Ectomycorrhizae* have intercellular hyphal infections. The fungal hyphae form a complete sheath or mantle of tissue covering the tips of the rootlets. This produces a characteristic pinnate or dichotomously branched root form (Fig. 10-15). The hyphae (Fig. 10-15) penetrate the root and grow between the cells of the cortex to form a network (the Hartig net) within and around the rootlet. Ectomycorrhizae are also widespread and occur on many important tree species, including the *Pinaceae* (which include Douglas-fir, hemlock, spruce, larch, and pine) as well as *Salicaceae* (poplars and willows), *Betulaceae* (alder and birch), and *Fagaceae* (beech and oak).

3 *Ectendomycorrhizae* have both intercellular and intracellular hyphal infections. This group is found only in tree species that are usually ectomycorrhizal (Hacskaylo, 1972). The fungal mantle in this group is not readily apparent (Fig. 10-15a).

Mycorrhizae were first recognized and named as a symbiotic combination of fungus and tree root on beech by the German botanist Frank (1885). He later attributed to the ectomycorrhizae a capacity to absorb organic nitrogen from the soil and pass it on to the host—the nitrogen theory. A mineral theory was proposed by Stahl (1900) in which the fungal hyphae were considered to function like root hairs to absorb all soil nutrients. Time has proven both hypotheses to be correct.

The function and behavior of mycorrhizae are complex. The association between plant roots and the fungus usually results in increased growth and vigor of the plant host. This is due to a complex of factors including increased nutrient uptake, increased water uptake, increased mineral solubility, protection of plant roots against pathogens, production of plant growth hormones, and the

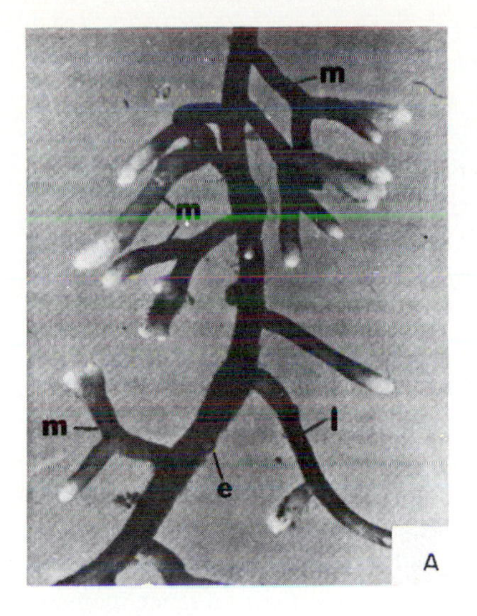

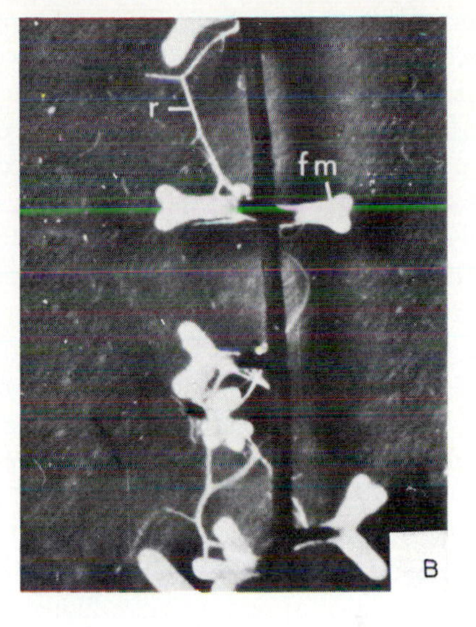

Figure 10-15 Comparison of ectendomycorrhizae and ectomycorrhizae in red pine seedlings and mycorrhizae in Douglas-fir. A: Ectendomycorrhizae in red pine seedlings, × 6.5. No fungus mantle evident. B: Ectomycorrhizae in red pine seedings, × 4.5. Conspicuous white mantle with connecting rhizomorphs; e, emerging lateral roots; fm, fungal mantle; m, mycorrhizae; l, long root branches. C and D: Douglas-fir mychorrhizae, *Pseudotsuga menziesii* and *Lactarius sanguifluus.* First dull orange, the smooth elements gradually become deep verdigris. C: Closeup of compound pinnate fan, × 3.5. D: Cluster of mycorrhizae along root, × 0.70. (*From Wilcox, 1971; Zak, 1971.*)

movement of carbohydrates from one plant to another (Sutton, 1969; Hacskaylo, 1972). The most commonly reported characteristic of ectomycorrhizal associations is the effect of enhancing nutrient uptake, particularly of the ions of low mobility such as phosphorus. This effect seems to be particularly important in soils of low site quality. Extensive fungal hyphae in the soil absorb ions released from soil minerals or by other microorganisms and translocate them through the fungal mycelia to the host root (Melin and Nilsson, 1950). The increased nutrient uptake of trees with mycorrhizal associations is therefore partly attributable to the greatly extended absorption system provided by the fungal mycelia. This is particularly important in trees, especially conifers, which have relatively sparse root-branching characteristics.

Mineral availability and uptake is also enhanced by ectomycorrhizal fungi in the capacity of their hyphae to excrete exudates that increase the solubility of ions bound to soils. The presence of the rich variety and quality of exudates in the rhizosphere (Slankis et al., 1964) also apparently creates an environment favorable to the development of nitrogen-fixing bacteria. Thus, although there is no evidence of ectomycorrhizae fixing nitrogen directly, their presence increases nitrogen-fixing action.

A further function of the ectomycorrhizae is their protection of the root from invasion by pathogenic fungi. This could be achieved through a number of different ways (Zak, 1964): (1) by utilizing the root carbohydrates and other chemicals that would attract the pathogens, (2) by providing a physical barrier to their attack, (3) by secreting antibiotics to inhibit or kill the pathogens, (4) by supporting a protective rhizosphere population of other organisms, and (5) by stimulating the cells of the roots to elaborate chemical inhibitors. It is the last idea which is suggested as the reason why the ectomychorrhizae remain symbiotic instead of damaging the host (Melin, 1955). A test of the hypothesis that ectomycorrhizae actually protect the roots from pathogens has been made on shortleaf pine seedlings using *Phytophthora cinnamomi* as the pathogenic infecting agent (Marx and Davey, 1969). Their results, shown in Table 10-5, together with subsequent analyses (Marx, 1971), indicate that a strong protective barrier is created by the ectomycorrhizae against the pathogen.

Ectomycorrhizae decrease in numbers as the site becomes richer. A number of hypotheses have been advanced to explain this, the most plausible being that the flow of auxins from rapidly expanding crowns stimulates root elongation and this results in an increased production of fungal inhibitors. The concept is debatable, but it is considered by some pathologists that on a poor site where slower-growing trees produce relatively less auxin, the development of ectomycorrhizae produces auxins which inhibit the extension or maturation of short roots.

Importance of Mycorrhizal Associations in Silviculture

As stated earlier, mycorrhizal associations with plant roots are exceedingly common. Their presence is a prerequisite to normal growth of many forest trees.

Table 10-5 ***Phytophthora cinnamomi*** **Infection of Roots of Aseptic Shortleaf Pine Seedlings with and without Mycorrhizae.**

	Inoculation with zoospores of *Phytophthora cinnamomi*						
		Mycorrhizae		Short roots		Lateral root tips	
Mycorrhizal fungus	No. of seedlings	No. inoculated	Percent infected	No. inoculated	Percent infected	No. inoculated	Percent infected
Laccaria laccata	9	7	0	29	100	8	100
Leucopaxillus cerealis var. *piceina*	6	27	0	32	25	6	100
Pisolithus tinctorius	9	42	19*	13	100	6	100
Suillus luteus	8	9	0	34	77	6	100
Nonmycorrhizal control	7	—	—	23	100	8	100

*Mycorrhizae with incomplete mantle and Hartig net development.
Source: Marx and Davey, 1969.

Certain trees, such as in the genus *Pinus*, have an obligate requirement for ectomycorrhizae and cannot grow normally without them (Marx, 1975). The importance of mycorrhizal associations to tree growth first became apparent in afforestation programs in many parts of the world where exotic trees were planted in the absence of naturally occurring symbiotic fungi. Examples of this kind have been reported for many species of pines in Puerto Rico (Vozzo and Hacskaylo, 1971), Peru (Marx, 1975), Australia (Kessel, 1927; Bowen et al., 1973), and also in Chile and Korea. In these cases, attempts to establish exotic pines were commonly complete or partial failures until ectomycorrhizal infection was obtained. This was commonly achieved by applying a mulching layer of soil containing the mycorrhizal fungus taken from around the base of successfully growing pines. A good example of the dramatic effects of such an inoculation is described by Hacskaylo (1972), where uninoculated slash pine trees planted in Puerto Rico grew only to about 30 cm tall in 5 years, with only a few tufts of needles at their tips. In contrast, the plants which were inoculated 3 years previously were up to 2.5 m in height and fully needled. It can be seen, therefore, that an important practical application of mycorrhizal symbiosis in forestry is in the introduction of exotic tree species and in the establishment of planted forests.

Similar problems may also occur in forest nurseries, even in locations where the trees grown are native and where some mycorrhizal fungi must already exist. This is due partly to the common use of soil fumigants in nursery management to reduce losses due to root diseases and weeds. Prior to sowing seeds in the nursery, the beds are often fumigated using methyl bromide. This kills all fungi in the upper layer of treated soil, but reinvasion is rapid since most fungal spores, including those of mycorrhizal fungi, are readily spread by air movement. Reinvasion also occurs from the development of fungi which survived at depths below the treated upper layer of soil. The nursery beds are therefore usually rapidly recolonized. This usually occurs when the established seedlings have developed beyond the stage of maximum susceptibility to damping-off fungi but in time for the seedlings to benefit from some infection by mycorrhizal fungi. However, delays in reinfection are detrimental, particularly with current trends of growing seedlings in nurseries for as short a time as possible. Adequate infection is also increasingly important in intensive management, where it is desirable for the planted seedling to become established and grow as rapidly as possible. For these reasons, increasing attention is being placed on the desirability of mycorrhizal inoculation of nursery beds as a routine part of management (Mikola, 1973). Details of the considerations involved have been reviewed by Trappe (1977). Because of the importance of mycorrhizae in assisting seedling growth and development, considerable research is being done to identify and culture those fungi shown to develop a favorable mycorrhizal association. Work of this kind is being done in the Southeast and in the Pacific Northwest, and it is expected that commercial mycorrhizal inoculums will soon be available for forestry use.

Controlled burning of slash following harvesting commonly kills a high proportion of the soil fungal population in the surface layer of the soil. The heat usually penetrates only a few centimeters, and fungi which are deeper in the soil survive without injury and rapidly recolonize (Mikola et al., 1964; Mikola, 1973). In addition to the effects of heat, burning changes pH, which can render a soil unfavorable since ectomycorrhizal fungi have rather narrow pH requirements (Mikola et al., 1964). Although intensive burning can delay the commencement of mycorrhizal infection, the effect is not regarded as harmful (Mikola, 1973).

Application of weed killers or other pesticides has not been found to be harmful to mycorrhizal fungi when applied at rates which are customary in forestry practice (Mikola, 1973). Similarly, clearcutting does not seriously harm mycorrhizal fungi unless the area remains treeless for a very long time (Mikola, 1973).

An important aspect of the role of mycorrhizae in forestry is in the revegetation of overburden or spoils from strip-mining operations. The surface of these waste areas is commonly dark, rocky, and devoid of humus and biological systems. They may also contain only small quantities of the essential elements for normal plant growth. Schramm (1966) showed that on mining spoils early ectomycorrhizal development was essential for seedling establishment of *Betula*, *Pinus*, *Populus*, and *Quercus*. Similarly, Marx (1975) has shown that inoculation with *Pisolithus*, an ectomycorrhizal fungus, increased survival of *P. virginiana* seedlings on strip-mined coal sites in Kentucky from 1.5 to 45.5 percent. Inoculation stimulated height growth by 30 percent. The key role played by mycorrhizae has been further demonstrated by Daft and Hacskaylo (1976), who reported that most herbaceous plants growing on anthracite and bituminous coal wastes in Pennsylvania, Scotland, and England were infected with endomycorrhizae. These findings are of paramount importance in reclaiming these large areas of adverse sites. The problem is complex since, although these sites are nutritionally poor, heavy fertilization of spoil material which contains little humus may actually depress the development of the mycorrhizae (Daft and Hacskaylo, 1976). Revegetation programs, consequently, must be developed which recognize the interactions of mycorrhizal development with mulching and fertilizing.

Until recently, there has been no evidence of mycorrhizal problems in developing and mature stands; however, there is isolated evidence that monocultures of pines may lead to a mycorrhizal deterioration of the soil. It has been reported (Robinson, 1973) that in some instances where a decline has occurred in productivity in the second rotation of *Pinus patula* on a variety of soil types in South Africa, both the number of mycorrhizal roots present and the degree of infection were less than during the first crop. This deterioration of mycorrhizal structure may not directly cause a decline in productivity and may be symptomatic of a gradual change in organic matter content, accumulation of nutrients within the buildup of litter, and possibly an increase of fungitoxic or phytotoxic materials.

The importance of mycorrhizae in growth and development of trees is being increasingly recognized, particularly with the introduction of intensive forest management and programs to revegetate difficult sites. The future role of mycorrhizae in silviculture has been aptly stated by Trappe (1977), who suggests that we may eventually "have mycorrhiza-improvement programs that are just as comprehensive and well financed as the large-scale tree-improvement programs that have been developed in various countries around the world."

Nutrient Requirements

Nutrient requirements of trees depend on species, stage of development, and season. There are three prime ways to evaluate these requirements: soil analysis, foliar analysis, and by directly testing the effects of applications of nutrients on biomass production.

Soil Analysis Of the total pool of elements present in a soil, only a small fraction is "available" for plant uptake. The remainder is relatively unavailable in crystalline or organic forms. The objective of soil analysis is to use chemical extractive methods which, as closely as possible, provide a measure of the amount of each element that is "available." Different extractive methods produce different results, consequently standardization of procedures is important. A further difficulty in soil analysis is the extreme heterogeneity of the soils themselves. Differences in location of a few meters and in depth of a few centimeters can make considerable difference in the estimates of nutrient content obtained. A large sample size is therefore required if reasonable estimates are to be obtained. A further difficulty is that fertilizer treatments based on soil analyses do not always produce changes in plant productivity expected, partly because of the interactions between elements which influence the availability and performance of fertilizers in plant growth. For these reasons, soil chemical analyses must be viewed as a helpful diagnostic tool but by no means as an entirely adequate means of forecasting plant growth or prescribing nutritional treatments.

Foliar Analysis An alternate approach to soil analysis is to analyze the quantities of elements actually taken up by the plant. In agronomic crops, much use has been made of the concept of *critical concentration*, which is defined for a given form of nutrient and plant part as that concentration at which the growth rate of the plant begins to decline significantly (Ulrich and Hills, 1973). Calibration curves can be constructed using solution culture techniques which show the effects of nutrient concentration on growth (Fig. 10-16). In diagrams of this type, three zones can be identified. The first zone is the zone of de-

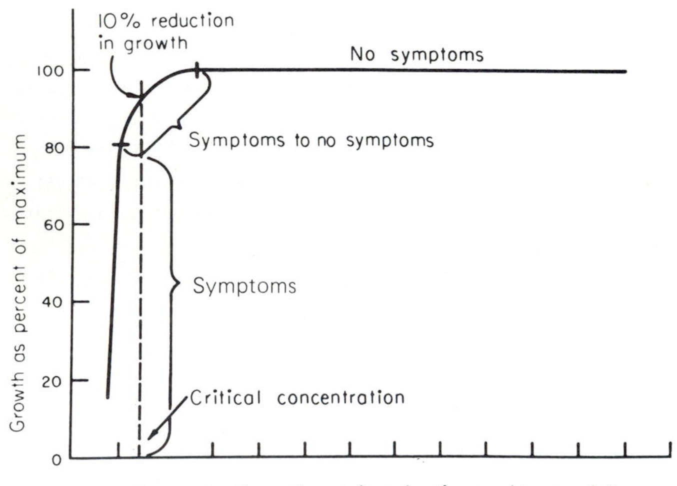

Figure 10-16 The critical concentration of a nutrient is taken at the point where growth is 10 percent less than normal. Symptoms generally appear below this concentration and fail to appear above this concentration. The transition zone, where symptoms may or may not appear, separates the zones of symptoms and no symptoms. The sharper the transition zone, the more useful the calibration for purposes of plant analysis (*From Ulrich and Hills, 1973.*)

ficiency, where plant growth increases sharply as more nutrient is absorbed. In this zone, there is little change in concentration of the nutrient in the plant part analyzed. The second zone is a transition zone, in which both nutrient concentration and growth increase as more nutrient is absorbed. Within this zone there is a critical concentration at which the plant will develop to 90 percent of maximum, providing other factors are not limiting. The third zone is an "adequate" zone, or zone of "luxury uptake," where addition of nutrient raises nutrient concentration without a corresponding increase in growth (Ulrich and Hills, 1973). With forest trees, foliar analysis usually involves sampling foliage from the upper portion of the crown in the fall or in early winter. This timing is selected since foliar nutrient concentrations are most stable at this time of dormancy. Good arguments can be made, however, for the desirability of sampling early in the growing season when growth is active and demands for nutrients are high. This is especially important since the contents of the mobile elements in foliage tend to increase during the first half of the growing season and decrease during the latter portion (Leaf, 1973). For reasons of this kind, results from foliar analyses are difficult to interpret and the method is not suitable for determining the presence of multiple deficiencies.

A different approach to foliar analysis which shows considerable promise has been developed in Canada (Durzan, 1974). This approach involves the determination of the concentration of particular amino acids. Since these acids are fundamental to nitrogen metabolism and to the physiological functioning of the plant, it is believed that a determination of their relative concentration will provide a better diagnostic basis for nutrient needs.

Because of difficulties of interpreting results from either soil or foliar analyses alone, combined analyses are often used in determinations of plant requirements.

Field Tests Actual prescriptions of fertilizer applications are most commonly made from field trials where a particular soil and tree species are treated with increasing amounts of a particular compound. This can be done in the laboratory and "critical levels" of nutrients determined for maximum growth performance. In forest practices, the "trial and error" approach of actually testing different rates of nutrient application still seems to be the most useful way of developing prescriptions for fertilizer treatment.

Nutrient Deficiencies and Deficiency Symptoms

In most productive forest regions of the world, nutrient supplies are adequate for growth. Consequently, it is uncommon to observe visual symptoms of deficiencies such as chlorosis. However, although nutrient supplies may be adequate, particularly in terms of total nutrient content, productivity of even high-site-quality forest land can generally be enhanced by fertilization because the availability of nutrients in the rooting zone commonly may be too low to permit maximum growth rates. In the afforestation of areas that have not previously had a forest cover, spoil areas from mining operations, sands, or drained areas, the available nutrient capital may be inadequate to sustain the demands made by a rapidly growing tree crop. In these situations, deficiency symptoms may become apparent.

Table 10-6 Nutrient Deficiency Symptoms in Conifers

Deficient nutrient	Symptom
Nitrogen (N) mobile	Stunting, yellowing, short needles. Young seedlings may remain in primary needle stage with little or no branching. Older trees have poor needle retention, stunted needles. Lower crowns may be yellow, while upper crowns remain green. Common on burned or scalped sites or in coarse-textured soils, especially in frigid zone.
Phosphorus (P) mobile	Young seedlings show purple needles, starting at the tips of lower needles and progressing inward and upward. Few or no secondary needles may appear. Buds may set early, or seedlings may remain dormant longer. Older trees take on dull blue- or gray-green color. Roots are sparse, with no evidence of mycorrhizae. May occur on very acid or alkaline soils or following fumigation.
Potassium (K) highly mobile	Dark, blue-green color in oldest foliage, progressing to yellowing, and finally to red-brown at needle tips. Needle retention is poor, and needles are often stunted. Short, thick stems in seedlings with abundant buds. Frost injury is frequent. Rare in conifers, but may occur in sandy soils or glacial till.
Calcium (Ca) extremely immobile	Primary needles may develop naturally in seedlings, but secondary needles may be badly stunted or dead. Growing points stunted and needles may "hook" at tips. Most severe in youngest leaves in upper crown. Very rare in conifers, except on ultramafic or heavily leached, acid soils.
Sulfur (S) mobile	Similar to nitrogen deficiency. Needle tips may be red or mottled, particularly on older needles. Poor needle retention. Deficiency may occur on heavily burned or scalped sites.
Magnesium (Mg) mobile	Tips of needles are orange-yellow, or rarely red. In young seedlings, cotyledons and primary needles remain blue-green. But in older trees, older needles and lower crown show symptoms first. In affected needles, transition to green may be sharp. Rare in conifers. May occur on calcareous, or badly leached soils.
Iron (Fe) immobile	New growth very stunted and chlorotic. Older needles and lower crown remain green. In seedlings, cotyledons remain green. Rare in conifers, except on calcareous or otherwise alkaline soils.
Boron (B) extremely immobile	J topping, splitting of meristematic tissue of main leader. Necrotic blotches on magnified cross sections of buds, terminating in death of terminal and some lateral buds. Rare, but may occur on sites high in organic matter (10 percent), or coarse-textured, droughty soils.

Source: R. F. Powers, USFS, PSW Station, Redding, California.

Considerations of nutrient use by trees requires an understanding of the concept of relative ion mobility. Nitrogen and potassium are very mobile and can therefore be readily leached from foliage or translocated from old tissue to young tissue. Phosphorus, chlorine, and sulfur are moderately mobile; zinc, manganese, calcium, iron, and magnesium are essentially immobile or only partially so under certain circumstances; and boron and cobalt are extremely immobile.

Table 10-6 indicates the visual symptoms that can develop in conifers when particular nutrient elements become limiting. Deficiency symptoms for mobile nutrients appear first in lower crown and innermost, older needles. Symptoms for immobile nutrients appear first in upper crown, growing points, and youngest, outermost needles. Details on the importance of individual nutrients for the nutrition of forest plants and descriptions of deficiency symptoms in both hardwoods and conifers are given in the text by Baule and Fricker (1970).

Nutrient Content of Trees

In recent years, considerable attention has been given to determining the actual amounts of nutrients taken up by forest crops. Generally in conifers, about half the above-ground nutrients is in the needles and branches and the remaining half is in the stem wood and stem bark. Data for loblolly and lodgepole pine are provided in Tables 10-7 and 10-8. Despite differences in species, ages, and soil types, it can be seen that the general magnitude of the quantities of nutrients in forest stands is generally 300 to 700 kg ha^{-1} of N, 40 to 60 kg ha^{-1} of P, 70 to 200 kg ha^{-1} of K, and 300 to 400 kg ha^{-1} of Ca. These quantities appear relatively small in comparison with the nutrient capital of good forest sites.

Effects of Silvicultural Treatments on Nutrient Capital

Silvicultural practices are designed to provide for reproduction of a stand or to modify the stand's composition, structure, or distribution of growth. In the

Table 10-7 Distribution of Tree Biomass and Nutrients in a 16-Year-Old Loblolly Pine Plantation in the North Carolina Piedmont

Component	Biomass, t ha^{-1}	N, kg ha^{-1}	p*, kg ha^{-1}	k*, kg ha^{-1}
Trees				
Needles	8.0	82	10	48
Branches	23.2	60	6	28
Stem wood	109.6	79	11	65
Stem bark	15.2	36	4	24
Root total	36.3	64	17	61
Total tree	192.3	321	48	226
Forest floor	—	307	30	28
Mineral soil, 0–70 cm	—	1,753	371	404
Site total	—	2,381	449	658

*Total in vegetation and forest floor and extractable in mineral soil.
Source: Jorgensen et al., 1975.

Table 10-8 Nutrient Storage in a 100-Year-Old Lodgepole Pine Forest on Ward Creek Drainage, Lake Tahoe Basin, California

Part of Site	Carbon*†	Nitrogen	Phosphorus	Calcium	Magnesium	Potassium
Foliage	5,550	250	12	40	10	16
Branches	7,340	110	7	38	12	12
Stem						
Wood	62,300	150	3	140	36	20
Bark	5,160	70	4	46	8	4
Roots	25,700	140	16	156	43	23
Tree Total	106,050	720	42	420	110	74
Litter	44,900	2,210	76	600	144	23
Soil						
(1-m depth)	259,900	6,360	1	12,000	1,920	1,950
Site Total	406,850	9,290	119	13,020	2,173	2,048

*Not a nutrient, but indicative of organic matter status of the site.
†Weights in kg ha^{-1}.
Source: P. T. Zinke, 1950, California Agricultural Experiment Station Project 1762.

process of applying these practices, there are potentially serious impacts on the nutrient capital. An awareness of the nutrient cost of a particular practice is an essential background for making silvicultural decisions.

Burning The interrelationships between fire and soil physical, chemical, and biological properties are extremely complex. Consequently, the effects of fire on soil cannot be generalized in an entirely satisfactory manner. Detailed information on this topic can be obtained from Kozlowski and Ahlgren (1974), Armson (1977), and Ahlgren and Ahlgren (1960). Fire can cause substantial losses of nutrients from a site. The magnitude of this loss depends on whether the fire is a wildfire or a prescribed burn (a *prescribed burn* is one in which the intensity of the fire is controlled). Burns in light slash or prescribed burns cause only small losses of nitrogen in the order of 10 to 15 percent (Grier, 1975) and enrich the soil with ashes. A potential for rapid leaching of these ashes is created on steep slopes with heavy rainfall unless these minerals are withheld by the cation exchange capacity of the soil or taken up by soil organisms or vegetation. Heavy slash fires or wildfires have such intense heat that 60 to 80 percent of the nitrogen is volatilized (Knight, 1966; De Bell and Ralston, 1970). A study in a Douglas-fir stand in Washington showed that the amount of nitrogen in the forest floor was 181 kg ha^{-1}; to this was added the slash from a harvesting operation which contained 160 kg ha^{-1} of N, resulting in a total of 341 kg ha^{-1} of N. The subsequent slash fire volatilized some 200 kg ha^{-1} of N (Cole et al., 1967). The importance of a loss of this magnitude is largely dependent on the nutrient capital of the site and the percentage of nitrogen removed. In many forest soils of high site quality, the proportion of total nitrogen volatilized may be quite small, whereas at the other extreme sands and tropical forests may have nearly all the nitrogen in the above-ground biomass.

Burning of slash or prescribed burns tend to raise soil pH due to depositions of alkaline ash (Klemmedson et al., 1962). The ash mainly contains the elements calcium, magnesium, potassium, and phosphorus. This raised pH tends to increase both phosphorus availability and the proportion of nitrate nitrogen, which is more readily leached. The burn reduces the population of soil microbes, but the raised pH enhances their development and it is usual to find that populations rapidly increase to previous levels. This enhanced microbial activity stimulates nitrification, which results in burned areas commonly having higher nitrate nitrogen levels than before the burn.

Hot burns also change the physical properties of soils by reducing porosity and consuming organic matter to a depth of several centimeters. This may be seen where slash piles have been burned. However, the total effect of these hot burns is commonly favorable to growth, as evidenced by the worldwide reporting of the "ash-bed effect." When piled slash is burned and the area is subsequently planted, observations are commonly made of the generally greater height growth of trees planted in the ash beds compared with those planted in unburned areas. This was shown to be the case in an analysis of 3-month to 17-year-old stands of *Pinus radiata* in Australia, where the 12- to 17-year-old trees on ash beds had produced about 50 percent more merchantable volume (Cromer et al., 1970). Ash-bed soils were shown to have a higher pH and higher total salts of N, P, K, and Mn in the top 25 cm. Foliar analyses showed the ash-bed trees to have higher content in P and K. In another study, a simulated ash bed was created by burning logs so that the soil was maintained at 100°C to a depth of 20 cm for 6 h (Renbuss et al., 1973). Four weeks after burning, more bacteria were in the ash bed than in the unheated control. *Actinomycetes* and other fungi recolonized more slowly. Early bacterial and fungal colonies included many species not detected in the control soil, but after 1 year the microbial populations in both soils were much the same. Eucalypt seedlings potted in ash-bed soil taken from various depths, after 3 years' growth, were about twice the height of seedlings potted in unburnt soil (Renbuss et al., 1973).

Windrowing In the mechanical removal of brush or slash from a site it is common practice to bulldoze the material into small piles or long extended piles called *windrows*. Unless the operator is very careful, even with the use of a "raker blade" or teeth on the blade of the tractor, it is inevitable that some topsoil is scraped into the piles or windrows. The importance of removing this top layer of soil depends on the proportion of organic matter and cation exchange capacity that this thin layer represents. On high site lands the removal of some small amount of material can be less important than on impoverished sites, where it may be highly critical.

Thinning and Pruning These treatments increase the rate at which the transfer of nutrients takes place within the system. Precommercial thinning and pruning where the cut material is left on the site essentially fertilizes the site by making the nutrients from suppressed trees and lower branches available for the growth of the more vigorous crop trees. However, the addition of organic matter to the soil, particularly when chipped, can cause temporary reductions in available nitrogen due to changes in the C/N ratio and increased consumption of nitrogen by decomposers.

Harvesting The amounts of nutrient elements removed from the site depend on the proportion of the stand that is harvested and the standards of utilization. The importance of this removal depends on the proportion of the total available nutrients that are removed from the site.

In conventional harvesting, where only the stem wood plus stem bark is removed, approximately 40 to 50 percent of the nutrient content of the tree is lost to the site. The actual quantities involved have been given previously in Tables 10-7 and 10-8. Further data are provided in Table 10-9. The element removed in greatest quantity in both hardwoods and conifers is calcium. On an annual basis, the removal of a forest once every 100 years is equivalent to a loss of approximately 0.5 to 3.0 kg ha^{-1} $year^{-1}$ of N, P, or K, which, on a good site, may not be a significant amount since similar quantities are commonly brought to the site in rainfall.

Some examples where studies have been made of nutrient depletion accompanying harvesting will be illustrative. In Oregon, two watersheds were compared with a control for 2 years before and after harvesting (Brown et al., 1973). The watershed which was patch cut showed no change in yields of nitrate nitrogen, phosphorus, or potassium after logging. After clearcutting and burning,

Table 10-9 The Drain of Nutrients from a Site Resulting from Removal of Stem Wood and Stem Bark

Species, species group, or site	Age, years	Nutrient drain in bole and bark* Ca	K	P	N	Investigator
Picea abies	120	—	24	6	45	Kvist, 1964
Pines	100	250	95	19	—	Rennie, 1957
Other conifers	100	506	269	37	—	Rennie, 1957
Hardwoods	100	1 266	279	56	—	Rennie, 1957
Pinus silvestris	55	135	58	9	101	Ovington, 1957; 1959
Pinus radiata	35	170	230	30	200	Will, 1964
Pinus resinosa						
Good site	31	166	35	9	87	Madgwick, 1962
Poor site	31	90	9	5	38	Madgwick, 1962
Pinus resinosa†	20	—	18	6	52	Heiberg et al., 1959
Tanbark oak	25	328	109	38	—	Von Schroeder, 1890

*In pounds per acre; 1 pound per acre = 1.12 kilograms per hectare.
†Midpoints of the ranges reported by the authors.
Source: Maki, 1966.

the yield of nitrate nitrogen was markedly increased (from 0.70 to 2 to 10 mg l^{-1}) in the first year after treatment and gradually returned to preharvest conditions by the sixth year. The yields of phosphate on this watershed also increased but returned to normal within 2 months after the clearcut and burn. Potassium levels remained unchanged. The changes in nutrient yields following clearcutting and burning were not regarded as a significant deterioration of water quality.

In another study of clearcutting and burning slash of Douglas-fir in Oregon (Fredericksen, 1971), the losses of nitrogen in stream flow was reported to be approximately 2 kg ha^{-1} $year^{-1}$, which again is less than the annual inputs from precipitation. Similarly, in Washington (Cole and Gessel, 1965), clearcutting of Douglas-fir produced immediate releases of nitrogen, phosphorus, potassium, and calcium. However, over a 10-month period nearly all the released nutrients were retained within the rooting zone of the soil. Absolute losses from the system were very small, approximately 0.5 kg ha^{-1} for all elements except calcium.

The potential loss of nutrients following harvesting is consequently influenced to a marked degree by the proportion of vegetation removed, the manner in which the slash is treated, slope, the nature of the soil, and precipitation. Losses can be considerable if vegetation on a watershed is removed and the area is deliberately kept denuded of vegetation, as was well illustrated on the Hubbard Brook experiments in New Hampshire (Bormann et al., 1968; Likens et al., 1970). Here, where all trees, saplings, and shrubs were cut and limbed, no products or slash removed from the forest, and regrowth of all vegetation was inhibited by the use of herbicide sprays, increased runoff and losses of nutrients from the site, particularly nitrate nitrogen, were considerable. During the year after cutting and spraying, the treated area showed a net loss of 52.8 kg ha^{-1} of N compared with a net gain of 4.5 kg ha^{-1} of N in the undisturbed system.

Intensity of Management and Nutrient Supply For conventional harvesting, where only the stem wood and stem bark are removed from the site, about half the nutrients in the above-ground portion of the tree is exported. The fate of the other half of the nutrients in the slash is dependent on the method of slash treatment, as has been discussed earlier in this chapter. If the crown of the tree is also utilized, the potential rate of depletion of nutrients from the site is doubled. It has been estimated (Cole et al., 1967) that chipping and exporting branches from the site would remove two-thirds of the nitrogen in the standing tree. Whole-tree utilization would remove 288 kg ha^{-1} of N, 50 kg ha^{-1} of P, 196 kg ha^{-1} of K, and 296 kg ha^{-1} of Ca in a Douglas-fir forest. Removal of roots would increase the loss of nutrients by 12 percent.

The impact of varying intensity of utilization on nutrient base is illustrated for black spruce in Canada in Fig. 10-17. This diagram clearly shows the magnitude of both total and "exchangeable" nutrients and the potential losses due to conventional harvesting compared with whole-tree harvesting.

The length of rotation also has an impact on rate of nutrient depletion.

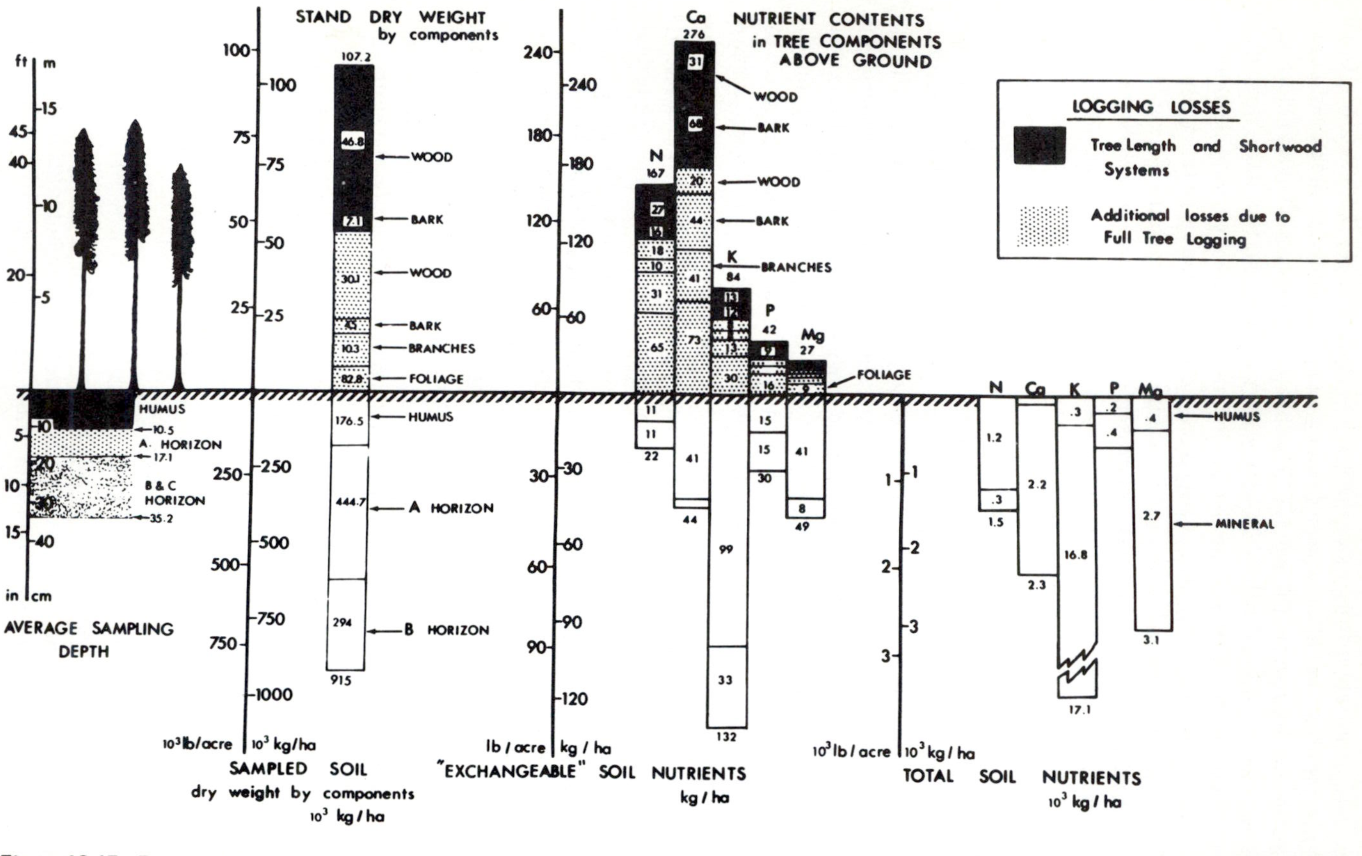

Figure 10-17 Dry weight and nutrient content of the soil and above-ground parts of the trees (kg ha^{-1}), black-spruce stand, 65 years old, Baic Comeau, Quebec. (*From Waetman and Webber, 1972.*)

Long rotations remove more nutrients in any one harvest because of the larger size of tree removed. However, since the most rapid accumulation of nutrients occurs in the early stages of the rotation, prior to maximization of stand foliar biomass, short rotations result in a greater removal of nutrients on an annual basis (Armson, 1977). Boyle (1975), for example, estimated that in managing stands of aspen on a whole-tree basis, a change from the single 30-year rotation to three 10-year rotations would increase nutrient removal of N, P, K, and Ca to 345, 239, 234, and 173 percent, respectively. The impact of short rotations is probably greatest in the case of growing hardwoods such as sycamore on rotations of 3 to 5 years in a system called "sycamore silage" (Steinbeck et al., 1970; Dutrow and Saucier, 1976). Here, rapidly growing hardwoods are grown at spacings of 1.2 to 1.8 m (4 to 6 ft) and "mown," producing yields on good sites of 17,000 to 56,000 green kg/ha^{-1} at the end of the short rotation. Yields at the lower end of this range are for crops at the end of 3 years' growth which were established from seedlings; yields at the higher end of the range are obtained from crops established from sprouts or coppice at the end of a 5-year rotation. Continued cropping of a high-producing crop of this kind could be expected to lead to a decline in nutrient availability similar to that experienced by intensively managing agricultural crops.

The intensity of management and extent of utilization raise important questions as to the capacity of forests to maintain their productivity and fertility. These general questions are of considerable international interest and have been brought into focus in this country by the pioneering work of Young, who established the Complete Tree Institute at the University of Maine. Young and others (Young, 1964; 1968; 1971; 1975; Chase and Young, 1976; Kimmins, 1977) are investigating trees and shrubs and quantifying their biomass and nutrient contents from root tips to leaf tips inclusive. The aim of such investigations is to determine the maximum amount of material potentially available for use in all components of trees and shrubs regardless of size. Traditional uses of forest biomass have included lumber and pulp, but these workers are exploring possibilities for new uses such as solid and liquid fuels, cellulose and textile fibers, chemicals, and food and fodder.

The amount of biomass in forests, including trees, shrubs, and herbs, can be quite considerable. Art and Marks (1971) have compiled a large amount of published data on forest biomass by species, location, and age. Their review shows, in general, that in temperate forests, biomass ranges from 14,000 to 30,000 g m^{-2}, and net primary production ranges from 1200 to 2800 g m^{-2} year^{-1}. This large amount of potentially usable material can be expressed in traditional terms of timber, slash, stumps, and roots and compared with a forest's allowable cut for timber production. For example, Hakkila (1971) estimates that in Finland, which has a short-term allowable cut of 50 million solid cubic meters per year, the amount of branches and needles left in the forests after harvesting is approximately 6 million tons of dry matter, which is equivalent to 15 million solid cubic meters of timber; and the amount of material left in stumps and root wood is approximately 4 to 5 million tons of dry matter, which is equivalent to 7 to 10 million solid cubic meters of timber. Taken together, slash, stumps, and roots left in the woods after harvesting can therefore, in the Finnish case, constitute the equivalent of some 45 percent of total harvest. The proportion in other situations would depend on tree species and size, the amount of live crown, the proportion of cull material, and utilization standards.

The complete-tree concept is concerned with evaluating the extent to which this amount of fiber can be used, competitive with costs of fiber from alternative sources, without causing site deterioration. Keays and Hatton (1971) discuss four critical questions relating to complete tree utilization: (1) how much of each component is available for utilization; (2) what quality of pulp or other product could be obtained from each tree component; (3) how would the various tree components be extracted, transported, and processed; and (4) what would be the effect of complete tree utilization on forest growth, natural and artificial regeneration, silvicultural practices, and ecosystem interactions. Included in this last question are factors such as the effect on nutrient cycles, soil compaction and erosion, need for and extent of site preparation, reduction in fire hazard, and silvicultural prescriptions for all intermediate treatments including thinnings. Another concern is the energy cost involved. Consideration of complete tree utilization will become increasingly important as fiber resources become more scarce. Of prime concern to us here is the potential impact on forest productivity and fertility. From a nutritional standpoint, as intensity of management and utilization increases, there is an increased need for sensitive nutrient monitoring to determine whether ameliorating treatments become necessary to maintain forest fertility. This will probably be particularly important in young stands in which there is generally a greater potential loss of nutrients in the changeover from conventional to whole-tree harvesting than will be the case with older stands (Kimmins, 1977; Armson, 1977).

Monocultures and Productivity Decline Similar to agriculture, it should be expected that the continued harvesting of a single crop, especially with short rotations and high levels of cultivation, should potentially result in a depletion of the nutrient capital of a site. Further, it should be expected that evidence of nutrient depletion should be noticed first on lower-quality sites with a more limited nutrient base. Rapid loss of growth in forest crops has occurred where plantations have been established on areas which have been subsequently shown to be deficient in a particular element, particularly a trace element. What we are concerned about here, however, is the question of whether there has been a decline of productivity following intensive management of conifers on apparently suitable forest sites.

There are two prime examples which are purported to illustrate productivity decline, but both of these raise questions as to whether the actual decline was due directly to the excessive nutrient demands of monocultures. The first commonly quoted example is the planting of monocultures of spruce in Saxony in Germany. Here, after apparently successful early growth, the plantations declined in productivity. This case, however, is believed to have resulted from the off-site planting of spruce in areas that were not suitable for spruce growth. In Europe there is no documented case of decline in productivity following intensive management of conifers (Professor D. Mülder and Director E. Holmsgaard, personal communications).

The second example of decline in growth is in some Southern Hemisphere countries where exotic conifers are managed intensively. In south Australia where *Pinus radiata* is grown on low-quality sandy soils, and in South Africa (*Pinus patula*), there are examples where the second rotation of pine is performing at one site class lower in level of productivity than the first. These examples all occur on low-quality sites. On plantations that have been established on medium- and good-quality sites, second-rotation productivity is equal to or better than the first rotation. The problem is complicated and the reasons for the lowered productivity on some low-quality sites are not clear. It does not seem to be a simple case of lowered nutrient capital alone but is probably a complex of this plus additional factors that have modified the soil's physical, chemical, and biotic characteristics following the removal of native vegetation and the planting of conifers (Bednall, 1968; Keeves, 1966; Robinson, 1973). Difficulties in growth comparisons between rotations are made more difficult by gradual improvement in establishment techniques. Improved genetic stock, control of weeds, and use of fertilizers all tend to give enhanced growth rates in current plantations.

CONCLUSIONS

Forest ecosystems are normally in a steady state, where, once the foliar biomass of the stand has maximized, the amount of nutrients taken up by the vegetation approximately equals the amount returned by leaf fall and mortality. The inputs to the system from precipitation and mineralization approximately equal the losses through deep seepage and leaching. All silvicultural practices will tend to interrupt this balance, and the task of the silviculturist is to predict the likely magnitude of the interruption. The greater the change to vegetative cover, the more likely will be the change in the nutrient balance. The higher the level of utilization, the more nutrients are exported from the site and the more likely will be the need for replacement of nutrients by fertilization.

The potential problem is most acute on low-quality sites, and it is here that the silviculturist must pay considerable attention to the likely impact of intensive management practices. On high-quality lands, providing vegetation is either maintained or rapidly replaced and no erosion losses occur, the impact of management of conifers on a 30- to 50-year rotation on the nutrient base will probably be small. However, our knowledge of the long-term effects of vegetation manipulation on the nutrient capital of sites is limited. The situation needs to be more carefully assessed as intensity of management increases.

Chapter 11

Site Quality Evaluation

Productivity estimates for wildlands have been made primarily in terms of single renewable natural resources such as timber, forage, water, and game. Because of the lack of information on total productivity of an area and on tenable balanced relationships among the renewable resources, a concentrated effort is demanded in an analysis of all the components of an ecosystem.

Productivity of forest lands is largely defined in terms of *site quality*, which is measured by the maximum timber crop the land can produce in a given time. Within a given macroclimate, site quality determines the kind and magnitude of problems and opportunities that a forester has in managing a stand. Site quality is an essential consideration when stands are managed for various combinations of possible forest products: timber, water, forage, recreation, and game. No valid silvicultural decision can be made without reference to site quality and other site conditions.

Site quality is the sum of many environmental factors: soil depth, soil texture, profile characteristics, mineral composition, steepness of slope, aspect, microclimate, species, and others. These factors, in turn, are functions of geologic history, physiography, macroclimate, and successional development. Pro-

Figure 11-1 White-pine stumps indicate the productivity of a former northern hardwood site that was converted to pasture. The area grows poor-quality jack pine now that the soil has lost its humus and reverted to fine sand (near Cadillac, Michigan). (*By Daniel, 1949.*)

ductivity potentials develop slowly in areas with residual soils and more rapidly on sites subject to wind- or water-borne deposition of soil particles.

In contrast, productivity can be rapidly lowered by poor management, and particularly by various kinds of erosion. Aspen in Minnesota loses a full site class with repeated burns on soils with 10 to 60 percent silt and clay (Stoeckeler, 1960). A broad outwash plane of sand near Cadillac, Michigan offers an example of how rapidly site quality can be reduced. This area was covered in 1949 with eastern-white-pine stumps that measured 24 to 30 in in diameter (Fig. 11-1). The stumps of the hardwood components of the northern-hardwood type that had previously occupied the area had rotted rapidly and had been removed in the late 1800s to allow easier cultivation. After about 60 years of cultivation and wind, the soil profile had lost the accumulated organic material that had helped retain a rich mineral supply and high water-holding capacity; by 1949 the soil was completely impoverished. A young plantation of jack pine in the area had only stunted, twisted trees because of this extreme reduction in site quality.

Since site quality is measured by the maximum timber crop (volume) produced within a given period, it can vary with tree species and the time element chosen. Thus, if the period is held constant at 100 years, then a particular area could have a different site quality depending on whether it supported Douglas-fir, western hemlock, or western red cedar. The problem of the forester, therefore, is to determine the site quality of a particular area relative to a certain tree species. The area may have a pure or mixed composition and be even-aged or uneven-aged and partially or fully stocked. Yet the standards used to define site quality (normal yield tables[1] for different species) provide data based on

[1] Normal yield tables are derived from unmanaged stands and represent net volumes for any particular age, in contrast to managed-stand yield tables, which are based on the sum of thinnings plus residual stand.

pure, even-aged, normally stocked stands. Unfortunately, such stands are rarely encountered, so an indirect method for estimating site quality from a yield table was developed: site index.

The indirect approaches for estimating site quality have been divided into three categories (Jones, 1969): site index, vegetational, and environmental. It is the purpose of this chapter to examine these methods and briefly analyze their strengths and weaknesses. The ultimate criterion for an indirect method is how effectively it can be translated into an accurate estimate of site quality. In other words, no matter how well an indirect method may reflect variation in the environment, its worth depends on its being convertible to an accurate estimate of the maximum size of the timber crop at a given age.

DIRECT METHOD OF SITE-QUALITY EVALUATION

A precise method of determining site quality is to grow a fully stocked stand of the desired species on a site for a designated period. This is what the Europeans have done to produce their latest standards or yield tables. Obviously, this procedure could not produce guidelines for American foresters for many years. An alternative is to measure fully stocked natural stands. This was done, but to simplify the interpretation and use of the resultant tables, stands were used only if they were pure, even-aged, and fully stocked. Of these criteria, only full stocking was subjective. For species with wide distribution, it was more practical to develop a yield table for a limited portion of the range, i.e., yield of Douglas-fir in the Pacific Northwest (McArdle et al., 1949). For any species, stands of all age classes from the whole range of site-quality classes were needed as well as a sufficient number of stands to give a substantial base for interpretation. To ensure locating sufficient numbers of stands within a reasonable amount of time, the criterion for full stocking was not maximum stocking but a full canopy of normally developed crowns, which could include small holes (Meyer, 1937).

In each selected plot, every tree above a specified minimum diameter at breast height (DBH) was measured for diameter and height. The volume of timber on an acre and the many stand values that constitute a normal yield table were then calculated from these data. A series of anamorphic curves was fitted to the data, such as for volumes (Fig. 11-2). Data thus derived have provided today's silviculturist with knowledge about the growth and development of a stand, and about the relationships that establish the principles of stand development as affected by site quality.

When the direct method is used on a particular stand, its characteristics must match those of the stands used in the construction of the standard. Even if such a stand existed where needed, it would be costly to make the necessary volume measurements if the only purpose was to determine site quality. A number of technical difficulties are also inherent in the way the volume-age curves were constructed. Fitting anamorphic curves to the volume data assumed the same growth patterns on all sites. In addition, there is no allowance for the prob-

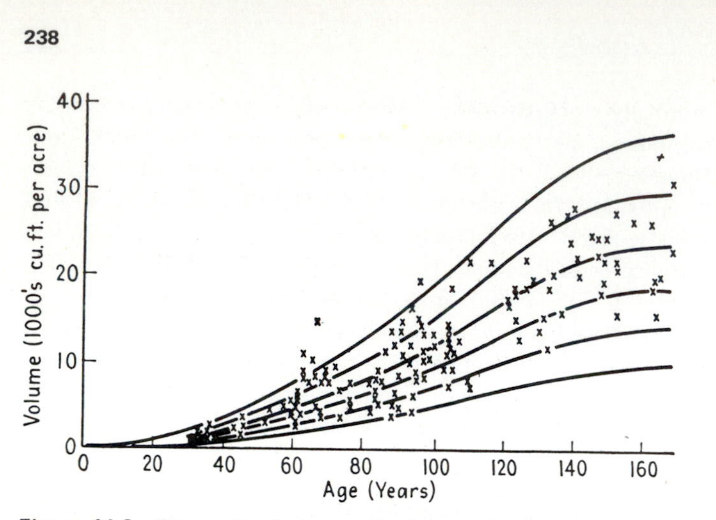

Figure 11-2 Site quality in terms of cubic-foot volume growth at a particular age with anamorphic curving.

able imbalance among younger stands which occur in disproportionate numbers on the best sites and older stands which occur more frequently on the poorer sites. This may have resulted because early loggers worked the best and most accessible stands first, or because young growth on the best sites was logged off at an earlier age. In addition, yield tables are based on the average of the stands measured. A site with a stand approaching maximum productivity at a particular age could therefore be classified under the direct method as being of better quality than its actual production might justify.

INDIRECT METHODS OF SITE-QUALITY EVALUATION

Site-Index Approach

Site-Index Method The data collected from each of the plots used in constructing a normal yield table always includes the age of the stand and the total height of each tree. The authors of the various normal yield tables have used the average height of dominants or of dominants and codominants in determining the site index. *Site index* is the height to which a tree will grow under forest conditions in 100 years for most old-growth Western species, or in 50 or 25 years for Eastern and Southern species and young-growth stands managed on shorter rotations. [*Site index* has also been expressed as the height of a dominant tree at a particular diameter (McLintock and Bickford, 1957; Pflugbeil, 1960). This designation is preferred when age is meaningless because the trees have had a long period of suppression, as in uneven-aged spruce stands.] The average height of all trees has not been calculated because it would be too easily affected by the removal of smaller trees in low thinnings, by differences in stocking, and by fire. Average ages have generally been based on the age at breast height, although some authors add an estimated age to grow to breast height to give a total age. The data on age and height for all the sample plots provided a basis for another family of anamorphic curves from which site index can be read. The typical yield table combines the site index with age in the analysis of the basic plot data on volumes, number of trees, distribution of trees among the diameter classes, basal area, and average stand diameter.

Site index is a more sensitive indicator of site quality than the yield table expressions of volume productivity because the height of a dominant tree at a particular age represents the full height attainable. In contrast, the normal yield table volume represents only the average volume of what has been found at one period in time for many stands with a particular site index. In addition, the site index can be quickly and economically determined for a species in a mixed, uneven-aged, or poorly stocked stand. It is essential, however, that trees measured for site-index determination have been dominant throughout their lives.

Yield table site indices have been developed usually by anamorphous curve fitting of height-over-age data, which assumes that the form of the height-growth curve is the same on all site-quality classes. Polymorphic curves developed from

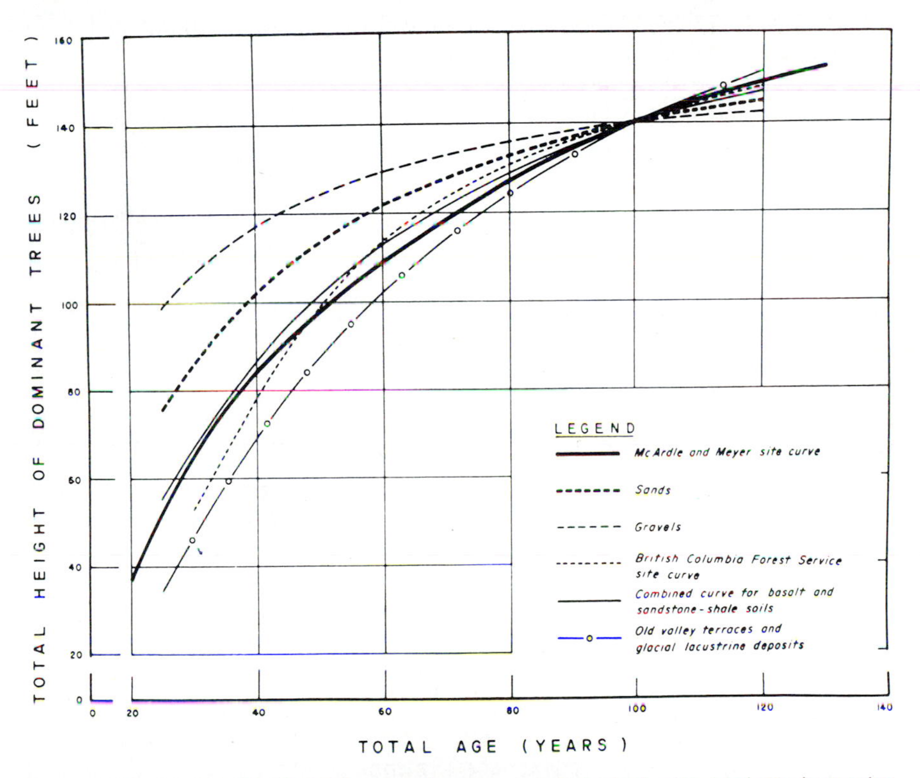

Figure 11-3 Douglas-fir (SI 140) curves for young, well-stocked, even-aged stands growing on major soil groups of southwestern Washington. The soil-site curves express the change of tree height with age when the site index is held constant. (*From Carmean, 1956.*)

stem analysis (Beck, 1971*a*; Beck and Trousdell, 1973) indicate that height-growth curves do not have the same form on all sites. He found that a difference in site index of 14 to 20 ft could exist between anamorphic and polymorphic curves of eastern white pine if the site index was determined at 20 years. Even where the heights of trees on two areas are the same at the site-index age, the trees have not necessarily arrived at that height by following the same growth pattern. Carmean (1956) found that trees on sands and gravels grow rapidly at first and more slowly later, while trees on imperfectly drained soils grow more slowly initially but do not slow down appreciably with age (Fig. 11-3). The effects of soil texture, slope position, and aspect on the site index are illustrated by the growth patterns derived from stem analyses of black oak on site index 60 in southeastern Ohio (Fig. 11-4) (Carmean, 1970). These sources of error in site-index determinations raise the question of the applicability of a set of regionwide site-index curves that incorporate data from the whole range of sites over which the species occurs.

The effect of stand density on height growth (i.e., the effect on site index) depends on the species. An awareness of density effects on a particular species' height growth would be necessary for an adequate interpretation of site-index measurements (Lynch, 1958; Alexander et al., 1967).

Applying the site-index method to a species requires the presence of that particular species on an area, since any site may have several site qualities depending on which species is being considered. If the species of interest is absent

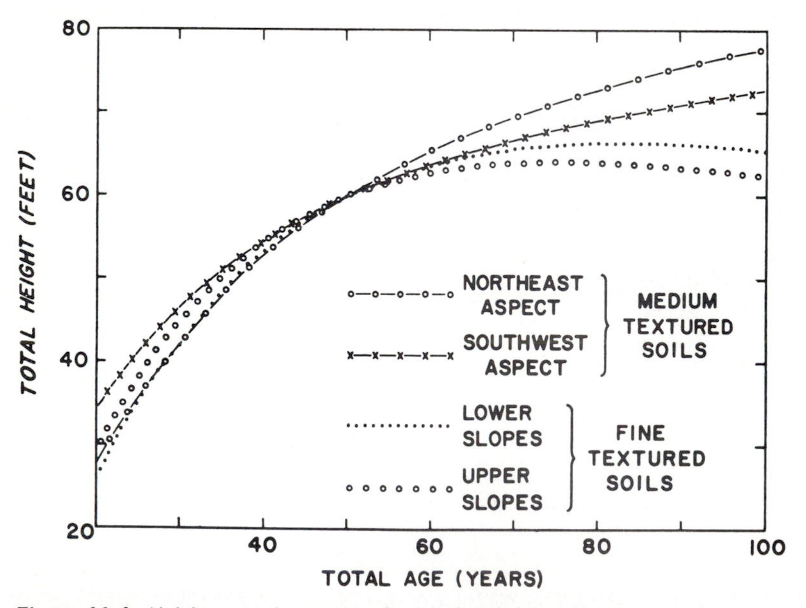

Figure 11-4 Height-growth patterns for black oak in southeastern Ohio show differences due to aspect and slope position. (*From Carmean, 1970.*)

from an area, its site index may be predicted from the presence of a second species of tree, if the relation of their site indices has been established (Carmean and Vasilevsky, 1971).

Intercept Methods Because most normal yield tables make no effort to predict site quality for stands younger than 20 years, site-index determinations for young stands have been poor. The trend toward short rotations, such as the 25-year rotations in the South, prompted development of the *5-year intercept method* (Ferree et al., 1958; Wakeley and Marrero, 1958), which uses the total length of the first five annual internodes above DBH. The 5-year intercept method has a number of advantages for use relative to crowded young stands, especially if the height-age relationship is obscured by pre-breast-height growth hazards (browsing, suppression, and insects) or by differences in planting stock and planting techniques. The method does not require a knowledge of the age of the stand, and it is easier to use in closed stands than total height. It has the disadvantages of being influenced by the effects of short-term climatic fluctuations or of not recognizing sites where the early growth rate does not correlate with later growth rates. The measurement of the 5-year intercept from the first node above 8-ft height, instead of above breast height, doubled the accuracy with which the site index was estimated for red pine (Alban, 1972). It has been used as a basis for site classification rather than in determining site quality. Thus the method does not allow direct conversion from index to productivity unless a relationship between the 5-year intercept index and site index can be established (Beck, 1971*b*; Alban, 1972).

The *polymorphic site-curve method* (Stage, 1963) is based on the intercept that represents the growth for the 10 years after the dominant tree has reached a certain height (such as 55 ft). The rationale is similar to that of the 5-year intercept method, i.e., site quality is a function of the rate of growth of a tree at a particular height. However, it assumes that the growth rate in the upper portion of the height-age curve is a better indicator of site capability than height at a particular age. Site index required a new definition according to this method, i.e., "sites will be rated according to the increment attainable by a dominant tree of a standard height." Stage used data from the stem analysis of the tallest tree on each of 66 plots, and the families of site-index curves were generated from a formula for the shape of the curve. Each family of site-index curves was based on a particular degree of suppression in the early life of the trees from which the family curves were derived. The degree of suppression was determined from the number of rings at DBH in a 1.5-in radius from the pith and was used as an index to a family of curves. A family of site curves was constructed for variations in the number of rings in 1.5 in from the pith when N (the number of rings) was unknown, 10, 15, 20, and 25. While the site-index conforms to the height of the tree at the index age if the degree of suppression is known or is unnecessary, the height may be greater or less than the site index at the index age depending on the degree of suppression. The method was developed for grand fir, which is a tolerant species, and it provides a means for analyzing the dynamics of uneven-aged stands, as well as other stands in which suppression is prevalent.

Vegetational Approach

Early folk knowledge recognized that certain plants occur in reasonably definite local environments; thus, as a corollary, one can draw conclusions about the environment from the plant cover. The weakness of folk knowledge built up in one environment and transferred unmodified to a different area is demonstrated by the pioneers in their western migration. They erred in believing that the high-yield conifer lands of the Lake states and Pacific Northwest were as good for agricultural crops as the high-yield hardwood lands of the East. A similar weakness afflicted attempts to directly transfer the use of ground cover in site-quality evaluations from cold to warm latitudes. No account was taken of the fact that species in the cold latitudes have rather narrow ecological amplitudes (species can grow only in a limited range of situations), while species in the warm latitudes have wide ecological amplitudes. At all times, however, vegetation has been recognized as useful in delimiting site changes (Farrar, 1962) or setting boundaries of certain physical conditions. The difficulties have been encountered in trying to assess site quality from the species in the plant cover. The history of the vegetational approach centers on a very successful pioneer study in Finland. Early attempts to apply this approach to other areas failed, until it was recognized that the *principle* was universal but the *method* had to be adapted to local conditions. Discussion of this pioneer work and some successful adaptations of the principle will illustrate how well vegetation can be utilized to establish the site quality in an area.

Finnish Forest-Site Types The first to recognize and classify the relations between plant cover and site productivity for tree growth was Cajander (1926) in Finland in 1909. His method assumed that the presence of certain climax species in the ground cover (i.e., forest-site type) under a mature stand indicated the site quality. When certain species are consistently present in association with a certain site quality and lacking on any other site, they are called *indicator species*. The main criticism of the method, even in Finland, is that ground cover changes when its overstory is removed. This change is not serious in northern Finland, but some difficulty in interpretation occurs in southern Finland because of a richer ground flora and more species of trees on deeper soils. A major criticism has been that ground vegetation is affected only by surface soil conditions, while trees are influenced by the greater depth of root penetration in deeper soils in warmer climates. However, the climax species originally observed do remain evident unless fire or serious disturbance eliminates them, and the differences in rooting depth are apparently not significant enough to influence

Table 11-1 A Partial Set of Forest-Site Types and Their Productivities on Upland Soils within Different Regions of Finland
(Annual growth figures in solid cubic meters without bark per hectare are calculated from the results of the National Forest Inventory)

General characteristics of site and vegetation	Site	Forest-site types		
		South Finland*	Central Finland	North Finland†
Driest and most barren upland site, usually coarse sand	V	*Cladina* type 1.4 m^3	*Cladina* type	*Cladina* type 0.4 m^3
Moderately dry site, sandy to morainic soil	III	*Vaccinium* type	*Empetrum-Vaccinium* type	*Empetrum-Myrtillus* type
Most fertile upland site, with more humus	I	*Oxalis-Myrtillus* type, 5.1 m^3	*Geranium-Oxalis-Myrtillus* type	*Geranium-Myrtillus* type, 1.7 m^3

*Region 1, most southerly.
†Region 10, most northerly.

the forest-site-type classification. Cajander based his method on ground cover only, without concern for the soil type, because he found widely varying ground-cover types (forest-site types) on the same soil type and the same ground-cover type on different soil types. Some of the Finnish upland forest-site types that have been categorized (and their productivities) are given in Table 11-1. The table illustrates that the same relative site quality (Site III) supports different species, depending on locality, and that its productivity also differs. The necessity for evaluating the plant community rather than merely noting an indicator plant is indicated in the table by the presence of the same species (*Myrtillus*) in two forest-site types in the same region (Sites I and III in northern Finland). A complete list of the Finnish forest-site types would reemphasize the need for evaluation of the community of plants in deciding on site quality. Thus it is not just presence or absence of an indicator species that determines each forest-site type, but the amount of imbalance among the species in the site community. Inventories, taxation of forest lands, and even bank loans are based on the productivity of the forest sites, which are classified according to forest-site type as determined in the field.[2]

Forest-Site Types in North America Various studies have shown that Cajander's ideas are applicable in Canada and Northern United States, but the accuracy of the site evaluations have commonly proved uncertain. One study

[2]Personal communication from Professor P. Mikola, University of Helsinki, Finland, December 1971.

(Spilsbury and Smith, 1947) was quite successful in identifying forest-site types in British Columbia and the Pacific Northwest for young-growth Douglas-fir. Their five forest-site types, based on dominance and codominance of only a few species (*Polystichum*, *Achylys*, *Oxalis*, *Gaultheria*, and *Berberis*) (Fig. 11-5), correlated very well with site index and site quality in various-aged stands. It is the relative abundance and vigor or dominance of one species over the others under various conditions that is significant in classifying a site and not the frequency of occurrence. The species within the plant communities representing the forest-site types changed in numbers, dominance, and development in a continuous fashion, so the number of plant communities could have been infinite (Fig. 11-5). The forest-site types and the site qualities they represent are given in Table 11-2. Spilsbury and Smith's dominance principle was adequate in many instances for determining the forest-site type. When the dominance concept could not differentiate between two possible forest-site types, other species whose presence or absence was decisive to a given type were evaluated. Thus, within the particular climatic province for which Spilsbury and Smith developed their forest-site-type classification, where a few dominants and codominants played decisive roles, the entire community of ground-cover species had to be noted in determining the forest-site type. In the limited area in British Columbia to which it was applicable, it had the advantage of being useful with a limited amount of training. It has been superseded by Krajina's much more complex system because Krajina's system can be applied to the whole province.

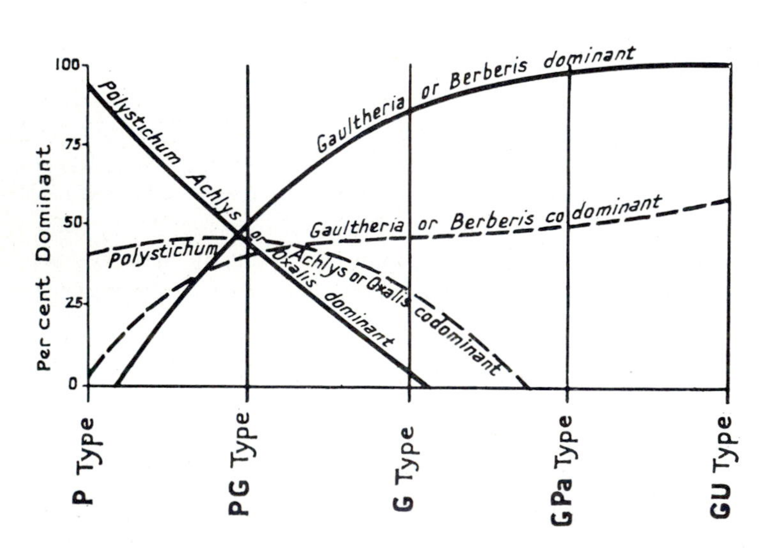

Figure 11-5 Forest-site types for young-growth Douglas-fir in British Columbia and Washington based on the dominance species with P type about equivalent of Site I and GU type between Sites IV and V by volume according to U.S. Forest Service yield table. (*From Spilsbury and Smith, 1947.*)

Table 11-2 Forest-Site Types for the Coastal Douglas-Fir Type in the Pacific Northwest

	Forest-site type	Average site index, ft	Production, ft^3
P	Sword fern (*Polystichum*)	170	19,000
PG	Sword fern–salal (*Polystichum-Gaultheria*)	150	15,000
G	Salal (*Gaultheria*)	125	11,000
GPa	Salal–"pale green" lichen (*Gaultheria-Parmelia*)	95	8,000
GU	Salal–"bearded" lichen (*Gaultheria-Usnea*)	70	6,000

Daubenmire's Habitat Types An active program of habitat typing has been inaugurated over most of the Rocky Mountain region based on Daubenmire and Daubenmire's (1968) habitat types (Table 11-3), which include dominant cover as well as ground cover. A *habitat type* is an aggregate of land areas having the same potential for growth. This potential is indicated by its supporting a single-climax plant-community type which is an amalgam of the compositions of a number of climax plant communities. A habitat type is identified by a binomial made up of a series species (the climax tree species on an area) and the dominant or characteristic species of the ground cover in the climax-community type. This land-classification system provides the information needed for silvicultural and management decisions. An early comparison of habitat types with site indices [interpolated from the ponderosa-pine yield table (Meyer, 1938)] indicated that habitat-type designations had very little value for predicting growth rate of ponderosa pine. However, if the habitat types were plotted against the raw data of actual ages and heights of dominants, the habitat types formed four groups on the basis of ponderosa-pine height growth that were in agreement with productivity (Table 11-3). Site-index values from Meyer's curves diverged greatly from site indices determined by stem analysis, which in turn were in close agreement with predictions made from habitat types (Daubenmire, 1961).

In addition to being well correlated with growth, habitat types provide a basis for estimating regeneration, invasion by other species, and disease problems. *Pinus ponderosa-Purshia* and *Pinus ponderosa-Agropyron* habitat types usually have dwarf-mistletoe infestation and regeneration problems. High-potential growth sites such as *Abies grandis-Pachistima*, *Pseudotsuga-Physocarpus*, and *Pseudotsuga-Calamogrostis* represent areas where ponderosa-pine stands are subclimax and where Douglas-fir invasion of ponderosa-pine stands is usual.

Daubenmire's habitat types were developed from an analysis of the taxonomic composition of climax stands. The habitat types could then be recognized under disturbed conditions resulting from fire, logging, and insects. Disturbances do not normally eliminate all the species that are characteristics of a given habitat type. Therefore, despite an abundance of seral species, an analysis of the residual

Table 11-3 Daubenmire and Daubenmire's (1968) Habitat Types for Northern Idaho and Eastern Washington

Series	Habitat type	Ponderosa pine growth potential
Abies lasiocarpa	*Abla/Pinus albicaulis*	–
	Abla/Vaccinium scoparium	–
	Abla/Xerophyllum tenax	–
	Abla/Menziesia ferruginea	–
	Abla/Pachistima myrsinites	–
Tsuga mertensiana	*Tsme/Xerophyllum tenax*	–
	Tsme/Menziesia ferruginea	–
Thuja plicata	*Thpl/Athyrium filix-foemina*	–
	Thpl/Oplopanax horridum	–
	Thpl/Pachistima myrsinites	–
Tsuga heterophylla	*Tshe/Pachistima myrsinites*	–
Abies grandis	*Abgr/Pachistima myrsinites*	I
Pseudotsuga menziesii	*Psme/Symphoricarpos albus*	–
	Psme/Physocarpus malvaceus	II
	Psme/Calamagrostis rubescens	III
Pinus ponderosa	*Pipo/Symphoricarpos albus*	II
	Pipo/Physocarpus malvaceus	II
	Pipo/Purshia tridentata	IV
	Pipo/Stipa comata	–
	Pipo/Agropyron spicatum	IV
	Pipo/Festuca idahoensis	IV

climax species can provide a basis for determining the habitat type. Every habitat type has many climax species characteristic of it, and with few exceptions, the same species are also characteristic of several habitat types. Species that are restricted to one habitat type may not always be present, and the habitat type must then be determined from the associated species. Thus, while the characteristic species of a habitat type may be called indicator species, it is again the analysis of the community of plants (overstory and ground cover) that is the determinant in establishing the habitat type.

In a field examination of soils in relation to habitat types, Daubenmire (1952) found that the same habitat type occurred on several soil series. However, several habitat types noted in one area were found to be on the same soil series. Habitat-type mapping appears to be a satisfactory method of classifying land according to its growth potential for timber and helps meet the needs of decision makers. However, in applying the habitat-type method beyond the area where it was developed, the relationships among the characteristic species undergo gradual changes. Thus Daubenmire's habitat types are not directly transferable to any considerable distance, but the idea and technique have been. Habitat typing has spread from Daubenmire's original area into Montana (Pfister et al., 1974), central Idaho (Steele et al., 1975), southeastern Idaho (Steele et al., 1974), Wyoming (Steele et al., 1977), northwest and northeast Utah (Henderson et al., 1976, 1977), and will be extended to include the whole of the Rocky Mountains. The classifications have been accepted as the basis for management decisions and compartmental analyses. In addition, habitat types now provide the basis for inventory surveys. Some efforts have been made to carry the system into West Coast forests, but no habitat types have been developed for specific areas. It should be noted that two so proximal areas as the Uinta Mountains of northeastern Utah and the Wasatch Mountains of northern Utah and southern Idaho have only three habitat types in common. This indicates that a habitat type integrates the environmental factors in a rather narrow range, thus providing an excellent basis for determining the potential productivity of the site. Considerable field application and research are necessary before the potential growth of a habitat type can be given quantitative values, but the silvicultural implications for regeneration and competition problems are quickly associated. Approximate quantitative values for productivity have been developed for Montana habitat types.

Hodgkin's Vegetal Site Index The preceding examples (forest-site type or habitat type) of determining site quality are categorized in terms of the climax species which are characteristic of the site. These methods frequently depend on a visual analysis of a plant community's components, which could inject subjectivity into naming the type. In addition, the procedures were developed in the cooler latitudes which have the advantage of ecological amplitudes of species that are on the whole rather narrow.

The value of the plant community as an objective basis for site-quality determination has also been demonstrated by the development of a vegetal site index for longleaf pine (Hodgkins, 1960) in Alabama, where there is the disadvantage of numerous plant species and wide ecological amplitudes. The species used in calculating the vegetal site index were those available on a year-round basis. In listing potential indicator species, the initial 13 sampling plots were inventoried and each species within a plot was given a dominance rating. The inventory of a plot included not only the plants within the well-stocked, even-aged, uncut stands of longleaf pine from which the site index could be obtained, but also the plants (trees and understory species) in the openings and adjacent cutover areas. A mean site index was calculated for each species, which were then arranged in the order of their mean site index along the ordinate of dry to wet. After the mean site indices were adjusted according to the known site index of a plot, a vegetal site index could be calculated from the list of species on each plot and their dominance, as shown in Table 11-4 (Hodgkins, 1960). The standard error of estimate of the vegetal site indices of the initial plots was ±2.61 ft, which was improved by later evaluations of additional plots.

Table 11-4 Calculations of a Plot Site Index Using a Trial Plant-Indicator Scale for Longleaf Pine

Indicator species	Site index		Cover abundance rating							Coded index X cover abundance rating
	Actual	Coded	1	2	3	4	5	6	7	
Quercus laevis		1								
Chrysobalanus oblongifolius		2								
Quercus incana		3								
Tephrosia virginiaia	60	4								
Quercus stellata v. marganetta	62	5								
Ceanothus americanus	64	6								
Schrankia microphylla	66	7								
Cornus florida	68	8	X							8
Quercus marilandica	70	9	X							9
Quercus falcata	72	10		X						20
Quercus stellata	74	11		X						22
Andropogon tener	76	12			X					36
Axonopus afinis	78	13		X						26
Prunus serotina	80	14		X						28
Sorghastrum nutans	82	15	X							15
	15.3									
Liquidambar styraciflua	84	16			X					48
Ilex glabra	82	17					X			85
Nyssa sylvatica	80	18				X				72
Quercus nigra	78	19			X					57
Rhexia spp.	76	20		X						40
Magnolia virginiana	74	21								
Arundinaria tecta		22								
Aronia arbutifolia		23	X							23
Xyris spp.		24								
Number of *X*'s times column head			4	10	9	4	5			
Total										489

$4 + 10 + 9 + 4 + 5 = 32$. Mean coded site index $= \dfrac{489}{32} = 15.3$

Site index = 83

Source: Hodgkins, 1960.

Later application of the developed vegetal site index to a different physiographic province showed much less agreement with the site index despite the floristic similarity of the two physiographic provinces. In other words, a vegetal site index is directly applicable only to the area conditions within which it is developed. Each change in province requires a reevaluation of the relationship between species and site quality.

The vegetal site-index method has primarily demonstrated that if properly interpreted, the vegetation can reflect site quality of an area even in the warmer latitudes.

Environmental Approach

A variety of causes directed research into establishing environmental bases for estimating site quality. Site-index determinations required the availability of height-age curves for a particular area and species, and areas without suitable trees could not be classified. The vegetational approach, with its dependence on indicator plants, had not proven as satisfactory elsewhere as Cajander had found it in Finland. Indicator plants were useful in the northern countries, including Canada and Northern United States. Farther south, however, greater insolation, evaporation, surface drought, and human disturbance, plus more species and a greater ecological amplitude per species, made correlation of site quality with a few indicator species improbable. The inability of indicator understory vegetation to provide reasonable estimates of site quality (Coile, 1938) led to efforts to explore the use of locally significant physical factors in an environment.

Soil-Site Index Site quality within a given macroclimate is associated with the soil's capacity to provide moisture and nutrients. A voluminous literature has developed concerning the evaluation of relationships between site-index and soil-site variables such as parent material, profile characteristics, slope, and aspect. On some limited areas, estimates of site index based on soil and site factors have been able to account for only 50 to 60 percent of the total variation (Covell and McClurkin, 1967). The various prediction formulas have had standard errors of between ±4 to 9 ft for the average site index. This means that even at the point of highest accuracy (i.e., the average site index), only 67 percent of the determinations fall within the limits of the standard error (Hodgkins, 1959).

Two soil-site studies done in one region did not agree on the relative significance of various soil characters. Coile and Schumacher (1953), working with loblolly and shortleaf pines in the Piedmont region of the Carolinas, Georgia, and Alabama, found that the depth of the A horizon and the imbibitional water value of the subsoil (B horizon) had a highly significant correlation with site index. Position on the slope was found to have no significant relationship to site index apart from its effect on the soil characteristics. Their results were presented in a table: one axis consisted of five subsoil classes based on consistency when moist, and the second axis was the depth to subsoil. The standard error of estimate was 11 percent of the calculated site index for loblolly pine and 12 percent for shortleaf pine.

In contrast to the preceding study, Della-Bianca and Olson (1961), working with hardwood and pine-hardwood types also in the Piedmont region, found no significant correlation between site index and either depth of A horizon or imbibitional water value of the B horizon. Their most significant factor was position on the slope, but they were unable to develop a prediction formula. The two studies do not necessarily invalidate each other, since the order in which a factor is evaluated in a regression analysis may easily change the significance of that factor. Tests (Hodgkins, 1956) showed a poor relationship between earlier soil-site regressions and site index because the studies had attempted to cover too broad a range of conditions. The tests indicated that a directly derived site index was the best measurement of site quality, and that the most useful alternative to this was a site-index measurement from a comparable local site.

The soil-site index studies have improved understanding of the ecological interactions of all site factors, as the following studies illustrate. Factors are generally shown to be significant because of their relationship to water availability during drought periods (White, 1958; Zahner, 1958). Slope, aspect, and textural and depth differences primarily reflect a soil's capacity to provide moisture. At the same time, mottling, texture, and surface drainage reflect limited aeration in wet soils. Increased elevation, with its usual concomitant increase in rainfall, may favor increased site quality until the rainfall exceeds the capacity of the soil profile to retain it for use in a critical period (Gessel and Lloyd, 1950). Individual soil series, the mapping units of the soil scientist, can incorporate a wide range of site indices and soil-site indices depending on variations in soil depth, rainfall, and elevation (Zinke, 1959). While most reports emphasize the critical aspects of the moisture supply for its effect on site quality, one should remember that the moisture-supplying capacity of a soil is usually highly correlated with its nutrient level.

Rarely have large areas been mapped in soil-site index research. Steinbrenner's mapping of 3 million acres in Western Washington and Oregon was an exception. He developed a site-prediction method (unpublished) based primarily on soil that accounted for 83 percent of the variations in site index. His method included as significant factors the depth of the A horizon (providing information on nutrient conditions), effective soil depth (60-in soil depth with 50 percent rock equals a 30-in effective depth), texture of the B horizon (reflecting water-holding capacity), and elevation (reflecting amount of rainfall). He correlated site quality with landforms so he could use aerial photographs for mapping site quality, and then he checked the maps along road cuts and with special check surveys. He has also developed an unusual site map—one that designates areas according to profit potential.

When soil-site index is adequately correlated with site index, site-quality classes can be mapped with unprecedented speed. The failures of the soil-site index method appear to stem from (1) weakness in the site-index curves with which correlations are made, and (2) the incorporation of data from too many physiographically different areas.

Physiographic Site Types The physiographic site type as developed in Ontario, Canada was conceived as an ecosystem (Hills, 1953, 1958, 1960, 1962;

Farrar, 1962) within which productivity was uniform. Differentiation among various physiographic site types was based on factor gradients within a fairly uniform macroclimate or "site region." Within each site region, any differences between site types as to productivity were attributed to differences in temperature, soil moisture, soil fertility, and exposure to drying. Moisture characteristics were defined according to topographic position, parent material, depth of soil, depth of water table, and profile characteristics. The moisture gradient within a site region could often be established from topography, with accuracy improved by a check of the profile, so aerial photographs and geology were utilized. Local temperature differences and potential for evapotranspiration were recognized by vegetation changes. In mapping a site region, the number of different physiographic site types was held to 12 to 17.

The basic approach to developing physiographic site types is to simultaneously develop the physical and biotic factors in a site region. The various physiographic site types within a site region will differ in their productivity, but no estimate of site quality is inherent in the establishment of specific site types. After a site region has been site typed, a special survey of productivity would seem advisable. Comparable physiographic site types in two site regions do not necessarily bear any relationship to each other since each is based on the situation within its own region.

Ordination Ordination is a method of investigating the responses of vegetation to factors in the environment. Each factor is considered to be a continuum, i.e., a smooth transition from hot to cold or from dry to wet rather than a series of discrete conditions, e.g., hot, very warm, warm, cool, cold. *Ordination* is defined as "an arrangement of units in a uni- or multi-dimensional order" (Goodall, 1954), as opposed to a classification "in which units are arranged in discrete classes."

The method is best understood by the use of examples such as the one described in the vegetal site index developed by Hodgkins (Table 11-4). The plant species in a vegetal site index is the unit referred to in Goodall's definition. These species are arranged in a unidimensional order according to the moisture condition of each one's optimum competitive site, e.g., from dry to wet on a scale of 1 to 24. If the plants had been arranged in discrete classes [e.g., six classes: dry (1 to 4), moderately dry (5 to 8), below-average moisture (9 to 12), above-average moisture (13 to 16), moderately wet (17 to 20), and wet (21 to 24)], the method would have lost much of its sensitivity. A two-dimensional order could be developed by arranging plant species in the order of nutrient conditions as well as moisture for each species' optimum competitive site. The vegetal site index would then have had to be determined by some function of the two axes.

A two-dimensional order of site factors was used for determining the site index of balsam fir in Minnesota (Bakuzis and Hansen, 1962). The fre-

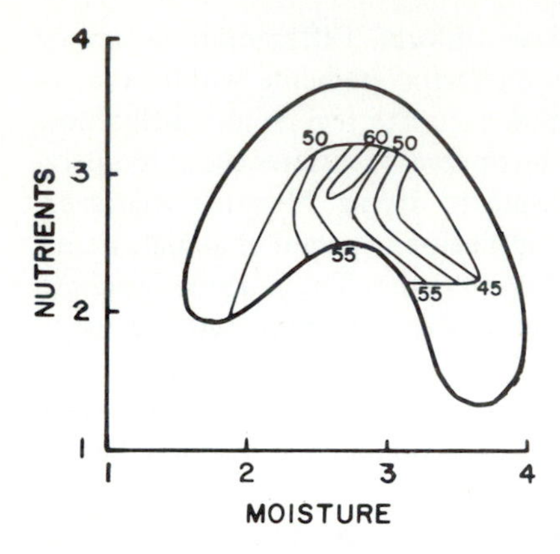

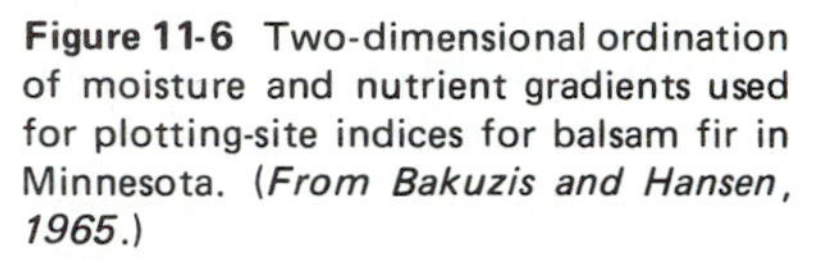

Figure 11-6 Two-dimensional ordination of moisture and nutrient gradients used for plotting-site indices for balsam fir in Minnesota. (*From Bakuzis and Hansen, 1965.*)

quency of occurrence of the species along the moisture and nutrient gradients established the effects of each factor. Within the two-dimensional figure, the iso-site-index boundaries could be plotted (Fig. 11-6).

Krajina's Biogeocoenoses Krajina came to British Columbia in 1949 after years of teaching at the Charles University in Czechoslovakia; so he brought a wealth of experience in the application of Braun-Blanquet's plant sociology coupled with Russian plant community and soil terminology to the study of the ecology of the province. In British Columbia he had what was lacking in Europe—the availability of climax vegetation—so his system of ecological classification of forest sites has an original interpretation from that of the European Braun-Blanquet schools. The basic unit of his system (Krajina, 1972) is the biogeocoenosis[3] (Fig. 11-7) within a biogeoclimatic zone, and it is a distinct entity on the ground. While it constitutes a unique ecosystem, it is not synonymous with ecosystem because an ecosystem has no finite boundaries except as artificially established by the investigator. There is a resemblance between a biogeocoenosis and a habitat type in that Krajina has been limited to describing the phytocoenosis for lack of expert help in the areas of associated fauna and microbiota. Krajina's phytocoenoses have a narrower scope than habitat types in that Daubenmire has described a single habitat type on an area composed of

[3]The functional structure of a biogeocoenose is diagrammatically represented by Kojima and Krajina (1975) from Sukachev and Dylis (1964):

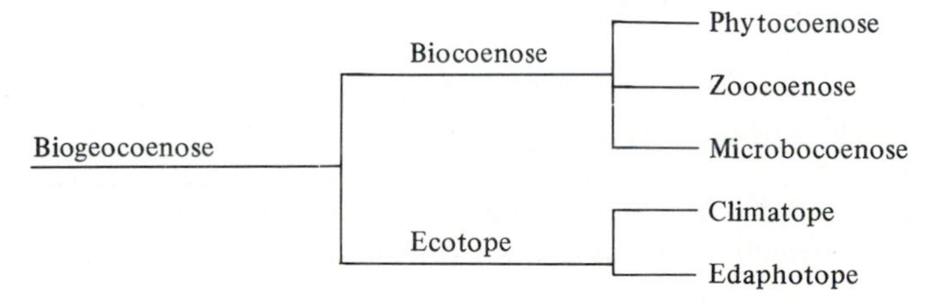

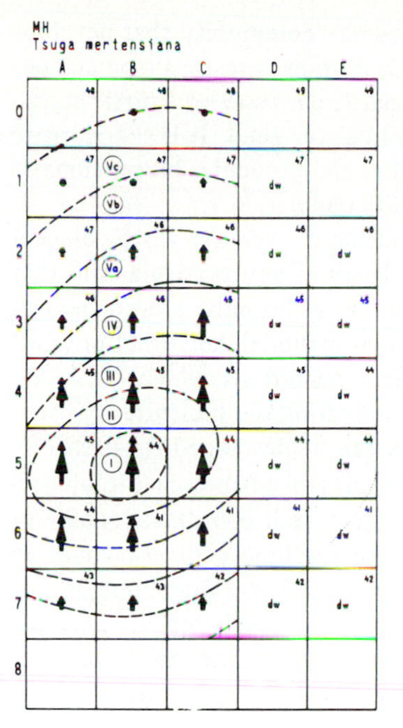

Information given in the diagram:

MH — abbreviation of the name of a biogeoclimatic zone [in this case: the Mountain Hemlock Biogeoclimatic Zone]

Tsuga mertensiana — the tree species ecological function, shown in the panel through edatopes derived from a combination of two composite parameters: *hygrotopes* and *trophotopes* (Pogrebniak, 1930)

Hygrotopes (vertical axis):

- 0 — very xeric
- 1 — xeric
- 2 — subxeric
- 3 — submesic
- 4 — mesic
- 5 — subhygric
- 6 — hygric
- 7 — subhydric
- 8 — hydric

Trophotopes (horizontal axis):

- A — oligotrophic
- B — submesotrophic
- C — mesotrophic
- D — permesotrophic
- E — subeutrophic to eutrophic
- F — hypereutrophic

dw — decayed (or decaying) wood as a substrate on which species is established

Figure 11-7 A biogeoclimatic zone (mountain hemlock) that is ordinated for moisture (hygrotopes) on the *y* axis and for nutrients (trophotopes) on the *x* axis is divided into edatopes. Each edatope is associated with a particular biogeocoenosis or plant-community type (small numbers in northeastern corner indicate which type). The tree symbols and their sizes indicate the growth potential. The dashed-line curves of site-quality zones extend into D and E nutrient zones but need rotted wood as a seedbed. (*From Krajina, 1972.*)

two or more soil series, but the similarities are the most prominent features since both use the union of the climax tree cover with climax ground cover to describe the unit. Krajina has created a zonal classification system based on the division of British Columbia into 11 biogeoclimatic zones (Krajina, 1972; but 12 in Kojima and Krajina, 1975) that are integrated to form seven biogeoclimatic regions and four biogeoclimatic formations. The zones have various subzones based on the amount of precipitation, latitude, or duration of snow cover. The ruggedness of the British Columbian topography prevents the same accuracy in differentiating biogeoclimatic zones that Hill was able to develop in Ontario for his site regions. The regional climates are described in terms of Köppen's classification. Each zone is characterized by a particular macroclimate, a zonal soil that occurs under mesic conditions and a climatic climax community that develops on the mesic soil. Among the zones the mesic condition is really a median condition relative to the extremes which exist in a particular zone, so a mesic condition would be much drier in a dry zone than in a wet zone. It is the climatic climax community on the zonal soil that typifies the particular biogeoclimatic zone, and the dominant species of that community name it.

Since the macroclimate is the common element over the whole biogeoclimatic zone, it is the moisture and nutrient levels of any particular site that determine variations in the vegetative composition of a particular biogeocoenosis from the climatic climax. Thus slope, aspect, parent material, and other physical and chemical factors of the environment create conditions which, with the vegetation, determine the biogeocoenosis. By ordinating the moisture regimens and nutrient regimens, an *edatopic* grid matrix can be developed for each biogeoclimatic zone (Fig. 11-7). Krajina (1969) has described the specific biogeocoenosis (plant-community type) that occurs within each coordinate square of the grid. The same biogeocoenosis may occur in several biogeoclimatic zones. In one zone, it may be the climatic climax on the zonal soil; in a drier zone, it may occur on hydric habitats; and in a wetter zone, on more xeric habitats.

Another area of integration in the system is in the hierarchical ordering of the biogeocoenoses into plant associations, alliances, orders, and classes. Each higher level in the hierarchy loses some of the homogeneity of lower levels, but the emphasis is on the floristic or ecological relationships within and across the borders of the biogeoclimatic zones.

A map of the biogeoclimatic zones of British Columbia has been developed, but the mapping of the vegetational units has been limited to the university research forest. In a province as diverse as British Columbia, many of the biogeocoenoses will be much too small to serve as a management unit, which introduces the problem of getting interpretations activated on the ground. The system has the potential of being able to predict site quality within a zone, as indicated in Fig. 11-6, but like the variation in productivity of similar physiographic sites between site regions in Hill's classification, the relative productivity of the same biogeocoenosis in different biogeoclimatic zones is uncertain. The system should be useful, however, in focusing attention on the constraints or advantages of applying silvicultural treatments to a particular area. On the other hand, the language poses a hurdle for anyone who is not one of Krajina's advanced students.

CONCLUSIONS

Site-quality evaluations are essential for decision making in the management of any forest property. This brief review of some of the means available for classifying lands according to site quality permits the following conclusions:

1 Most methods of site-quality evaluation depend on correlation with site-index values as the measure of their accuracy or leave the correlation for future work.

2 A site index indirectly measures wood volume productivity on a site, thus restricting the concept of site quality. Site quality has not been correlated with the production of other forest products.

3 All the methods examined are reasonably accurate if applied to restricted areas. In extending the use of any one method beyond the area in which it was developed, the principles can be retained but the practices should be modified to fit each situation.

4 Yield tables provide the bases for site-quality standards, and site index provides the access to them.

5 A carefully derived site index is the most accurate indirect measure of site quality available, but the site index for most yield tables should be checked by stem-analysis methods for their accuracy in any one particular locality.

6 Density and suppression can often influence a site-index rating, while vegetational and environmental approaches are free of these influences.

7 When forest-site types, habitat types, biogeocoenoses, and physiographic site types are accurately correlated with site quality, site qualities can be more rapidly recognized and mapped.

8 Any introduction of genetically superior trees will render obsolete the site-quality mapping based on site-index determinations and correlations with wild populations.

9 The rapidly expanding use of fertilizers in forest management will also upset past site-quality evaluations and cause revisions, although the Finnish forest-site type can reportedly predict the change in a forest-site type after fertilizing.

Part Four

Stands

Chapter 12

Stand-Density Determination

Stand density is the second most important factor after site quality in determining the productivity of a site. This is important because stand density is the major factor that the forester can manipulate in developing a stand. Through manipulation of stand density the silviculturist is able to influence species establishment during the regeneration period, as well as modify stem quality, rate of diameter growth, and even volume production during the stand-development period. When a series of alternative results dependent on stand-density manipulation is available, it is essential from the point of view of informed decision making to be able to express stand density accurately. Because accurate evaluation of stand density is so important, mensurationists have proposed various methods of obtaining it. Thus the methods of measuring stand density are many; yet no agreement as to the most reliable way of describing it has been reached.

Before developing the principles of stand-density measurement, however, it is necessary to clarify two terms most frequently used in this context: stocking and stand density. *Stocking* is defined as "a more or less subjective indication of the number of trees as compared to the desirable number for best results." *Stand density* is defined as "quantitative measure of tree stocking expressed either *relatively* as a coefficient, taking normal numbers, basal area or volume as unit, or *absolutely* in terms of number of trees, total basal area, or volume, per unit area" (Ford-Robinson, 1971).

Stocking expresses in terms such as understocked, fully stocked, or overstocked the adequacy of the number of trees in a given stand to meet the objectives of management. It is possible that a stand with a particular stand density could be understocked or overstocked depending on the management objectives. Stocking has been called a "loose term" (Bickford et al., 1957) because of the many senses in which it has been used, but it will be used here only to express adequacy of growing stock. Stand density, however, of necessity is expressed in quantitative terms and should represent a stand measurement that is independent of management objectives and consistent throughout the developmental stages of a stand. Other characteristics for a desirable stand-density measure would be independence from age and character of the stand and from site quality (Spurr, 1952). A desirable stand-density measure should also be easy to apply and easy to correlate with volume increment (Bickford et al., 1957).

METHODS FOR DETERMINING STAND DENSITY

In developing silvicultural prescriptions, it is fundamentally important to be able to determine stand density. These measures are actually one way of describing the biological development of stands. This chapter will explore the weaknesses and strengths of the principal stand-density measures and assemble the justifications for their use. Finally, a comparison will be made of the stand densities of three stands by applying the most useful methods. It is imperative to realize that stand density is not a direct measure of volume. Any stand-density determination requires additional information about a stand before the volume can be estimated.

Ocular Method

The pioneer European foresters maintained maximum density compatible with maximum growth by ocular estimates of crown closure and crown development. American foresters used ocular estimates to determine full stocking in selecting plots from which to construct normal yield tables; as a consequence, there has been variation in the criteria of normality among them (Gevorkiantz, 1944).

Normal Yield Table Method

The normal yield table developed from even-aged stands has been the most generally used base for measuring stand density. In the discussion of site quality, it was emphasized that a normal yield table gives an average value of many stand characteristics for fully stocked, even-aged, and pure stands of the same age and site quality. The "normal" concept and its usefulness have been criticized as no longer tenable for modern silvicultural or management practices (Nelson and

Bennett, 1965). The criticism loses sight of the fact that the development of a stand is toward "normal" (Table 12-1). The table shows that hemlock stands approach normal (100 percent) with age whether they start as understocked or overstocked. A knowledge of the equilibrium condition of stand density is a basic silvical requirement if the dynamic processes of stand development are to be understood (McArdle et al., 1949; Barnes, 1962; Smith, 1965; Gingrich, 1967).

Stand density of a *specific stand* under this method is expressed as the relationship of its basal area, number of trees, or volume to the normal yield table values for the same age and site index. Basal area is the most widely used criterion because it is readily determined in the field with instruments utilizing the Bitterlich angle principle.

Two criteria for a good density measure are met by the use of basal area, i.e., ease of application and capacity to convert to volume if a yield table is available. However, the method is dependent on a knowledge of the age and site quality of the stand. Errors in determining age and site index would limit the accuracy of the density measurement. Basal area also fails to provide any information on tree size. The ratio of the basal area of a specific stand to the normal yield table basal area for the same site index and age correlates well with the ratio of cubic-foot volume of the specific stand to the cubic-foot volume of the normal stand. The correlation is poorer using board-foot volumes. Stand densities based on normal yield table values are often in error because the

Table 12-1 Estimated Stocking* at Future Intervals, for Specified Present Density, in Percent of Normal for Western Hemlock

	Stocking at					
Present stocking	10 years	20 years	30 years	40 years	50 years	60 years
20	42	58	69	77	83	87
30	49	63	73	80	85	89
40	56	68	77	83	87	90
50	63	73	80	85	89	92
60	71	79	84	88	91	93
70	78	84	88	91	93	95
80	85	89	92	94	95	96
90	93	95	96	97	98	99
100	100	100	100	100	100	100
110	107	105	104	103	102	101
120	115	111	108	106	104	103
130	122	116	112	109	106	104
140	129	121	116	112	109	106
150	137	127	120	114	110	107
160	144	132	123	117	112	109

**Stocking*, as used here, is the percentage relationship between actual and normal number of trees for a particular average stand diameter (Barnes, 1962).

biology reflected in varying development patterns of stands tends to make all stands approach normal. As Table 12-1 indicates, a stand starting with 20 percent normal number of trees will in time reach 100 percent normal number, but the trees will be much bigger in diameter, they will be more tapered by having grown as open-grown trees for so long, and the stand height and basal area will differ from normal. An estimate of volume for such a stand would have considerable error. If, however, the stand had 160 percent the normal number of trees and it was measured when natural mortality had reduced the number of trees to normal, then its stems would be smaller in diameter and much more cylindrical, the basal area would be less, and the height would be shorter than normal. An estimate of volume would also be in error. Thus normal yield tables give best estimates in stands with average development.

Reineke's Stand-Density Index Method

Reineke's stand-density index (SDI) has the potential of being a most useful tool when intensive stand management requires a refined method for regulating stand density to fit prescribed goals. It is being more intensively developed in this chapter because of that potential and because it also has many neglected uses for better management under present stand conditions. Some of these will be enumerated after the method has been developed.

Reineke discovered that any pure, fully stocked, even-aged stand of a given average stand diameter (where the average stand diameter [ASD] is the diameter at breast height of a tree with the average basal area) has approximately the same number of trees per acre as any other pure, fully stocked, even-aged stand of the same species that has the same average stand diameter (Table 12-2) (Reineke, 1933). This table indicates that the most important factor in estimating stand density is the average stand diameter, and Bruce's yield table, discussed later, will support this. This characteristic is not significantly affected by age and site quality (Table 12-2).

Table 12-2 Number of Trees per Acre at an Average Stand Diameter of 10 in for Western Hemlock in British Columbia and Alaska for Trees over 1.5 in in Diameter.

Age	Average stand diameter	Site Index	Number of trees per acre
160	10.0	70	510*
100	10.0	90	510
60	9.9	130	510
50	10.0	170	510

*Other yield tables do not give such uniformity, but all demonstrate that the relationship holds.

Source: Barnes, 1962.

These elements—(1) same number of trees at a given average stand diameter, and (2) the fact that age and site quality have no effect on the number of trees—are fundamental contributions. Since the parameters which describe the shape of the diameter-distribution curve for an even-aged stand (coefficients of skewness, kurtosis, and variation) are also more closely related to average stand diameter than to age and site quality (Meyer, 1930), Reineke described the differences in stand density among stands of the same average stand diameter in terms of the number of trees. Since the number of trees per acre in normal stands varies with the average stand diameter, then when comparisons of density between stands were to be made, it was necessary to have a curve showing the number of trees per acre for any average stand diameter. For example, if one stand has 200 trees at an ASD of 9 in and a second stand has 40 trees with an ASD of 20 in for an effective comparison of densities, the two density measures have to be reduced to equivalent densities at the same ASD. Reineke, in order to provide a means for reducing these diverse density measures to a common base, plotted on log-log paper the original plot data for the yield table of a species. A curve of maximum number of trees per acre for any average stand diameter was obtained (Fig. 12-1). Note that the maximum curve for a species is *not* a least-squares fitted curve, but a curve that would represent the maximum density expected for a particular average stand diameter. For most of the species examined, the maximum curves have the same slope ($\log N = -1.605 \log D + k$, where N = number of trees per acre, D = average stand diameter, and k = the adjustment in elevation of curve for species differences). (Shortleaf and slash pines have maximum curves that do not parallel the curve of red fir, and there is no explanation for this fact unless it lies in their fire history.) The maximum curves for red fir (Fig. 12-1) and also for redwood passed through the point of 1000 trees per acre at an average stand diameter of 10 in. Since red fir and redwood had the highest stand densities of any species examined, their common maximum curve was called the *reference curve*. By use of a series of curves parallel to the reference curve (Fig. 12-2), the number of trees per acre at any average stand diameter could be converted to an equivalent density at an average stand diameter of 10 in. Thus Reineke's *stand-density index* (SDI) is the number of trees at an ASD of 10 in. Using this approach, 4000 trees per acre at 2 in ASD is equivalent in density to 300 trees per acre at 10 in ASD, and the stand-density index is 300. In the previous example, 200 trees at an ASD of 9 in is equivalent in density to SDI 185, and 40 trees with an ASD of 20 in has an SDI of 121. So the first stand is the denser; yet the basal areas would indicate that the density is almost the same: 88.3 and 87.3 ft^2.

Stand-density index is always expressed as a number of trees. Presumably the maximum stand-density index is 1000 at an ASD of 10 in. However, there may be a species, and certainly some stands, which could exceed this figure (Fig. 12-1). While stands of red fir and redwood have been found with stand-density indices of 1000, the maximum density curve for Douglas-fir (Fig. 12-3)

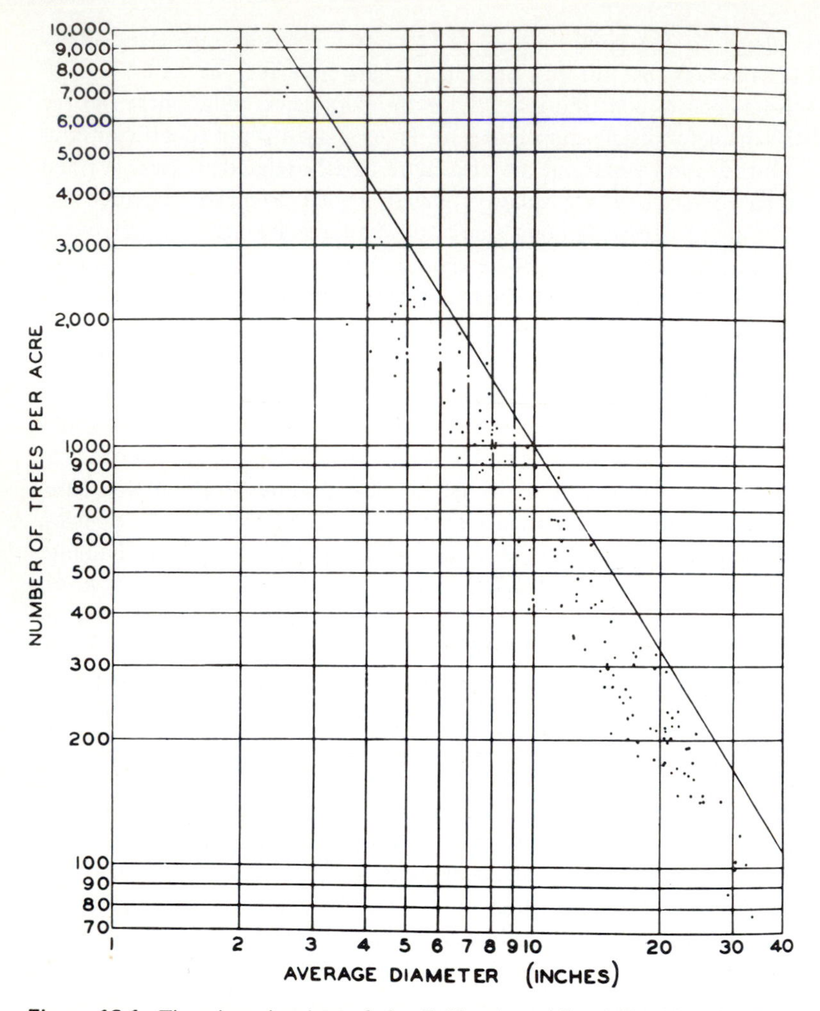

Figure 12-1 The plotted points of the California red-fir yield table's plot data show the maximum curve for red fir passing through the point 1000 trees per acre at 10 in average stand diameter. (*From Reineke, 1933.*)

⟶

Figure 12-2 Reference curve (solid line). The stand-density index of each of the broken-line parallel curves is the number of trees indicated by each at 10 in average stand diameter. (*From Reineke, 1933.*)

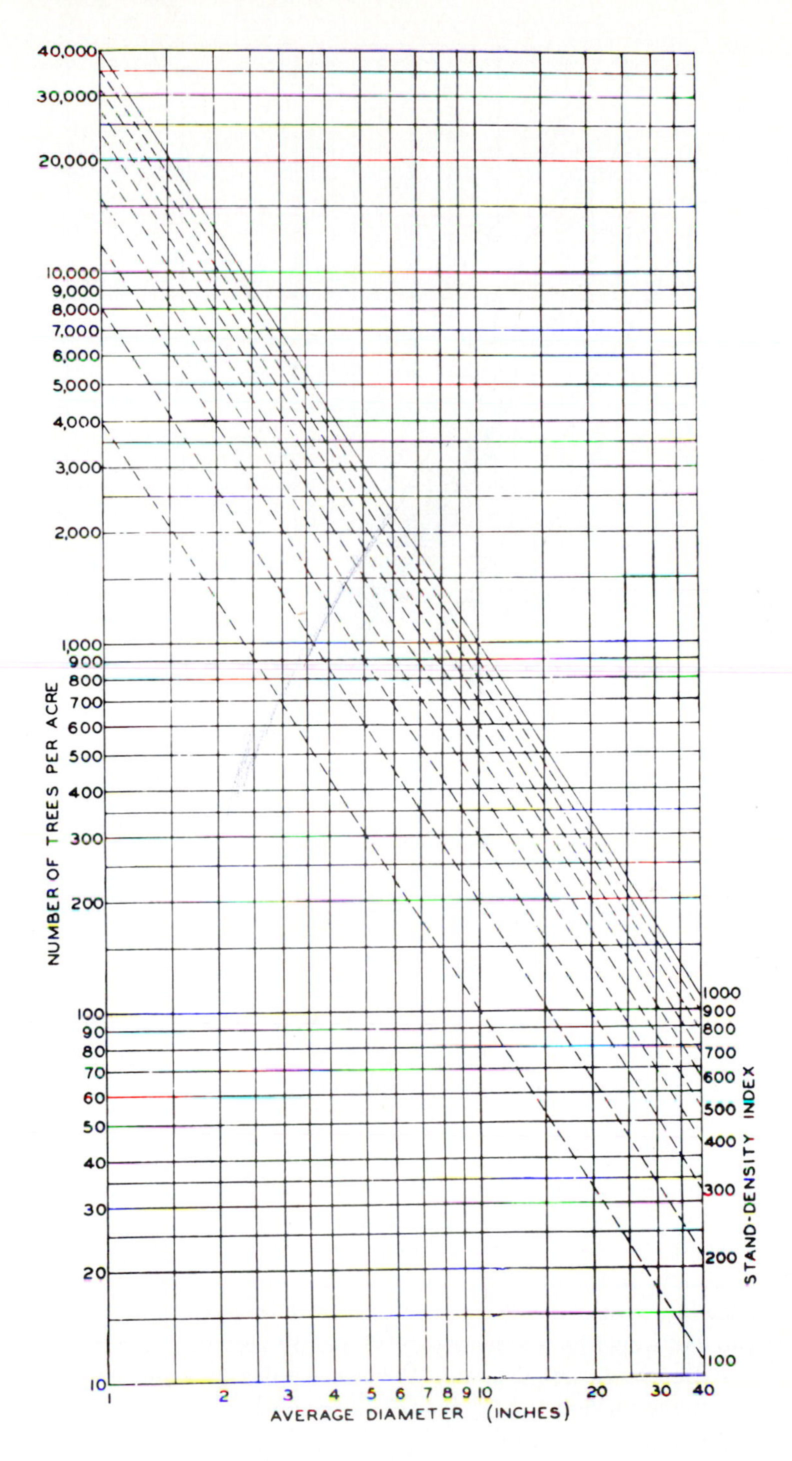

NUMBER OF TREES PER ACRE
40,000
30,000
20,000
10,000
9,000
8,000
7,000
6,000
5,000
4,000
3,000
2,000
1,000
900
800
700
600
500
400
300
200
100
90
80
70
60
50
40
30
20
10
1
2
3
4
5
6
7
8
9
10
20
30
40
AVERAGE DIAMETER (INCHES)
STAND-DENSITY INDEX
1000
900
800
700
600
500
400
300
200
100

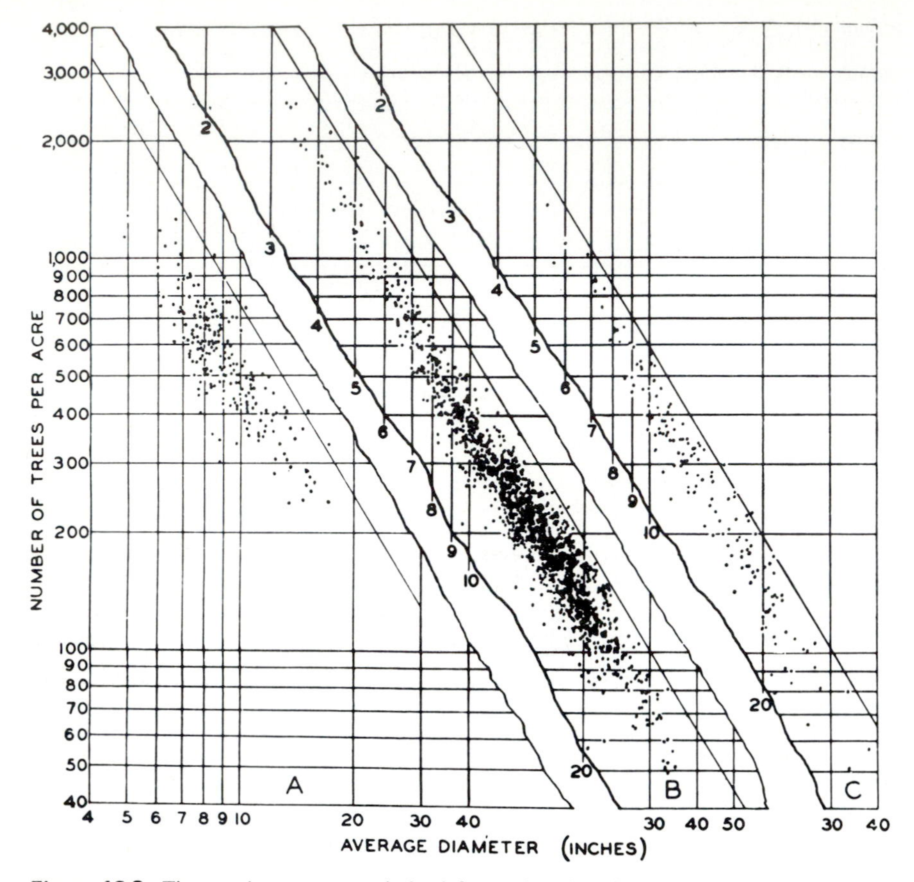

Figure 12-3 The maximum curves derived from plot data for the mixed-conifer type of the Sierra Nevada (A), Douglas-fir of the Pacific Northwest (B), and California Douglas-fir (C). Note that the maximum stand-density index for both groups of Douglas-fir is approximately 595. (*From Reineke, 1933.*)

has a stand-density index of 595, and other species have had different maximum stand-density indices determined from plotting their yield table plot data (white fir, 830; eucalyptus, 495; loblolly, 450; longleaf pine, 400; and ponderosa pine, 830) (Reineke, 1933). It is, however, unnecessary to have the maximum curve for a species in order to use the stand-density index. The maximum curve is useful primarily for comparisons of densities among stands of the same species. Reineke has noted that densities within a species could be compared as a percentage of the species' maximum stand-density index; i.e., Douglas-fir has a maximum stand-density index of 595, so two stands of Douglas-fir with SDI values of 197 and 298 would have a density at 33 and 50 percent of the maximum density that Douglas-fir could attain. The added value of expressing an SDI as a percent of the maximum of a species as opposed to a straight comparison of SDI values is minimal.

The stand-density index method exceeds the requirements suggested for a good stand-density measure. It is free of the need to consider the effects of site and age, and it is easily obtained by the use of a Bitterlich angle or Bruce's wedge for measuring basal area (BA) and by tallying the diameters of the trees counted at each point.[1] In addition, it provides a density measure that is independent of species; longleaf pine's maximum SDI of 400 gives the simplest contrast of its difference in density potential to any other known maximum SDI, such as red fir or redwood's SDI of 1000. Thus it provides the ability to compare the densities of two stands of any species in any location.

The difficulty of interpolating the SDI and accurately locating a particular average stand diameter and number of trees on Reineke's log-log diagram (Fig. 12-2) has been relieved by developing an SDI table (Daniel et al., 1978). Table 12-3 provides accurate readings except at very high numbers of small trees. Inasmuch as the metric system will be adopted in forestry, Reineke's SDI has been converted to a metric stand-density index (Table 12-4). Instead of the reference curve passing through the 10-in average stand diameter at 1000 trees per acre, the metric stand-density index reference curve passes through the 25-cm average stand diameter at 2500 trees per hectare. The reference curves for Reineke's SDI and the metric SDI are not precisely at the same location, with the discrepancy being about 1.2 percent lower for the metric reference curve at maximum stocking (metric SDI 2500 when ASD is 25 cm has an actual

[1] The following is an example of the calculations for determining average stand diameter and number of trees per acre (wedge factor 10.0). SDI is from Reineke SDI Table (Table 12-3). Even-aged stand:

Plotless cruise, sample point *A*:

Diameter in.	BA ft^2 $acre^{-1}$	BA per tree	Trees per acre
12	10	0.789	12.68*
12	10	0.789	12.68
16	10	1.396	7.16
9	10	0.442	22.62
14	10	1.069	9.35
8	10	0.349	28.58
15	10	1.227	8.14
11	10	0.660	15.15
14	10	1.069	9.35
13	10	0.922	10.85
	100		136.56

Tree of average BA $= \frac{100}{137} = .7326\ ft^2$

ASD = 11.59 in
SDI = 174

*Number of trees per acre equals $\frac{\text{BA per acre}}{\text{BA per tree}} = \frac{10}{0.789} = 12.68$

Numerous sample points:

BA	100	No. of trees	137
BA	80	No. of trees	184
BA	110	No. of trees	150
	290		471

Tree of average BA $= \frac{290}{471} = .616\ ft^2$

Average no. of trees $= \frac{471}{3} = 157$

ASD = 10.62 in SDI = 173

Table 12-3 Reineke Stand-Density-Index Table Based on the Number of Trees per Acre When the Average Stand Diameter is 10 in and the Reference Curve Passes through the Point 1000 Trees per Acre

No. of trees per acre	Average stand diameter, in																			
	2	4	6	8	10	12	14	16	18	20	22	24	26	28	30	32	34	36	38	40
5	–	1	2	3	5	6	8	10	12	15	17	20	23	26	29	32	35	39	42	46
10		2	3	6	10	13	17	21	25	30	35	40	46	52	58	64	71	78	85	92
20	1*	4	8	13	20	26	34	42	51	60	70	81	92	104	116	129	142	156	170	185
40	3	9	17	27	40	53	68	85	102	121	141	163	185	208	233	258	285	312	340	370
60	4	13	26	44	60	80	102	127	154	182	212	244	278	313	349	388	427	468	511	555
80	6	18	35	55	80	107	137	170	205	243	283	326	370	417	466	517	570	625	681	740
100	7	22	44	69	100	133	171	212	256	304	354	407	463	522	583	646	712	781	852	925
120	9	27	52	83	120	160	205	255	308	365	425	489	556	626	699	776	855	937		
140	10	32	61	97	140	187	240	297	359	425	496	570	648	730	816	905	998			
160	12	36	70	111	160	214	274	340	410	486	567	652	741	835	933	1034				
200	15	45	88	139	200	267	343	425	513	608	708	815	926	1044	(1166)					
300	22	68	132	209	300	401	514	637	770	912	1063									
400	30	91	176	279	400	535	686	850	1026											
500	37	114	220	349	500	669	858	1063												
600	45	137	264	419	600	803	1029													
800	60	183	352	559	800	1071														
1,000	75	229	440	698	1000															
1,500	113	344	660	1048																
2,000†	151	459	880																	
2,500	188	574	(1101)																	
3,000	226	689	(1321)																	
4,000	302	919																		
5,000	377	(1149)																		

*The columns form a straight-line curve; the rows form a quadratic curve that is approximately a straight line between any two adjacent average stand diameters.

†At high densities for small trees, more accurate readings can be obtained using an enlarged table from Daniel et al.

Source: Daniel et al., 1978.

$SDI = 10^{(\log n + 1.605 \log D - 1.605)}$

Table 12-4 Metric Stand-Density-Index Table Based on the Number of Trees per Hectare When the Average Stand Diameter is 25 cm and the Reference Curve Passes through the Point 2500 Trees per Hectare

No. of trees per hectare	Average stand diameter, cm																			
	5	10	15	20	25	30	35	40	45	50	55	60	65	70	75	80	85	90	95	100
10		2	4	6	10	13	17	21	25	30	35	40	46	52	58	64	71	78	85	92
25	1*	5	11	17	25	33	42	53	64	76	88	101	115	130	145	161	178	195	213	231
30	2	6	13	20	30	40	51	63	77	91	106	122	139	156	174	194	213	234	255	277
50	3	11	22	34	50	67	85	106	128	152	177	203	231	261	291	323	356	390	426	462
100	7	22	44	69	100	134	171	212	256	304	354	407	463	522	583	646	712	781	852	925
150	11	34	66	104	150	201	257	318	385	456	531	611	695	783	874	970	1069	1172	1278	1388
200	15	45	88	139	200	268	343	425	513	608	709	815	927	1044	1166	1293	1425	1562	1704	1850
250	18	57	110	174	250	335	429	531	642	760	886	1019	1158	1305	1457	1617	1782	1953	2130	2313
300	22	68	132	209	300	402	514	637	770	912	1063	1222	1390	1566	1749	1940	2138	2344	2556	
350	26	80	154	244	350	469	600	744	899	1064	1240	1426	1622	1827	2041	2263	2495			
400	30	91	176	279	400	536	686	850	1027	1216	1418	1630	1854	2088	2332	2587				
500	37	114	220	349	500	670	858	1063	1284	1521	1772	2038	2375	2610						
600	45	137	264	419	600	804	1029	1275	1541	1825	2127	2445								
700	52	160	308	489	700	938	1201	1488	1798	2129	2481									
800	60	183	352	559	800	1072	1372	1701	2055	2433										
900	67	206	396	629	900	1206	1544	1913	2311											
1,000	75	229	440	699	1000	1340	1716	2126	2568											
1,100	83	252	484	768	1100	1474	1887	2339												
1,200	90	275	528	838	1200	1608	2059	2551												
1,300	98	298	572	908	1300	1742	2231													
1,400	105	321	616	978	1400	1876	2402													
1,500	113	344	660	1048	1500	2010	2574													
1,600	120	367	704	1118	1600	2144														
1,700	128	390	748	1188	1700	2278														
1,800	135	413	792	1258	1800	2412														
1,900†	143	436	836	1328	1900	2546														
2,000	151	459	881	1398	2000															
3,000	226	689	1321	2097	3000															
4,000	302	919	1762	2795																
5,000	377	1148	2202																	

*The columns form a straight-line curve; the rows form a quadratic curve that is approximately a straight line between any two adjacent average stand diameters.

†At high densities for small trees, more accurate readings can be obtained using an enlarged table from Daniel et al.

Source: Daniel et al., 1978.

Metric SDI = $1.0147\ (10)^{\log n\ +\ 1.605 \log D\ -\ 2.250}$

density of SDI 988). At SDI 100, the discrepancy is reduced to 0.1 percent. However, metric SDI can be obtained from a Reineke SDI by multiplying the Reineke SDI by 2.5. (There is some variation around 2.5 probably due to using the tables for conversion of ASD and number of trees to a metric or Reineke SDI). The slight difference in density between the Reineke and the metric stand-density index for a particular stand is insignificant at normal operating densities. In addition, the difference becomes of interest only in converting from one to the other index. To determine metric SDI from field data using a wedge prism (with a 10-factor), each tree counted represents approximately 2.3 m^2 basal area per hectare instead of 10 ft^2 per acre or use formula in Table 12-4. The tree diameters must be measured in centimeters and basal area expressed in square meters.

While SDI is designated as applicable only to even-aged stands, it could be used to describe the density of uneven-aged stands better than the present dependence on basal area (Stage, 1968). SDI measurements can always be converted back to basal area, but they give the added dimensions of average stand diameter and number of trees, as well as by species if desired.

Reineke's stand-density index has many practical applications in evaluating stand development. For example,

1 It permits the densities of stands to be compared regardless of site and age differences. Even densities of stands of different species or mixtures can be compared using SDI.

2 Given a mature stand which is regarded as ideal to suit management objectives, using SDI the silviculturist can project backwards to determine the appropriate number of trees which should be maintained in an immature stand to develop at that same density.

3 Studies on thinning and stocking control establish the upper and lower levels of desirable basal area (Figs. 14-22 and 14-26). These levels can then be best defined in terms of SDI because (a) this eliminates the need for complex diagrams of optimal stocking range, and (b) it permits determination of the appropriate number of trees for that stocking level for any size class.

4 SDI permits the description of thinning regimens which are unambiguous, as opposed to regimens expressed only in terms of basal area. A stand maintained at constant basal area results in a gradual reduction in density because when this is done, basal area as a percent of normal basal area falls off with time. Stand-density index confirms this relationship and is therefore more useful as a measure of density since it also decreases with time when basal area is held constant.

Bruce's Yield Table Method

The inability of normal yield tables to provide a reliable measure of the normality of stands that have had different development patterns led to the development of a Douglas-fir yield table based on average stand diameter, number of trees, and height (McArdle et al., 1949). The basic plot data from normal stands of number of trees per acre, volume, and normal height were plotted as a function of average stand diameter. It was found that stands with the same average stand diameter were more nearly alike than stands of the same age and site index, which is a corollary of Reineke's discovery. A simple table (Table 25, McArdle et al., 1949) gives for each average stand diameter the normal number of trees per acre, the average height of trees of average stand diameter, and volume per tree. Stand density as a percent of normal by this method is greatly simplified and is more applicable to subnormal stands than the standard yield table. To measure density, a stand of a given average stand diameter and number of trees per acre is compared to the normal number of trees for the same average stand diameter. In order to determine stand density on a volume basis or the volume of the stand, the volume per tree was found to need a correction because of the variation in the height/diameter ratio within stands of the same average stand diameter. The variation in the ratio was loosely associated with site quality. However, the needed adjustment of the volume per tree was found to be proportional to the ratio of the average height of trees with average stand diameters to the normal height. The normality of a stand with this method can be expressed in terms of number of trees or of cubic-foot volume.

The variables in this method are easily and accurately measurable in a stand. In addition, stand density can be evaluated independent of a knowledge of age or site quality. A lack of comparable yield tables for most species reduces the usefulness of the method, and in any case, the usefulness is limited to a comparison of stand densities within a species and a region. The normal number of trees per acre by average stand diameters in Bruce's Douglas-fir yield table is the approximate equivalent of SDI 381 (range from 355 at 5 in ASD to 396 at 30 in ASD) and is approximately equivalent to the least-squares-fitted line through the data points of Douglas-fir given in Fig. 12-3. The average SDI for Bruce's yield table (SDI 381) indicates that the normal yield table for Pacific Northwest Douglas-fir is 64 percent of the maximum (SDI 595) for the species.

Crown-Competition Method

A second method (inasmuch as Bruce's has such limited use) for measuring stand density is based on a biological principle only recently recognized–that is, the high correlation between the width of crown of an open-grown tree and its diameter. It is proving useful in estimating the reduction in height caused by varying degrees of stagnation in lodgepole pine (Alexander et al., 1967). It contributes a means for recognizing the amount of side pressure a species can endure and gives a valuable insight into why some species are able to grow in denser stands then others. The method measures another biological characteristic of a species that is independent of age and site.

The crown-competition method was developed to provide data on the maximum amount of growing space a tree could utilize and on the minimum necessary for a tree to maintain a place in a stand (Krajicek et al., 1961). Open-grown trees must be used to collect data on the vertical projection of crown area by tree diameter. This is so because it is only with open-grown trees that the crown-area-per-DBH relationship is not influenced by competition. Crown area was found to be closely related to diameter at breast height, and the relation was nearly constant within a species regardless of site and age. There was a highly significant difference in the relationship of crown diameter to diameter at breast height between spruce and hardwoods, a significant difference between hickory and red oak, but no significant difference among the oaks. Thus a tree with infinite space could not occupy more than a certain maximum area which was proportional to its breast-height diameter (Fig. 12-4).

Maximum crown area (MCA) is expressed as the percent of an acre an open-grown tree can occupy at a particular stem diameter, and the determination of

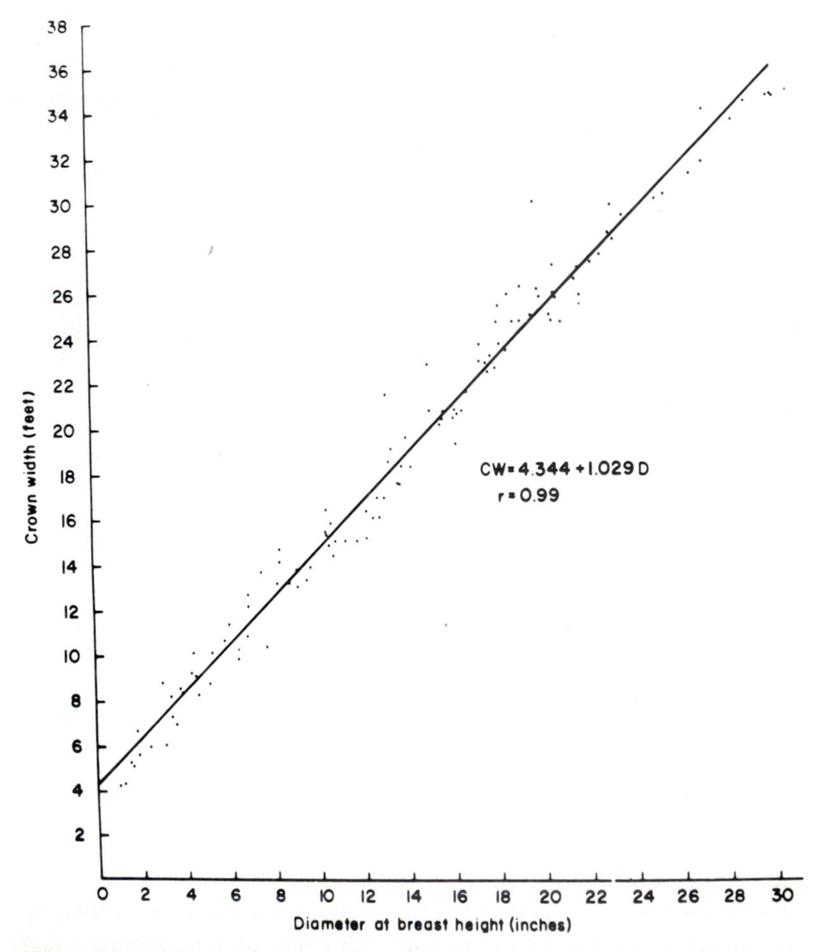

Figure 12-4 Relationship of crown width to stem diameter at breast height for open-grown Engelmann spruce. (*From Alexander, 1971.*)

the area requires a curve similar to Fig. 12-4 for each species. The *crown competition factor* (CCF) is the sum of all MCA values on an acre. If every tree on an acre has its full crown development and if all ground space is occupied, the CCF would be 100. However, among species, CCF 100 may be highly variable in terms of number of trees per acre in an even-aged stand. In fully stocked, even-aged oak stands in eastern Iowa, the CCF averaged 200, and some lodgepole-pine stands have CCF of 500. Thus, if the CCF of a stand is 200, then each tree has half the space it requires in the open; and if the CCF is 75, then the trees are unable to fully utilize the space available. However, if the maximum CCF for a species is 200, then a stand at CCF 200 can continue its development only through a continuing loss of trees. Other species could have different maximum CCF values because of differences in area development of the crown in the open, in shape of the crown, in tolerance, and in the ability to endure lateral pressure. No direct comparison of stand density from CCF values of different species is possible, except as a value provides the relative degree of crowding. The method does not necessarily measure crown closure (Curtis, 1970) because stand manipulations such as thinnings can artificially disrupt a canopy's integrity. However, under usual stand conditions, the method measures the degree of competition present by comparing the space available to the trees in a stand to the space that the same trees need if each has a fully developed crown.

Rule-of-Thumb Methods

A rule-of-thrumb method depends on a fixed function of diameter or height as a control for the density of a developing stand. Rule-of-thumb methods usually have a background of field studies. These methods have the advantage of being readily understood and used by work crews assigned to do the intermediate treatments. Inasmuch as ease of transmission and execution of instructions are important, these methods have a function. A Reineke stand-density index can be readily translated into a percent-of-height method or a *D*-plus method to be applied in the field.

Percent-of-Height Method Wilson (1946, 1955) introduced the idea of maintaining uniform density in a developing stand by treating spacing as a function of height; i.e., with height of trees at 50 ft and percent-of-height at 22 percent, then distance beween trees is 11 ft. He emphasized that height combines the effects of site and age and is reasonably independent of density. Diameter, on the other hand, is strongly affected by stand density. The particular percentage of height used for spacing depends on the species, especially its tolerance, and the management objectives. Site does not affect the percentage. Age has an effect only after the tree has appreciably slowed down in height growth. The method has the potential of maintaining a uniform density in a stand, but it has been used with the height of dominants as the controlling height. The effect of using the height of dominants is a reduction in density: a gradual reduction in stand-density index and a reduced or constant basal area per acre (Day and Rudolph, 1971). However, the use of the average height of the trees with the average stand diameter maintains the stand-density index and allows normal basal-area increase with increasing average stand diameter.

The question as to whether the distance between trees applies to all trees in the stand or only to the upper crown classes is moot. After several low thinnings, the question is answered by the removal of the lower crown classes. The implication is that the smaller trees may be ignored initially in spacing the crop trees. The application of the percent-of-height method requires only the measurement of the tree heights and the ability to assess the average distance between trees in spite of considerable irregularity of spacing. With a known distance between trees, a second criterion of density is available, i.e., number of trees per acre. Wilson has developed a special log-log paper through Keuffel and Esser that simplifies the recording of the treatments and their effects in a stand. It also permits direct reading of numbers of trees per acre as well as percent of height. The percent-of-height method has the major advantage of being completely interchangeable between English and metric units.

***D*-Plus Method** In the *D*-plus method, the distance in feet between trees should be the same as the mean diameter in inches plus a constant; i.e., a mean diameter of 12 in plus 2 equals a distance of 14 ft between trees. The weakness of the rule is that adding a constant value to the diameter does not maintain a particular density. Usually a stand is too open when young and too dense when mature, but an initially dense stand would become more open if the diameter-plus rule were followed. However, Mitchell (1943) has shown for a number of species (southern pines, northern hardwoods, and upland oaks) that the number of dominant and codominant trees in normal even-aged stands fits a $D + 4$ reasonably well.

***D*-Times Method** In this method, spacing should be the mean diameter times a constant; i.e., a mean diameter of 12 in times 2 equals a distance of 24 ft between trees. The rule favors very dense, young stands and open, mature stands. The results parallel the maintenance of a constant basal area throughout the life of a stand (Averall, 1945).

Growing Space

Growing space refers to the area occupied per tree and is expressed by the distance between trees. In the spacing of actual stands in the course of their development, one of the rules of thumb given in the previous section would be satisfactory as a field guide. However, there are occasions such as in establishing stands by planting or in releasing crop trees in cleanings or thinnings where definite growing spaces need to be set.

Table 12-5 Spacing Distance Calculations

	Square		Circle	
	Measurement, m	Measurement, ft	Measurement, m	Measurement, ft
Spacing distance (d)	$d = \sqrt{\frac{10000}{N}}$	$d = \sqrt{\frac{43560}{N}}$	$d = \sqrt{\frac{10000}{N} \cdot \frac{4}{\pi}}$	$d = \sqrt{\frac{43560}{N} \cdot \frac{4}{\pi}}$
	$= \sqrt{\frac{10000}{300}}$	$= \sqrt{\frac{43560}{121}}$	$= \sqrt{\frac{12732}{300}}$	$= \sqrt{\frac{55462}{N}}$
	= 5.8 m	= 19 ft	= 6.5 m	= 21.4 ft
Area available per tree	33.4 m^2	360 ft^2	33.4 m^2	360 ft^2

There are two main methods for calculating spacing: one assumes that the root and crown area occupied by a tree is square, and the other assumes that the area is a circle. Both methods are inefficient in calculating space utilization. Tree roots and crowns are usually circular in area, at least prior to competition; consequently, a square spacing does not permit initial full utilization of the site. Similarly, the simple circular spacing method commonly used to calculate spacing or tree numbers overestimates the number of trees because the calculation does not allow for the fact that, physically, there must be gaps of unoccupied land between adjacent circles. These two ways of allocating space are illustrated in the following example.

Spacing distance (d) or number of trees per unit area (N) are calculated in the manner shown in Table 12-5. In this example, spacing distance is calculated given $N = 300$ per hectare (121 per acre). Thus, given the number of trees per hectare as 300, spacing distance would be 5.8 m assuming square spacing and 6.5 m assuming circular spacing. In both cases, the area theoretically available per tree is 33 m^2. However, using a spacing of 5.8 m, a tree with a circular crown area would occupy only 26.4 m^2. Thus the total site is not initially utilized. Dividing this value of 26.4 m^2 per tree into a hectare results in a potential stocking of 380 trees per hectare, which is considerably more than the value of 300 per hectare used initially to calculate square spacing.

Regardless of whether square or circular growing space is used as a basis for calculating distance between trees, the actual layout of plants in the field using both methods would be generally in the form of squares. In practice, spacing is most commonly calculated assuming square or rectangular growing space. This is so because of simplicity and because trees usually develop into the space available. Attempts to allocate growing space designed to maximize efficiency in space utilization are not likely to be of much concern.

Other Methods of Measuring Stand Density

Alternative approaches to the measurement of stand density have been developed. Although of interest, they are not used in silvicultural practices because they provide less precise estimates of stand density.

Tree/Area Ratio Method Tree/area ratio method (Chisman and Schumacher, 1940) allocates the ground-surface area occupied by the canopy of an individual tree in a normal stand according to its stem diameter. The ground area for each stem diameter was calculated by means of a quadratic equation fitted by least squares to the data from a series of normally stocked loblolly-pine stands that had been used in preparation of a yield table. When the derived equation was applied to the diameter distribution within a plot, the tree/area ratio represented the ground area occupied by the trees as a ratio of the plot area. A normal stand would have a ratio of 1.0. It has been determined that the tree/area ratio and the ground area of a particular diameter tree are independent of site and age when calculated from normal stands.

The application of the method requires the use of a fixed plot and the diameters of all trees in order to calculate the ratio. Each species would require the calculation of a new set of constants for the quadratic equation. However, the principal weakness of the method is the concept that the ground area occupied by a tree of a given diameter is the same irrespective of stand density. It is an assumption that has been disproved (Krajicek et al., 1961). Thus the interpretation of a tree/area ratio would be difficult.

Stage's Volume-Increment Method Stand density has been associated with growth potential; yet two stands may have the same stand density on the basis of number of trees or volume and not have the same growth potential. Stage (1969) has developed a method for describing stand density on the basis of growth potential. The volume increment per acre of trees at a sampling point divided by the growth capacity per acre of the site gives the fraction of the site's resources utilized at that point. The variation in site utilization among sampling points gives an estimate of stocking uniformity. The method estimates growth at a point by relating volume increment of a tree to the cube of its diameter and requires a knowledge of the growth capacity of a site at any particular stage in the development of a stand. While the ratio of current volume increment to growth capacity may be independent of age and site, its derivation is not. In addition, volume increment may be constant for a considerable stand density range (Briegleb, 1952; Gingrich, 1967), so the ratio would reflect only the degree of understocking and overstocking rather than being an objective measure of stand density.

Lexen's Bole-Area Method One reason for a stand density measure is the need to estimate stand growth potential. Stand growth potential has been equated with a number of stand parameters such as basal area, board-foot volume, and cubic-foot volume, but objections have been raised for each. Lexen (1943) introduced the concept of bole area as a criterion for xylem growth potential because this takes place at the stem cambium. He used hypothetical plots with trees of varying sizes to show that constant board-foot volume gave a highly variable amount of growth. Bole area is a function of circumference, height, and form. Lexen found bole area could be reasonably approximated in ponderosa pine by summing the diameters in inches times height in feet for all trees and then dividing by 7. No intensive evaluation of this method as a stand density measure has been made. It would seem particularly adapted to uneven-aged stands since all trees would contribute in proportion to size. The effect of the more deliquescent form of hardwoods on the calculations and interpretation of bole area is an added difficulty in its use.

EVALUATION OF METHODS

Each method discussed has a unique approach to measuring stand density, but Curtis (1970) has shown the mathematical relationships among most of the methods and found them differing mainly in algebraic form. However, density measures comparing diameter, height, or age and site with a standard will give different numerical values that may not be equally representative of relative stand densities. Since the differing methods produce different estimates of density, silvicultural, management, and communication problems are created unless, of course, one or more provide reasonably accurate and consistent results under a wide variety of conditions. In addition to being accurate and consistent, a method should be readily understood, easily visualized, and able to aid in the interpretation of the stand-density contributions of a variety of authors. Finally, a suitable method should be easily applied in the field.

Briegleb (1952) has used three stands of Douglas-fir for comparing various methods of measuring stand density (Fig. 12-5). His data and most of his density comparisons are presented in Table 12-6 along with some additional methods. Data of stands A and C are derived from light to moderately thinned Danish plots, and stand B data are from a Pacific Northwest plot after a light thinning. The site indices of the stands and the normal yield table values for basal areas and volumes were taken from the Douglas-fir yield table (McArdle et al., 1949). The stand data could justify the conclusion that stand B has the greatest density, although values of board-foot volume indicate that stand C is the most dense. However, the relative density of stands A and C is open to question. The density measures fall into two distinct groups: (1) those which indicate stands A and C have a difference in stand density and also show considerable fluctuation with respect to the relative density of stand B to stands A and C (Table 12-6, bracketed numbers 1, 2, 3, 8, 9, and 10), and (2) those which indicate stands A and C have the same density and have stand B maintaining its relative density (Table

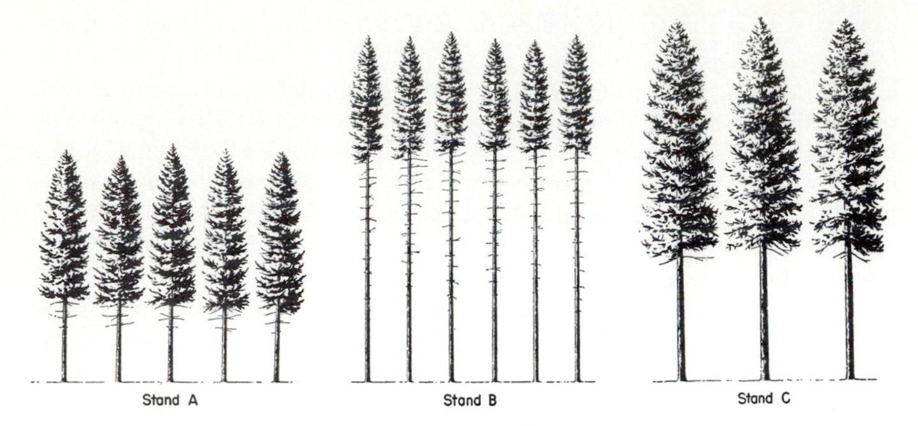

Figure 12-5 Three contrasting Douglas-fir stands after thinning. Data in Table 12-6. Average stem dimensions, crown dimensions, and spacing all drawn to the same scale. (*From Briegleb, 1952.*)

12-6, bracketed numbers 4, 5, 6, and 7). In low-thinned stands with an ASD of 10 in, it would seem that the basal-area table for stems 7 in and over would be appropriate for measuring relative density. However, the results are completely out of line (Table 12-6, bracketed number 2), since the normality of stand A far exceeds that of stands B and C. Bruce's normality by number of trees at a given average stand diameter has the same relation between stands A, B, and C as that of Reineke's stand-density index since stands A and C are approximately 55 percent of stand B (Table 12-6, bracketed numbers 4 and 6).

In balance, the first group of density measures, by disagreeing narrowly as to whether stand A or stand C has more density, tend to confirm the findings of the second group. Of the second group, Bruce's yield table number of trees or cubic-foot volume data provides a reliable estimate of stand density, but such yield tables have not been developed for most species, so the application of the Bruce's yield table method is limited. Percent-of-height method has the potential for simplicity and accuracy without requiring information on site quality. Relative densities of even-aged stands could be readily determined regardless of species. However, a number of problems hinder its use: (1) the method's failure to provide any information on tree size except as height is incorporated into records; (2) the lack of knowledge as to what percentage of the height should be used for a particular species and management objective; and (3) the question as to the choice of trees from the height of which the spacing should be determined if the density is to be maintained.

Reineke's stand-density index is the method that provides an accurate estimate of stand density for even-aged stands irrespective of species, site quality, or age. The data needed for determining the SDI of a stand are readily acquired and provide the average stand diameter and number of trees as well as the SDI. Bruce

Table 12-6 Density, by Various Measures, of Three Contrasting Douglas-fir Stands after Thinning

	Stand A*	Stand B†	Stand C‡
Stand data			
Average stand diameter (in)	10	10	17.5
Average height (ft)	60	90	93
Average crown width (ft)	12.6	8.6	19.2
Average crown length (ft)	37.3	31.7	58.4
Average space between trees (ft)	14.6	10.8	22.6
Site index	113.6	122	126
Stand data (per acre basis)			
No. of trees	206	372	85
Basal area (ft^2)	112	203	142
Volume (derived from Bruce's yield table)			
Cubic feet (M ft^3)	2.9	7.9	4.9
Scribner (M board feet)	6.4	17.3	23.5
International 1/8 (M board feet)	9.8	26.6	34.0
Crown projection (M ft^2)	25.7	21.6	24.6
Crown surface (M ft^2)	304	319	299
Density measures			
BA (percent normality, total stand) [1]	80	102	70
BA (percent normality, 7 in and larger) [2]	181	116	78
Volume (percent normality, total stand, ft^3) [3]	94	124	73
Bruce (percent normal, no. of trees) [4]	55.5	100	54.2
Bruce (percent normal, volume ft^3 total stand) [5]	40	109	38
Stand-density index (Reineke) [6]	206§	372	204
Spacing (percent of height) [7]	24.4	12.0	24.3
Spacing (D plus k) [8]	+4.6	+.8	+5.1
Spacing (D times k) [9]	1.46	1.08	1.29
Bole area (M ft^2) [10]	17.7	47.8	13.8

*First thinned at 20 years; 6th thinning just completed at age 35 years.
†Overdense at 55 years; first thinning just completed at age 55 years.
‡First thinned at 20 years; 12th thinning just completed at age 55 years.
§A = 55.4% B; C = 55% B.
Source: Briegleb, 1952, with additional measures of density.

(1965) has found that the data of several of the stand-density methods parallel the SDI curve between 4- and 30-in DBH, except the constant basal-area method. By adjusting the SDI from diameter at DBH to diameter at ground line, he found that the effect of nursery density on seedling size closely approximated the extrapolation of SDI 500, whereas the other methods of measuring density did not fit.

Reineke's principle has been much more widely accepted and used than generally realized. In fact, any study in which average stand diameter is the independent variable and some other stand parameter is the second variable has only modified the use of an SDI. As indicated in the discussion of Bruce's method, the average SDI in his yield table is 381, and the SDI values gradually increase with increasing average stand diameter. This trend reflects the tendency of the sample plots in the basic data (from which Bruce built his table) to increase in SDI with age (McArdle et al., 1949, Table 2; site index 90, age 30 years, SDI 318, and age 160, SDI 394). Briegleb (1952) did a study of growth for intensively managed stands of Douglas-fir in Denmark and in the Pacific Northwest and compiled a table giving a range of values for average stand diameter, average stand height, and number of trees within which gross growth would be constant. The complex study can be reduced to the fact that any managed stand of Douglas-fir with a density between SDI 187 and 382 will be equally productive. The particular density chosen would depend on the economics involved in setting up the management objectives. Another study which establishes the limits of stand density for full production is Gingrich's (1967) work on upland hardwood forests in the Central states. He used a CCF of 100 to set the lower limit for a stand that fully utilized a site and the tree/area ratio developed from Schnur's (1937) yield table plot data and other plots for upland oaks to set the upper limit. He provides a chart (Fig. 14-22) for determining percent stocking based on average stand diameter, number of trees, and basal area. His presentation can be simplified by the recognition that the lower limit of stand density capable of fully utilizing the site is SDI 117, while the upper limit for full site utilization is set by SDI 209. Stands below or above these limits are understocked or overstocked, with the tendency to approach normal stocking from either condition. A third paper using Reineke's principles with the objective of establishing stand density is that of Stahelin (1949) for loblolly pine. He has expressed his densities in percent of normal basal area for a given average stand diameter. Yet the normal basal areas for loblolly pine at various average stand diameters are very close to the basal areas for SDI 300, so specific densities could be better designated by SDI values or percent-of-height spacing.

There is a general tendency to conceive every forest type as having nothing in common with any other forest type as far as utilization of site is concerned. Reineke's concept of the average stand diameter as the key to density, rather than site and age, provides a base for unifying the many studies of density and its effects on development and growth of stands. Each study should contribute to a general understanding of density in addition to the limited objectives of the study itself. Thus, when new studies utilize the average stand diameter and number of trees relationship, the calculation of the slope of a new base curve by the least-squares balance of plots assumes that a similar set of plots would have the same slope. In view of past variations among individuals gathering data on normal stands, the assumption is not well-founded. The use of the term *stand-*

density index for measures of density obtained from such curves adds unwarranted confusion to a subject already overloaded with diverse approaches. The fact that the slopes of these base curves calculated by the least-squares method [such as, Schnur (1937) for upland oaks, -1.499; Stoehr (1955) for aspen, -1.515] so closely approximate Reineke's -1.605 reference curve slope indicates how well the slope of Reineke's reference curve (Fig. 12-2) may fit the slope of the maximum-density curve for the various species. If for some species there was some discrepancy between its maximum-density slope and the slope of Reineke's reference curve, it is unlikely that densities expressed as SDI values would be seriously influenced. In addition, there is the considerable value of being able to reduce all expressions of density to a common base.

CONCLUSIONS

1 While normal yield tables are essential for the determination of site quality and for estimating the potential yields of an area for a particular species, Bruce-type yield tables based on average stand diameter, normal number of trees per acre, average height of trees of average stand diameter, and normal volume should be developed for their increased simplicity and accuracy. The Bruce-type yield table provides accurate stand-density estimates by number of trees or volume in cubic feet.

2 Reineke's stand-density index provides an accurate measure of the full range of stand densities. It is easily determined and is not dependent on a knowledge of site quality, age, or species. The term *stand-density index* should not be plagiarized by use in other senses than the number of trees in a stand at an average stand diameter of 10 in or its metric equivalent.

3 Percent-of-height method is also an accurate, easily applied measure of stand density that is not dependent on a knowledge of site quality or species. It may be the most flexible measure in that densities can be compared or maintained in which nonsignificant elements in the stand can be eliminated from consideration in the stand-density values, i.e., suppressed and intermediate trees in a high thinning schedule. In fact, percent-of-height method may be the most practical way of applying a particular SDI value on the ground. However, the percentage should be applied to the height of trees of average stand diameter so as to maintain a given stand density.

4 Stand-density index and crown-competition methods may have the potential of measuring the stand density of uneven-aged stands more reliably than the basal-area and merchantable-volume methods now in use.

Chapter 13

Stand Dynamics

The dynamics of stands are based on ecological principles which have contributed to the nature of the stand, such as succession, competition, tolerance, and the concept of zone of the optimum. These factors directly influence growth and development of an established stand, which is the subject of Chapter 14. Knowledge of the interactions of these factors in stand dynamics enables one to predict the way in which vegetation will develop and is thus basic to the development of sound silvicultural prescriptions. Holdridge identified the macrofactors of the environment which create the situation in which a particular vegetation can exist (See Fig. 3-7). Knowledge of soils and site quality refines the expectations in terms of productivity, but the reason why particular species are found in a particular area from the many species suited to exist there or in several of Holdridge's zones has not been fully explained. It is a common experience to move through what appears to be the same macroenvironment and to have the cover types change, gradually or abruptly, for no apparent reasons. However, within one particular area, the driving force that causes changes in the vegetation is systematic and predictable. This driving force is called *plant succession.*

PLANT SUCCESSION

Plant succession is the replacement of one plant community by another. This can occur in slow integrating stages where a site is initially so harsh that few plants can survive on it, or it can happen very rapidly where one community is destroyed by an agent like fire, flood, or insect epidemic and is replaced by another.

Successional changes in plant communities over time have been recognized for a long time, and the observations were finally conceptualized by Clements (1916) into a theory of process with a complete and intricate vocabulary of its own. Essentially, the theory postulates that within a given macroclimate and site, the interactions among soil, plants, and climate will lead to a plant community so attuned to its environment that it will maintain itself in competition with any outside species–this community is called a *climatic climax.* A climatic climax requires a mature soil, stable climate, and no disturbance. These three requirements, however, are unlikely to exist over any considerable space or time because (1) the soils of an area form a mosaic in development depending on time, topography, aspect, and parent material; (2) climate stability occurs only within fairly wide limits; and (3) disturbances occur whether they result from an overmature tree falling in the forest or a family of beaver damming a forest stream. Climatic climaxes do occur, e.g., ponderosa pine at its lower elevations and sizable areas of western hemlock, but it is only a matter of time before a disturbance will set an area back to an earlier stage of succession. Some site conditions cannot directly support a climatic climax but are suspended in their succession at some intermediate point. For example, a "permanent" bog may be designated a physiographic climax, or a serpentine ridge with its high magnesium content would support an edaphic climax, and both could be called subclimaxes. The complete cycle of changes in an area from barrenness to climax formation is called a *sere.* It is a common practice in mapping stages of secondary succession for inventory or management purposes to give cover-type designations based on the predominance of a few species. The cover types take on the status of discrete entities; and by virtue of the rigidity inherent in a type name, the continuum existing in the species distributions, which reflects the past successional forces, is lost. Species from many seral stages are intermingled in a cover type, and the recognition of a continuum would more nearly reflect future expectations than reliance on a cover-type designation.

Primary succession begins on a raw previously unvegetated surface like a sand dune, an earth slide, a lava flow, a rock surface, or even a sterile pond formed by a retreating glacier. Each raw environment would develop a sere appropriate for its climatic conditions, and the rate of community change would depend on how fast the site became favorable for higher seral stages. In the case of a lava flow, eons would pass before the lava would decompose sufficiently to support higher plants, whereas an earth slide would have a relatively rapid transition from raw earth to a climax. Each plant community changes its environment to the degree that is favorable for the pioneers of the next seral stage. This situation continues until finally the site may reach a climax condition and the rate of change is slowed.

A sere does not progress in an unbroken series of advancing stages. Any one of a wide variety of disturbances may set the succession back at frequent intervals. In fact, the disturbance may happen so frequently that the succession is halted at a level below the climax. Douglas-fir in the Pacific Northwest is a fire climax on many sites because fire has to occur only once every 300 to 500 years (the usual life span of Douglas-fir) to prevent succession from moving on to the development of stands of the more tolerant hemlock. The short-grass prairie of the "Great American Desert" once considered a climatic climax, now appears to have been a grazing climax in many places because of the grazing pressure of the enormous buffalo herds the area once supported. When the course of succession is set back by some perturbation, the recovery toward its original stage is called a *secondary succession.*

Foresters are generally managing stands that are developing as part of a secondary succession. In fact, many of the most valuable forest species are components of seral stages below the climax, and the forester's major efforts are frequently directed toward controlling the tendency of a community to move toward the climax species. Each region has its own ecological distinctiveness, and foresters use silvicultural practices to maintain the stands in the seral stage that meets society's objectives most closely. The following examples of natural succession illustrate some particular seral developments.

Engelmann-spruce-subalpine-fir type in Utah forms a climax with the spruce often dominant because it is long-lived. This type is uneven-aged and often interspersed with aspen and is characterized by a mosaic of timber and small meadows. The succession is from meadows to timber, but the destruction of a stand by fire or logging frequently sets the area back to meadow. This succession, which is usually secondary from meadow to timber, takes two forms. Where aspens occur on the margins of a stand, they advance into the meadow by means of root suckers, provide shade, reduce gopher concentrations, and give a favorable seedbed and growing conditions for the establishment of the conifers (Fig. 13-1 and Table 13-1).

A slower process takes place where aspens are missing but occurs only on northern margins of stands. Snow drifts are deep on the lee side and last well into the warmer weather. Consequently, vegetation is sparse. Conifers establish on the margin of the drifts and a few create a favorable environment for further conifer establishment. The advance stops until the new invaders are tall enough to influence the location of the drift.

Douglas-fir in the Pacific Northwest depends on disturbance to maintain itself, so the type has usually been clearcut and the slash burned. The erosion hazard is generally low because the vegetation moves in almost immediately.

Figure 13-1 The succession of meadow to spruce-fir showing the aspen seral stage that provides the microclimate and seedbed conditions favorable to the establishment of Engelmann spruce and subalpine fir, with T. W. Daniel, College Forest, Wasatch National Forest, Utah. (*By Henderson, 1977.*)

The sere usually begins with annuals and perennials, but on badly burned sites it might commence with mosses. The annuals dominate for a few years, then the perennials take over. If Douglas-fir seeds are available, the shade of the invading vegetation favors its establishment, and the trees finally dominate the site. On huge burns like the Tillamook fire in Oregon (Fig. 13-2) or the Yacolt fire in Washington, or where repeated burns give the perennial ground cover or brush the advantage, the Douglas-fir is prevented from becoming established because of the absence of a seed source or the development of a dense stand of vegetation. Salal, bracken fern, vine maple, and red alder may control large areas for years before the conifers invade from the sides. On protected slopes, the more tolerant hemlock forms a component of the invading conifers or it develops as an understory in more mature Douglas-fir. Its presence assures that the succession will go to climax unless people or nature disrupts it. In Clark County, Washington, a few widely scattered Douglas-fir which were more than 1000 years old were all that intruded on the culmination of a climatic climax (Fig. 13-3).

In the northern Lake states, the climatic-climax community would normally be dominated by sugar maple and hemlock, but continual small disturbances create opportunities for the intermediately tolerant yellow birch and the tolerant basswood to remain important constituents. A large disturbance like clear-

Table 13-1 A Transect (10 m Wide) from Meadow into an Aspen Fringe of a Spruce-Fir Stand Illustrating the Rate of Succession from Meadow to Spruce-Fir

Age distribution by species		Distance from a random point in meadow, m								
		0–3	4–6	7–9	10–12	13–15	16–18	19–21	31–33	34–36
		Number of trees								
Populus	1–5 years		9	10	17	13	7	1		
tremuloides	6–10			1	18	2	1	1		
	11–20				2	2	6	1		
	Over 20				1(25)*	20(56)	15(52)	12(72)	8(170)	1(124)
Picea	1–5 years			2	19	57	27	2		
engelmannii	6–10					6	11	1	1	
	11–20				2	13	19	3	1	
	Over 20					3(23)	15(35)	1(43)	5(102)	2(141)
Abies	1–5 years					2	3	2	12	1
lasiocarpa	6–10					2	3	9	10	6
	10–20					7	30	5	7	7
	Over 20					4(25)	6(30)	3(37)	8(128)	16(135)
Pseudotsuga menziesii							1(17)			

*Numbers in parentheses indicate the maximum age for trees in group.

Source: J. A. Henderson, unpublished observations.

Figure 13-2 The Tillamook burn of 1933 after the Douglas-fir had been salvaged, showing hemlock snags and brush as a consequence of two subsequent burns. (*By Daniel, 1950*.)

Figure 13-3 Succession almost at the climatic climax of a western hemlock cover except for a few surviving Douglas-fir (approximately 3 m dbh) in Mount Rainier National Park, with W. Moir and J. Franklin. (*By Henderson, 1970.*)

cutting or heavy partial cut starts the secondary succession, with most of the tolerant species already established, but enables invasion by the intolerant aspens and paper birch. The community takes on the appearance of the invaders, with the climax species in the understory, but the climax community becomes dominant as the fast-growing, short-lived invaders drop out. If one or more fires sweep over the site, secondary succession may be set back to brush (Fig. 13-4), intolerant hardwoods, or conifers, and the sere has a long way to go to reach climax. On sandy soils (Fig. 11-1), fires may reduce the site quality to where a primary sere may almost be the beginning point of the secondary succession.

In the South and Central states, oaks are the climax over the greater part of the region. There are, however, many other shrub and tree species as well as grasses and forbs capable of invading cutover areas or becoming established under an overstory of the seral pines. It takes considerable and continuous effort to maintain the seral stage composed of an intolerant coniferous community. A major problem in holding the community at this stage is the capacity of competing hardwoods to grow more rapidly as a result of their capacity for vigorous, persistent sprouting (Fig. 13-5). Since the pines are quite intolerant, the environment favors the continuous encroachment of the higher seral stages. Thus foresters must commonly use control measures such as fire to hold back successional development.

Figure 13-4 Invasion of brush, mostly hazel, following clearcutting or fire in Minnesota, with Larry Neff. (*By Daniel, 1949.*)

Figure 13-5 An overstory of longleaf pine with an understory of oak in Alabama that will retain possession of the site unless fire or people interfere. *(From photo by Daniel, 1949.)*

Descriptions of plant succession frequently ignore the contributions of the zoological life forms to the environmental changes. It is the burrowing animals and soil fauna that mix the organic debris through the soil profile to improve its nutritional status and water-holding capacity. The whole complex of flora and fauna and physical factors of an environment is called the *ecosystem.* Any disturbance of one element of the ecosystem has an effect on the rest of the ecosystem. This recognition of the ecosystem has concentrated attention on the energy balance, nutrient cycling, diversity, and biomass production along a seral gradient. Currently, it is believed that there is lower nutrient cycling, greater diversity, and higher biomass but lower production as the sere progresses to the climax (Odum, 1969).

The concept of succession has important silvicultural implications for the forester. One must be aware of the eons of time responsible for the forest development in which silviculture is practiced. Grass and brush are very competitive in terms of direct displacement through the seeding-in of trees, so an area may be in trees because trees move from the edges to slowly crowd out the lower vegetation. Thus some ecosystems are delicately constructed, and the recommendation of a poorly chosen treatment may result in a loss of the site to less desirable vegetation. Or even worse, a misapplied clearcutting may result, 5 to 10 years later, in the entire slope being stripped of its soil to bedrock by mass sloughing because the original roots holding the soil in place are finally rotted (Burroughs and Thomas, 1977). Less traumatic than land sloughing but equally illustrative of the lessons of succession is that most tree species are representative of higher seral stages and are not true pioneer plants which can produce their seed and die, or go dormant, ready to come back the next year. These trees must survive throughout the year and be capable of withstanding the unnaturally harsh environments created by inappropriate stand treatments. However, some trees do have pioneering capabilities. For example, what is a suitable site for lodgepole-pine or jack-pine establishment may invite failure for spruce regeneration. When in doubt as to what to plant because of site severity, a step down in the pioneering hierarchy may be an answer. When the desired tree species happens to be a pioneering species, of course, the principle cannot operate.

Secondary seral stages create forest-cover types of intolerant species that are mostly even-aged, while climax-cover types are frequently uneven-aged. There are exceptions for a species like lodgepole pine which may be pure over areas of thousands of hectares. In this case, the lack of competition may give an overmature stand the opportunity to develop a series of age classes. Another example is a spruce-fir stand destroyed by beetle damage that may develop into a relatively even-aged stand through release of an understory. Clearcutting in old-growth northern hardwoods gives a climax composition ultimately without age-class distribution because the removal of the overstory releases the seedling understory.

COMPETITION

Succession is dynamic because species differ in their capacities to develop within a given environment, and the less adapted species are eliminated from that community. To be a successful competitor, a species must have an adequate seed source, a suitable seedbed, adequate growing conditions, and no major susceptibility to disease, insects, or animal damage that would hamper survival. Species differ in the effectiveness of their seed distributions. Species like groundsel and fireweed in the Pacific Northwest appear almost immediately after an area is burnt and may dominate until less mobile species arrive. Both species, in spite of their light weight and plentiful seed, retain their dominance only a short time for probably different reasons: groundsel because the seedbed is no longer favorable and fireweed because the high nutrient level of a fresh burn is no longer available. Seedbed sensitivity is illustrated by Indian mustard, which made an ideal plant for quick erosion control in the Pacific Northwest without becoming a weedy nuisance, because it establishes quickly and is then unable to reestablish on its own litter. Pioneer seral species are particularly sensitive to seedbed conditions in their demand for mineral soil, while climax species are less sensitive to seedbed conditions but more demanding of growing conditions.

Light seed has the competitive advantage of reaching distant areas, but the situation in which the species can become established is frequently limited to the more mesic sites. A heavy seed does not disseminate well, but the larger amount of food reserve permits a rapid root development for establishing under xeric conditions, which explains the success of the oaks and pines like longleaf, piñon, and digger. In terms of competition, a large-seed species can avoid drought by enabling deep root development to tap an adequate moisture supply.

Species differ in the range of growing conditions under which they grow best and in the adaptations which give them a competitive advantage over other plants. Perennials take over from annuals or even new perennials because they become well established and begin growth as soon as weather permits. The competing seedling is dominated by the lush growth of the perennial above ground and a soil mass preempted by an established root system. Competition among perennials is expressed in terms of height dominance–in the Pacific Northwest salal eliminates native blackberry and Douglas-fir can eliminate salal. Among trees, the intolerant species like larch depend on their fast height growth to take advantage of an opened area, while tolerant species like a true fir grow more slowly but can survive under competiton that closes in and eliminates the larch.

Other competitive factors frequently determine the domination of one species over another, such as the relative susceptibility to rot. In an Engelmann-spruce–subalpine-fir community, the subalpine fir dominates the early stages of stand development, then in the pole stage root rots begin to decimate the subalpine fir until at maturity the Engelmann spruce dominates with a few fir. Northern red oak with its aggressively expanding crown tends to increase its dominance in a stand at the expense of such species as white ash and yellow birch, which have easily damaged crowns. Epidemics of larch sawfly in the past have killed the larch and turned the site over to the understory of balsam fir. Competition usually provides that the survivor is the most suited to the site and growing conditions unless factors other than the ability to compete influence the final competition.

TOLERANCE

Succession advances because invading species are better able to compete in the changing environment. *Tolerance* is the forestry term for expressing the relative capacity of a tree to compete under low light and high root competition. *Tolerant trees* reproduce and form understories beneath canopies of less tolerant trees or even beneath their own shade. *Intolerant trees* reproduce successfully only in the open or where the canopy is greatly broken. A knowledge of tolerance and its implications for competitiveness and growth is fundamental to good silviculture and should support every management decision.

Historically, Heyer (1852) was the first to describe the various European tree species according to their need for light and their ability to endure shade. He arranged the species into a tolerance table, and contrary to his meaning, the practicing foresters spoke of them as light-species and shade-species as though the shade-species had to have shade. It was not until Fricke (1904) put some trenched plots in a stand and found that trees growing without root competition grew much better than trees with root competition that the dominance of light in controlling growth was questioned. Without further testing, he ascribed reduction in growth with decreasing light to the increasing root competition associated with lower light intensities. Fricke's school of silvics did not last very long, since Cieslar (1909) grew seedlings in a well-watered garden under different degrees of full sunlight (produced with lath coverings). He maintained the soil moisture of the open-grown seedlings by covering the soil with a sphagnum mulch. Cieslar used the green volume of the open-grown seedlings as a base for comparing the effects of light on growth. Figure 13-6 illustrates his results. All species lost volume at 75 percent full sunlight, but pine and larch were reduced the most. Silver fir exceeded its growth in the open at 66 and 60 percent, but no other species reached its open maximum. Silver fir and spruce were still doing well at 25 percent of full sunlight, while pine and larch did poorly at the low light intensities. In the second year, Scotch pine died at 25 percent and larch at 33 percent. The study proved that light intensity had an effect on the growth of all species, and the impact of a particular intensity varied with the species. It took Fabricius (1929) to quantify the effects of root competition by growing seedlings in trenched and untrenched plots in a stand and in an adjacent area in the open without competiton. The pine seedlings in the trenched plots were 2 to 3 times as big as those in the untrenched plots, but the seedlings in the open were 10 times as big. Toumey (1929) confirmed Fricke's trenched-plot experiments in white-pine stands in New Hampshire. Craib (1934), who worked with Toumey, recognized the importance of root competition and developed spacing and thinning schedules for plantations of *Acacia* and exotic conifers in South Africa. He introduced exceptionally wide spacing and heavy thinning procedures in anticipation of competition rather than to relieve it.

Controversy over the definition of tolerance persisted until the late 1950s, making it one of the longest debated issues in forestry. Consensus is that toler-

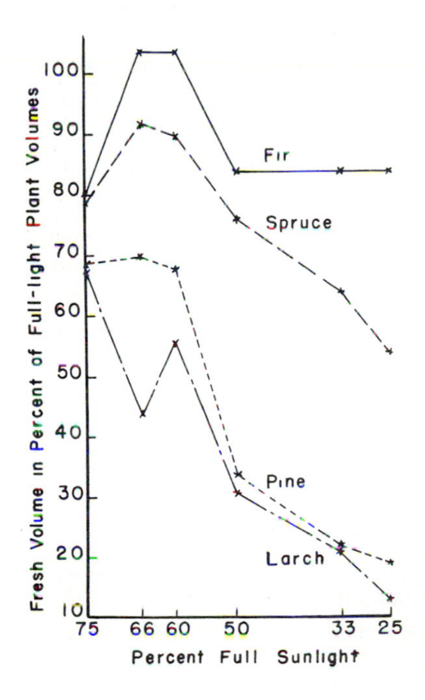

Figure 13-6 The effect of light intensity on the growth of tolerant and intolerant species using green volumes of seedlings grown in the open as a base. (*From Cieslar, 1909.*)

ance cannot be defined as the result of the single factors of light or water. Rather, it is the relative genetic and physiological capacity of the plant to develop in a given environment, with the capacity to withstand low light intensities being generally the most important characteristic.

Factors Affecting Tolerance

Tolerance is not a constant for a species under all circumstances, although two species with a particular tolerance difference will continue to reflect that difference if conditions permit them to grow together in the new situation.

Tolerance is affected by light intensity and root competition, although it is difficult to separate the two. Longleaf pine shows the effect of root competition very well, where fully exposed seedlings established in the root zone of the seed trees are conspicuous for their lack of development. In ponderosa pine on the Mogollon Rim, Arizona, the 1919 seedlings established themselves in dense thickets in the openings, but only scattered seedlings established in areas of root competition. In some circumstances, the root systems of the parent trees exercise no influence on seedling development, such as for loblolly pine in Arkansas (see Fig. 14-6), where rainfall and soil nutrients are adequate for all. Factors like dry, south slopes will decrease the apparent tolerance, and poor soil does the same; thus a species' tolerance will be greater on a rich, moist soil.

Tolerance is markedly affected at higher latitudes; for example, beech requires only 1.8 percent full sunlight to establish in the Vienna woods but requires 20 percent in Norway. The difference represents the need of beech for the warmth at the northern edge of its range.

Age affects tolerance, i.e., species get more intolerant with age. Young balsam fir is more tolerant than red spruce, but as the seedlings get 20 to 30 years old, the balsam fir dies under conditions where red spruce can survive. White-fir seedlings are tolerant but quickly become more intolerant once established.

Tolerant species have a lower light-compensation point and perhaps utilize photosynthate more efficiently than intolerant species. There is certainly more efficiency in retaining leaves for 6 to 8 years in the case of tolerant conifers as opposed to 2 to 3 years for intolerant conifers. The efficiency may also result from an ability to function at a lower temperature.

Criteria of Tolerance

No direct measure of relative tolerance is available, since tolerance is an expression of total genetic and physiological response to an environment. Simple measures of light-compensation point are unsatisfactory since these are commonly made when moisture and nutrients are adequate. Also, the level of light compensation measured is influenced by the immediate past environment to which the plant was exposed. Thus the ranking of trees according to their relative tolerances depends on indirect criteria with considerable subjectivity involved because the criteria, as applied to a comparison of two species, do not always indicate the same rankings. Baker (1950), in summarizing the tolerance tables of individuals from all sections of the country (Table 13-2), found widely divergent placing of some species.

Many of the criteria for estimating tolerance allow the ready separation of the extremes, but the placement within the extremes or within the broad area of intermediate tolerance requires the sensitive balancing of species responses. The following list of criteria for estimating tolerance of trees is given in the approximate order of their usefulness in practice.

1 *The condition of the reproduction* of a species under varying canopy covers and competitive conditions is probably the most widely used and reasonably reliable way of evaluating its tolerance. It is deficient in that adult requirements are judged from those of its regeneration, and it is ineffective for comparing the relative tolerances of two nonassociated species. However, by this criterion, Douglas-fir can have a relative tolerance varying throughout its range from tolerant to intolerant depending on whether it is associated with ponderosa pine and larch or hemlock and western red cedar. On balance, therefore, Douglas-fir is classed as intermediate.

2 *The density of the crown* readily separates the extremes, since intolerant species have relatively thin foliage and open crown and canopy (or low leaf-area index) and tolerant species have thick, dense crowns and canopy cover (high leaf-area index). Some hardwoods such as aspen, yellow poplar, and vine maple have "staggered" placement of leaves to maximize light interception. This concept is discussed by Horn (1971).

3 *Juvenile height growth* tends to be more rapid in intolerant trees than in associated tolerant species when they are both growing in the open. After the seedling stage, tolerant trees grow faster in the open than they do in the shade.

4 *Self-pruning* is useful in separating the extremes of tolerance if allowances are made for stand density. Tolerant trees clean their boles of side branches relatively slowly, since the leaves remain functional in low light and keep the twigs and branches alive. Intolerant species clean their trunks more rapidly, sometimes even when growing in an isolated position in full light, like longleaf pine. They tend therefore to yield a higher proportion of clear lumber when grown to an old age in wild or poorly managed stands.

5 *Number of branch orders* or the length of time the leaves are retained on conifer trees is a difficult measure to apply since the range in number of orders, usually 4 to 8, between tolerant and intolerant species is so narrow. The tolerant species have more branch orders than do intolerant species. Baker (1950) lists bristlecone pine as very intolerant, yet it retains its leaves for 15 to 20 years or more.

6 *Natural thinning* of a stand is faster for intolerant species if the stand has not stagnated. An overly dense stand of most species will stagnate in growth for a period, and the more tolerant the species the sooner some individuals will gain dominance. Stagnation can endure for long periods, and some species like lodgepole pine seem unable to generate dominance once stagnated. For example,

Table 13-2 Tolerance of American Forest Trees

Eastern conifers	
Very tolerant:	Intolerant:
Eastern hemlock	Eastern red cedar
Balsam fir	Red pine
Atlantic white cedar	Pitch pine
Tolerant:	Shortleaf pine
Red spruce	Loblolly pine
Black spruce	Virginia pine
White spruce	Very intolerant:
Northern white cedar	Tamarack
Intermediate:	Jack pine
Eastern white pine	Longleaf pine
Slash pine	
Bald cypress	
Eastern hardwoods	
Very tolerant:	Intermediate:
Eastern hop-hornbeam	Yellow birch
American hornbeam	Sweet birch
American beech	American chestnut
American holly	White oak
Sugar maple	Red oak
Flowering dogwood	Black oak
Tolerant:	American elm
Red maple	Rock elm
Silver maple	Hackberry
Box elder*	Magnolias*
Basswood	White ash
Tupelos	Green ash
Persimmon*	Black ash
Buckeyes	Very intolerant:
Intolerant:	Willows (as a class)
Black walnut	Quaking aspen
Butternut	Bigtooth aspen
Pecan	Cottonwoods
Hickories	Grey birch
Paper birch	Black locust
Yellow poplar	Osage orange
Sassafras*	
Sycamore*	
Black cherry	
Honey locust	
Kentucky coffee tree	
Catalpas	

Table 13-2 Tolerance of American Forest Trees (*Continued*)

Western conifers	
Very tolerant:	Intolerant:
Western hemlock	Limber pine
Alpine fir	Piñon pines*
Western red cedar	Ponderosa pine
Pacific yew	Jeffrey pine
California torreya	Lodgepole pine
	Coulter pine
Tolerant:	Knobcone pine
Sitka spruce	Bishop pine
Engelmann spruce	Big-cone spruce
Mountain hemlock	Noble fir
Pacific silver fir	Junipers
Grand fir	
White fir	Very intolerant:
Redwood	Whitebark pine
Incense-cedar	Foxtail pine
Port Oxford white cedar	Bristlecone pine
Alaska yellow cedar*	Digger pine
	Western larch
Intermediate:	Alpine larch
Western white pine	
Sugar pine	
Monterey pine	
Blue spruce	
Douglas-fir	
Red fir*	

Western hardwoods	
Very tolerant:	Intermediate:
Vine maple	Red alder*
	Golden chinquapin*
Tolerant:	Oregon ash
Tan oak	California white oak*
Canyon live oak	Oregon white oak*
Big-leaf maple	
Madrone*	Very intolerant:
California laurel*	Quaking aspen
	Cottonwoods

*Cases of great uncertainty.

the densest plot found in a stagnated lodgepole-pine stand had 264,000 trees per hectare at 70 years with a height under 1.5 m (Mason, 1915).

7 *Capacity for release* is another important difference between tolerant and intolerant trees. When tolerant trees form an understory, they are very persistent, clinging to life in spite of very small growth for many years. When finally released, they develop very well unless the suppression has been very long and severe. Intolerant trees die out rapidly; and if released before death, they often respond sluggishly to the release.

8 *Density of stems* in fully stocked stands of tolerant species tends to be greater than in stands of intolerant trees of equal age and height.

9 *Stem taper* is less in intolerant species owing to the fact that they lose their lower branches and develop a more cylindrical form. Tolerant trees under equal conditions of stand density would tend to have a more conical form.

10 *Leaf structure* differs between tolerant and intolerant species, with the tolerant leaf containing more spongy parenchyma and the intolerant leaf having more palisade paranchyma. The two forms tend to intergrade if both are in the sun or in the shade.

FACTOR EVALUATION FOR SILVICULTURE

In the broad view of vegetation management, silviculture looms as excessively complex because of the multiplicity of factors involved. Physical factors of the environment in the infinite permutations and combinations of temperature, light, soil properties, soil organisms and nutrients, and moisture as well as physiographic impacts complicate the understanding of growth and reproduction of trees and competing vegetation. Adding to this the large number of species to be considered, each with its own particular requirement for each of the factors, then the complexity of the problem obviously grows. Finally there is the complication of dealing with aggregations of trees of either pure or mixed species on top of which is superimposed the variety of purposes that silviculture may serve, i.e., recreation, timber, pulp, watershed management, wildlife management, and/or pollution control.

In practice, the factors affecting decisions may be a less formidable array of variables because the concern is for a particular area and combination of species and purposes. However, the situation is complicated enough to warrant guidelines as to where to focus attention for maximizing returns from a minimum of inputs. In looking at a particular problem area, the important factors fall into three broad categories: first, factors universally present (such as incoming solar radiation, oxygen, carbon dioxide, and soil) which have limited opportunities for manipulation; second, factors which the forester can modify (such as composition of species, stand density, light within the stand, perhaps wind movement, nutrients through fertilization, temperature, and soil moisture) and by their manipulation better attain management objectives; and third, factors at a critical level or concentration. Almost any factor may, on occasion, fall within this category and limit tree growth (such as light, air impurities, biotic hazards, seedbed, temperature, moisture, or nutrients).

In terms of silviculture, even in broader contexts, there are few situations where improvement would not be acceptable, and there are many where something must be done. In most cases, improvement depends on the solution of problems raised by the critical levels of a few factors. If a deliberate effort is not made to look at a situation and to identify the actual critical factors, one is risking poor or inadequate solutions. This concept of identifying limiting or critical levels of operating factors is expressed by "laws of the minimum," of which there are two.

Liebig's law of the minimum (1855) was expressed by Blackman (1905) as follows: "When a process is conditioned as to its rapidity by a number of separate factors, the rate of the process is limited by the pace of the 'slowest' factor." The law has no validity in a biological situation except at a point where the minimum factor is so low as to stop the process entirely. Liebig recognized that it did not fit the results of his fertilizer trials because when nitrogen was the most deficient factor, he could get an improvement in growth by adding potassium or phosphorus. He postulated the "law of compensating factors," through which the phosphorus could "substitute" for the nitrogen. One element will not substitute for another except possibly in some cases where sodium can replace a little of the needed potassium. Liebig's law of the minimum applies very well to certain mechanical situations, like pouring water into a barrel with varying length staves. Water can rise only to the height of the shortest stave. If more water-holding capacity is wanted, then the short stave must be lengthened. When now a second stave becomes the shortest, it does no good to continue lengthening the first stave.

Mitscherlich's law of the minimum (1921) is stated as follows: "The increase in crop production by unit increment of *any* lacking factor is proportional to its decrement from its optimum." This fits everyday experience in that a person improves growth in a garden by adding commercial fertilizer or manure, not by supplying the minimum nutrient shown by a soil analysis. The law implies that increasing any factor that is below its optimum will improve the rate of process, but increasing the factor furthest from its optimum will give the greatest increase in production. If an added unit of a factor gives a certain increase in production, a second unit will give a lower response, and a third unit will give an even lower response, which corresponds to the economic law of diminishing returns.

The comparison between the two laws of the minimum is illustrated in Fig. 13-7. In this situation, Liebig's law says that production will be improved only by increasing the CO_2 content of the air. Mitscherlich says that bringing any factor closer to the optimal level will improve production, but that the biggest increase, for a given percentage increase of a factor, would be gained by adding to the CO_2 content of the air. It is only in greenhouses that the CO_2 content can

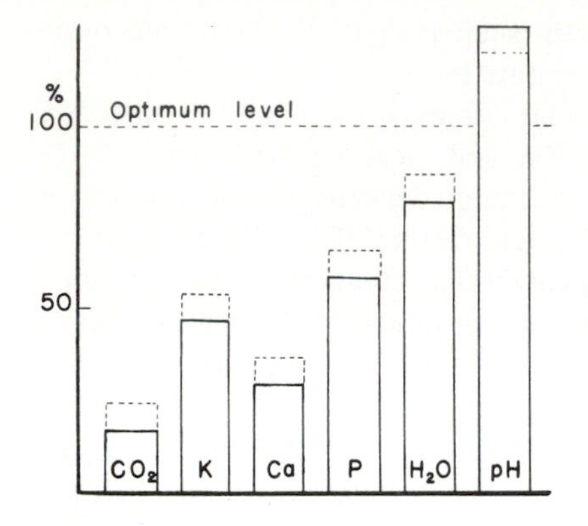

Figure 13-7 A comparison of Liebig's and Mitscherlich's laws of the minimum, where Liebig requires more CO_2 to affect production, while Mitscherlich indicates that addition of any factor below its optimum will add to the yield or a reduction in an overabundant factor.

be increased, but it is quite feasible in many situations to lime the soil. Of course, if only one factor is below its optimum, then both laws predict the same results. A wide variety of decisions about problems, including silvicultural decisions, would be made in a less haphazard fashion if people attempted to identify and ameliorate the effects of obvious minimum (or overabundant) factors in light of Mitscherlich's principles, e.g., in analyzing the factors affecting seedling mortality.

ZONE OF THE OPTIMUM

Mayr (1909), from his years of observations in the United States, Japan, and Europe, proposed that every species has an area where it is best attuned to the climate. Within the natural range of every species, there is an area where it is better attuned to the environment than anywhere else. Mayr divided the potential range of a species into a series of zones based on a temperature gradient from north to south, or altitudinally, with moisture gradients affecting the shape of the zones. He also had a list of characteristics for recognizing the location of a species relative to its optimum zone but emphasized that the characteristic differences assumed comparison between similar sites.

Mayr's zones are the zone of optimum, the zone cooler than the optimum within the natural range, the zone warmer than the optimum within the natural range, the zone cooler than the natural range where the species will survive if planted and beyond that cultivated, and the zone warmer than the natural range where the species will also survive if planted (Fig. 13-8).

Mayr described the *zone of the optimum* as the place where the species is commonly encountered on a wide variety of sites and soils; where on the best sites it reaches its maximum size, age, merchantable length, and fullness of bole (which fall off toward the cooler and warmer zones); where the trees retain their sprouting and seedling ability to an older age than in the cooler or warmer zones; and where regeneration is easiest and more certain. A tolerant tree is most likely

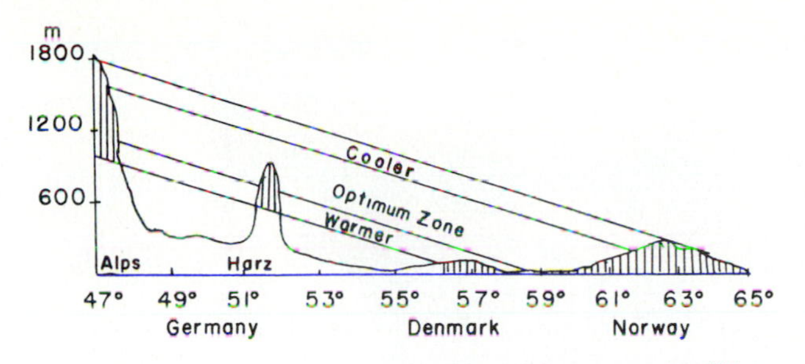

Figure 13-8 Mayr's zones of the optimum for stone pine (*Pinus cembra*)—a timberline species in the Alps—along the 10° East longitude line from the Alps to Norway. (*From Mayr, 1909.*)

to form a climax in the zone of its optimum. There is a gradual transition in these characteristics as the species approaches the cooler boundary of its natural range. The species is encountered only on the warmer and usually drier sites—generally south slopes—and the average leaves are smaller, thicker, and fewer than on similar sites in the optimum zone. It is more intolerant; it grows more slowly; it is shorter-lived and does not seed until later in life and only with small, infrequent crops. Its sprouting is weaker, and regeneration is difficult. The bark percentage increases from the cool zone through the warm zone, while litter accumulation decreases.

As the species approaches the warmer boundary of its natural range, it is limited to the coolest, moistest sites available—generally north slopes—where it has thinner, and more numerous leaves and is more tolerant. Growth is fastest for the first third of its life than in the zone of the optimum, then it slows down. Its seed crops begin early and are heavy and frequent. Sprouts initially develop very vigorously, but they are less likely to mature.

Some of the understanding to be derived from Mayr's observations pertain to competition among species in a mixture—a species in the southern part of its distribution will initially outgrow and perhaps suppress one that is in the cooler part of its range if both have the same potential in their optimum zones. This explains some of the findings in provenance trials where northern provenances seed earlier than local strains and even grow faster initially. The principal value to be obtained from this concept is in sharpening an awareness of whether the climate and site are working with or against management objectives for a particular species. In Eastern United States, transitional areas may cover hundreds of kilometers with gradual changes in the climatic factors; but in the West, a few hundred meters in elevation and a few kilometers in distance cover the range from cool to warm for a species' local distribution. Sargent (1884), in mapping some of the distributions of tree species in North America, indicated the zone of optimum growth for some of them. In Fig. 13-9 he indicates the zone of the optimum for white ash. His map shows white ash on the horizontal plane;

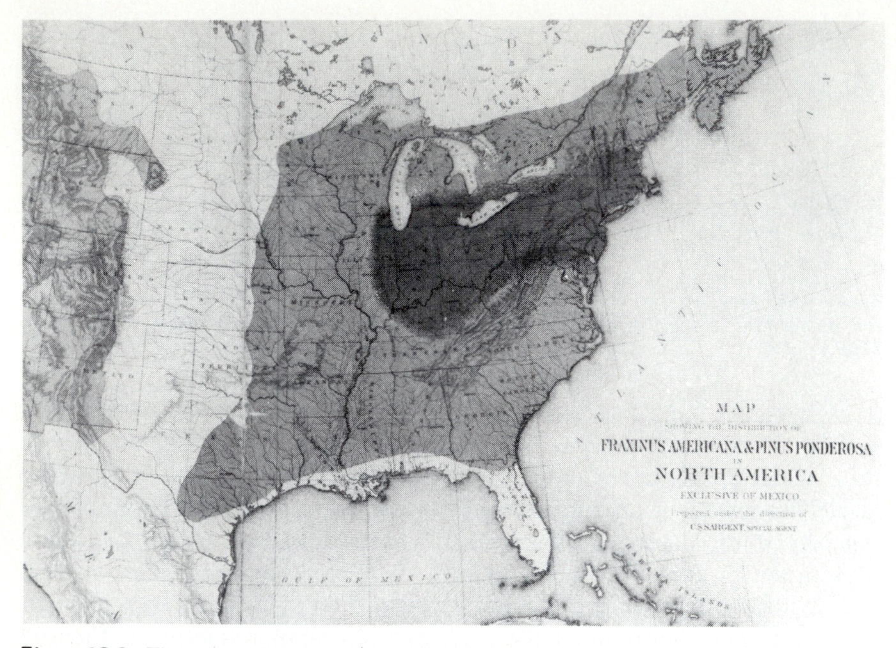

Figure 13-9 The natural range of white ash with the zone of the optimum indicated. (*From Sargent, 1884.*)

yet within the horizontal zone of the optimum, a vertical dimension has its own warmer, optimum, and cooler zones for the species. Serious errors can and have been made, by neglecting Mayr's observations, such as clearcutting where a species is already at an upper or lower boundary in its distribution. While Mayr's observations still provide a basis for understanding what pressures are operating on a species, the development of habitat types and the management implications to be derived from them have a greater potential for focusing the forester's thinking on the needs of a particular stand where habitat types have been de-

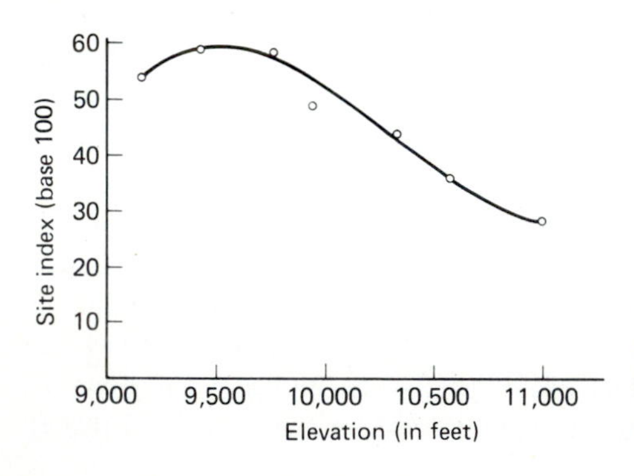

Figure 13-10 The change in site index of Engelmann spruce in the Uinta Mountains of Utah with elevation confirms one aspect of Mayr's zone of the optimum. (*From Henderson, 1977.*)

veloped. Henderson (1977), in habitat typing in the Uinta Mountains in Utah, found that there was a vertical optimum for Engelmann spruce, and its site index was reduced on the warmer and cooler elevations (Fig. 13-10). Any point along the curve represents the average site index and average elevation of all the plots with the same habitat type. Beech has an optimum zone in the Allegheny Mountains and is considered the climax. However, ponderosa pine in the West forms a climax in the warmer, drier part of its range, and the zone of its optimum occurs at higher elevations where the climatic conditions are cooler and moister (see Table 11-3). Ponderosa pine illustrates another element of an optimum-zone concept in that the ultimate optimum occurs at the middle elevations on the west side of the Sierra Nevada. Thus a species with such a tremendous range has local optimum zones too, as Henderson's Engelmann spruce in the Uinta Mountains and Mayr's map of stone pine along longitude 10° East in Europe illustrate very well.

Chapter 14

Growth of Stands

Silvicultural manipulation is on a tree-by-tree basis or by aggregations within a stand, but the major concern of forestry is the effect of treatments on volumes, values, or structure on a stand basis rather than on the behavior of individual trees. The mensurational literature is voluminous on how to make predictions of growth in stands, but biological understanding of growth and of how silvicultural practices affect growth is not well developed.

Total growth in a stand, including possibly stumps and roots, can be expressed in a variety of different ways depending on the purpose of management. These may include total yield over a rotation including volumes removed in intermediate cuts, mean annual increment, periodic annual increment, gross growth, and net growth. Each of these terms can be expressed in a variety of units (cubic feet, board feet, cubic meters, cords, tons, or metric tons). Analyses of growth are influenced by the choice of units, and unit selection depends on the management objectives and utilization potential.

The concentration of interest on xylem production and its measurement ignores the production of other forest products that offer appreciable economic returns, such as Christmas trees, ferns and other floral trade items, medicinal barks and herbs, and resin. Nor do these measures provide a means of evaluating nonmarketable values, such as recreation, water production, and aesthetics. In addition, more emphasis is needed on use and measurement of the total tree (Young, 1974), including bark, branches, leaves, and stumps, if the full growth of a stand is to be made useful. Increasingly, bark is being used as mulch, as a constituent of particle board, as a source of waxes and chemicals, and as a source of energy. In addition, work being done for very-short-rotation pulp production using the silage concept of silviculture (Steinbeck et al., 1970; Dutrow and Saucier, 1976) is focusing attention on above-ground biomass production.

While growth of stands in terms of xylem production has been studied for almost 200 years, the work has been almost limited to even-aged, reasonably pure stands. From this early work, the first yield tables for the principal local species were constructed by the German forester Paulsen in 1795 (Assmann, 1961). It has been the collection and analysis of yield table data that have developed the principles underlying the growth of stands. More recent studies have shown why individual trees behave the way they do in response to stand treatments. Stand growth is influenced by such a complex of interacting factors that the choice of even-aged, pure stands for study to reduce the complexity made progress easier. Uneven-aged stand management has a long history too, but no one has been able to reduce growth expectancy of these stands to tabular form because separate tables would be required for each combination of species mixture and site. However, computer simulation modeling may overcome this problem. Some yield tables do exist for mixed species in even-aged stands, such as for hemlock and Sitka spruce (Meyer, 1937) and spruce and fir (Meyer, 1929).

STAND DEVELOPMENT

Growth of even-aged, pure stands is affected by stage of stand development, age, site quality, species, density in terms of basal area and number of trees per acre, stand treatments, and the units in which growth is expressed. Inasmuch as the factors influencing growth of stands are commonly interdependent, the principles of stand growth usually have to be developed by observing factor interactions.

Even-Aged Development

Stand development is a function of stand age, but its character is dependent on species and site quality. As the site quality improves, a stand of a given species reaches a particular stage of development at a younger age. The stages of development generally recognized are: seedling, 1 m (3 ft) tall; small saplings, 1 to 3 m (3 to 10 ft) tall; large samplings, 3 m (10 ft) to 10 cm (4 in) dbh; small poles, 10 to 20 cm (4 to 8 in) dbh; large poles, 20 to 30 cm (8 to 12 in) dbh; thrifty-mature, 30 to 60 cm (12 to 24 in) dbh; mature-overmature, over 60 cm (24 in) dbh. The specific diameters appropriate to a given stage for a particular species may require modification according to the characteristics of the growth pattern. Figure 14-1 illustrates the stages of development as a function of the growth pattern of a species, with the pole stage corresponding to the "grand period of growth." Diameter at breast height (dbh) is used on the ordinate axis in Fig. 14-1 because the age corresponding to particular stages of development vary with site quality and species. In the case of Douglas-fir in the Pacific Northwest, the "grand period of growth" probably ends with an average stand diameter greater than 30 cm on the better sites, but the general values of height and diameter for the stages of development are useful divisions.

The development of even-aged stands is therefore structurally very simple. Using one of the three classical reproduction methods (clearcutting, seed tree, or shelterwood, described in Chap. 18), stands are regenerated. They may be manipulated by intermediate treatments (Chap. 17), and grown to harvest. Except during the relatively short regeneration period, the structure of even-aged stands for most of the rotation is identical, and one could not distinguish between them once the seed trees or shelterwood cover is removed. Of major

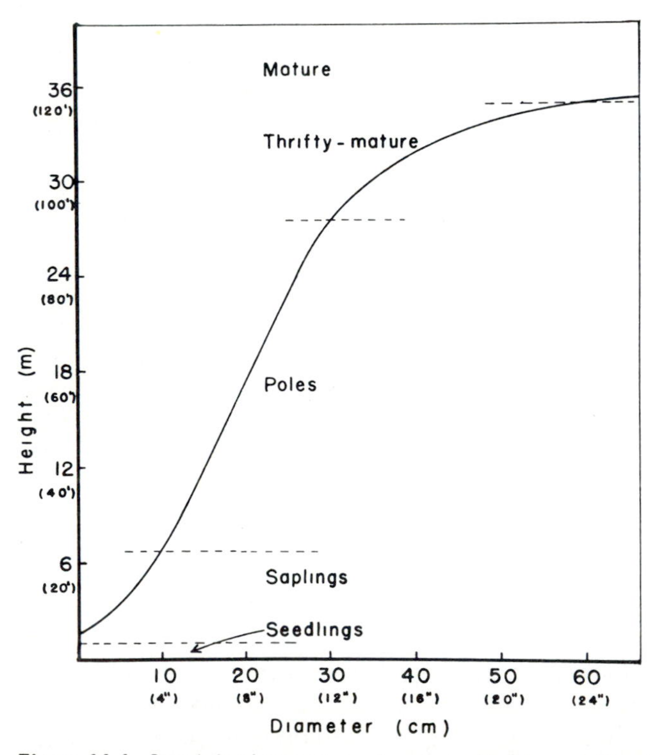

Figure 14-1 Stand development as a function of height and diameter growth, with the pole stage generally associated with the "grand period of growth" in the stand.

concern to the silviculturist in developing even-aged stands is the control of density (see Chap. 12). By controlling density, the silviculturist not only controls growth but is able to control the presence or absence of understory. This is accomplished through a familiarity with the silvical characteristics of the plants involved (Chaps. 5 through 9) and with the characteristics of the site which control growth (Chaps. 10 and 11). Microclimates can be developed under the even-aged tree structure such that the temperature, light intensity, and other site factors are suitable for growth of the particular understory plants desired. In this way, even-aged stands can be developed which are both functional and attractive and have the desired structure and composition to suit most diverse management objectives, including wood production, watershed, wildlife habitat, and aesthetics.

Uneven-Aged Development

In uneven-aged stands, there are a number of stages of development present on a given area. The structure of the stand depends on the emphasis placed by the silviculturist on the growth of particular size classes in that structure. Yields are regulated through control over growing stock, and this is done by controlling (1) residual stocking level to be left after harvesting, (2) the diameter of the largest tree, and (3) the number of trees desired in each diameter class.

First, as suggested earlier, total stand growth for most species is relatively constant over a wide range of stocking levels. Stocking goals, usually expressed in basal area or volume, are commonly set near the lower end of this range so that growth is concentrated on the fewest larger trees without losing growth through understocking. This minimizes the time taken to grow trees to a given diameter and also minimizes the capital investment represented by the standing timber or growing stock. Second, the maximum diameter of trees to be left after harvesting depends on management objectives. For timber production, this maximum diameter is commonly regarded as being about 60 cm (24 in) dbh, since beyond this size, trees are unlikely to be capable of growing at an acceptable rate of about 5 to 6 percent per year. Third, to control the numbers of trees in each diameter class requires some method of defining the characteristic inverse-J-shaped curve of uneven-aged stands. The position of the curve on the abscissa is fixed by the choice of the tree of largest diameter; the residual stocking level determines the position of the curve between the axes; and the slope of the curve is determined by the desired distribution of diameter classes. This slope is defined by the diminution quotient q, which expresses the ratio of the number of trees in any diameter class to the number in the next higher diameter class, resulting in a stepwise curve. Usually q ranges between 1.3 and 2.0 for 2 in diameter classes; however, changes in width of diameter class result in different values of q. Low values of q result in a flat curve for the frequency distribution of diameters, which produces a stand with a relatively high proportion of the growing stock allocated to the larger diameter classes. High values of q result in a curve which produces a stand with a greater proportion of small trees. It can be seen, therefore, that for a given level of growing stock a q value can be chosen to develop the uneven-aged stand which has the most suitable structure for management objectives. Table 14-1 (Alexander, 1977) shows the distribution of numbers of trees in each diameter class for a residual growing stock of 100 ft^2 basal area and maximum dbh of 60 cm (24 in) for various q values.

Previously unmanaged, uneven-aged stands are usually not brought under complete regulation to the desired curve defined by q in one harvest. This would commonly open the stand too rapidly, and even tolerant species would not be able to immediately fully utilize the site, thus resulting in loss of productivity. The silviculturist has to evaluate the current vigor and growth potential of species in all diameter classes and judge the extent to which the stand can be opened up. In mixed stands with mosaics of vegetation, this analysis must be made for each aggregation of vegetation separately. Consideration must be given to the potential impact of harvests on windfall, insect and disease susceptibility, impacts on wildlife habitat, slash problems, and in particular, the creation of appropriate microsites for tolerant and/or intolerant regeneration. After these appraisals, determinations can be made on the heaviness of the cut and the number and interval between cuts in order to bring the stand to a fully regulated condition.

The various structures that can be created or maintained in an uneven-aged stand can be defined by de Liocourt's law (de Liocourt, 1898; Popescu-Zeletin,

Table 14-1 Residual Stand Structures for 100 ft^2 of Basal Area and Maximum Tree DBH of 24 in for Various q Values
(All data on a per-acre basis)

	q = 1.3		q = 1.5		q = 1.8		q = 2.0	
Diameter class	Trees, no.	Basal area, ft^2	Trees no.	Basal area, ft^2	Trees, no.	Basal area, ft^2	Trees, no.	Basal area, ft^2
4	38.90	3.38	79.08	6.89	156.01	13.62	210.18	18.35
6	29.90	5.87	52.72	10.34	86.68	17.01	105.09	20.63
8	23.02	8.04	35.14	12.26	48.15	16.81	52.54	18.35
10	17.96	9.65	23.43	12.78	26.75	14.59	26.27	14.33
12	13.62	10.69	15.62	12.26	14.86	11.67	13.14	10.32
14	10.47	11.20	10.41	11.12	8.26	8.83	6.57	7.02
16	8.07	11.26	6.95	9.70	4.59	6.41	3.28	4.58
18	6.21	10.97	4.63	8.18	2.55	4.50	1.64	2.90
20	4.77	10.41	3.08	6.73	1.42	3.09	0.82	1.79
22	3.67	9.68	2.06	5.42	0.79	2.08	0.41	1.08
24	2.82	8.86	1.37	4.30	0.44	1.37	0.20	0.64
Total	159.14	100.01	234.49	99.98	350.50	99.98	420.14	99.99

Source: Alexander, 1977.

1936). The general equation describing this law, which produces a continuous negative exponential curve, is

$$N = ke^{-aD}$$

where N = the number of trees per acre
k = a coefficient representing the number of trees per acre at 0 dbh
e = the Napierian constant
a = the slope of the distribution, which controls the rate at which the number of trees changes between successive diameter classes
D = the diameter

De Liocourt's equation is used to generate the inverse-J relationship for any uneven-aged distribution. An example of its use to calculate the diminution quotient q, and the relationship between q and a, is given in the footnote.[1]

Moore (1964) analyzed the structure of an Engelmann-spruce-subalpine-fir forest in Utah and evaluated the potential structures that existed in the stands

[1] The formula describing de Liocourt's law is used to determine the number of trees in a particular diameter class.

Assuming 2 in diameter classes, then the fractional increase in number of trees per acre (N) in a given class (14 in class) over the number in the next higher diameter class (16 in class) is

$$\frac{N_{14} - N_{16}}{N_{16}}$$

From de Liocourt's law, $N = ke^{-aD}$

$$\frac{N_{14} - N_{16}}{N_{16}} = \frac{ke^{-a \cdot 14} - ke^{-a \cdot 16}}{ke^{-a \cdot 16}}$$

$$= \frac{ke^{-a \cdot 14}}{ke^{-a \cdot 16}} - \frac{ke^{-a \cdot 16}}{ke^{-a \cdot 16}}$$

$$= \frac{ke^{-a \cdot 14}}{ke^{-a \cdot 16}} - 1$$

$$= \frac{ke^{-a \cdot 14 + a \cdot 16}}{k} - 1$$

$$= e^{a \cdot 2} - 1$$ (The *fractional increase* in number of trees from the next higher 2 in dbh class)

Then $\frac{N_{14}}{N_{16}} = e^{a \cdot 2}$

= diminution quotient q for 2 in dbh classes (The *actual ratio* for the number of trees in a 2 in dbh class to the number in the next higher diameter class)

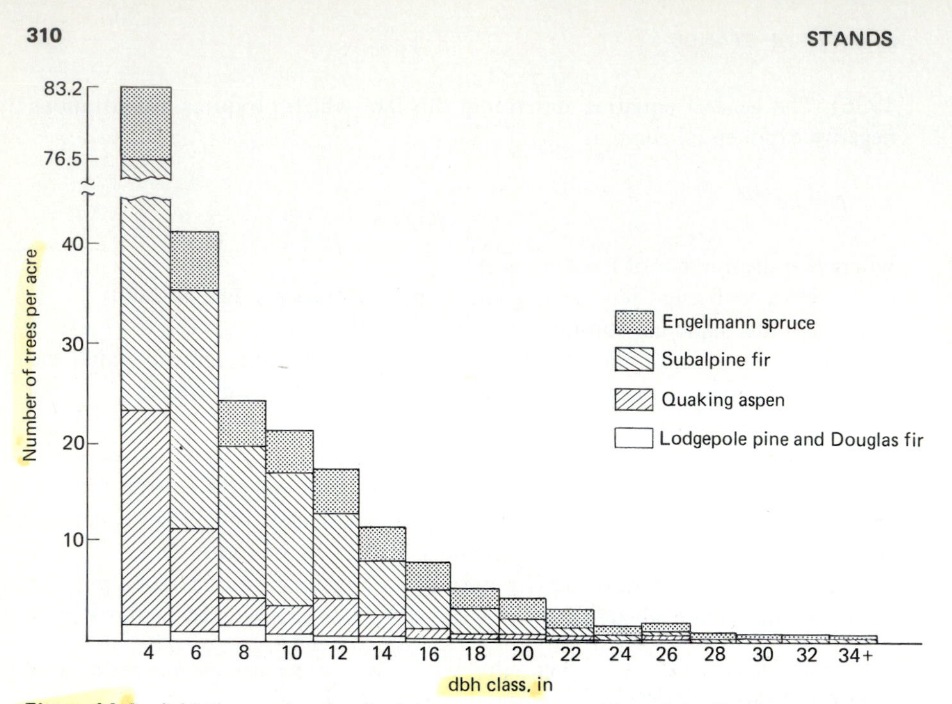

Figure 14-2 A histogram showing the structure and composition of the Engelmann-spruce-subalpine-fir uneven-aged uncut stratum on the College Forest, Logan, Utah, based on the 1950 inventory using 274 plots (¼ acre each). (*From Moore, 1964.*)

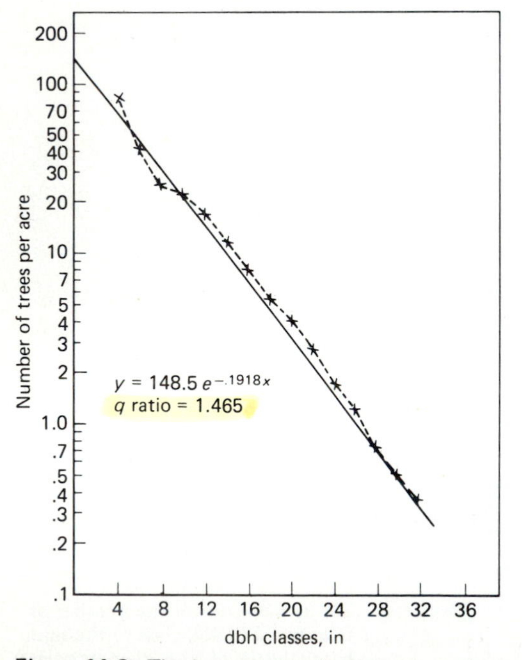

Figure 14-3 The least-squares fitted curve to data provided by the 1950 inventory of the College Forest, with the de Liocourt formula's constants. (*From Moore, 1964.*)

from the 1950 inventory of 274 quarter-acre plots. The distribution of species and diameter classes for these plots is given in Fig. 14-2. A least-squares fitted curve for the structure with its de Liocourt's constants and q value is given in Fig. 14-3. Using the frequency distributions by 2 in diameter classes of the trees on 21 samples of five contiguous quarter-acre plots, a family of 21 structures was found to be present in the stand, with q values varying from 1.2 to 1.5 (Fig. 14-4). This illustrates the variability of structure within a natural stand and the fact that there is considerable choice available in selecting different structures to suit management objectives. Similarly, within a single forest-cover type, different stands may be managed using different diminution quotients. In introducing management, the same diminution quotient does not have to be

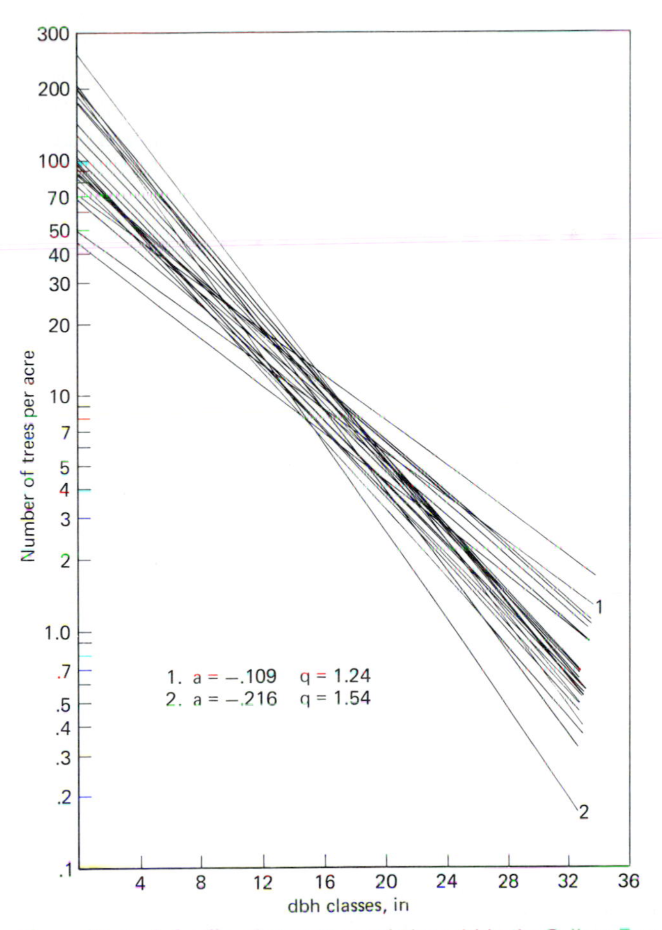

Figure 14-4 A family of structures existing within the College Forest based on 21 samples, with each sample containing five contiguous ¼ acre plots and using 2 in diameter classes. The minimum and maximum *a* values and corresponding *q* values are indicated. (*From Moore, 1964.*)

applied throughout the whole diameter range. One value of q could be used for the larger diameter classes and another for the smaller classes.

The diameter distribution in all uneven-aged stands over the country tends to follow the q concept (Meyer, 1952); in other words, natural uneven-aged stands tend to be balanced. This is shown in the examples from several forest-cover types (Table 14-2).

It should be remembered that q values are derived from de Liocourt's law and do not give any estimate of density; diminution quotients are strictly related to stand structure.

The growth of various diameter classes in an uneven-aged stand can be strongly influenced by the silviculturist through treatment. Through selective harvesting, the structure of the forest can be made to change to suit management objectives. This has been well demonstrated by Solomon (1977) who, working in the northern-hardwood type in New England, determined the influence of density (measured by basal area) and the percent of sawtimber on the growth of poles and sawtimber. In establishing the study he created 12 different stand structures, and after 10 years, the low-density stands had appreciable structural changes while the dense stands had only small changes in structure.

The data showed that if high saw timber *production* was desired, then the present sawtimber should be appreciably reduced in basal area to allow the pole-sized trees to rapidly grow into sawtimber. To obtain maximum *growth* on sawtimber, a high percentage of sawtimber should be retained in conjunction with maintaining a high basal area. In this latter case, growth on pole timber is not appreciable, and a longer time is taken for pole timber to become "in-growth" and develop into sawtimber.

In both the northern-hardwood example (Solomon, 1977) and the Engelmann-spruce-subalpine-fir type (Moore, 1964), the species were for the most part tolerant, with some admixture of intolerants, but the same control over structure is possible for intolerants, as demonstrated by Reynolds (1969) in mixed loblolly and shortleaf pines in Arkansas. Figure 14-5 shows the in-

Table 14-2 Range in q Values for Various Unmanaged Forest-Cover Types

Type	q Value (approximate)	Author
Ponderosa pine in Southwest (densities from 30 to 100 ft^2 per acre)	1.24	Schubert, 1974
Northern hardwoods in New England	1.2–1.8	Solomon, 1977
Engelmann spruce–subalpine fir, Utah	1.24–1.54	Moore, 1964
Engelmann spruce–subalpine fir, Rocky Mountains	1.3–1.5	Alexander, 1977
Loblolly pine–shortleaf pine, Arkansas	1.25	Reynolds, 1969
Mixed-conifer type, California	1.3–1.5	Grah, 1977

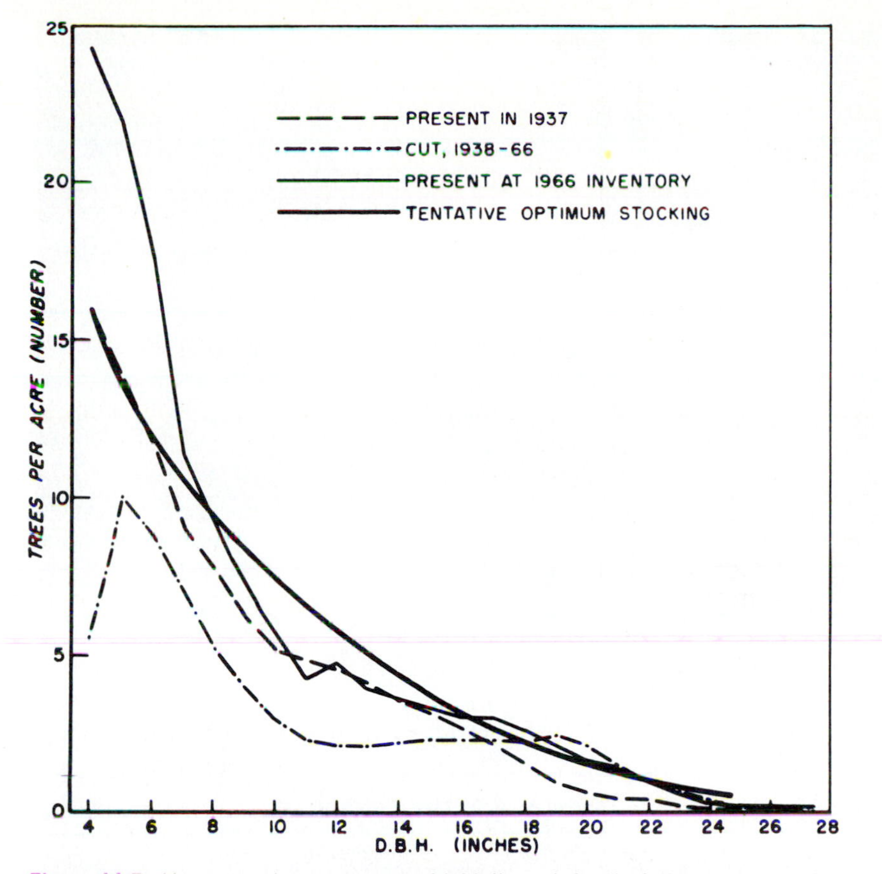

Figure 14-5 Uneven-aged management of loblolly and shortleaf pine mixture at Crossett, Arkansas, showing the production and change in numbers of trees in the various diameter classes over the 29 years of the study with a q value of approximately 1.25 for the optimum stocking curve. (*From Reynolds, 1969.*)

verted-J curve with an approximate q value of 1.25 and changes in stand structure accompanying harvesting. The selection method can be very productive under intensive management. Almost as much material was removed in 29 years as was in the original stand, while the present stand has an appreciable increase in volume over the original. Figure 14-6 shows a view of one of the stands at Crossett, Arkansas.

BASAL AREA IN EVEN-AGED STANDS

Basal-area increment is a measure of the capacity of trees or stands to produce xylem elements. This production is partly controlled by genetics but is also influenced by all biotic, physical, and chemical factors of the environment. As

Figure 14-6 Loblolly and shortleaf pine mixture growing under uneven-aged management at Crossett, Arkansas. (*From Reynolds, 1969.*)

stands develop, basal area accumulates and is maximized under conditions of full stocking. Silvicultural treatments redistribute basal-area increment but may reduce total production in even-aged stands.

Basal Area as a Function of Site Quality and Age

Understanding silvicultural treatments usually requires working with yield tables. For a given species, entrance into a yield table requires a knowledge of site quality and age of the stand. Since basal area is one of the most significant characteristics in a stand's description, a knowledge of the effects of site quality and age on basal area is necessary for a discussion of stand growth. The rate at which basal area accumulates with age is the most dramatic growth rate in forestry. A stand of 10,000 seedlings per acre at the end of their succulent stage will have an approximate total basal area of 0.14 ft^2, but at 20 years western white pine will have 46 ft^2 and western hemlock at the same age on a low-medium site will have 151 ft^2. The effect of site quality and age on basal-area accumulation is illustrated for Douglas-fir (McArdle et al., 1949) in Fig. 14-7. This figure shows three general characteristics: (1) that Douglas-fir has an initially high rate of growth, especially on the better sites; (2) that there is a tendency for basal-area growth to level off when the stand is mature; and (3) that the difference in basal-area production between the higher site qualities as age increases is relatively small compared with that for the lower sites. A small improvement in site quality of a poor site consequently gives a large increase in basal-area production. This is one element of a decision as to whether or not to fertilize poorer sites.

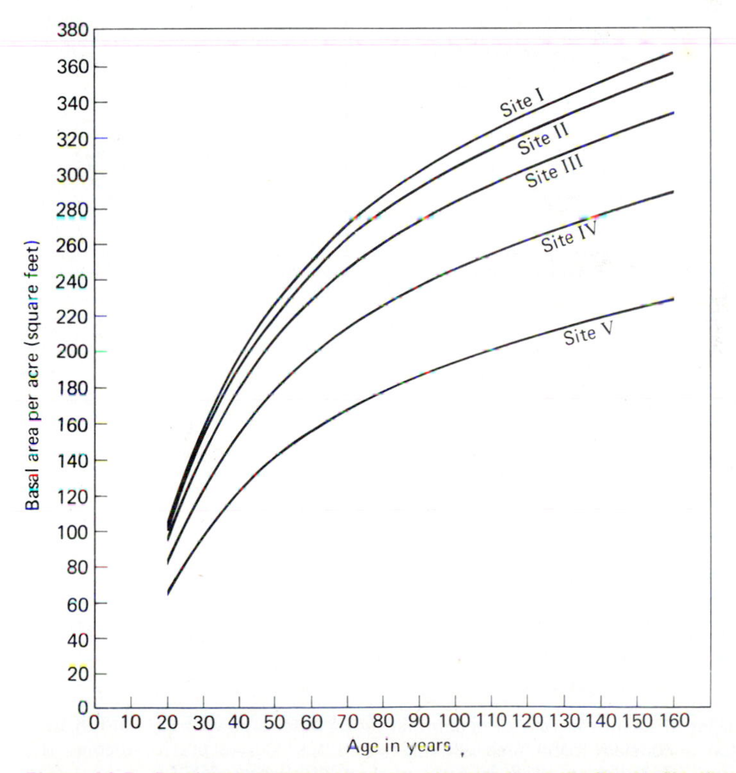

Figure 14-7 Basal-area accumulation for Douglas-fir in the Pacific Northwest as influenced by site quality and the age of the stand. (*From McArdle et al., 1949.*)

Basal Area as a Function of Species and Age

An extreme example of the tendency for well-stocked, unmanaged stands to level off with age is given by ponderosa pine (Fig. 14-8). Ponderosa pine levels off at 60 years of age for all its site qualities. The site qualities differ in the steepness of the early growth curve and the total basal area at which the curve flattens. This uniformity of age among the site qualities for the break in the rate of growth curve for basal area is less evident in other species, but a significant slowing of growth does occur consistently between 110 and 130 years for Douglas-fir sites and between 80 and 100 years for western-hemlock sites. Within the limits of available yield table data, no other species flattens out in basal-area production like the intolerant ponderosa pine. In contrast, the tolerant red fir, even with a 285 ft^2 basal area on its best site at 160 years, added 12 ft^2 in the last 10 years (Schumacher, 1928).

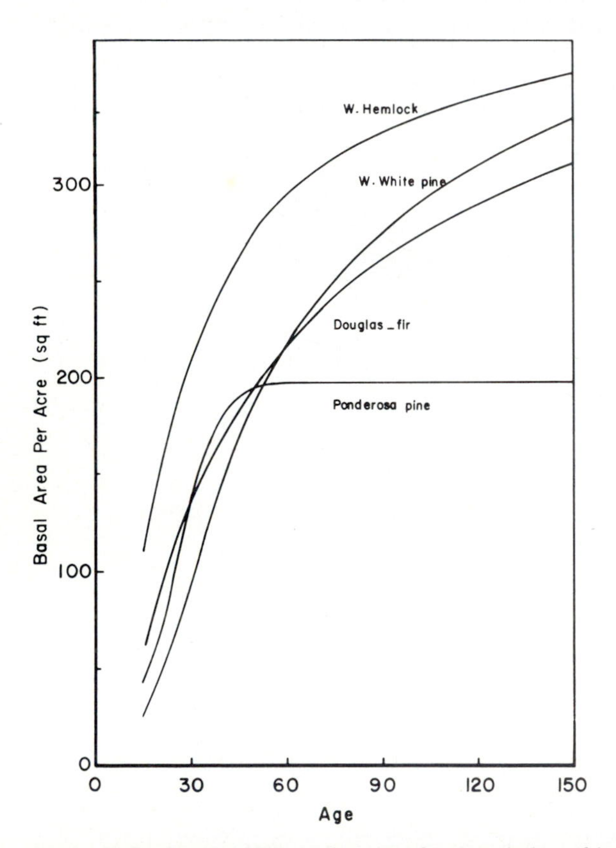

Figure 14-8 Species differ in the rate of accumulation of basal area with age and in their tendency to reach a constant basal area with advanced age. Curves are for median site quality. (*From Haig, 1932; Meyer, 1938; McArdle et al., 1949; Barnes, 1962.*)

Table 14-3 Basal-Area Variation with Age, Site Quality, and Species, ft²

	Western white pine*				Western hemlock†				
Age	40	50	60	70	100	130	160	190	210
20	45	46	46	47	127	151	160	166	172
60	215	218	221	223	276	295	306	315	319
100	286	289	292	296	315	334	346	355	359
140	323	327	331	335	335	355	367	377	382
160	338	342	346	350	342	363	376	386	392

*Site-index age is 50 years.
†Site-index age is 100 years.
Source: Haig, 1932; Barnes, 1962.

Species vary in the extent to which their basal area differs at a given age from one site quality to another. Thus western white pine demonstrates a constancy of basal area at a particular age irrespective of site quality (Table 14-3); yet it shows the least tendency to reduce basal-area growth with age of the species in Fig. 14-8. However, the poorest site has 97 percent of the basal area of the best site at a given age. Ponderosa pine is extreme in this instance too, since the poorest site has less than 50 percent of the basal area of the best site. However, species as different as western hemlock and jack pine have their poorest site producing 87 percent as much basal area as their best site at maturity.

It would be a nice simplification if the variation in basal area could be tied to a species' tolerance. There is enough correlation to ensure that tolerance does have its influence, but erratic behavior with respect to a species' tolerance, probably due to the fact that tolerant trees become more intolerant with age, indicates that many other genetic and site differences influence basal-area growth in a stand.

Effect of Number of Trees on Basal Area and Volume

As the number of trees per hectare increases, the basal area will increase with each tree, adding an equal amount of basal area until the trees begin to compete with each other (Baskerville, 1965). After between-tree competition begins and intensifies as tree numbers increase, the sum of the tree basal areas will increase, but the basal area per tree will decrease. Finally, with an excessive number of trees, the sum of the tree basal areas will decrease, as will the average basal area per tree. The reasons for the reduction in total basal area with increasing tree numbers include the possible reduction in photosynthate production, the increased proportion of respiration to photosynthesis, and, what is most likely the principal cause, the constriction of the supply of moisture and nutrients because of poor root development.

Langsaeter (1941), in discussing the effects of thinning on volume growth, described the effects of density (with density given in terms of volume in cubic meters) on volume growth (Fig. 14-9). While the actual curves in quantitative

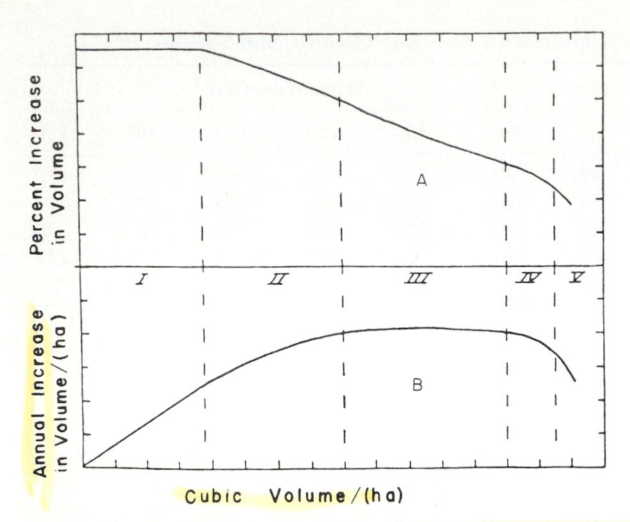

Figure 14-9 Langsaeter's curves showing the effect of increasing cubic volume per hectare on the annual increment (current year's growth) and on the percent of growth. Actual values for the axes are not known, but the principle is important in the silvicultural manipulation of a stand. (*From Langsaeter, 1941*.)

values are unknown for different soils, species, or ages, the form of the curve for a particular soil and age is considered to be fairly well established. He divides the curves into five segments based on the response of the stand to increases in volume. In zone I, which represents trees growing independent of each other, the yearly growth increases directly as the cubic volume increases (curve B), with growth a constant percentage of stand volume (curve A). In zone II, which begins when the growth rate is reduced by the beginning of competition between the trees, the annual growth continues to increase with increasing density but at a decreasing rate (curve B), while the percentage growth decreases as the stand volume increases (curve A). In zone III, within which the growth rate changes very slowly in spite of increasing stand volume (curve B), the percentage growth of stand volume continues its steady decline and is nearly inversely proportional to the stand volume increase (curve A). In zone IV, the annual growth and the percentage growth decline at an increasing rate as the stand volume increases (curve B). In zone V, the annual growth rate (curve B) and the percentage growth of the stand volume (curve A) decline very rapidly. In this zone, the resistance of the trees to insects and diseases is low. Langsaeter concludes that zone III would be much the same for silver fir and Norway spruce and thinning would be productive only in zone III and part of zone II.

Since Langsaeter specifies that his curves are applicable only to a *particular site and age*, and if it is *assumed* that height growth is relatively constant throughout the zones, then an increasing volume can be attained only by increasing the number of trees. However, volume can increase only if an added tree increases the total volume. Figure 14-10 indicates that the total volume increases with number of trees up to a certain density (*critical density*), after which increasing the number of trees will cause a reduction in total volume. The critical density will vary with species, also probably with site quality and age, but may be constant for a given average stand diameter. Curves derived from yield table data for red fir (Schumacher, 1928) show the critical density for basal area and cubic volume (Figs. 14-11 and 14-12) at about 130 percent of the normal yield table values for number of trees per acre.[2] The close correlation between increasing basal area and cubic volume is illustrated in Fig. 14-13. This correlation depends on the height being reasonably constant. The point A in Figs. 14-11, 14-12, and 14-13 illustrates the effect of an excessive number of trees on basal area, volume, and height; basal area was reduced by only 5 percent and cubic volume by 23 percent, so height must have been seriously reduced since any change in the form factor would be insignificant and tend to favor greater volume.

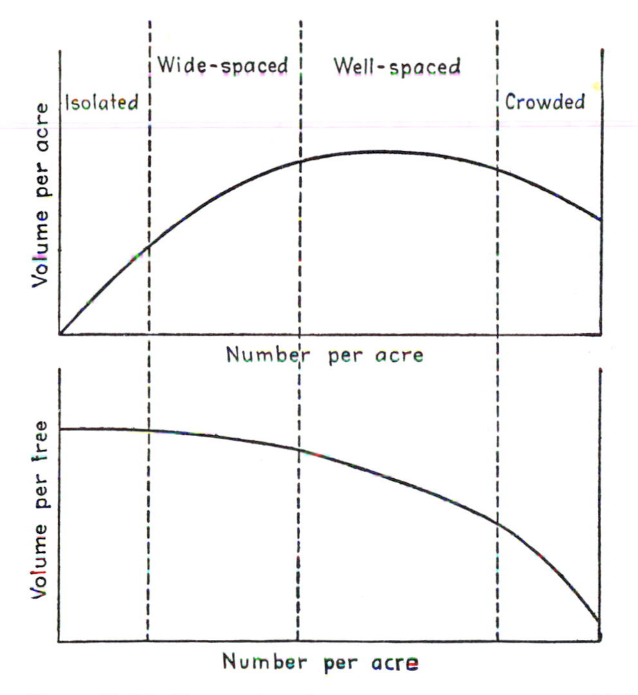

Figure 14-10 Changes in volume per acre and per tree with increasing numbers of trees. (These general relationships were originally derived for agricultural crops and have been shown to be valid for trees.)

[2]By expressing field-plot data as a percent of the normal yield table values for a given plot's site index and age, the resulting percentage data are independent of site quality and age, as plotted in Figs. 14-11, 14-12, and 14-13.

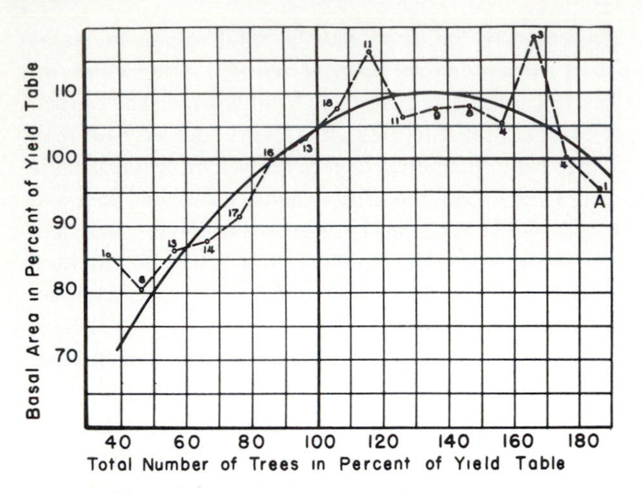

Figure 14-11 Relation between basal area and number of trees per acre from red fir yield table. (*From Schumacher, 1928.*)

The significance of the critical density is that Langsaeter's curves apply to stand development up to the critical density and not beyond it because his curves are based on continually increasing volumes at a particular age on a given site. His zone III begins at about 60 percent full stocking (see Möller's curve, Fig. 14-23) and ends at or slightly beyond full stocking. His zone IV corresponds to the continuing rise in basal area and volume with increasing numbers of trees beyond full stocking, while zone V is the flattened portion of the curve before the critical density is reached. In Fig. 14-12, if full stocking is 100 percent nor-

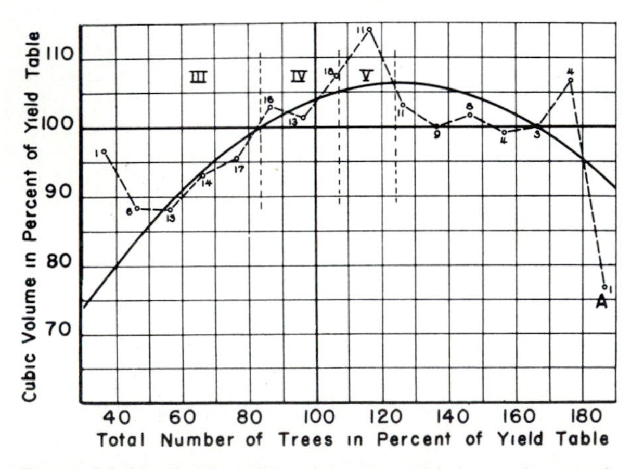

Figure 14-12 Cubic volume increases with increasing number of trees for red fir until the number of trees reach the critical density, then cubic volume decreases with increasing number of trees. Langsaeter's zones (III to V) are superimposed on the curve on the assumption that volume increase is obtained only by increasing tree numbers (with constant site, age, and species). (*Revised from Schumacher, 1928.*)

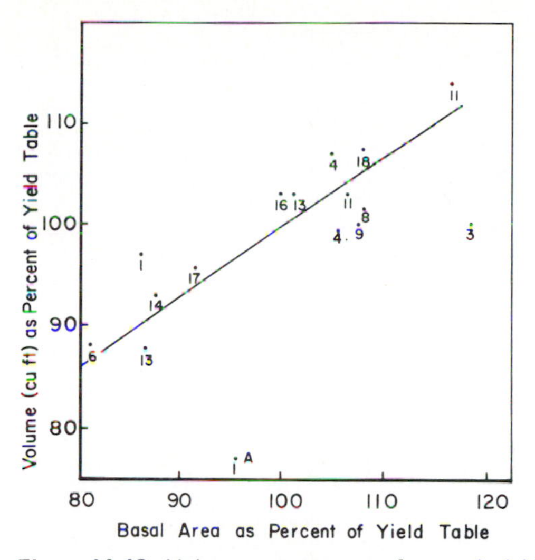

Figure 14-13 Volume as a percent of normal yield table values increases as the basal area increases. (*From Schumacher, 1928.*)

mal volume on the curve and represents the end of zone III, then zone IV would end at 105 percent full stocking and zone V would end at 107 percent (critical density). The interpretation of Langsaeter's curves gives a third dimension to Fig 14-12, in that not only is the average tree getting smaller in spite of increases in total volume in zones IV and V, but the annual increment is falling off rapidly. If the annual increment has dropped badly at the critical density, stagnation must be really severe at higher numbers of trees. It is possible for two stands of the same species, age, and site quality to differ appreciably in volume, and for the stand that has the higher volume per hectare (therefore, the higher *mean* annual increment) to have the lower *current* annual increment. This could occur, for example, when one stand is in zone III of Fig. 14-9 and the other in zone V. The growth mechanisms that produce this situation are not well understood, but Fig. 14-12 seems to confirm Langaseter's observations.

In the red fir examples of basal-area and cubic volume increases with number of trees, the failure of the curves to pass through the intersection of the two 100 percent scales is confusing [100 percent cubic volume (Fig. 14-12) is reached with 83 percent of the normal number of trees]. It indicates that normal yield table estimates of density cannot be accepted as entirely precise, although density estimates based on volume or basal area would agree reasonably well. This is probably true because the tables are constructed from samples of stands that are considered full stocked.[3]

[3]The failure of the curve to go through the intersection of 100 percent for both axes results from normal values being data averages, and in Fig. 14-12, 80 of the plots (points on the curve) had below-normal numbers of trees as compared with 69 that had above-average numbers. It should be noted that for red fir, volume and basal area continue to increase with more trees until tree numbers are approximately 50 percent greater than at 100 percent normal volume or basal area.

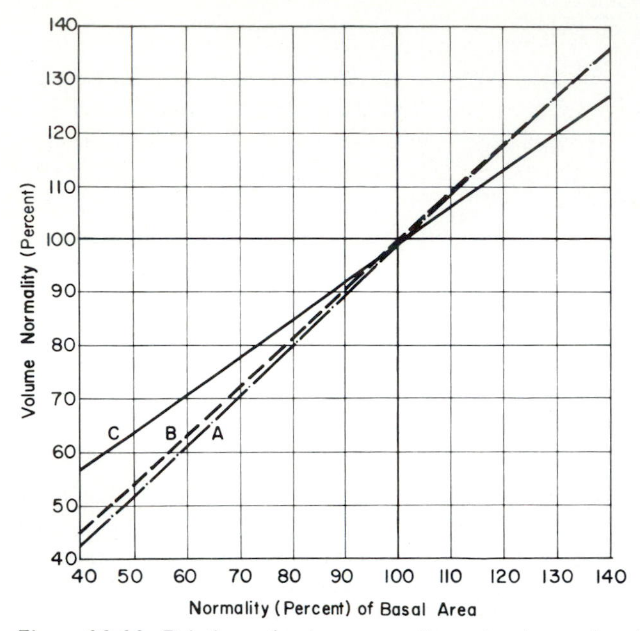

Figure 14-14 Relations of volume normality to basal area: A, cubic-foot volume; B, board-foot volume, International rule; and C, board-foot volume, Scribner rule. (*After Meyer, 1938.*)

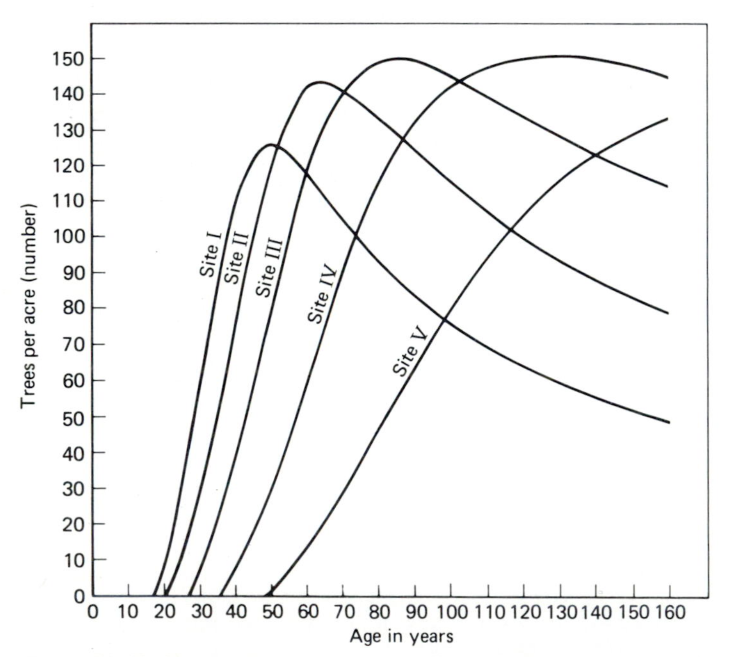

Figure 14-15 Number of trees per acre 12 in in diameter and larger for various ages and site qualities for Douglas-fir. (*From McArdle et al., 1949.*)

It is common for the maximum basal area (critical density) to occur when the number of trees per hectare is above yield table normal values. For Douglas-fir, the maximum basal area occurs at 130 percent of the normal number of trees, and the added 30 percent in tree numbers contributes only 5 percent more basal area. With red spruce (Meyer, 1929), at 180 percent of the normal number of trees the basal area has increased by only 10 percent. Red spruce has a balanced yield table since 100 percent of the normal number of trees will provide 100 percent normal basal area.

Basal area in closed stands has been shown to have an excellent correlation with volume for a given species, site quality, and age. This relationship holds best for cubic-foot volume (Fig. 14-14). When volume is expressed in board feet or other units where material below a certain utilization standard is disregarded, the correlation with basal area improves as the trees in the stand grow beyond the limiting diameter. If the merchantable diameter for board feet is 12 in, then a stand accumulates basal area with no board feet until some trees reach 12 in in diameter (Figs. 14-15 and 14-16). This is in contrast with the accumulation of cubic-foot volume in Fig. 14-17. Minimum log size is now much less than 12 in dbh in most parts of the country, which reduces appreciably the differences in yields, as represented by Figs. 14-16 and 14-17.

The interactions between basal area and number of trees and their effects

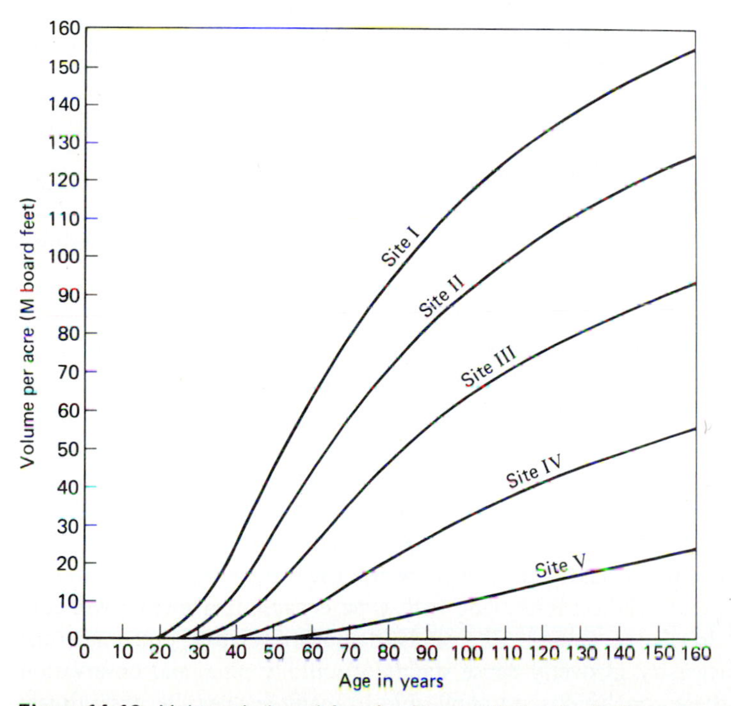

Figure 14-16 Volume in board feet for Douglas-fir by the Scribner log rule for trees 12 in in diameter and larger for various ages and site qualities. (*From McArdle et al., 1949.*)

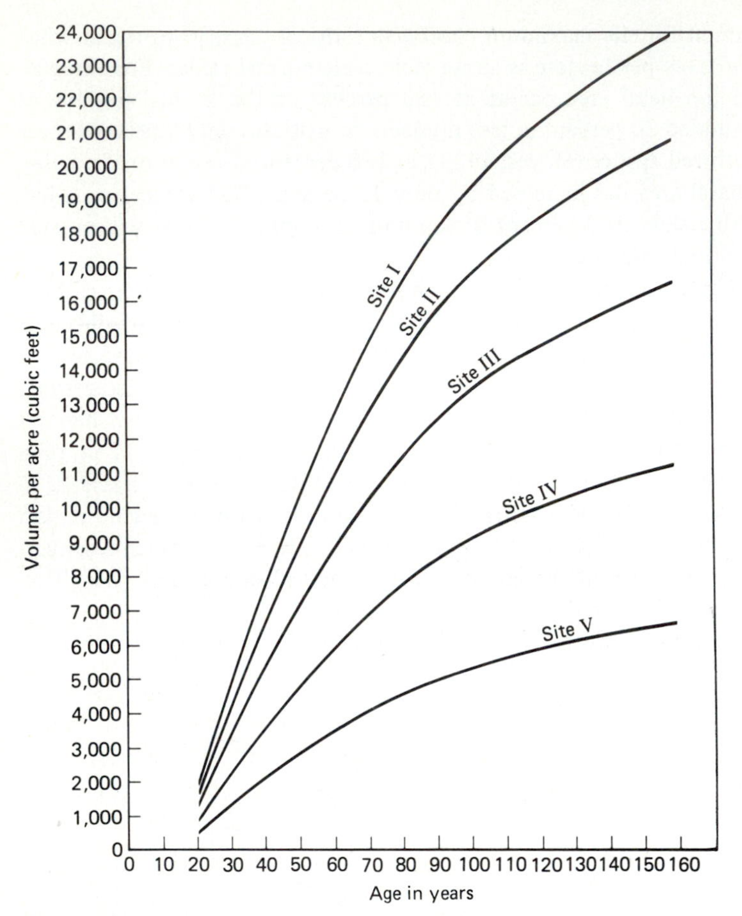

Figure 14-17 Total yield in cubic feet for Douglas-fir with age for various site qualities. (*From McArdle et al., 1949.*)

on growth can be illustrated in a precommercial-thinning study (Table 14-4) in 7-year-old loblolly pine (Grano, 1969). This study showed that 12 years after treatment, as thinning intensity increased, both height growth and diameter growth increased. Basal area appears to be inversely related to volume growth until it is recognized that these data reflect the situation at one particular age during stand development. At age 19, treatment 1 had accumulated basal area very rapidly as compared with treatment 3, almost a 10 to 1 difference. Any future measurements of basal area will show the usual relation of basal area to volume, with the correlation improving with time because treatment 1 will continue its rapid basal-area growth. The study demonstrates that precommercial thinning is a necessity in overly dense stands of loblolly pine, and observation indicates that these conclusions are applicable to many species. Because a mea-

Table 14-4 Loblolly Pine Growth as the Result of Two Precommercial Thinnings*

	After thinning: age 7			Age 19					
Treatment	Number of trees	DBH, in	Basal area, ft^2	Number of trees	DBH, in	Basal area, ft^2	Tree height, ft	Volume, ft^3	SDI
1	1,850	0.06	5.3	547	5.9	103	52	1,188	224
2	5,400	0.05	11.5	1,687	3.5	110	48	694	322
3	30,520	0.05	51.1	2,717	2.8	120	44	301	373

*Treatment 1: machine plus hand thinning; treatment 2: machine; treatment 3: control.
Source: Grano, 1969.

sure of density based on tree numbers and basal area is a product of the levels of these factors, and because both tree numbers and basal area change as the stand develops, the interpretation of any given measure of density must change as the stand grows. The use of Reineke's stand-density index (SDI) overcomes this problem, and has been included in Table 14-4, since at any given SDI the competition among trees is the same regardless of the number of trees present.[4] Knowing that the maximum SDI for loblolly pine is 450, the relative densities of the 3 plots in Table 14-4 can be more readily compared. The yield table for loblolly pine (U.S. Forest Service, 1929) shows the SDI varying from 353 to 291 on site index (SI) 60 and from 300 to 265 for site index 100, so treatment 1 has a density well below normal. This illustrates that in the management of young-growth stands, densities should be maintained at much lower levels than in "normal, fully stocked" stands, i.e., the lower end of Langsaeter's Zone III.

Basal-Area Increment per Tree

It is a surprising fact that for a normal, fully stocked stand on any given site, basal-area increment per tree[5] remains relatively constant over the range in age from 40 to 160 years (Table 14-5). This is in contrast to the almost doubling of total

[4] Reineke's SDI applied to even-aged pure stands of a species.

[5] Calculation of basal-area increment per tree is possible because the stand tables in the Douglas-fir yield table (McArdle et al., 1949) have the same total number of trees as given in the yield tables for various site qualities at different ages. (This is in contrast to some yield tables which show a different number of trees from the number in the stand tables.) Since a single stand table is given for each site class, the median site index within a site class was chosen as most likely to fit the site-class stand table. The method of calculation assumes that the trees lost to competition between ages were of the smallest diameters in the stand. By progressively subtracting these smallest-diameter trees identified in the stand table at age 60 years from those listed present at age 40 years until the total equaled the trees lost between 40 and 60, the diameter distribution of the trees lost could be determined and their total basal area calculated. By subtracting the total basal area of trees lost from the stand basal area at 40 years, the residual basal area is the basal area of the trees that survived to be 60 years old. The difference in average basal area per tree for the same trees at 40 and 60 years gave the growth of the average tree in 20 years.

Table 14-5 Basal-Area Increment per Tree with Age on Two Douglas-Fir Sites as Compared to Total Basal-Area Increment from Normal Yield Table Data

	Site index 140		Site index 170		Percent of SI 170 to SI 140	
Age	Basal area	Basal-area increment per tree	Basal area	Basal-area increment per tree	Basal area	Basal-area increment per tree
40	177	–	189	0.346	107	–
60	226	0.235	241	0.360	107	153
80	259	0.271	276	0.428	107	158
100	283	0.250	302	0.431	107	172
120	301	0.263	322	0.426	107	162
140	317	0.260	338	0.352	107	135
160	331	0.229	353	0.385	107	168
Average		0.251		0.390	107	155

Source: McArdle et al., 1949.

basal area over the same period. The table shows that for Douglas-fir the mean annual increment per tree on SI 170 of 0.390 ft^2 is 55 percent greater than the mean annual increment on SI 140 of 0.251 ft^2. This contrasts with the 7 percent increase in total basal area for SI 170 over SI 140. Anamorphic curving of plot data accounts for the uniformity of the 7 percent increase in basal area and may contribute to the uniformity in the increments per tree, but the performance of the average tree within and between sites gives a better picture of stand development.

VOLUME GROWTH OF A STAND

The basic implication in site quality measurements is that they measure the productivity of the site for a particular species over a given period of time. It is debatable whether a forester can influence the maximum volume that a site can produce without introducing a superior genotype or changing the environment by fertilization or watering. The cubic-foot volume of a stand is a function of its basal area and height, while board-foot volume depends on the average stand diameter, number of trees, and height. Therefore, if volume production is to be improved, a forester's manipulations must be able to change one or more of these dimensions to favor added growth.

Height Growth of Stands

The height to which a tree will grow on a given site by a certain age is assumed to be fixed by the soil and climate, except as affected by stand density. Height growth of trees is adversely affected in both excessively open and very dense stands. Since height development in closed stands remains reasonably constant, volume has a straight-line relation to basal area. Insofar as increasing the number of trees per hectare adds to basal area and maintains the straight-line basal area–volume relation, density has no significant effect on height growth. When basal area diminishes with increasing numbers, volume growth and probably height growth are also reduced. In general, therefore, good management of stands which are well stocked has no impact on height growth.

Basal Area

Basal area varies widely among plots selected as fully stocked for normal yield table construction because these measurements are taken at a discrete point in time in the development of the stand. The circumstances responsible for the current level of stocking on a plot at the time of measurement cannot be fully

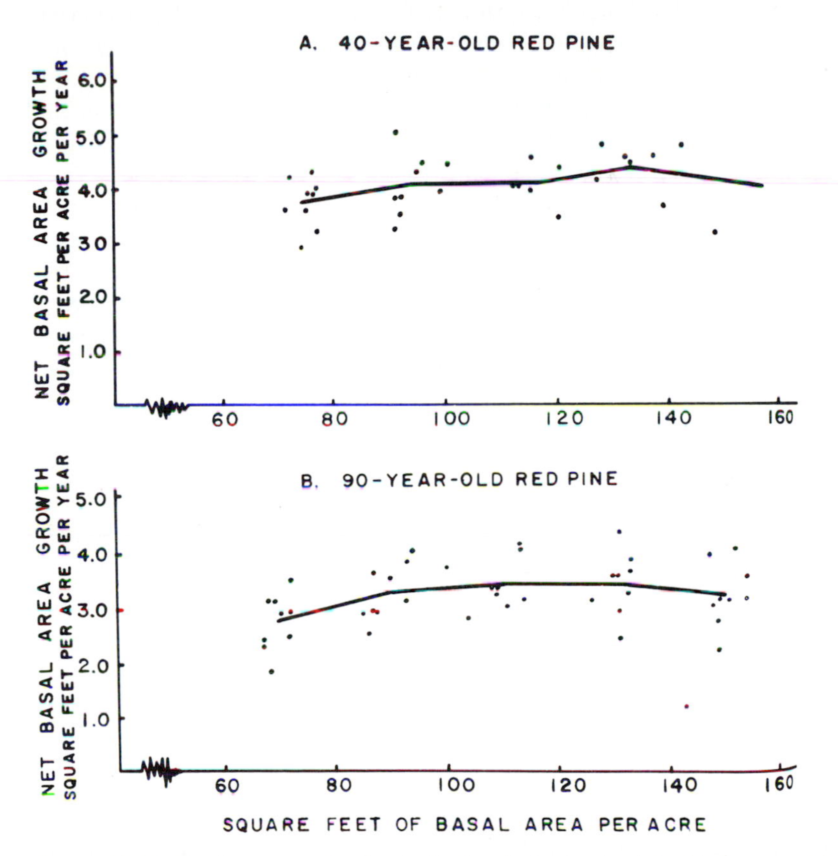

Figure 14-18 Basal-area growth of red pine in relation to stand density (measured by basal area) for two experiments on the Cutfoot Experimental Forest, Minnesota. (*From Buckman, 1962.*)

evaluated. One stand may have reached full stocking very recently from a previously understocked condition and another may have been overstocked for a long period. As indicated in Chap. 12, the tendency is for stands to approach normality. Therefore, it would be difficult in managed stands to hold the basal area at any point much above the normal level, so production could not be increased by increasing basal area to higher levels. In addition, net basal-area growth remains almost constant over a wide range of stand densities, as is shown in Fig. 14-18 for red pine. However, there is evidence that young red pine stands with rapid height growth may grow more cubic-foot volume at the higher densities (Buckman, 1962).

Average Stand Diameter

Average stand diameter is defined as the diameter corresponding to mean basal area. It is *not* the average diameter of the trees present because diameter is a linear function and basal area is a quadratic function of the diameter. The mean diameter is also poor as a stand parameter because it is badly distorted by any extra-large or extra-small trees in a population.

Diameter growth is one of the important factors over which the forester can exercise considerable control. The effect of a small increase in number of trees has been shown to give appreciable reduction in average stand diameter. While the increase in basal area results in greater cubic-foot volume, the reduction in average stand diameter will reduce the board-foot volume of the stand, with a consequent reduction in stand value. The reduction is particularly serious at the margin of merchantability. The change in board-foot/cubic-foot ratio with changes in the average stand diameter in California Douglas-fir is illustrated in Fig. 14-19 (Schumacher, 1930). A stand with an average stand diameter of 6 in provides only 2.5 board feet per ft^3; at 32 in the ratio is 7 board feet per ft^3 (Scribner decimal C scale).

The number of trees per acre will influence the average stand diameter. In red fir in California, an increase in number of trees to 180 percent of normal yield table values will reduce the average stand diameter by 25 percent; a decrease in number of trees to 40 percent of yield table values will increase the average stand diameter by 20 percent (Fig. 14-20). The actual increase or decrease in average stand diameter with number of trees will vary with the species, but the influence of density on diameter is evident in every species.

Care must be taken in evaluation of the increase in average stand diameter over time. Measurements taken after a period of growth tend to be inflated by the mortality, usually of the smallest size classes, which has occurred during the period. For mixed stands of Sitka spruce and western hemlock, almost half the yield table increases in average stand diameter are the result of mortality among small trees (Fig. 14-21). The amount of false growth is negligible in open stands of low mortality, but it is appreciable in overly dense stands with high mortality of small trees. Estimating stand diameter growth by increment cores underesti-

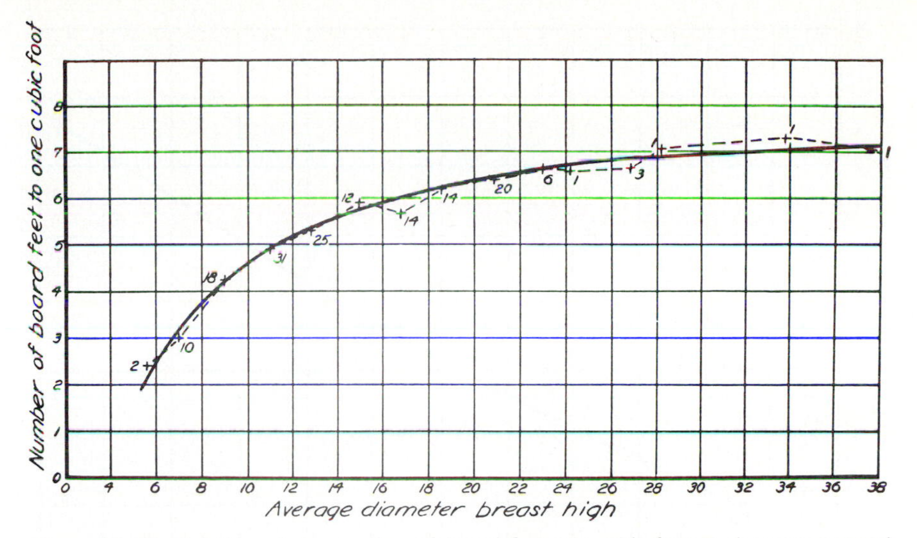

Figure 14-19 Relation of the number of board feet per cubic foot to the average stand diameter in California Douglas-fir. (*From Schumacher, 1930.*)

mates to some extent the future average stand diameter of a stand unless some correction can be applied for this false growth. Low thinning has the same effect on average stand diameter as normal mortality has in an unmanaged stand, and the heavier the thinning the more pronounced the increase in the average stand diameter of the residual stand.

From this discussion of the factors that influence the volume growth of stands, it becomes evident that there is little a silviculturist can do to increase *total* cubic-foot growth. If a stand has the desired composition and remains fully

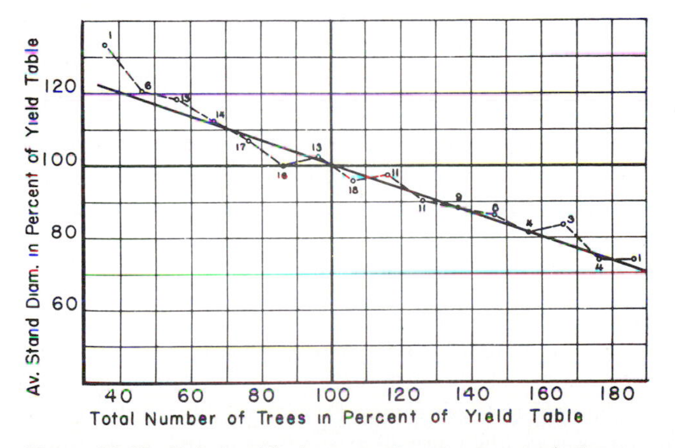

Figure 14-20 Relationship between the average stand diameter and the number of trees per acre in red fir. (*From Schumacher, 1928.*)

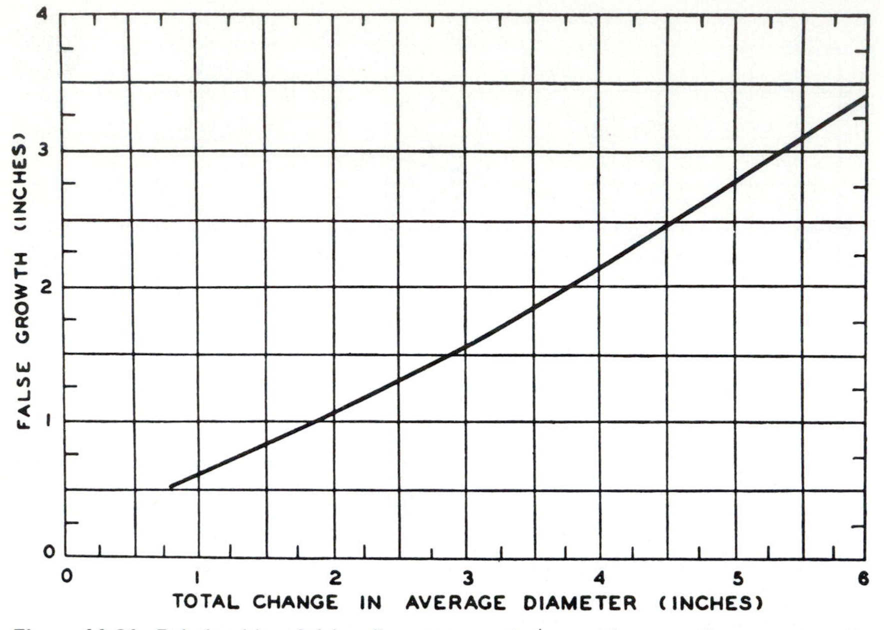

Figure 14-21 Relationship of false diameter growth (caused by mortality among smaller trees) to total change in average stand diameter over a 20-year period for fully stocked, even-aged, mixed stands of Sitka spruce and western hemlock. (*From Meyer, 1937.*)

stocked throughout the rotation, the final basal area and height are predetermined by the rotation age. But the problems of ensuring desired composition and full stocking with quality trees throughout a rotation demand the utmost skill on the part of a silviculturist. Only two areas seem open to manipulation beyond controlling composition and obtaining full stocking: (1) the average stand diameter can be influenced by manipulating density, but the manipulation would have to be carefully done unless some reduction in yield at final harvest is acceptable; and (2) an increase in *total yield* can be accomplished by salvaging the potential mortality.

Volume production and value production are not likely to be synonymous in many circumstances, so any stand manipulations have to balance these two objectives. In the case of thinnings which reduce crop basal area, the improvement in stem quality and greater diameter increment of the crop trees may make value production outweigh the loss in final volume.

TREATMENT EFFECTS ON VOLUME GROWTH OF STANDS

Generally, the effect of thinning is to lower total yield (compared to a stand that is unthinned but where its mortality is salvaged at frequent intervals) but to increase *merchantable* productivity. Some studies, however, have shown that total yield can be increased under some circumstances.

In European thinning trials, it is common practice to salvage the mortality on the unthinned control plots and to include this volume in the measure of total production. By not including this salvaged mortality in measures of production on the unthinned control plots, which would certainly be the practice in this country, thinning can be shown to increase total yield if stocking is maintained. The fact that 50 percent of the annual cut in a country like Sweden has been from thinnings, in spite of some loss in total volume production, indicates the importance of the value increment that results from the improved quality and increased diameter growth of the final harvest. Currently in Scandinavia, as elsewhere, the desirability of thinning stands to increase yields is being reevaluated in light of the increasing costs of labor and mechanization.

Leaf Surface Area with Treatment and Site

It has been the general assumption that the various intermediate cuts and partial cuts in stands managed on an uneven-aged basis increased crown depth and opened up more leaf surface to light so that growth was enhanced. In summarizing his own and other European research, Möller (1947) found that stand leaf area was almost constant for a species from initial control of site to maturity (for beech from age 20 to 120 years in Denmark) over a wide range of site qualities and stand treatments. If a thinning caused a stand to be opened to where it could not close in a reasonable period, the leaf area was reduced. Changes in climatic conditions, elevation, and species would affect the leaf area per square meter of a stand (leaf-area index), while year-to-year changes in local climate had no observable effect. In the Bowmont study of Norway spruce in Scotland (Mackenzie, 1962), crown surface area was also found to be much the same for all low thinnings. Since the leaf area was constant, the lower production with decreasing site quality was attributed to the lower efficiency of the leaf surface caused possibly by a poorer supply of nutrients or water.

Effects of Thinning on Volume Production

Associated with the concept of constant leaf area over a range of densities is the general reporting of constant increment on a given site over a range of densities. Solomon (1977) found the average annual gross growth in basal area to be relatively constant over a wide range of residual basal areas and size classes in the northern hardwoods of the Northeast. There was some decline as the residual basal area and the percentage of sawtimber increased. Stand-density experiments for red pine on the Cutfoot Experimental Forest show relatively constant basal-area increments for varying densities (see Fig. 4-18).

The oak-dominated upland hardwoods of the central region (Gingrich, 1967) have the same volume increments over a range of average stand diameters and densities (Fig. 14-22). These diagrams of the upland hardwoods indicate the range of basal areas for each average stand diameter that will give uniform growth increments without regard for age and site quality. Although interesting

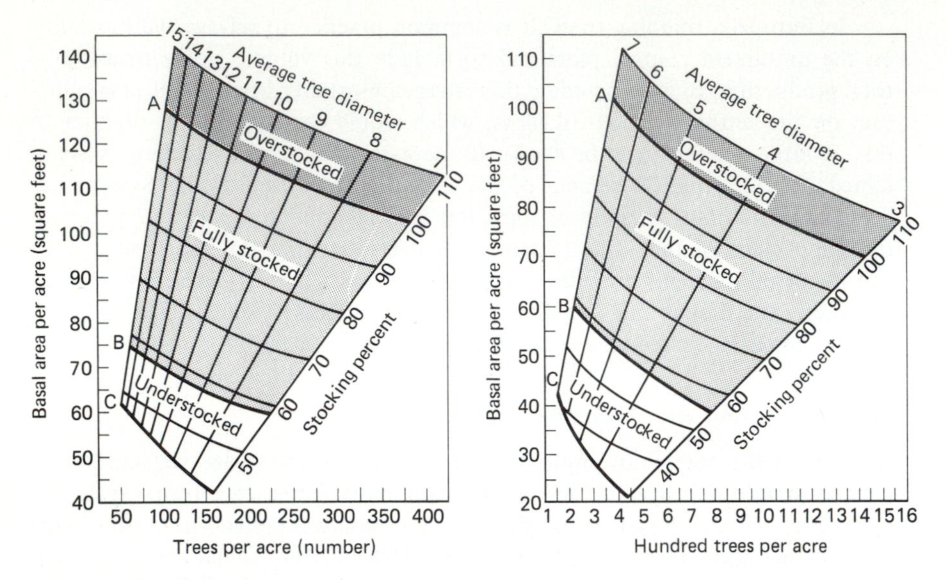

Figure 14-22 Relation of basal area, number of trees, and average stand diameter to stocking percentage for upland hardwood stands (chiefly oak-hickory and mixed species) of average uniformity. The area between curves A and B indicates the range of stocking where trees can fully utilize the site. Curve C shows the lower limit of stocking necessary to reach the B level in 10 years on average sites. (*From Gingrich, 1967.*)

and very useful in setting the limits of stocking, the relative complexity of these diagrams demonstrates the value of Reineke's stand-density index in that the SDI of the lower level of optimal basal area is 117 and the upper level is 209. Thus one needs only to use the same parameters of diameter, basal area, and tree number to determine the SDI of a stand to know its position in or out of the optimal range.

Experiments in Germany and Denmark in beech and spruce show that thinnings of varying grades have no influence on average increment over a long period of time (Fig. 14-23). Heavy thinnings do give a short-lived rapid increase in growth which slows down to give the average growth. Cuts that produce excessively open stands or stands where densities are too high do not produce as well as stands held within an optimal range (Langsaeter's zone III, see Fig. 14-9).

Results of 50 years of thinning, in which cuts began at 31 years and were repeated every 5 years with an occasional 4- or 6-year period in stands of Norway spruce in southern Sweden, are given in Table 14-6 from the report of Carbonnier (1957). This table shows that *total* production and average dominant height for all thinning plots and the control were approximately the same. However, if the unthinned stand's total yield (plot 1) is used as a base, then thinning reduces total yield by 2.0 to 7.3 percent; but plot 1, while not thinned, had had the natural mortality salvaged and added to its standing volume to give the total

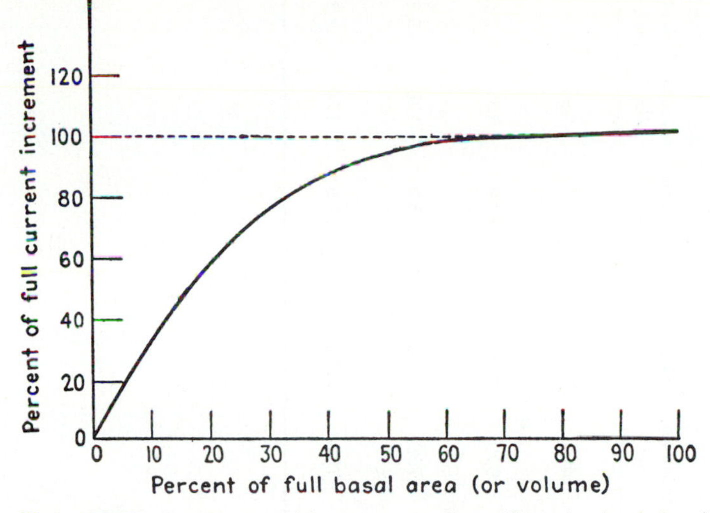

Figure 14-23 Current annual increment as affected by reduction in basal areas (or volume). (*From Möller, 1945.*)

yield. Under American conditions, the salvaging of mortality would be an unlikely practice, so thinnings based on the standing volume of plot 1 would increase total yield by 24 to 31 percent. Average stem diameter and volumes obtained from thinnings increased with thinning intensity. It can be seen, therefore, that the increase in value and reduction in handling costs represented by doubling or tripling the volume of the individual tree, coupled with the early returns on the capital investment resulting from thinnings, favors the inclusion of thinnings to maximize total merchantable yields.

Table 14-6 also shows that periodic annual basal-area increment[6] increased with thinning. However, since there is no real difference in this value between the two most heavily thinned plots, the heaviest thinning probably resulted either in understocking or in loss of growth as a result of a heavy infection of red rot. On the other hand, periodic annual volume increment decreased with thinning. This may be from the fact that with heavy thinnings, the relation between basal-area and volume growth becomes poorer because of increased stem taper. However, in the first 20 years of the study, the heavily thinned plots surpassed the growth of the unthinned plots by 6 to 9 percent; rot became a factor at 25 years of thinning (55 years old); and growth dropped 18 percent in the very heavily thinned plot from the performance of the unthinned plot in the last 30 years. The seriousness of the effect of red rot on the root system is evident from the fact that the trees in the heavily thinned plots were blown over or

[6] *Mean annual increment* is the volume or basal area (in any unit) at a particular age divided by the age in years. *Periodic annual increment* is the difference in volume or basal area between two ages divided by the difference in the ages in years. The most common periods used are 5- or 10-year intervals.

Table 14-6 A Summary of Thinning Results in Norway Spruce Stands in the Dalby National Forest in Southern Sweden for Different Degrees of Thinning between Ages 31 to 81 Years*

										Average periodic annual increment		Percent of	
Plot	No. of trees	SDI	Basal area, m^2	Average stand diameter, cm	Height of dominants, m	Standing volume, m^3	Volume per tree, m^3	Volume thinning, m^3	Total yield, m^3	Basal area, m^2	Volume, m^3	Standing volume, plot 1	Total yield, plot 1
1. Mortality salvaged	1396 (565 per acre)	531	64.9	24.3 (9.6 in)	29.0	917	0.657	(305)	(1222)	1.00	19.1	–	–
2. Moderate thinning	600 (243 per acre)	321	43.2	30.3 (11.9 in)	29.4	608	1.013	590	1198	1.11	18.8	131	−2.0
3. Heavy thinning	336 (136 per acre)	239	34.3	36.1 (14.2 in)	29.3	437	1.301	739	1176	1.30	18.0	128	−3.8
4. Very heavy thinning	200 (81 per acre)	188	28.8	42.9 (16.9 in)	30.2	367	1.835	766	1133	1.31	17.5	124	−7.3

*Data per hectare; stand data at 81 years.
Source: Carbonnier, 1957.

Table 14-7 A Summary of Bowmont Thinning Results in Norway Spruce Stands in Scotland between the Ages of 20 and 50 Years*

Treatment	No. of trees, acre (ha)	SDI	Basal area, ft^2 (m^2/ha)	Average stand diameter, in (cm)	Height of dominants, ft (m)	Standing volume, ft^3 (m^3/ha)	Volume per tree, ft^3 (m^3)	Volume thinning, ft^3 (m^3/ha)	Total yield, ft^3 (m^3/ha)	Percent of: Standing B volume	Percent of: Total B yield	Percent of: Volume over 11.8 in dbh (30 cm)
B†	1032 (2550)	520	236.8 (54.4)	6.5 (16.5)	55 (16.8)	6414 (448.5)	6.2 (.18)	(915) (64.0)	(7329) (512.5)	—	—	0
C	583 (1441)	377	185.4 (42.6)	7.6 (19.3)	55 (16.8)	5172 (361.8)	8.9 (.25)	2710 (189.4)	7882 (551.2)	122.9	+7.5	0
D	172 (425)	213	122.1 (28.0)	11.4 (29.0)	57.5 (17.5)	3390 (237.0)	19.7 (.56)	4524 (316.4)	7914 (553.4)	123.4	+8.0	89
LC	397 (981)	278	136.1 (31.2)	8.0 (20.3)	56 (17.1)	3553 (248.4)	9.5 (.27)	4088 (285.9)	7641 (534.3)	119.1	+4.3	55

*Data per acre and per hectare.

†B-grade thinning: a light, low thinning with only badly suppressed, dead, and dying trees removed; C-grade thinning: moderate, low thinning intermediate between B and D; D-grade thinning: heavy, low thinning leaving mainly dominants with spacing for full crown development; LC-grade thinning: a light crown thinning that resembles the D-grade thinning except that suppressed and subdominant trees are left to fill spaces between best dominants. The poor dominants and codominants are cut.

Source: Mackenzie, 1962.

badly damaged by a storm, and that terminated the study. No storm damage was done to the unthinned stand, and it was unaffected by red rot.

The Bowmont Norway spruce plots in Scotland provide additional support for the principle of increased merchantable yields from thinning. The Bowmont study is the earliest use of replicated thinning plots in forestry with four treatments and four replications which were laid out in a latin square. Each plot was only 0.04 ha (0.1 acre), with narrow buffer strips around the plots. This study was on a broad, level ridge top with 76 cm (30 in) of sandy soil, good drainage, and a droughty growing season. Rainfall of 700 mm (28 in) occurred primarily in winter, with evaporation exceeding rainfall in the growing season. Planted in 1910 with 7660 (3100 per acre) accurately spaced seedlings per hectare, the thinning regimen was begun in 1930 with thinnings every 5 years until 1960. A summary of the results (Table 14-7) shows that the heavily thinned plots had diminishing standing volumes with heavier thinning, but contrary to the Swedish study (Table 14-6), heavy thinning increased the total yield by about 8 percent. If the salvaged material in the B thinning is again ignored as inapplicable in the United States, total production is increased by 19 to 23 percent in a stand only 50 years old.

After total volume production was adjusted by means of covariant analysis for the initial inequalities in the plots in 1930, the precision of the comparison is greatly increased (Table 14-8). The results show a significant difference between B plots and the heavier thinned stands. At age 35 years, the D thinning showed a significantly greater production than the C thinning, but there was practically no difference between the two treatments at 50 years. This early rapid accretion of volume took place in the Swedish plots too, after which differences between light and heavy thinnings became less apparent. Another significant difference in the treatments shown in Table 14-7 is in the percent of volume greater than 30 cm (11.8 in) dbh. The D-grade thinning had 89 percent of its volume in trees over this diameter, while all trees in the B- and C-grade plots were smaller than this size. The LC-grade plot had only 55 percent of its volume in trees above 30 cm in diameter because of the inclusion of suppressed and subdominant trees.

Table 14-8 Total Adjusted Volume Production by the Bowmont Thinning Plots of Scotland for Norway Spruce Stands at Age 50 Years with Thinnings between Ages 20 and 50 Years

	B	C	D	LC	Standard error
Adjusted volume at age 35	3540	3830	4430	4140	±113
Adjusted volume at age 50	7132	7909	7912	7673	±61

Source: Mackenzie, 1962.

In contrast to the heavier thinnings, the B-grade thinning had 54 percent of its volume in trees less than 19 cm (7.6 in) in diameter.

MANAGED-STAND YIELD TABLES

The relationship between gross and net increment[7] in an even-aged stand, which is the basis of thinning theory (developed further in Chap. 17), is one where total *net* cubic volume production increases linearly with age throughout the thinning period if potential mortality is harvested and the stand remains fully stocked. This relationship has been well demonstrated for a stand of Norway spruce in Denmark which was thinned every 5 years from age 34 to 94 years (Hall, 1955). The final cut provided only 43 percent of the total yield. In this stand, the mean annual increment culminated at the age of 90 years, and the stand was harvested at age 94 years (Fig. 14-24). Data such as these permit the development of generalized relationships describing the growth of managed stands (Fig. 14-25). In reading the two figures, *net production* indicates the actual harvesting of potential mortality in Lovengreen's data, while *gross production* is what could be produced by a stand whether or not the potential mortality is cut. These concepts are used in the development of managed, or variable-density, yield tables.

Managed-stand yield tables have a long history in other parts of the world, but they are now becoming available in this country with the introduction of young-growth management. To manipulate young-growth stands, the silviculturist needs such yield tables to predict yields with any reasonable accuracy. The red pine bulletin is an example of a managed-stand yield table (Buckman, 1962). The yield table provides estimates of growth, in terms of cumulative and periodic net annual increment, that can be expected for a wide variety of stand manipulations in terms of the effects of residual basal areas as well as thinning schedules. The independent variables used are age, site index, and stand density. No improvement in prediction was added by considering intensity of cutting, number of trees per acre, cutting method, or variability of tree diameter. The differences in yield for the three thinning methods (high, low, and a combination of the two) were not significant, which confirms the Danish findings. Table 14-9 is taken from the red pine bulletin showing, for site index 50, the yields expected from a thinning schedule in which basal area is maintained between 90 and 120 ft^2 per acre. This range in growing-stock levels may define the limits of Langsaeter's zone III of the production curve shown in Fig. 14-9, and ranges very close to this are commonly quoted as optimal for other moderately intolerant trees grown on short rotations, such as the southern pines, ponderosa pine, Monterey pine, and Douglas-fir. Table 14-9 also illustrates the basic principle

[7] *Gross increment* is the total growth that has taken place since the last measurement. *Net increment* is the gross increment minus the mortality that has been lost since the last measurement.

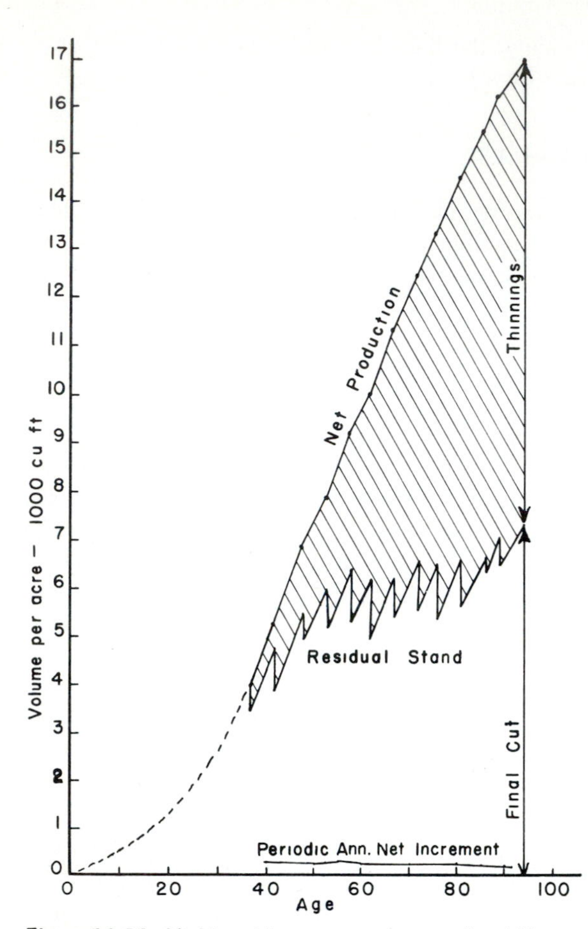

Figure 14-24 Yield and increment on a completed Norway spruce plot in Denmark. (*From Lovengreen, 1951.*)

that when stands are thinned, the culmination of mean annual increment is delayed—in this case, culmination of mean net annual growth in cubic feet occurred at 85 years of age. Unthinned stands would have culminated much earlier.

To illustrate the dynamics of thinning schedules in the growth of stands, data from Table 14-9 have been presented in diagrammatic form (Fig. 14-26). This diagram shows the application of two thinning schedules where the objective is to maintain stand densities between 90 and 120 ft^2 per acre and also between 150 and 180 ft^2 per acre basal area. The diagram clearly shows the commonly observed uniform rate of net basal-area growth over a wide range of stand densities and the manner in which growth diminishes with age.

A significant weakness of short rotations is the high percentage of the rotation that occurs immediately after establishment, in which the site growth

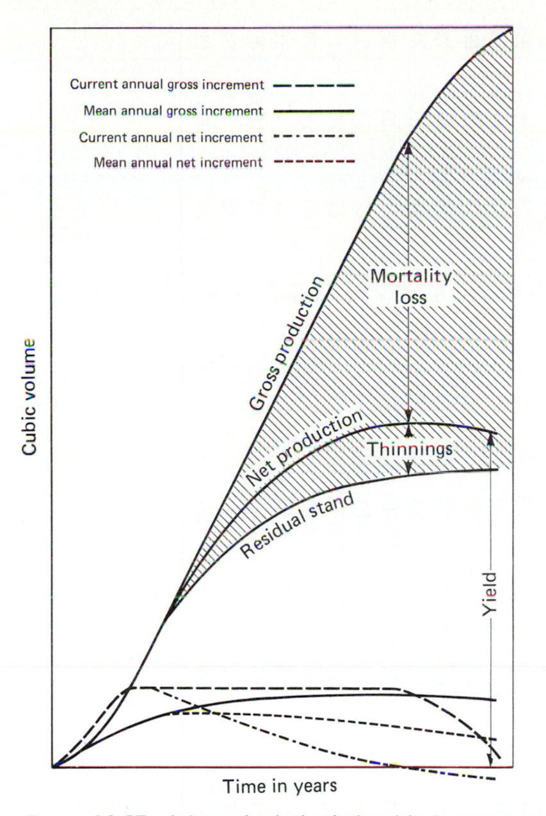

Figure 14-25 A hypothetical relationship between gross and net increment in an even-aged stand. Note that the net production curve is the production of an unthinned stand, and the cross-hatched area between the residual stand and gross production curves is the volume of thinnings. (*From Hall, 1955.*)

capacity is not fully exploited. Most foreign managed-stand yield tables report periodic *gross* annual increment, but periodic *net* annual increment is more appropriate to many of the present management practices in this country.

A set of optimum thinning intensity tables for thinning yield control of Douglas-fir in the Pacific Northwest (Hoyer, 1966) provides the young-growth manager with a means of predicting yield. It defines *thinning intensity* as the volume cut divided by the *thinning cycle* (i.e., the time between successive thinnings), and the recommended optimum thinning intensity (for stands 60 to 70 percent of normal basal area or greater) is given by site quality for 5-year thinning cycles. After years of required thinnings on Washington's state lands, thinnings were made optional (1976) because the total yields obtained from thinned stands were little better than the yields from unthinned stands; however, the optimum thinning intensity tables would have had too little opportunity to prove themselves.

Table 14-9 Managed-Stand Yield Table for Site Index 50 Red Pine in Minnesota* (Thinning Schedule: Maintain Given Stand Density at Beginning of Each Thinning Cycle)

	Height of dominants	Remaining stand				Cut stand				Cumulative total volume—cut stand plus remaining stand				Periodic net annual	Mean net annual	Proportion of stand basal	
Age, years	and codominants, ft	Basal area	Cubic feet	Cords	Board feet	Basal area	Cubic feet	Cords	Board feet	Basal area	Cubic feet	Cords	Board feet	growth, ft^3	growth, ft^3	area cut, %	Age, years
(1)	(2)	(3)	(4)	(5)	(6)	(7)	(8)	(9)	(10)	(11)	(12)	(13)	(14)	(15)	(16)	(17)	(1)
25	24.5	90.0	900	8.7	—	—	—	—	—	90.0	900	8.7	—	—	36	—	25
35	35.5	90.0	1,300	12.6	—	53.7	780	7.5	—	143.7	2,080	20.1	—	118	59	37.4	35
45	15.5	90.0	1,670	16.2	8,500	47.4	880	8.5	—	191.1	3,330	32.2	8,500	125	74	34.5	45
55	53.5	90.0	1,970	19.1	10,000	41.9	920	8.9	4,700	233.0	4,550	44.0	14,700	122	83	31.8	55
65	60.0	90.0	2,210	21.4	11,200	36.4	890	8.6	4,600	269.4	5,680	54.9	20,500	113	87	28.8	65
75	66.0	90.0	2,430	23.5	12,400	31.4	850	8.2	4,300	300.8	6,750	65.2	26,000	107	90	25.9	75
85	71.0	90.0	2,610	25.3	13,300	27.0	780	7.6	4,000	327.8	7,710	74.6	30,900	96	91	23.1	85
95	75.0	90.0	2,760	26.7	14,100	23.2	710	6.9	3,600	351.0	8,570	82.9	35,300	86	90	20.5	95
105	78.5	90.0	2,890	28.0	14,700	19.6	630	6.1	3,200	370.6	9,330	90.3	39,100	76	89	17.9	105
115	81.0	90.0	2,980	28.9	15,200	16.6	550	5.3	2,800	387.2	9,970	96.5	42,400	64	87	15.6	115
125	83.0	90.0	3,050	29.6	15,600	13.8	470	4.5	2,400	401.0	10,510	101.7	45,200	54	84	13.3	125
135	84.5	90.0	3,110	30.1	15,800	11.9	410	4.0	2,100	412.9	10,980	106.2	47,500	47	81	11.7	135
145	85.5	90.0	3,140	30.5	16,000	10.0	350	3.4	1,800	422.9	11,360	110.0	49,500	38	78	10.0	145
155	86.5	90.0	3,180	30.8	16,200	9.0	320	3.1	1,600	431.9	11,720	113.4	51,300	36	76	9.1	155
165	87.0	—	—	—	—	98.0	3,480	33.7	17,800	439.9	12,020	116.3	52,900	30	73	100.0	165

*All data per acre.

Source: Buckman, 1962.

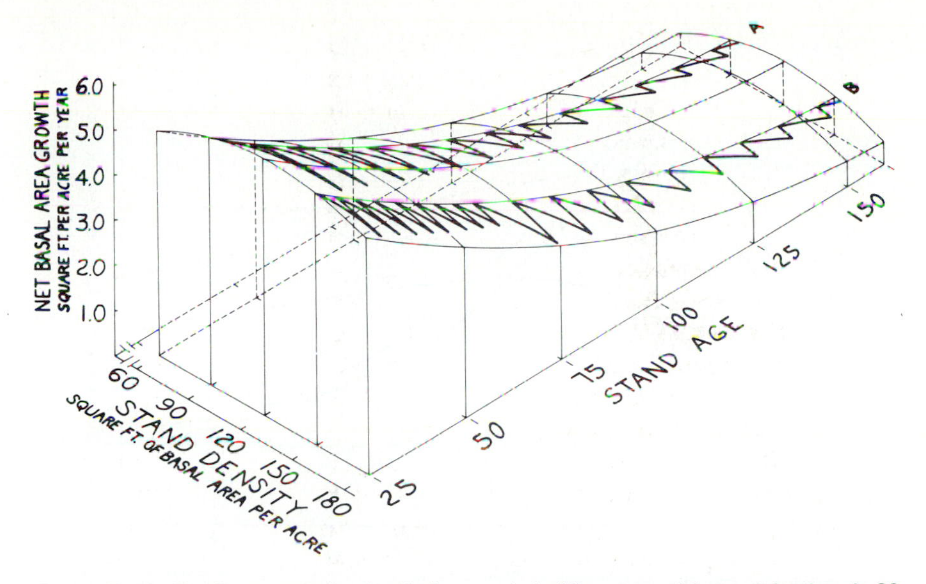

Figure 14-26 Basal-area growth of red pine stands in Minnesota with stand density: A, 90 ft^2 of basa area at beginning of each thinning cycle; B, 150 ft^2 of basal area at beginning of each thinning cycle; site index 50. (*From Buckman, 1962.*)

GROWTH AND ROTATION LENGTH

Total production from a given site depends on the length of the rotation. The biological basis for determining the rotation is defined by the *culmination of mean annual increment*, i.e., where the periodic annual increment curve crosses the curve for mean annual increment. This culmination is used by the U.S. Forest Service for defining rotation. The most common alternative to this approach is to define rotation as the age at which annual growth rate (in either physical or economic terms) falls below some acceptable level, say 5 to 6 percent. A long rotation that extends beyond the culmination of the mean annual increment usually sacrifices volume growth for quality, as the curves showing the relations between mean annual increment and periodic annual increment in unmanaged stands illustrate (Fig. 14-27). Periodic annual increment rises rapidly and falls off sharply, while the mean annual increment increases more slowly and falls off slowly. For a given site quality and species, the age at which growth culminates depends on the utilization restrictions and intensity of management. For unmanaged stands of Douglas-fir at SI 170, total cubic-foot volume culminates at about 65 years, whereas total board-foot volume (international $\frac{1}{4}$-in kerf rule for volume over 7 in dbh) culminates at 80 years (Fig. 14-27). If board-foot volume was restricted to that part of the stand over 12 in in diameter using the same international rule, the volume would not culminate until the stand was

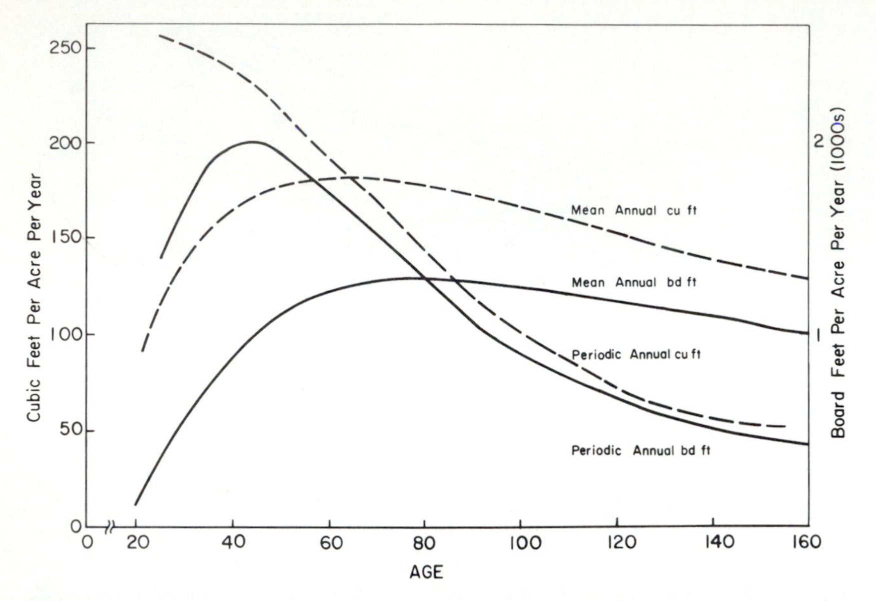

Figure 14-27 Periodic and mean annual increments in cubic feet and board feet for Douglas-fir (SI 170) showing the difference in time for the culmination of the mean annual increments of the two measures of volume in unmanaged stands. (*From McArdle et al., 1949*.)

90 years old. A poorer site quality takes a longer time to reach culmination than a better site quality.

Maintenance of rapid growth by thinning may either increase or decrease rotation length depending on the criteria used to determine it. Since maintaining good growth delays the culmination of mean annual increment, rotations are lengthened. However, if rotation length is determined by the time taken to achieve a given average stand diameter, then rotations are shortened.

GROSS AND NET VOLUMES

Biomass production might be thought of as the gross growth of a stand, but the forester in this country usually limits measures of gross growth to the xylem growth of the main stem. As fully stocked stands develop, there are usually trees that have died from natural competition between any two periods of measurement. The difference in volume between these two measurements is called the *net increment* for the period. The net increment is made up of two parts: the volume laid down on trees that were measured earlier and the volume of trees that have grown to exceed the minimum diameter limit and are measured for the first time (*ingrowth*). Net increment underestimates the real growth of the period by the volume represented in the trees that died. Gross growth measures the actual growth of the period. The European forester mea-

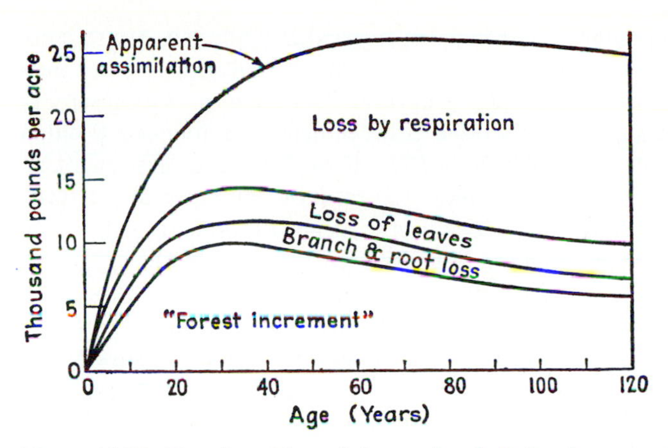

Figure 14-28 The disposition of the total assimilation by a stand of beech in Denmark. (*From Möller, 1945.*)

sures gross growth in estimating the productivity of a site, while the American forester's thinking has often in the past been limited to net growth. Normal yield tables are net-growth tables, although mortality can be estimated from one time period to another with some limiting assumptions as to which trees are most likely to be the ones that died, if stand tables are provided.

In considering the gross or net growth of a stand, it is interesting to have an estimate of how little of the total assimilation produced is actually laid down as stem xylem. By far the greater proportion of the assimilation is consumed in respiration and the production of leaves, branches, and roots (Fig. 14-28).

ACCELERATED GROWTH

Accelerated growth has a special meaning in forestry: it refers to the marked increase in growth rate following the release of a tree and is a response due to delayed management. Accelerated growth in height occurs in the case of understory and overstocked trees that are released and have the capacity to respond to that release. However, even when height growth does accelerate, it is less conspicuous than the dramatic change in diameter growth. In general, accelerated growth refers to the sudden increase in diameter growth due to the reduction in stand density of overstocked stands. As discussed earlier, height growth is commonly little affected in stands that have more desirable spacing.

Accelerated growth is an effect associated with the increased light to existing foliage, the rapid buildup of foliage, and the increase in root space, with its increase in moisture and nutrient availability. An extreme case of acceleration is noted by Mason (1915), where a lodgepole pine as a suppressed tree in a closed stand had a growth-rate change from 134 to 8 rings per inch. Such a response to release is not "immediate," i.e., the tree took about 4 years to build up crown and root surfaces to produce the increase in ring width. While it is easy to rationalize why a tree responds to release, it is difficult to explain why other trees fail to do so. The reason undoubtedly is associated with the tree's incapacity to utilize increased growing space as a result either of being in a dominant position already or having a poor root system and low live-crown ratio that cannot respond to the treatment. When trees whose crowns have developed in the shade are released, their shade-adapted leaves are suddenly exposed to higher light intensities and must become adjusted to this new environment. Within a canopy, the less vigorous trees of moderate intolerance, such as intermediate Douglas-fir in a mixed-conifer forest, may not be able to adjust rapidly to this change. The foliage may become chlorotic or yellowish in appearance and height growth may be retarded. In extreme cases, released suppressed trees may die. The slowing down of growth after release is called *thinning shock*.

Species differ in their response to release. Generally, tolerant species like hemlock, red spruce, beech, and sugar maple give quicker and more striking responses than do more intolerant species like yellow birch, yellow poplar, or longleaf pine. Intolerant species have their whole pattern of growth limited under extreme competition because branch orders decrease to a low level and buds fail to develop, so a structure develops in which there are few growing points upon which a new crown can be built. Some intolerants respond excellently to release, such as ponderosa pine, which frequently survives in dense thickets and when thinned, grows rapidly (Teeguarden and Gordon, 1959).

Capacity for accelerated growth tends to decline with increasing age. Yet some species are very persistent, such as ponderosa pine, Douglas-fir, and redwood, which have shown accelerated growth at about 200 years of age.

The duration of accelerated growth depends on the rate at which the released tree is again crowded by its neighbors. The growth rate peaks, then the rings become more narrow, partly as a result of renewed competition and partly because of the effect of increasing diameter. A regular thinning schedule is designed to give continuous release, resulting in uniform ring development and high-quality logs. Naturally, to produce a stem with uniform ring width as the stem becomes increasingly larger in diameter would require a continual increase in growth rate. This may be possible up to a point while the crown is expanding, but clearly there is a limit—in the life of all trees, ring width will eventually decline.

Part Five

Forest Manipulation

Chapter 15

Tree Improvement

Tree improvement is the objective of silviculturists and geneticists as they cooperate to attain faster-growing, more disease-resistant, higher-quality trees. In the South, private companies spend millions of dollars to develop seed orchards that will increase stand yields from 10 to 25 percent. In other regions, the investment has been primarily by the U.S. Forest Service and various states. The initial efforts and a clearer understanding of future needs as well as improved techniques have accelerated pressures to produce more improved seed. The Tree Improvement Master Plan for the California region (Kitzmiller, 1976) of the U.S. Forest Service, which outlines short- and long-term goals for the supply of superior seed for various species within the different seed-collection zones, illustrates the penetration of the superior-tree idea into administrative thinking.

Tree improvement has motivated silviculturists to leave the best-looking trees to regenerate stands before geneticists ever knew why it worked. Although genetic principles have been applied from prehistoric times in the domestication and breeding of plants and animals, the knowledge of the underlying principles dates from the rediscovery of Mendel's laws of gene segregation in 1900. Some tree breeding was done in Europe as early as the 1850s, but the first full-time breeding programs began in the United States in 1925 at the Eddy Tree Breeding Institute in Placerville, California (specializing in conifer breeding) and the New York Botanical Garden (specializing in poplar breeding). The Placerville Institute concentrated on interspecific crosses to gain hybrid vigor, while European forest geneticists were more interested in intraspecific crosses for the production of "elite" trees. Intraspecific crossing dominates the genetic work in this country today (Zobel, 1971), although interspecific breeding is worldwide and has produced some exceptionally valuable crosses.

The drive toward production of superior trees gathered momentum in the 1950s when the focus shifted from natural regeneration to planting logged-off areas. The dangers inherent in using seeds supplied by seed dealers where provenance or source information was often missing focused attention on seed quality. The failure of various external control methods to cope with the devastating attack of white pine blister rust in western white pine stands brought a reliance on disease-resistance tree breeding for a solution. Economics indicate that it costs no more to plant a superior tree than an average tree, although the development costs of breeding the superior tree would need to be spread over time and numbers of trees.

The forester should be aware of why and how the practices of silviculture and the science of genetics are able to help in the management of forests. Basically by providing more understanding, they contribute to greater production through the development and use of trees superior in growth rate and form as well as more resistant to diseases and insects.

PROBLEMS IN THE APPLICATION OF GENETIC PRINCIPLES TO TREE IMPROVEMENT

The factors which contributed to Mendel's success in developing laws of inheritance contrast with the factors which have made forest genetics a difficult science to apply and should clarify why it has matured slowly. Mendel was fortunate both in the selection of sweet peas and in the conditions for his first experiments. His advantages can be described as follows:

1 Sweet peas are annual plants with low space requirements and ease of handling.

2 Mendel's garden provided a uniform environment where the phenotype[1] reflected the genotype.[2]

3 Sweet peas are naturally self-pollinating, which enables the breeder to hold the genetic constitution constant.

[1] *Phenotype* is a classification based on the appearance of an individual or a particular characteristic, i.e., the sum of the effects of environment and gene makeup on the individual.

[2] *Genotype* is a classification based on the gene constitution of the individual and determined from breeding behavior.

4 The many specific varieties developed by gardeners over the years enabled Mendel to select distinct pairings of characteristics: short or tall plants, smooth or rough pea surfaces, yellow or green pods, and four other such pairs.

5 In the course of his experiments, Mendel established that each pair of characters was controlled by *one pair* of genes; yet there were two kinds of genes, dominant and recessive, which could give three different genetic constitutions (Fig. 15-1). If the flowers of a homozygous tall plant were fertilized by pollen from a short plant (or vice versa), plants grown from the seed (F_1 generation) would be tall. If the self-pollinated seed from F_1 plants were planted, the plants (F_2 generation) would segregate into one quarter homozygous short plants (both genes of the pair for short plants), one quarter homozygous tall plants (both genes for tall plants), and one half heterozygous tall plants (one gene of the pair for tall and the other for short)–a segregation ratio of 3:1. Mendel proved that a dominant gene (tallness) had its full expression in the height to which the plant grew even though it was paired with a recessive gene (shortness). In other words, a plant could be short only if both genes of a pair were for shortness. His work proved that heritable characteristics were inherited in a definitely predictable way and segregation ratios for any combination of genes in the parents could be calculated.

In contrast to the situation faced by Mendel, the forest geneticist's efforts have been handicapped by a number of difficulties.

1 Mendel's selection of peas was fortunate in that their self-fertilization resulted in no loss of vigor; however, self-fertilized seeds for most of the important forest species usually result in a depression rather than a maintenance or improvement of a characteristic such as growth. Fortunately, most forest species self-fertilize poorly and require two parents to produce a fertile seed (*outcrossing*, as opposed to inbreeding, which is the crossing of two closely related parents, of which self-fertilization is the extreme). However, the necessity for

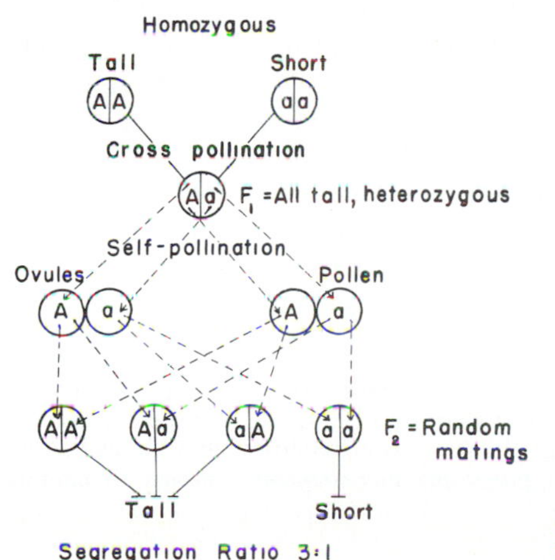

Figure 15-1 Mendelian segretation of genes in the F_2 generation.

outcrossing appreciably increases the difficulty of developing a superior tree that is certain to pass its superiority on to its progeny.

2 Only a wild population of trees, with no background work for selective breeding upon which to build, was available to the forest geneticist. In the long run, this is an advantage in that the breeder can work with the total gene pool.

3 Selection of a superior tree from a stand might provide a phenotype with the desired characteristic but no certainty that these characteristics are genotypically controlled, since the tree's appearance could be due to a large extent to environmental influences. Thus the heritability[3] of particular characteristics within a species had to be established.

4 A long interval exists between tree generations since trees do not become sexually mature until relatively late in life. The minimum time for seed production is about 4 years for a few species such as jack pine but may be 20 years and more for most others.

5 The forester is primarily interested in characteristics of mature trees, but methods to reduce the generation interval may mean selections based on sapling characteristics. However, selection for progeny in the sapling stage for further breeding assumes a high correlation between sapling and mature-tree characteristics. Too frequently that correlation has been poor (Toda, 1964).

6 Locating and evaluating "plus phenotypes" within a provenance is time-consuming, especially if the tree must be visited several times for conelet pollination and cone collection.

7 Progeny seed testing from many plus phenotypes requires extensive areas of reasonably uniform-quality land, particularly if a commercial seed supply is desired from the F_1 generation following selective elimination of the poorest of the F_1 progeny.

8 A planting program and nursery or container facilities are necessary adjuncts for intensive breeding work in a particular species (Wright, 1962).

9 Breeding work in tall trees is difficult and dangerous.

10 Administrative and research continuity are necessary for long-term studies like tree breeding.

11 Most characteristics of trees worthy of improvement seem to be controlled by multiple genes. Thus diameter growth could be affected by a wide variety of characteristics, each subject to considerable independent variation, such as size and number of stomata, efficiency of the photosynthetic process, size of cells and cell-wall thickness, amount of cambial activity devoted to phloem increase, temperature thresholds for initiation and cessation of growth of crowns and roots, efficiency of conductive tissue, tendency to develop compression wood, and many others. Each of these characteristics in its turn may also be controlled by a multiple-gene inheritance pattern.

[3] Heritability (narrow sense) is additive genetic variance divided by total phenotypic variance and is useful in predicting improvement that can be achieved in the offspring of a selected parent. Heritability (broad sense) is additive plus nonadditive variance divided by total phenotypic variance and is useful in predicting improvement achieved by cloning selected trees.

Figure 15-2 Progenies of sugar pine completely susceptible (right) and segregated (left) showing the effects of a single dominant gene for resistance to white pine blister rust. Dead and dying seedlings of the resistant line have been removed. *Ribes* bushes are in background. (*From Kinloch et al., 1970.*)

Variation in inheritance patterns is illustrated by two investigations of resistance to white pine blister-rust infection in western white pine and sugar pine. The studies indicate variations where one might expect more uniformity. Progeny tests by Bingham and coworkers (1960) of selected western white pines resistant to the disease indicated that several genes control that resistance. Kinloch and coworkers (1970), on the other hand, found that resistance of sugar pine to the disease is controlled by a dominant gene (Fig. 15-2). Selecting those individuals with multiple-gene resistance to blister rust has provided planting stock which in 1976 had an overall resistance to the disease of 50 percent. Consequently, western white pine is again being grown in the U.S. Forest Service nurseries for introduction into cutover areas with severe blister-rust incidence on low-resistance strains. Because of the prevalence of multiple-gene inheritance, the forest geneticist has to depend on principles of population and quantitative genetics for breeding. Consequently, the prediction of genetic gain[4] resulting from varying intensities of selection procedures over several generations requires sophisticated mathematics.

In spite of difficulties inherent in tree breeding, the tremendous planting expansion, coupled with the increasing costs of nursery stock and planting, has increased the incentives for genetically improved stock. It costs as much or more to plant and grow a poor tree as a superior one. In spite of the development costs for superior seed, Davis (1967) indicates that the cost of collecting wild seed is comparable to that of improved seed. Also, the changing demands of the wood market have shortened rotations, increasing the profits and the desire to augment productivity through breeding programs.

Because short rotations increase the importance of a few years' loss in production when natural regeneration fails, planting is the common regeneration method in countries practicing intensive forest management and is becoming common in the United States. A planting program, particularly when coupled with intensive practices such as thinning and fertilizing, increases the incentive to grow the most productive trees. In consequence, tree-breeding efforts have multiplied several times since 1950, and the patient work of pioneers in forest genetics to surmount difficulties and develop appropriate techniques for the new field has provided a strong foundation upon which to build.

BASIS FOR TREE IMPROVEMENT

A successful tree-improvement program depends on a tree's capacity to pass its characteristics on to its progeny. If no variation exists within a tree population, then no basis for tree improvement exists; consequently, the greater the varia-

[4] *Genetic gain* is the improvement in a given characteristic as a result of a breeding program; it is represented mathematically as the product of the selection differential and heritability (where *selection differential* is the difference between the mean of a particular characteristic from selected parents and the mean of the unselected population).

tion among individuals capable of successful breeding, the greater the opportunity for a tree-breeding program to show results. For example, *Pinus resinosa* is a remarkably uniform species with little genetic variation, whereas *Pinus radiata* is quite the reverse. Brief descriptions of the four levels of variation commonly utilized by forest geneticists follow.

Variation within a Genus

Within a genus, between-species breeding (interspecific hybridization) offers the widest variation and a high potential for tree improvement. A number of factors have contributed to the interest in hybridizing two species (Duffield and Snyder, 1958): (1) the economic or utilitarian gain from the resulting successful crosses, (2) the challenges presented to the researcher, (3) the development of technical skill in manipulating genetic materials needed in the emerging field of forest genetics, and (4) the need for more knowledge of species' interrelations to advance the field. Whether a crossing is a success or a failure, an analysis of the results provides an understanding of taxonomic relations.

The hybrid *Larix leptolepsis* × *L. decidua* has exhibited hybrid vigor in the environment of its European parent *Larix decidua*, as well as outside the native range of both parents. Species of poplar have been hybridized in Europe and the United States, and the many resulting successful crosses have been widely planted. The propagation of these successful crosses has been possible because poplars are reproduced by cuttings. The hybridization studies in white pine by Wright (1959) are illustrated by the following results using the 1950 seed crop:

Crosses	Average height in 4 years
P. strobus × *P. strobus*	39.4 cm
P. strobus × *P. griffithii*	51.4 cm
P. griffithii × *P. strobus*	51.4 cm
P. strobus × *P. ayacahuite*	44.0 cm
P. griffithii × *P. griffithii*	34.3 cm

Hybridization secondarily combines desirable properties of both parents. A classic example is the cross pitch pine × loblolly pine which produced a hybrid with improved growth and form that enabled Korea to have the largest planting of a conifer hybrid in the world. Pitch pine, naturalized to Korea, had outgrown the local pine, although it had a rough form. The successful crossing with loblolly pine (which had survived in Korea only along the south coast) combined the cold resistance of pitch pine with the cleaner-boled form of loblolly pine. The backcross pitch pine × (pitch pine × loblolly pine) (A × F_1) was even more cold resistant (Fig. 15-3), with little or no loss in vigor or form (Hyun, 1971). The use of the ecotypic variation in pitch pine was significant in the success of the hybrid, since the pitch pine from northern New Jersey succeeded and the ecotype from the southern Appalachians failed.

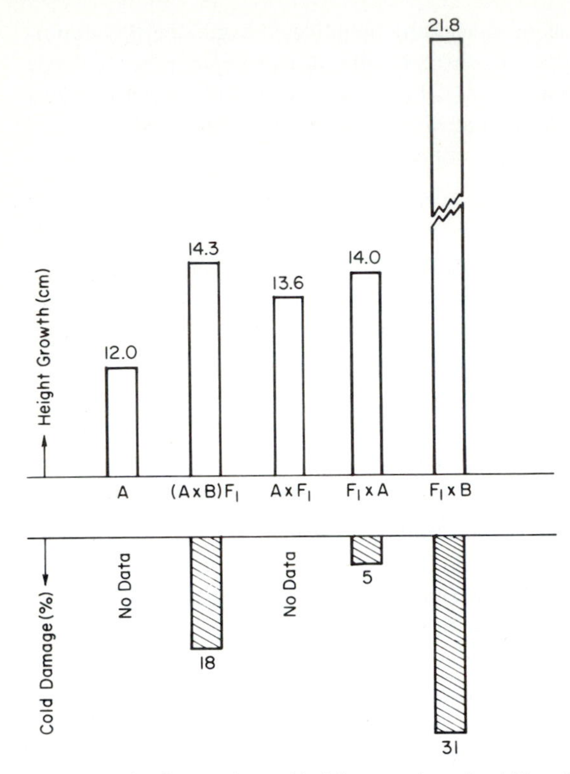

Figure 15-3 Comparison of height growth and cold hardiness between pitch pine × loblolly pine F_1 and backcross progenies at their transplanted nursery beds. A: Pitch pine. B: Loblolly pine. The backcross $F_1 \times$ B had superior height growth but unacceptably high cold damage. *No data* indicates little cold damage.

Hybridization of two species has been used to combine the insect or disease resistance of one species with the growth potential of another. A natural hybrid, Jeffrey pine × Coulter pine proved useful when the weevil (*Cylindrocopturus eatonii*) riddled the plantations of Jeffrey and ponderosa pine in northern California (Libby, 1958). Although Coulter pine was susceptible to cold and Jeffrey pine vulnerable to the weevil, the hybrid had the cold resistance of the Jeffrey pine and the weevil resistance of the Coulter pine. Coulter pine's weevil resistance can be transmitted to ponderosa pine, in spite of ponderosa pine's failure to cross with Coulter pine, by a triple cross: ponderosa pine × [backcross Jeffrey pine × (Jeffrey pine × Coulter pine)]. Another example of such a cross is the hybrid *Thuja plicata* × *T. standishii*, which has demonstrated resistance to the needle-spot disease caused by *Keithia thujina*; western red cedar is usually quite susceptible to the disease, while *Thuja standishii* is resistant.

The forest geneticist, anticipating problems, makes some imaginative crosses against possible future needs. Every successful cross may be one less riddle for the silviculturist, who in the final analysis is the ultimate judge of the success of the forest geneticist's work.

Variation within a Species

Species with wide distributions, such as ponderosa pine, Douglas-fir, and aspen, generally have many provenances (ecotypes or races) with great variation among them. High variation also develops in species having a discontinuous distribution with little or no exchange of genetic material among the groups, as in Monterey pine's small genetically isolated groups occurring along a short stretch of the California coast.

Provenance testing has been a widely used method of tree improvement. It consists of assembling many provenances and growing them in a variety of test localities to determine which provenance is the best for each locality. The common recommendation from these studies of ponderosa pine (Weidman, 1939), red pine (Rudolf, 1948), loblolly pine (Wakeley, 1944), and many other types in Europe is that the local provenance is the best. Apparently, the process of natural selection had developed local provenances which are best adapted to their environments. Consequently, it is almost axiomatic that the silviculturist should reforest using seed from a local area rather than seed from a distant ecotype.

However, the correctness of this finding has been subjected to doubt. Local ecotypes may not necessarily exhibit optimal growth (Namkoong, 1969), but rather the ecotype from a milder part of a species range may perform best. The climate may have changed faster than the local provenance's ability to adjust, or more likely, the local provenance may have evolved to survive in the extremes of a physically limited environment at the expense of growth rate. In spite of such possibilities, normal caution precludes the use of any provenance other than a local one unless the outside provenance has been tested over a period sufficient to "cover" the normal extremes in local climate.

Several provenance tests have shown the need for caution in shifting seed among provenances. In a test of nine provenances of loblolly pine on two sites in Georgia exposed to severe ice storms in 1963, more severe damage was evidenced in those provenances from the warmer, milder parts of the species range in southwestern Louisiana and southeastern Texas (Jones and Wells, 1969). For the Dooley County site, a correlation of $r^2 = 0.75$ was found between the amount of damage and the mean minimum temperature during January. Yet the increased growth rate of the provenances from the milder coastal areas more than offset the amount of damage by the ice storm. In another test in northern Idaho, two provenances of ponderosa pine from a relatively mild climate in northern California gave extremely different responses in a 25-year provenance test (Weidman, 1939). One was killed by a freeze after 9 years, and the other did as well as most of the more local provenances for the first 25 years reported.

Conversely, catastrophies can occur in which ecotypic variation has no influence, such as the spread of white pine blister rust and chestnut blight (*Endothica parasitica*).

Variation within an Ecotype

The most commonly used variation for developing a breeding program is the between-tree variation within an ecotype. Individual trees within an ecotype vary quite markedly in phenotype. The more desirable trees are selected as "plus trees," usually at a selection intensity of one individual out of every 100,000, and these are used in breeding programs.

Variation within a Stand

A stand is the smallest unit capable of providing a base for a breeding program. However, the variation in a characteristic from tree to tree within a stand is not necessarily small. In the Southeast, the same specific-gravity range of loblolly pine found between stands was also the range found between trees within a stand. However, the selection differential is greatly reduced when selection is made within a stand rather than over an ecotype.

METHODS OF TREE IMPROVEMENT

Tree-improvement programs vary widely in complexity and cost. Time is an element of cost for which money is unable to compensate, since trees require a certain minimum number of years for seed production and frequently additional years to ascertain degree of inheritance. The expanding demand for better seeds has induced reliance on some stopgap measures until forest geneticists have had time to develop highly improved trees and the necessary commercial seed supplies.

A review of forest-improvement programs and their status is provided by Libby and coauthors (1969), and a penetrating coverage is given by Wright (1962). A summary of tree-improvement methods follows.

Use of Wild Seeds Designated as to Origin

This is a minimal effort to ensure the use of seed of known origin. No check on the seed collector is usually made. If a price differential were to develop for seeds of different provenances, there is no guarantee that seeds of one origin would not be misdesignated. No seed improvement is indicated, and it is possible that the poorest sites provide the seed, especially if trees have to be climbed to gather cones.

As a minimum safeguard against the use of seed beyond its locality, seed zones have been described in great detail for the Pacific Northwest, California, the Southwest, and Georgia.

Use of Seed-Production Areas

Stands with a better-than-average phenotype are selected and set aside for seed production. Normally, these stands are heavily rogued (removal of the poorest phenotypes and opening of the stand) to raise the average quality of the stems and increase seed production. Fertilizers and other intensive management treatments are used to further increase seed production. Such a thinned seed-production area provides improved seed, with the degree of improvement depending on the selection differential as developed by the initial stand selection and subsequent thinning. The cross-fertilization with pollen from surrounding unthinned stands can diminish the average degree of genetic improvement, but this can be reduced by a wide buffer strip. However, seed-production areas are increasingly suspect because of the greater degree of self-fertilizing and too-close inbreeding in relation to the potential genetic gain. Seed-production areas have one advantage: the only added costs are locating the stand and fertilizing, unless thinning has to be done at a loss.

Use of Seed Orchards

Seed orchards are essential to a tree-improvement program designed to achieve maximum results. Seed orchards can be established and operated at modest cost for moderate improvement or at a much greater cost for rapid achievement of maximum genetic gain. Wright (1962) describes 12 different kinds of seed orchards from which increased levels of genetic gain can be obtained. A seed orchard is developed by assembling selected plus phenotypes into one or more orchards on carefully chosen sites. The plus phenotypes are planted in a prescribed design with replications within each orchard. Phenotypes may be limited to selections from within one ecotype, which is the most common practice, or from several ecotypes. Orchards should be located where topography and site quality favor easy maintenance, care, and collection and high seed production. Isolation of an orchard to prevent haphazard pollination from surrounding unselected pollen sources is determined by the percentage of contamination allowed in the breeding objectives.

Two major means of establishing seed orchards are by seedlings and by clones.[5] Seed orchards differ not only in the way they are established, but also in the type of breeding program utilized. Whether establishment is by seedlings or by clones, either of the following breeding methods may be used: general-combining ability[6] and specific-combining ability.[7]

[5] A *clone* is a group of plants derived from a single individual by asexual reproduction.

[6] *General-combining ability* is the ability of a parent to give superior progeny in combination with several different males.

[7] *Specific-combining ability* is the ability of a specific individual to give superior progeny in combination with another specific individual.

Seed Orchards Utilizing General-Combining Ability *Seedling Orchards* After field selection of the plus phenotypes, the open-pollinated seed of each is collected; the seed is commonly kept separated by families, and seedlings are produced from the seed. A seedling orchard is then planted, and when the trees are sufficiently mature for selection of the plus progeny, the poorer phenotypes are removed. Seed from this type of open-pollinated orchard is used in commercial planting. A higher genetic gain in a second orchard can be obtained by repeating the method, using the best phenotypes in the first orchard as a source of seed. This method has the silvicultural advantage of minimum cost since, except for selection cost, the orchard can be handled as an ordinary planting operation for commercial timber production. The long period required to determine the plus-phenotype trees and to produce seed is a major disadvantage.

Clonal Orchards Clones are taken from each of the selected plus phenotypes and grafted onto seedling rootstocks either in the nursery or in the field. If nursery-grafted, the clonal stock is planted as an orchard with clonal placement designed for maximum cross-fertilization. Field-grafting employs a similar scheme for the same purpose. Cone production may be much earlier in a clonal orchard than in a seedling orchard, especially if the seedlings produce cones late in life. Clonal orchards are most commonly used for this reason. Roguing is done to remove the poorest clones, while seed from the open pollination of the superior clones may be used for commercial seedling production. Care is necessary in collecting clonal material since different crown positions produce clones of differing vigor (topophysis). Some clones may retain branch characteristics instead of developing a tree form. When a clone's vigor is adversely affected by the crown position, the roguing of an orchard may remove a superior genotype because it is phenotypically poor. An improved seedling orchard can be developed from seed of the choicest clonal phenotypes, or an improved clonal orchard can be developed by use of clones from the best phenotypes.

The use of a general-combining-ability seed orchard provides a higher degree of improvement in the resulting progeny than the seed-production-area method. Finding and selecting the plus phenotypes and collecting the cones, in the seedling orchard, or grafting the clonal material on rootstocks, in the clonal orchard, are the only added expenses, although graft incompatability has caused problems in some species, such as Douglas-fir (Karlsson, 1970; Sweet and Thulin, 1973). Land and roguing costs can be treated as ordinary stand-management costs. Keeping performance records for each orchard may be a further cost.

Seed Orchards Utilizing Specific-Combining Ability *Seedling Orchards* After the plus phenotypes have been located in the field, pollen from several outstanding trees is collected and used to pollinate a number of cones on each plus tree. The cones of each family are segregated, and seedlings from each family are planted in replicated patterns in one or several orchards. Whole families are eliminated, and a more intensive program may also eliminate the poorest individuals among the best families. A family progeny test allows a calculation of genetic gain, while analysis of half-sibs permits identification of the genotype and its heritability. Genetic gain is a calculated value derived from the product of the selection differential and the heritability as determined by analysis of the variance of a two-parent progeny test (Wright, 1962). Improvement can be continued for a second progeny test by repeating the individual crossing of one or several superior males with superior females. The progeny-test area must be sufficiently removed to prevent significant quantities of wild pollen from reaching the stand.

Since the rogued first progeny test will provide improved seed, the breeding is allowed to revert to the general-combining-ability method, except for controlled crosses, which provide seed for the second progeny tests. The only way that a seedling orchard can qualify as a specific-combining-ability orchard is if only two tightly inbred lines are planted in the same orchard.

Clonal Orchards After the plus phenotypes have been selected in the field, a clonal seed orchard is established and pollens from exceptional plus phenotypes are used for controlled conelet fertilization on the clones. The families of these matings are isolated and a progeny test is run to calculate the genetic gain. The poor families in the progeny test are rogued and the clones with poor combining ability are also rogued. Both the clonal orchard and the progeny-test orchard will then produce improved seed by the general-combining-ability method, even if further hand-pollination is practiced for greater improvement. If the specific-combining-ability progeny test indicates that two clones have exceptional specific-combining ability, then a specific-combining-ability seed orchard could be developed by the clonal propagation of the two parents.

The specific-combining-ability breeding method is a high-cost method primarily because of intensive performance records needed for each family. Zobel (1971) estimated the cost of establishing and progeny testing clonal specific-combining-ability orchards at $5000 per acre. Two specific advantages are derived from this method: (1) genetic gain per generation of trees is rapid; and (2) it is a reliable way of calculating the heritability of various characteristics. These advantages have been used to justify such breeding programs.

Second-Generation Seed Orchards After the first generation of seed orchards has been rogued, it is used to provide improved planting. It is also the foundation for a second generation of seed orchards, developed either from seedlings or from clones. Genetic gain is enhanced by use of second-generation seed orchards. The degree of additional gain determines whether a further generation of seed orchards would be profitable. When breeding for disease resistance where the recombination of multiple genes is a factor, the higher order of seed orchards may well be profitable. However, each generation of seed orchard and the associated roguing reduces the gene pool and introduces more homozygosity in the gene structure with its concomitant threat of disaster.

An estimate of the total seed-orchard acreage is 7257 acres (U.S. Forest Service, 1971).[8] This expensive program is devoted almost exclusively to increasing wood production, since economic predictions have focused on increaseing production if future demands are to be met. It is unrealistic not to recognize that these same techniques can provide trees with a wide variety of characteristics —trees with reduced water need, greater resistance to soil compaction on campgrounds, adapted to abandoned strip-mine areas, or to other environmental or aesthetic needs. It is also feasible to plant improved seedlings along with seedlings from poor families, thereby ensuring well-spaced crop trees without the necessity of a noncommercial thinning. The poor-family seedlings would be more certain to be eliminated by natural competition. Although no demand has justified tree breeding for any purpose other than wood production, forest genetics has developed a technique and a scientific base applicable to many potential requirements.

Silen's Simplified Tree-Improvement Program

Silen (1966) has suggested a low-cost method of initiating a tree-improvement program in second-growth Douglas-fir that is applicable to other species. Owners or owner associations with as little as 100 to 250 sections of timber in a reasonably compact unit could initiate the program. Early results can be obtained without a professional staff of geneticists, and the program can be expanded to any degree of sophistication desired. The total gene pool is retained, and the program can be initiated with a low additional cost if a planting program is presently active. Two to three hundred or more plus-phenotype trees are recognized and reserved along the roads within the property. If a real estimate of genetic gain is desired at any future time, about 20 percent of the trees are randomly selected as a base. In a good seed year, the cones are collected from each tree and the half-sib families are separately maintained in the nursery beds. The 75 to 100 best-family trees are the seed source in future plantings. A seedling seed orchard could be easily established from the out-planted seedlings by roguing the poor families, yet the parents of all the families would still be reserved should specific crosses be wanted.

Rooted Cuttings

Tree-improvement methods have concentrated on developing seed sources for producing improved seedlings. Passing a plus phenotype through a sexual reproduction screen with its infinite number of potential recombinations of alleles slows the attainment of a strain that will breed true for the plus phenotype. However, vegetative propagation offers a more direct and certain method of

[8]The breakdown of this acreage shows 6421 acres in the South, with 51 percent on public lands, 252 acres in the Middle and Eastern States, with 97 percent on public lands, and 584 acres on the West Coast and in Idaho, with 70 percent on public lands. The non-public-land acreage is almost entirely on forest-industry land.

producing propagules. The Lombardy poplar and naval orange are excellent examples of a single mutant that has been proliferated. In addition, extensive plantations of poplar hybrids have been made in many countries including the United States.

Vegetative propagation among commercial conifers has met with varying success in the past, but all conifer species attempted have been rooted from cuttings. Juvenile shoots root the fastest and retain a capacity to duplicate the performance of the parent. The time required to produce a given number of propagules from a particular plus phenotype depends on the species, because the growth habits of some species interfere with maintenance of the juvenility of ramets (individual members of a clone). A Monterey pine plus phenotype could be brought to a 1 million rooted-cutting capacity in 4 years because ramet production can be quickly multiplied by hedging (pruning back to increase branching) each initial rooted cutting. A Monterey pine seedling has many side shoots developing in the axils of its fascicles. In ponderosa pine, the production of rooted cuttings would be much slower because of differences in growth habits.

The use of rooted cuttings has additional advantages. They provide an efficient way of capturing and making available total genetic variability, none being potentially lost through gene recombination. They provide a means of perpetuating genotypes of desired characteristics, particularly when hedged. Hedging is useful for managers wishing to maintain especially desirable individuals for future regeneration programs. Also, the use of cuttings provides a means of maintaining gene banks in gene-conservation programs designed to preserve sources of genetic material representative of a particular population.

GENETIC IMPLICATIONS OF THE REPRODUCTION METHODS

In the unmanaged original forests there is a large amount of variability in the existing gene pool for almost any characteristic consistent with an individual's survival. A mutation that produces an individual with an exceptionally fast rate of height growth but a slender stem is likely to fail in competition for dominance in an area where snow breakage is common. Up to where it becomes a liability, any improvement in height growth rate will increase the chances of a tree's survival in competition with other trees.

In a managed forest, the gene pool can be continually contracted through treatments which selectively remove individual trees. In regenerating forests using improved seed or cuttings from desired genotypes, it is critical to maintain genetic diversity in the out-planted stock. The number of clonal lines to be used in plantations developed from rooted-cutting programs or the number of superior trees used in seed orchards to ensure desirable diversity is a matter of debate among tree breeders. The objective, however, is to ensure that regeneration developed from tree-improvement programs has sufficient diversity to avoid unacceptable losses due to insect and disease pests, unexpected changes in factors such as temperature or precipitation, and changes in utilization.

There are, theoretically, some genetic consequences in the use of the various natural reproduction methods (W. J. Libby, personal communication).

1 *Clearcutting* has little effect on the genetic composition of the subsequent naturally regenerated stand. The regeneration will have a genetic composition similar to that of its parents, which are primarily located on the edges of the harvested block. If the original stand was composed of trees with serotinous cones, the seedlings might draw mostly on the gene pool of the harvested trees. Within any given stand or population, variability will tend to be increased by mutation and migration, whereas selection and genetic drift tend to lower variability.

2 The *seed-tree method* ideally selects the best individuals in the stand to be parents of the next crop in the ratio of approximately 1 in 20. This constitutes the selection pressure which, although low, is the highest of all natural reproduction methods. The offspring of seed trees have a low probability of being inbred because the relatively widely spaced seed trees are not likely to be near relatives, such as sibs or cousins, which will result in more vigorous progeny. Also, there is commonly a time differential in maturation of the male and female flowers on an individual tree which tends to limit the amount of "selfing." Many forest geneticists now believe that selfing is relatively unimportant, particularly since selfed individuals are normally unsuccessful competitors and have high rates of mortality in competition with the outcrossed seedlings at the stocking levels normally achieved with seed-tree regeneration.

3 In the *shelterwood method*, the selection may be at a level of about one tree out of every four, so the selection differential is reduced as compared with the seed-tree method. Because more trees are left in the canopy, neighboring trees are more likely to be related, and the degree of outcrossing is therefore less than in the seed-tree method. The possibility of inbreeding is further increased if the leave trees have been selected toward some common goal (whether positive or negative), since similar phenotypes tend to be relatives. This is also enhanced if the released crowns produce increased amounts of pollen, thereby increasing the proportion of pollen from near neighbors. This potential for the production of a higher proportion of inbred seed, which results in inferior seedlings, casts doubt as to the wisdom of collecting seed from shelterwood stands for use in nurseries. Similarly, it raises significant questions regarding the genetic implications of obtaining seed from seed-production areas. It has been noted, for instance, that growth rates of seedlings from seed-production areas are lower after the poorer trees have been removed from the seed-production areas, presumably because such roguing increased the proportion of near relatives in the stand and thus increased the amount of sib- and cousin-level inbreeding.

4 For the *selection method*, the uneven-aged condition of the stands, with its intermittent waves of seed production and regeneration, gives less opportunity to change seed quality. The proper application of the selection method should ensure a continuously upgraded gene pool, particularly with respect to bole form and tree health. It is difficult, however, to select for small differences

in growth rate in uneven-aged stands. A poorly managed stand, with high grading of the better trees and little care in selection among the lower age classes, is very likely to lead to a lowering of genetic quality (dysgenic selection).

Thus, in summary, natural stands tend to be comprised of family groups, some crossing among relatives is to be expected, and the reduced health and vigor associated with such inbreeding is tolerated as normal in most natural tree populations. The clearcutting method can be characterized as maintaining the status quo. The seed-tree method provides the greatest opportunity for change by selection and natural regeneration. Whether this selection is positive, negative, or random depends on the goals and skills of the forester choosing the seed trees. No matter how seed trees are selected, it is likely that the seedlings from seed-tree parents will have fewer problems resulting from inbreeding than seedlings from any other natural regeneration method. Genetic change, resulting from selection of individual trees for harvesting in the various reproduction methods, is less intense in the shelterwood method than in the seed-tree method, and least intense in a selection forest. The shelterwood method presents the greatest danger of a substantial increase in levels of inbreeding.

PROBLEMS FOLLOWING SUCCESSFUL TREE BREEDING

Every genetically improved seedling created by the forest geneticist has to be planted in the forest, where it will no longer be possible to isolate the introduced superior progeny from the wild pollen of the same species in the adjacent stands, except as these stands are replaced by seedlings raised from improved seed. Replacement of all stands implies planting good and poor sites—an economic improbability. The result is either a continual reliance on replanting or a gradual loss of superiority through outcrossing with wild pollen. If rotations are shortened to 25 years, then natural regeneration is probably out of the question, because most commercial conifers do not produce seed in any quantity at such an early age, because the loss of production due to failure of natural regeneration is too costly, or because of costs of controlling stocking. In long rotations, any planting, even of superior trees, may be financially unjustified, although future trends in wood prices may warrant planting. However, planting for disease, insect, or drought resistance may justify initial heavy costs.

A more serious problem hinges on the extent to which the forest geneticist produces a highly uniform strain, because uniformity increases susceptibility to diseases and insects (Adams et al., 1971). Widespread plantings of uniform strains could result in epidemics of catastrophic proportions. The example of the 1970 corn crop, when the southern corn-leaf blight (*Helminthosporium maydis*) caused a 20 percent crop loss in the first epidemic year and left 80 percent of the corn susceptible (Tatum, 1971), poses a warning to foresters. Knowing that corn with T cytoplasm was the susceptible strain, corn pathologists had other cytoplasm types available for emergency seed production for the 1971 crop. The producer of hybrid corn seed will lose the advantage of male sterility of T cytoplasm but will merely have to hand-detassle to prevent inbreeding. In a similar situation, the forester would be unable to make quick changes, even if the resistant stock were available, since trees are not an annual crop. The corn-blight disease evolved from relative unimportance to a major new disease that swept the country in only 1 year. Pure lines of horticultural trees, such as cherries and pears, have developed virus problems. As mentioned previously, however, the objective of tree breeding is to out-cross and maintain genetic diversity to avoid this problem.

PRACTICAL APPLICATIONS OF THE PRINCIPLES OF FOREST GENETICS

Despite the fact that at first glance the contributions of the forest geneticist appear to be future possibilities, it should be pointed out that much that is critical and useful has already been contributed. A large body of research has established the existence of ecotypes and the danger of haphazard seed transfer among them. Increasing concern with this danger is shown by the recent revision of seed-collection zones in California from the initial 13 (Fowells, 1946) to a division of the entire state into 6 physiographic and climatic regions, 32 subregions, and 85 seed-collection zones (Buck et al., 1970). Silvicultural refinements may continue to demand further breakdowns of broad seed-collecting zones, but not before the silviculturist knows more about microsites and microprovenances and their reflection in seed performance.

Enough work has been completed to indicate that many of the economically desirable characteristics of a tree are sufficiently heritable to be selected in a tree-breeding program. Some of the characteristics are growth rate in height (Fig. 15-4) and diameter (Fig. 15-5), wood specific gravity, tracheid length, crown form, stem crook, stem forking, oleoresin yield, resistance to disease, resistance to drought, branch length (Fig. 15-6), branch angle, and self-pruning (Dorman, 1962). Although the phenotype may be more strongly influenced by environment than by the gene constitution, the silviculturist must be a practicing geneticist. Since the increasing intensity of management allows more frequent stand treatments, the forester has available such highly selective tools as various thinning methods and other intermediate cuts, as well as choices among harvest methods. Increases from genetic gain may be added to increases in production through fertilization and density control.

Silviculturists or practicing foresters look at trees for their quality as part of their day-to-day assignments and, in the process, can identify the plus phenotypes that must be found before a tree-improvement program can be initiated. A method of evaluating plus phenotypes has been outlined (Brown and Goddard,

Figure 15-4 Longleaf pine progeny tests for uniform rapid early height growth. (*From U.S. Forest Service.*)

Figure 15-5 The two pine cross sections sum up the supertree story—same species, same age, grown on the same site under the same conditions. The smaller section was grown by natural pollination of normal forest parents, the other section is from a supertree. (*From International Paper Company*.)

1961). It is a guide to anyone who wants to advance the day when the best sites are producing at their maximum. A plus phenotype is most easily recognized in an even-aged stand that has not had an intermediate cut and is still growing in height at a normal pace. A plus-phenotype candidate is judged by comparison with the best dominants in the immediate vicinity for the following characteristics:

1 Significantly greater total height
2 Smaller average crown diameter
3 Naturally well-pruned bole with few persistent branch stubs or excessive bumpiness around limb scars
4 Average-sized limb diameters or less
5 Straight bole, as determined by a line from the top of the merchantable length, staying within margins of the bole
6 Bole free of short crooks
7 Less than a quarter turn of spiral in bole
8 Average limb angle greater than 45° from vertical

Figure 15-6 Short-branched trees in the right row of progeny test versus the long-branched trees, with George Stephenson. (*From U.S. Forest Service.*)

9 Free of ramicorn branches
10 No evidence of disease or insect injury

After the preceding standards have been met, the tree must fall outside the 5 percent confidence interval of the regression of crown size (length × average radius) against average annual basal-area increment (for the last 10 years). Ten dominant trees are enough to establish the regression. Other factors of importance identified by Kitzmiller (1976) include diameter, which is strongly influenced by stocking but is highly heritable in some instances (Silen and Rowe, 1971), low stem taper, desirable live-crown ratio, and rapid growth in the terminal internodes.

The difficulty in choosing superior phenotypes by the preceding characteristics depends on how many traits are being simultaneously selected for improvement. Selection for more than one trait at a time reduces the likely gain in any one trait.

Improved Seed

In view of the impending availability of genetically improved seed supplies, the Certification Subcommittee of the Tree Seed Committee of the Society of American Foresters (Libby et al., 1968) recommended that two levels of codification be applied to seed collected from superior-phenotype trees. The codifying terms are *selected* and *certified*, but each signifies a different level of improvement when applied to seed-production-area seed than when applied to seed-orchard seed. *Selected* seed from a seed-production area signifies that the stand has trees of above-average phenotype, but a stand grown from this seed may not show the same superior phenotype. *Certified* can be applied to seed from the same stand after the stand's progeny-performance records have been collected long enough to (1) ensure that its progeny will give a better-than-average performance and (2) specify the conditions required to attain that performance. Selected seed from a seed orchard has a high expectation of producing plus phenotypes, but is only an expectation by virtue of the assembling of plus phenotypes and the roguing of the poor selections; thus selected seed connotes unproved performance. Certified seed-orchard seed becomes available when progeny testing has determined the expected performance and conditions required. A third level of codification is *source-identified* (seed from wild stands), which implies no improvement in performance other than adaption of the seed to a particular environment by evolutionary processes and probably better performance in that environment than seed from another environment.

Uniformity within the progeny increases as the source of seed shifts from a source-identified area to a seed-production area to a seed orchard, but it is not predictable with reasonable certainty until records provide a basis. Thus predictability of performance with respect to uniformity, site requirements, and specified disease and insect resistance for a seed source is implicit in the label *certified*.

TREE IMPROVEMENT RETURNS

Perry and Wang (1958) attempted to determine the worth of genetically superior seed. The monetary values used are now out of date, but the principle is still sound. They assumed a $10.50 per acre profit on a 25-year rotation based on pulpwood at $7 per cord and 1 lb of $4 per pound seed planting 10.7 acres. An increase of 0.5 percent in growth rate through use of improved seed was estimated to give a net increase in the final crop value plus 5 percent interest on the investment which, if discounted back to the present, would permit an extra $4.52 to be paid for the initial seed. If seed were available to provide a 30 percent gain (a conservative estimate of the gain likely to be attained) (Zobel,

1971), then the net increase in crop value plus 5 percent interest on the investment, when discounted to the present, would permit paying an extra $271 per pound for seed. It is apparent, even if interest is limited to the increased wood production and wood quality, that the tree-breeding program will prove to be a profitable investment. Add the potential for breeding trees to meet other needs, such as Christmas trees, and trees adapted to specific situations, and the full possibilities of the forest geneticist's contribution will be realized.

Seed-crop failure in a provenance where the seed reserves have been exhausted and the planting schedule requires seedling production has resulted in the occasional substitution of seed from another provenance. A pertinent question in such cases is how long to wait for a local seed crop when the use of an outside provenance results in growth loss.

Calculations for 25-year rotations made by Perry and Wang (1958) indicate that forest productivity obtained by delaying planting for 1 year is equivalent to that obtained by immediately planting nursery stock which has 4 percent lower growth potential. Similarly, if the expected growth loss is 10 percent, the forester can afford to wait 2.5 years for local seed to be available. In Louisiana, a provenance test of loblolly pine was established using seed from Georgia and Arkansas (Wakeley, 1944). The Georgia seed resulted in a 60 percent lower productivity, which meant that within the 25-year rotation, the forester could have afforded to wait 15 years for local seed with no added loss.

Using conservative estimates of genetic gains and the time needed to produce improved seed, the California program (Kitzmiller, 1976) indicates an added present net worth (6 percent discount rate) of $27.37 per acre for a stand of improved trees. Current loblolly pine improvement programs have estimated internal rates of return ranging from 10 to 14 percent with progeny testing and roguing and from 8 to 13 percent without progeny testing and roguing (Porterfield et al., 1975). These authors conservatively estimate that total gain in volume of improved stands over unimproved plantations will be at least 20 percent. Consequently, there is considerable incentive to enter into tree-breeding programs not only to increase volume production, but also to enhance other attributes such as disease resistance, straightness, and specific gravity.

More recently, data have become available which allow for multiple-trait analysis of first-generation seed orchards. The technique is based on linear-programming techniques and provides a goal-programming model to guide and evaluate tree improvement programs using economic criteria such as benefit/cost ratios (Porterfield, 1976). Porterfield has been able to show that genetic gains of current tree-improvement programs in southern pines are 9.9 percent for volume, 9.3 percent for straightness, 6.2 percent for crown characteristics, 4.8 percent for specific gravity, and 4.2 percent for resistance to fusiform rust disease. Benefit/cost ratios at 6 percent interest were shown to be 9.2 percent, and internal rates of return were 12 percent. Use of goal-programming techniques enabled Porterfield to determine projected gains and benefit/cost ratios for a unit area of seed orchard at a specified interest rate. A major advantage of this approach is its flexibility in permitting the answering of questions on the effects of different inputs such as changes in selection intensity.

A hidden threat in using an outside seed provenance is the discovery of the seed's weakness (susceptibility to freezing, disease, or insects) after it has been widely used over a period of years. Using a fully regulated forest with a 60-year rotation and assuming that a freeze killed all plantings of the provenance introduced 10 years earlier, Staebler (1966) calculated that the value of the forest had been reduced by 8.3 percent for one rotation.

The science of forest genetics has come a long way since the early provenance tests were established to ascertain the best provenance for a particular locality. A lot of evidence has accumulated that demonstrates the potential of applying the principles of forest genetics to improve growth rates, form, and disease and insect resistance. Some of this potential has been realized on a commercial scale, and studies necessary for a more rapid expansion of the benefits from the application of genetics to trees have been initiated. Despite the relative youth of forest genetics (as shown by the debate over seedling or clonal seed orchards), the disagreement between authorities over calculation of some genetic gains (Johnsson, 1964), the confusion and proliferation of terminology, and the rarity of the second generation among the breeding programs, few question the importance of genetic work if forest production is going to keep pace with the needs of an expanding population confronted with higher costs and mounting environmental, disease, and insect hazards. The importance of forest genetics will expand as the need for trees suited to the special requirements of water conservation, erosion control, aesthetic values, and even superior smog resistance becomes sufficiently obvious to justify the use of private or public monies for their development.

Chapter 16

Regeneration

INTRODUCTION

Historically, one of the most important challenges and responsibilities of the forester has been to establish forests on unstocked lands or to regenerate after harvesting. Incentives for reforestation are low, however, while wood is in relative abundance. For this reason, reforestation programs on the West Coast have developed later than those in other parts of the United States. For similar reasons, one can find the most intensive reforestation practices in those countries of the world which have limited timber resources.

Incentives for reforestation are usually economic or social. Economic incentives arise directly from the interactions of supply and demand. As wood resources become more limited, management develops a direct incentive to maintain continuing low-cost supplies by ensuring rapid regeneration of lands after they have been harvested. Limitations on supply develop through harvesting, wildfire, urbanization, or by withdrawals of land from the timber base by classification of wilderness areas, scenic corridors, and other reserves.

Social incentives are commonly developed by actions of government at the federal, state, or local level. The federal government may approve an act such as the National Forest Management Act of 1976, which, among many other provisions, requires the reforestation of all understocked lands in the United States. Many states have forest-practice legislation which provides penalties should private lands remain understocked after a certain period following harvesting, and counties may enact ordinances which prescribe reforestation standards that exceed minimum state requirements. Governments can also provide very effective incentives to reforestation through programs that provide tax concessions, subsidies, education, services, and research.

Social incentives for reforestation can also be very effectively developed through the action of conservation groups and concerned citizenry. In recent years, in many parts of the world, pressures developed by these activities have added a considerable stimulus to the development of reforestation programs.

The worldwide activity in reforestation is enormous. The total annual world production of seedlings and cuttings for reforestation purposes is currently about 6 billion. Approximately 16 percent of these is in the form of container stock, and the proportion is increasing. Hardwoods comprise 20 percent of bare-rooted plants, about half of which are transplanted in the nursery (Balmer, 1974). This practice of transplanting hardwoods in the nursery is, however, not done in the United States except perhaps for specialty shade trees.

In the United States there is a large backlog of lands requiring reforestation. In California alone, it is estimated that there are 300,000 ha (750,000 acres) of public land and 100,000 ha (250,000 acres) of private land that need reforesting. In the South, it has been estimated (Hughes, 1971) that by the year 2000, that region must provide half the nation's timber products or $2\frac{1}{2}$ times its present production. To do this about 4 million ha (10 million acres) of bare land and about 12 million ha (30 million acres) in low-grade oak and hickory must be converted to pine. Approximately 0.4 million ha (1 million acres) are harvested annually, which means that 1.2 million ha (3 million acres) must be reforested each year if the job is to be done in 15 years. In the South, the planting of hardwoods is also increasing, which has caused an increase in the production of hardwood seedlings in the southern states from 12 million to 17 million over the period 1965-1970, the most important species being sycamore, cottonwood, black locust, yellow poplar, and sweetgum (Rowan, 1972).

Principles of Reforestation

In securing the availability of appropriate seeds and planting stock, and to develop specific reforestation programs in a given locality, the actual acres in need of reforestation must be identified and their current condition assessed. Specific strategies then must be developed which will ensure that the reforestation will be successful. In every case, the approach taken in development of reforestation strategies should include the following elements:

1 Identification of land-management objectives and the translation of these objectives into specific kinds of stand structure and composition. For example, the kind of stand most suitable for wood production may differ from that which is most satisfactory for multiple uses. Consequently, the approach taken to reforestation and the evaluation of seeding, planting, and need for control of species mix and density will vary accordingly.

2 Identification of general constraints, which may be *managerial* (nearness to market, treatment priorities, site quality), *societal* (visual considerations, water supply, legal, taxes), *economic* (analyses such as benefit/cost ratio, present net worth), and *ecological* (inherent site factors such as ecosystem stability, wildlife considerations, presence of pathogens, etc.).

3 Analysis of specific stand conditions of relatively uniform aggregations of vegetation or by habitat types. This analysis is aimed at identifying controlling variables such as drainage problems, insect and disease factors, soil-water availability, temperature, evaporative stress, and competing vegetation.

4 Prescribing treatments involving choice of species, age class, seed source, type of stock for planting, spacing, and appropriate site preparation which adequately integrate managerial, ecological, and physiological considerations.

Consideration of these four factors will more likely ensure satisfactory establishment of a new forest which predictably meets managerial objectives.

Choice of Method

In developing reforestation strategies the silviculturist usually has the choice of:

1 Natural seeding, which utilizes clearcutting, seed-tree, or shelterwood methods to reproduce even-aged stands or the selection method to reproduce uneven-aged stands.

2 Direct seeding, which involves artificially applying seeds to an area either by hand or by aircraft. The seed may be broadcast or applied in rows or spots.

3 Planting, which may involve the planting of bare-root stock, container stock, rooted cuttings, or material raised from tissue culture.

There are numerous advantages and disadvantages to each method, but these can be evaluated only within the context of the four previously mentioned factors which are basic to the development of practical reforestation strategies. As objectives, constraints, stand conditions, and opportunities for site preparation change, then the silviculturist should modify recommendations for reforestation. Different methods will require different levels of managerial input. In cases where some limiting factor is operating, such as in areas that are excessively wet or dry, that have aggressive weed species, or that are plagued by pests, the choice of method may be very limited. In other nonlimiting situations, the silviculturist will have a wide choice of methods which are equally acceptable ecologically. Usually, however, a careful analysis of ecological, managerial, and social considerations indicates one approach that is more likely to achieve a desired result than the others. The relative advantage of different methods depends on current stand conditions and kind and extent of site preparation. In the case of reforesting loblolly pine in Louisiana, for example, direct seeding, natural seeding, and planting have been shown to be equally successful, with planting being the most reliable (Campbell and Mann, 1973). Whereas on other Louisiana sites dominated by low-quality hardwoods which were killed by injection with herbicide, 9 years after regeneration the planted areas had a height superiority equivalent to 1 year's height growth over the seeded areas (Lohrey, 1973).

Generally, one has more flexibility in choice of method when the site in question is within the "zone of the optimum" for that species. Northward from the zone of the optimum, flexibility of treatments becomes increasingly restricted to the south-facing slopes. Conversely, southward from this zone, flexibility in choice of treatments becomes increasingly restricted to the north-facing slopes.

In every case, the silviculturist must identify the extent to which control is needed over gene pools and environment. Gene pools are controlled by controlling species, provenance, and species mix. Control of environment may be more complex, in that the environment most suited to germination or out-planting survival may be markedly different from that most suited to subsequent growth. An example is in the regeneration of tolerant true fir, which prefers shade for establishment but grows best in more exposed situations once past the establishment phase. This must be recognized and accommodated in the silvicultural prescription, otherwise satisfactory development of reproduction will not be obtained.

The basic necessity for ensuring satisfactory regeneration, then, is to develop a systems approach to appraising the interrelationships between all controlling variables both in space and in time. If this is done, there should be relatively few surprises, and the silviculturist should be able to attain a predicted level of performance.

SEED

Seed Morphology

The production of seed, including flower initiation, stimulation, maturity, and periodicity, have been discussed in Chap. 8. In this section we are concerned with the nature of forest seed and the principles underlying germination, dormancy, stratification, seed testing, and storage. Details of these topics for many species are provided in the USDA Handbook 450, "Seeds of Woody Plants In the United States," 1974.

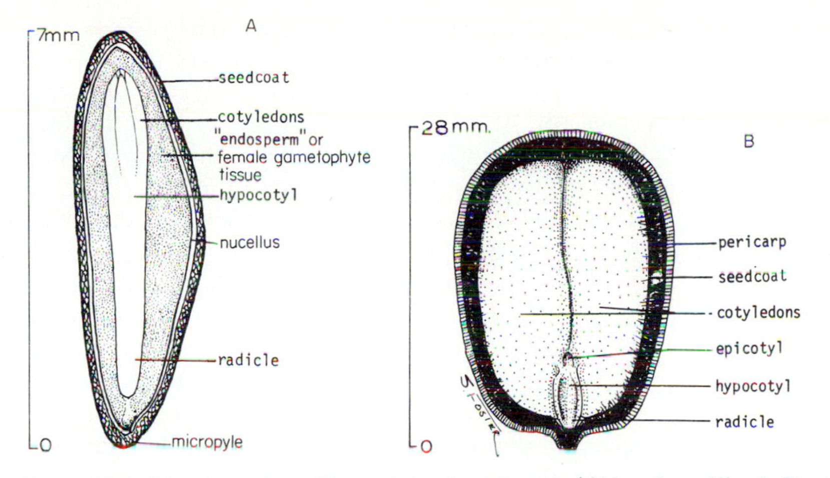

Figure 16-1 Diagrams of conifer and hardwood seed. (*Abies, from Wood, Plant Seed Manual, 1948; Quercus, from Seeds of the Wood, Plants in the U.S., 1974.*)

The basic components of conifer and hardwood seeds are shown in Fig. 16-1. In conifers, the *seed coat* is a hard outer covering which is relatively impermeable to oxygen, carbon dioxide, and water. The *nucellus* is a semipermeable membrane lining the inner surface of the seed coat. The *endosperm* (female gametophyte tissue) is the food reservoir of the seed which is utilized by the germinating embryo until it produces its own photosynthetic apparatus and capacity for producing carbohydrate. In conifers, the endosperm usually is composed of fats and oils. The central organ in the seed is the *embryo*, with *cotyledons* at one end, *plumal* or *epicotyl* in the center, and the *hypocotyl* and *radicle* at the other end. In conifers, the cotyledons are the initial photosynthesizing organs until the epicotyl develops to produce primary leaves. Pollen enters the ovule through the micropyle, which in the mature seed is a zone of weakness in the hard seed coat through which the radicle protrudes at germination.

Hardwood seeds have structures similar to those of conifers except that all the tissues are diploid; conifer seeds are also diploid, except for the endosperm. Hardwood endosperm commonly consists of starch, and the cotyledons do not act as leaves but are consumed by the germinating seed as a source of carbohydrate. The other major difference is that at the time of seed shed, the moisture content of mature conifer seeds is approximately 10 percent, whereas hardwood seeds may have a moisture content up to 40 percent.

Seed Collection

The success of most reforestation programs is dependent on the collection of seed. In recent years there has been a growing concern to collect better-quality seed and thus raise the quality of the resulting forest. Unfortunately, tree seed is difficult and sometimes dangerous to collect. It develops most profusely on large dominant trees and commonly is most abundant in the upper crown and toward the ends of the branches. Also, unfortunately, the trees which produce seed most prolifically are often those which are undesirable phenotypes having large, heavy branches. Tall, straight, light-limbed trees are difficult to climb and produce fewer seeds. Consequently, it is easy to see that when large quantities of seed are required, there is a danger that it may be collected from trees of relatively poor quality. Seed is commonly collected by following harvesting operations. However it is usually difficult to schedule tree harvesting over large areas so that it coincides with the time of seed maturation. Seed can also be obtained by raiding squirrel caches, but then one usually has no idea of the phenotypic characteristics of the trees from which the seed came. Some seed is obtained by tree climbing, but this is expensive and dangerous. However, since this is the major way in which the silviculturist can guarantee the quality of seed both as to the characteristics of the mother tree and the timing of the collection to ensure seed maturity and hence viability, there is currently a marked increase in seed collection by climbing.

To ensure success in survival and adaptability of seed to specific sites, silviculturists usually follow the tree-breeders' maxim that in the absence of tests proving the contrary, "local seed is best." For this reason, many states have divided their commercial forest areas into seed-collection zones. In the western states, these zones are delineated on the basis of seed collection criteria which specify that seed should:

1 Be collected within 100 mi north or south of the planting site
2 Differ in elevation by less than 300 m (1000 ft)
3 Be kept separate if areas have unusual climatic, topographic, or edaphic characteristics

The principle followed is that all seed collected is identified as to the zone from which it was collected. The seed carries this identification all the way through cleaning, storing, and sowing in the nursery bed. In this way, seedlings can be planted back in the same zone from which the seed originated.

As management becomes more intensive, these zones are often found to be too broadly defined. In these cases, and where industry or private owners have their own nursery facilities, seed may be collected from specifically designated "seed-production areas" or from "plus trees" which have been identified as having particularly desirable characteristics. Collecting seed from plus trees which are widely distributed over the stand is regarded as superior to collecting from seed-production areas. Both approaches are usually interim measures designed to upgrade the quality of seed used in reforestation in the period before seed becomes available from seed orchards. Descriptions of seed-production areas, seed orchards, and their use in upgrading the quality of forest seed are given in Chap. 15.

Seed Storage

When mature seed is disseminated from the tree, its storage depends on the species. Conifer seeds are shed with a moisture content of 10 percent and need drying to 6 to 8 percent if viability is to be retained over any extended period. Hardwood seeds vary greatly in moisture content when shed and some can be stored, like honey locust or black locust, and others, like oak, can usually be held only until spring planting. Dry conifer seed should be kept in sealed containers at a temperature just below freezing so that respiration will be kept to a minimum. Respiration should not be prevented entirely, since once it ceases any organism is dead.

Germination

There are two types of germination: hypogeous and epigeous.

Hypogeous Germination (Fig. 16-2*a*) This is the typical pattern of some angiosperms including *Quercus*, *Juglans*, *Aesculus*, and *Torreya*. Here, the cotyledons, which are storage organs, remain in the ground enclosed in the hull of the seed. The initial growth of the epicotyl develops into the primary stem and leaves.

Epigeous Germination (Fig. 16-2*b*) This is the typical pattern of most conifers. Here, the developing cotyledons lift the seed out of the ground. The cotyledons become the initial photosynthesizing organs, and the seed coat is dropped onto the surface of the ground.

Physiological Processes Involved in Germination

1 Absorption of water occurs, largely by imbibition.
2 Cell enlargement and cell division begins.
3 Enzymes are activated.
4 Insoluble carbohydrate, starch, fat, and protein are hydrolyzed into simpler water-soluble substances which are translocated from the endosperm to the embryo.
5 Rates of respiration increase rapidly and the energy liberated is used for growth and heat.
6 Increases occur in cell enlargement and cell division.
7 Rapid loss of weight occurs.
8 Differentiation of cells into the various tissues and organs of a seedling takes place.
9 Germination is essentially complete when the seedling has produced enough photosynthetic area to supply its own needs for carbohydrate.

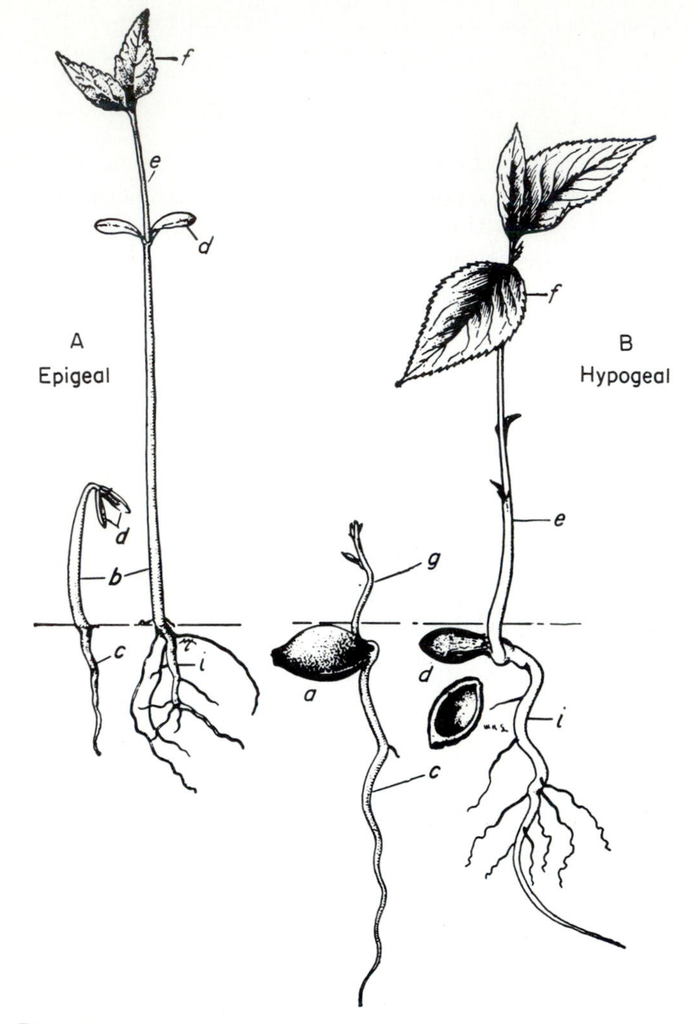

Figure 16-2 Seed germination. A: Epigeal germination of pin cherry (*Prunus pensylvanica*) seedlings at 1 and 10 days. B: Hypogeal germination of Allegheny plum (*P. alleghaniensis*) seedlings at 1 and 9 days: *a*, seed; *b*, hypocotyl; *c*, radicle; *d*, cotyledons; *e*, epicotyl; *f*, leaves; *g*, plumule; *i*, primary root. The plumule on both types of seedlings consists of the epicotyl and the emerging leaves. (*From USDA, 1974.*)

Internal Factors Affecting Germination

When seeds do not germinate promptly when exposed to favorable external environment, they are called *dormant*. There are a number of causes of dormancy in trees:

1 The embryo may be physiologically immature and may require a period of stratification before it will germinate well, e.g., sugar pine. Embryo dormancy is common.

2 The seed has an impermeable seed coat to water or oxygen or commonly both, e.g., honey locust. Seed-coat dormancy is less common.

3 The seed coat may be too strong for the developing embryo to break open. This cause of dormancy is rare.

4 The seed falls off the tree before the embryo has matured, e.g., *Gingko*. This situation is also quite rare.

The degree of dormancy varies between species and seed lots. Dormancy is a common characteristic in woody plants and has undoubtedly evolved as a mechanism to enhance survival by spreading the time that seeds germinate in natural conditions over a long period, thereby enhancing the probability that some seed becomes established.

Dormancy is broken in nature by exposure to cold temperatures in winter and by fluctuating moisture and temperature conditions. These changing conditions, plus the actions of bacteria and fungi in the soil, tend to break or weaken the hard seed coat. In some species with especially hard seed coats, such as *Ceanothus* brush, fire is commonly needed to scarify and break open the seed coat. Once the seed coat is made permeable to the entry of water, spring rains leach out any inhibitors to germination.

External Factors Influencing Germination

Fully mature seed can germinate only when external conditions are favorable. The major factors are as follows.

Water Abundant water is of primary importance because it is required to weaken the seed coat, hydrolyze food reserves, and raise the moisture content of conifer seed from about 10 percent when the seed is dispersed to a level of 45 percent, which must be attained before germination starts.

Oxygen Intensive respiration in germinating seeds requires abundant oxygen. Oxygen is usually readily available, but it becomes limiting when the seed is (1) in stagnant water, (2) buried too deeply, or (3) in an excessively fertilized soil which gives off large amounts of CO_2. Oxygen requirements vary widely, but the problems commonly occur in soils of poor porosity or high moisture content. Most species (with the notable exceptions of cottonwood, willow, sycamore, and rice) will not germinate under water. Even swamp species must be exposed to air and commonly become established during dry periods when the soil is exposed. Mangrove seed actually germinates when it is still attached to the parent tree and thus what falls into the tidal mud flats is actually a germinated seedling.

Carbon Dioxide High CO_2 levels retard germination because of the hindrance to respiration. For this reason, seed is stored in sealed cans so that the small amounts of CO_2 given off by the respiration of mature dry seed produce a CO_2-enriched environment inside the container. Moist soils with high organic content have high CO_2 levels. This is readily exchanged to the atmosphere if it is near the soil surface, but at deeper levels the soil atmosphere can contain high concentrations of CO_2. This is one reason why seeds planted too deeply may not germinate.

Temperature Germination of most tree seeds can occur over a rather wide range of temperatures. Most rapid germination takes place within the range of 25 to 40°C. Germination does not necessarily increase with increasing temperature, and it has generally been found that best germination occurs when temperature fluctuates rather than stays constant (Jensen et al., 1959). Knowledge of temperature requirements for germination has practical importance in timing the seeding of nursery beds and field sites. Dry seeds are quite resistant to detrimental effects of extreme temperatures. However, when seeds are in the imbibed state and are germinating, those same extreme temperatures can be lethal.

Light Some seeds such as peppergrass (*Lepidium virginicum*) do not germinate at all in darkness. Others such as Grand Rapids lettuce (*Latuca sativa*) are stimulated into germination by orange-red wavelengths (670 nm), whereas infrared wavelengths (760 nm) inhibit germination. Alternating exposure of seed to red and far-red light changes the stimulus back and forth, with the seed germinating or not depending on the wavelength to which the seed was last exposed.

Forest-tree seeds generally germinate equally well in the light or dark. There are some contradictory reportings on the effect of light and photoperiod on tree seeds, but this is not surprising since it can be shown that some seeds not usually light sensitive can be made so by holding them at a particular temperature in the imbibed condition.

Sonic Treatment Past investigators have sometimes claimed benefits in germination to yields of rice and tobacco from sonic energy. Sonic treatments of Douglas-fir and ponderosa-pine seed and seedlings have not substantiated these claims (Lavender and Atherton, 1964).

Stratification

Dormancy is a nuisance to nursery operators because of the particular need for uniform and even germination to ensure that all seedlings in the bed will be of uniform size at the time of lifting. Similarly, the silviculturist who wishes to direct seed in the spring finds dormancy a problem since it is desirable to obtain prompt germination while conditions are favorable. This is particularly important at high elevation where the time between snow melt and the onset of summer conditions may be only a few weeks. Prompt germination is also desirable in order to limit the time when seed is on the ground to be consumed

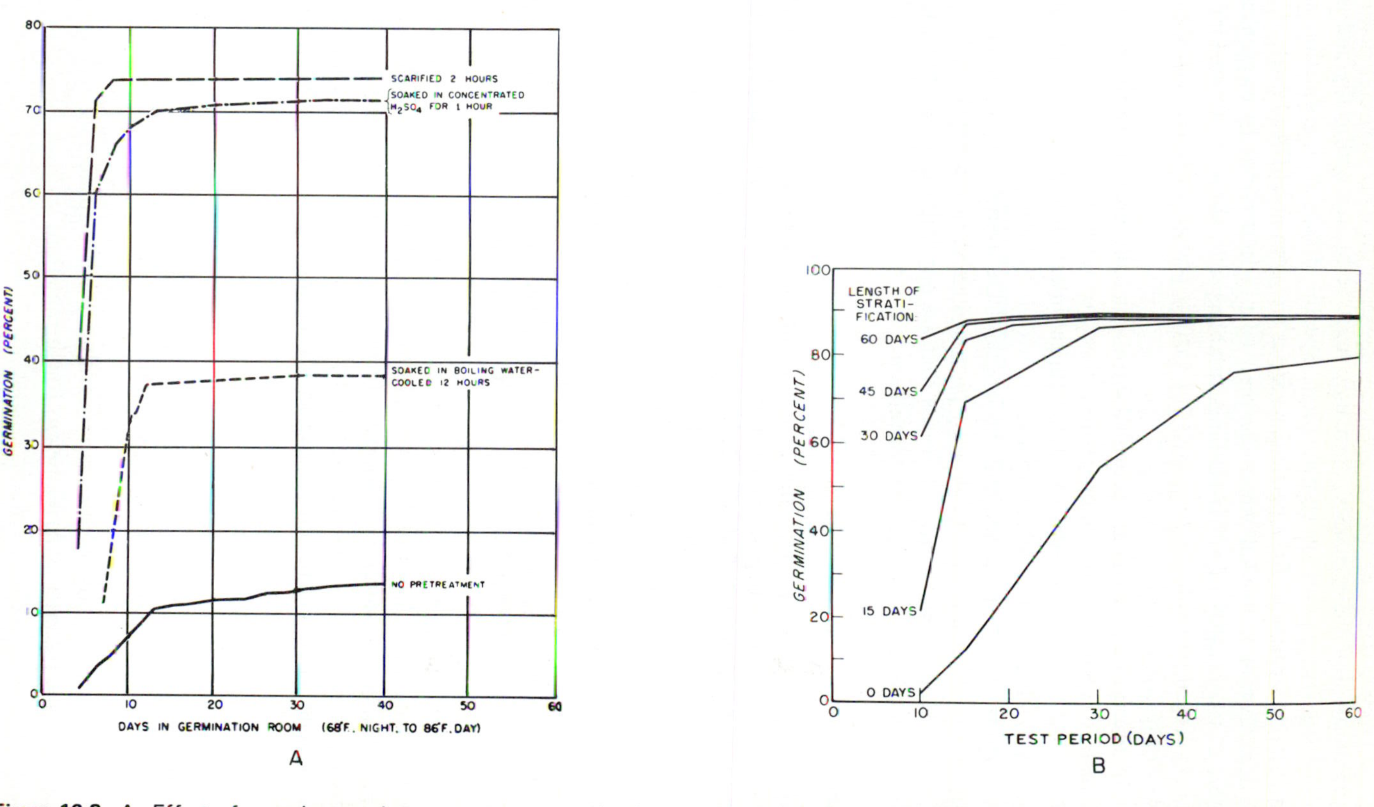

Figure 16-3 A: Effect of several presowing treatments on germination of black locust (*Robinia pseudoacacia*), a hard-seeded species. B: Benefits of cold stratification for speeding germination of loblolly pine (*Pinus taeda*). (*From USDA, 1974.*)

by predators. For these reasons, dormancy must be broken artificially and this is done by scarification or stratification. Specific details of scarification and stratification treatments for a particular species are given in USDA (1974).

Scarification is needed when the seed coat is impervious to oxygen and water, and it consists of either rubbing, tumbling with abrasives, or soaking in acid, hydrogen peroxide, or hot water for varying periods of time. Each species has its own requirements.

In *stratification*, seeds are provided with abundant moisture (well-drained), adequate oxygen, and low temperatures close to freezing for 1 to 4 months. During this time it is necessary to avoid fungal infection and this is commonly done by dusting the seeds with Captan.

The effect of scarification and the length of the stratification period on the germination of black locust and loblolly pine seed are shown in Fig. 16-3.

Some stratification is given to seeds that are not really dormant but whose germination period is too long for convenience.

Seed Testing

Seed must be tested to determine the amount that needs to be sown per unit area to provide a predictable number of healthy seedlings. In testing seed lots, the greatest care must be taken in sampling. Usually many subsamples are taken from different locations in the container and these are mixed together and sampled again until a convenient number of seeds is obtained. This number may be 50 to 100 seeds for an exploratory study or as many as 1000 to 2000 for research needs. Tests are usually carried out for the following attributes.

Genuineness Seed must be true to name and may need to be compared with a known sample for validation of its botanical identity.

Purity The amount of impurity in a batch of seed must be known. This is not normally a problem with large seed, but with small seed such as alder, birch, eucalyptus, and redwood, seed batches are relatively impure because of broken pieces of cone scales, wings, leaves, "chaff," and other debris. These impurities are difficult or impossible to remove in the cleaning process. The presence of impurities may even be desirable in that they "dilute" the seed and aid spreading. Purity is expressed in the following way:

$$\text{Purity (\%)} = \frac{\text{weight of "pure" seeds}}{\text{weight of seeds plus impurities}} \times \frac{100}{1}$$

Pure seeds are those which are outwardly normal in appearance even though they may lack the internal development for germination.

Number per Pound This is usually expressed as the number of seeds per pound of pure seed, and is important in determining the rate of sowing.

Table 16-1 Number of Cleaned Seed per Unit Weight of Common Tree Species

Conifers		Hardwoods	
Species	Cleaned seed per kg (lb)	Species	Cleaned seed per kg (lb)
Pinus ponderosa	26,400 (12,000)	*Quercus macrocarpa*	165 (75)
Pinus taeda	40,000 (18,200)	*Quercus nigra*	869 (395)
Pinus resinosa	114,400 (52,000)	*Juglans nigra*	88 (40)
Pseudotsuga menziesii	85,800 (39,000)	*Carya glabra*	440 (200)
Abies balsamea	131,120 (59,600)	*Betula glandulosa*	8,445,800 (3,839,000)
Larix occidentalis	301,400 (137,000)	*Populus deltoides*	770,000 (350,000)
Picea mariana	1,258,400 (572,000)	*Platanus occidentalis*	425,194 (193,270)
		Castanea dentata	286 (130)

Source: USDA, 1974.

Table 16-1 gives the approximate numbers of seeds per pound of common tree species.

Moisture Content This provides a guide as to the condition under which the seed was stored and an index of quality. Excessively dry or wet seed decreases the life of seed in storage. Conifer seed should be stored at a moisture content of 6 to 8 percent dry weight.

Viability Viability is influenced mainly by seed maturity at the time of harvesting, by damage due to handling during the cleaning process, and by the length of storage period. Viability may be tested physically, physiologically, or most directly and reliably by actual germination tests.

Physical Tests *Cutting:* this is the oldest and simplest test but it is least reliable and likely to give results which are often considerably higher than actual germination tests, especially for old seed. Cutting is often regarded as giving an estimate of the upper limit to viability. In this test, individual seeds in a sample are cut in half and the condition of the endosperm is observed. In mature viable seeds, the endosperm should be white and quite firm. In immature low-viability seed, the endosperm is still very fluid. Alternatively, the ratio of the length of the embryo to the total length of the seed plus seed coat can be observed. If this ratio is less than 80 percent, the seed has not ripened and is of low viability.

Transparency: small thin-coated seed such as birch and cedar can be passed over a strong beam of light. Empty seed is seen to be relatively transparent.

X-ray: the seed coat, endosperm, and embryo absorb x-rays to varying degrees. It is possible, therefore, to recognize seed with well-developed endosperm or seed with injuries resulting from mechanized processing or insects. To enhance the testing for injured seed, it is common to pretreat the seed with barium chloride solution, which cannot penetrate living tissue but which freely diffuses through necrotic parts. X-ray photographs clearly show injured seed by the marked contrast in density. X-ray treatments do not affect viability and may be used routinely to adjust seed-cleaning apparatus to deliver the highest proportion of pure and potentially viable seed. Figure 16-4 shows a pair of polaroid x-ray photographs of the output of a seed-cleaning device. One photograph is of Douglas-fir seed immediately after passing through the cleaner; the other is of the material that is being exhausted by the blower. Airflow in the cleaner is adjusted so that the minimum amount of impurity is included in the clean seed and the minimum amount of sound seed is discharged.

Physiological Tests Various biochemical and staining techniques have been used to test for enzyme activity. For example, tetrazolium chloride stains living tissue and indigo carmine stains dead tissue.

Germination Tests True viability can be determined only by actually germinating an average sample. In this test, seeds must be stratified in a prescribed manner and germinated under standard and controlled conditions. This is best done with a special germinator which uses a standard germinating medium

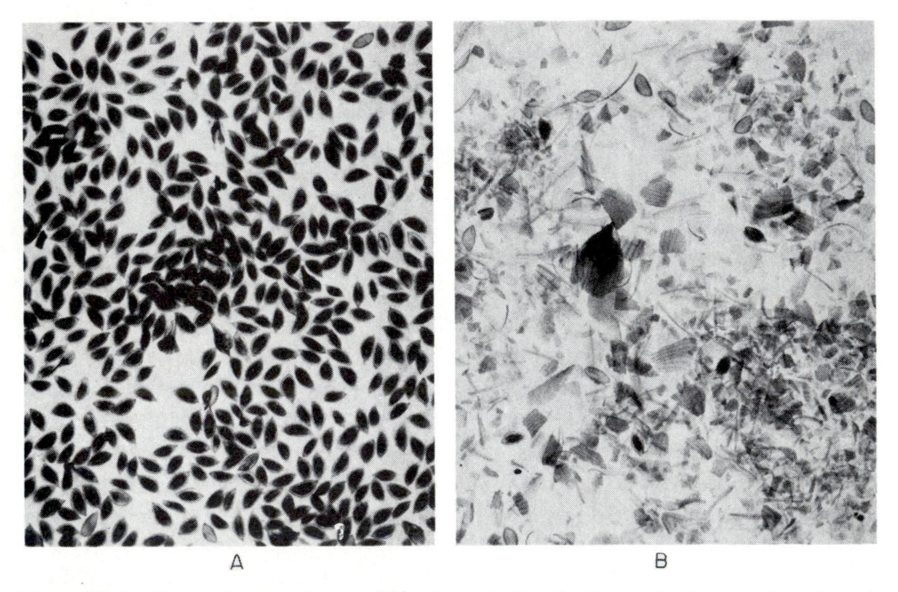

Figure 16-4 X-ray photographs of (A) cleaned Douglas-fir seed after passing through cleaner; (B) impurities exhausted from blower. The velocity of air in the seed cleaner is adjusted to deliver the maximum amount of pure seed and to minimize the amount of sound seed blown off with the impurities.

and provides controlled levels of temperature, moisture, light, and ventilation. If procedures are not strictly controlled, comparability between tests and samples cannot be obtained.

Germination tests provide estimates of *germinative capacity*, which is the total cumulative germination percent of the seed lot over a period of perhaps 1 or 2 months. Some seed in the sample will germinate quickly and some may never germinate even though it appears sound. In forestry operations we are commonly interested in prompt germination, particularly in nursery work, therefore we are more concerned with *germinative energy*, which is the percent germination at the point where the rate of germination slows appreciably. Since this will occur at different time periods for different seeds, germinative energy or rate is more commonly expressed as the number of days required for a certain proportion of total germination to occur (USDA, 1974). The concepts of germinative capacity and germination energy are illustrated in Fig. 16-5.

Seedling Development and Establishment

The development of seedlings from germination until establishment is the most critical and precarious period in the regeneration of a new forest. Most mortality occurs during this period. Consequently, if we are to enhance regeneration success, we must understand the basic characteristics of seedling growth. All plants pass through this period, but the time is short for many herbaceous species and hardwood trees. In conifers, this period is relatively long and the potential for mortality is higher. Consequently, conifer seedling development will be considered here in some detail.

There are two important periods which can be recognized:

1 The succulent stage, which lasts for only a few weeks, beginning as soon as the seedling appears above the surface of the ground and lasting until the hypocotyl has become hard and wiry.

2 The juvenile stage, which follows the hardening of the hypocotyl and lasts several years until the seedling is established and risks of mortality are low.

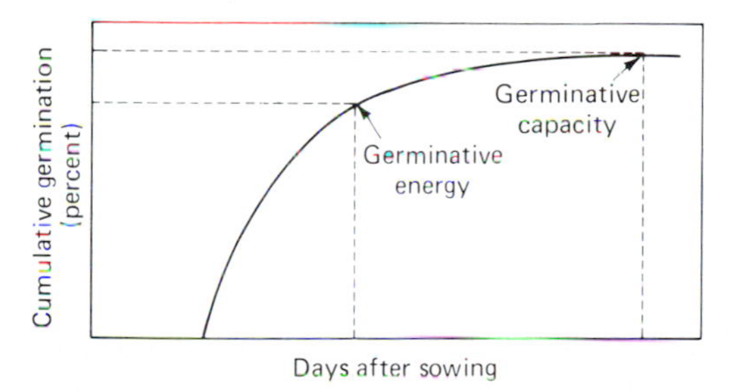

Figure 16-5 Distinction between "germinative energy" and "germinative capacity" used to describe seed viability.

The Succulent Stage Most coniferous seedlings will break ground and stand erect 1 week after the initial extrusion of the radicle. At this time the cotyledons are neither fully grown nor spread out, but will be cupped together. If the seed coat has been pulled out of the soil, the cotyledon tips will probably still be held together with the dry testa. At this time the hypocotyl will not have attained its final length. It is usually pale green, pink, or bright red, very succulent, and quite brittle. The root is still short, ranging from 13–50 mm in length depending on the initial size of the seed and species. Pine roots grow faster than spruce and fir, which in turn develop more rapidly than redwood. In all species the root is straight, white, and unbranched. The point at which the root becomes the hypocotyl is often impossible to distinguish, although it may be marked by a J-shaped bend.

Under favorable conditions, by the end of the second week, the hypocotyl will have attained its mature length and the cotyledons will be practically fully grown. At about this time the first of the true leaves appear from the bases of the cotyledons. By the end of the second week, the primary leaves may be as much as 6 mm long in fast-growing species or barely visible in species growing at high elevations. Root growth continues rapidly but slows down much earlier at high elevations and with short growing seasons than with species such as the southern pines and Monterey pine which grow in more favorable areas. Roots generally continue to be unbranched until about the sixth week, which is in striking contrast to herbaceous competitors.

By the end of the fourth week, the hypocotyl of rapidly growing species begins to shrink and wrinkle and becomes hard and wiry or leathery. With slower-growing species this change may not take place until the fifth or sixth week. At about the same time that these significant changes take place above ground, the first root branches begin to appear. The succulent stage is over.

Internal Development during the Succulent Stage During the first few weeks after germination there is a rapid differentiation of parts following a complex pattern which appears to be basically similar in all conifers. The gross development of the cotyledons and of the shoot appears to have little ecological significance, but the changes that take place within the hypocotyl are very important to the existence of the seedling.

Anatomically, the hypocotyl is a region of transition in which the collateral vascular bundles of the cotyledons change over to the radial bundles of the root. The change is made in the upper part of the hypocotyl, sometimes close to the base of the cotyledons; hence the greater part of the hypocotyl is essentially rootlike in structure, although it is above ground and has an aerial epidermal structure.

The development of the hypocotyl may conveniently be considered in four stages, which are illustrated in Fig. 16-6. This figure shows the pattern in pine as typical of the conifers that have several protoxylem and protophloem initials

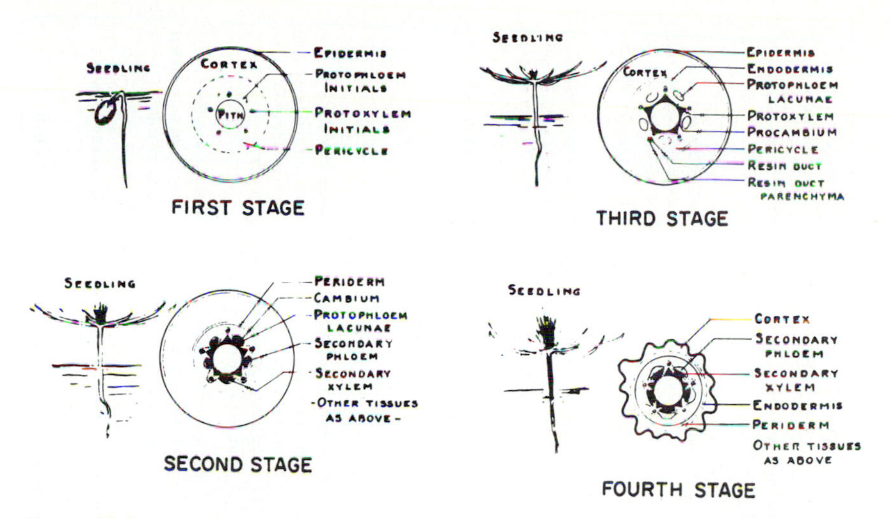

Figure 16-6 Developmental stages during the succulent period of stem development in conifers. (*From Mirov and Baker, 1942.*)

and a pith. The brief descriptions in Table 16-2 may best be followed by reference to Fig. 16-6. The ecological significance of many of the changes shown here is doubtful in the present state of knowledge, but several points are definitely significant.

The first is the weakness of the whole vascular system in the early days of the life of the seedling. The xylem is very small, and while it may be adequate to conduct sufficient water to the cotyledons, the ratio of xylem cross-sectional area to transpiring area is much lower than in the later life of the seedling, and inadequacy of conduction may be suspected. It is certain that these small strands contribute little to the mechanical strength and stability of the seedling, which maintains itself primarily by turgor of the parenchyma cells of which so much of the hypocotyl is composed. The development of the secondary xylem with its solid ranks of cells (nearly square in cross section) marks a definite change in mechanical strength if not in efficiency of water conduction.

Most important of all is the collapse of the cortex, which marks the end of the period of succulence and the development of the seedling into a small but wiry plant. The outer "bark" is firm and stringy in many species, resisting injury and providing some protection against fungal attacks. This, plus the heavy periderm, serves to prevent water loss from the stem. At the same time, the development of the vascular system adds further strength. The collapse of the cortex follows no set pattern. In the pines the collapse is sudden and the "bark" is crisp and hard. In some of the firs, notably white fir, the stem becomes gradually leathery because of the development of collenchyma in the cortex. In some of the conifers of the cypress family there is a development of a hypodermis of several cell layers immediately under the epidermal layer. Indeed, the

Table 16-2 Outline of Development of Tissues in Coniferous Hypocotyl in the Succulent Period

Seedling tissue	Stage 1. Elongation	Stage 2. Metaxylem and phloem	Stage 3. Cambial activity begins	Stage 4. Cortical collapse
Epidermis	Single row of cells moderately thickened on external side	No particular change; the outer wall apparently becomes more resistant to penetration by the hyphae of damping-off fungi		
Cortex	Wide zone, composed of large thin-walled parenchyma cells loosely fitted, containing chlorophyll	No particular change	No particular change	Sudden collapse and desiccation of cortical tissue into a dry "bark," in nearly all conifers
Endodermis	Site marked only by change in size of cells from large cortical to small pericyclic	Typical endodermis, single row of cells with Casparian strips	Cells beneath endodermis divided by tangential walls into several layers, periderm	Periderm becomes heavy and suberized (?) cutting off water supply to cortex
Pericycle	Composed of thin-walled cells usually containing chlorophyll; denser, smaller, and more closely fitting then the cortical cells	No essential changes; the area of pericycle tissues decreases with the gradual development of vascular tissues		
Phloem	Relatively large with poorly developed sieve tubes, thin-walled "precursory phloem"	True metaphloem develops on inner face of the phloem strand	Precursory phloem collapses and is resorbed leaving lacunae. True secondary phloem begins to be formed	Secondary phloem being formed and occupying precursory-phloem lacunae
Xylem	Small protoxylem strands of very small irregular cells	Metaxylem develops inwardly and tangentially. In two-poled species it may extinguish the pith	Metaxylem strands meet and the formation of secondary xylem begins	Secondary xylem is well developed with several rows of cells
Resin ducts	May be present or absent, variable in position and development; in *Pinus* they are prominent, standing radially outward from each protoxylem strand			
Pith	Variable in size, may be absent. When present, commonly of large thin-walled cells, usually without chlorophyll			

full details of this important stage of seedling development are as yet very little understood.

The precise time of cortex collapse varies a good deal with species and rate of growth. It has been noted that in the pines cortex collapse takes place when the length of the first true leaves is about one-half the length of the cotyledons and, at the same time, the secondary xylem has about the same radial extent as the protoxylem. In terms of time, it is usually 4 to 5 weeks from the initial germination and 3 to 4 weeks after the seedling appears above ground, although unfavorable conditions of growth may greatly extend the period.

This fundamental anatomy of the young seedling has been discussed in some detail in order to show clearly:

1 The long period during which the seedling has a succulent cortex
2 The slowness of development of the vascular system
3 The fact that all these developmental phases are interrelated processes—one cannot be accelerated while another runs at a slow pace. Hardening of the hypocotyl can therefore be speeded up only by giving the seedling optimal conditions for growth.

Injury during the Succulent Stage The high potential mortality in conifer seedlings in the succulent stage of development is largely due to direct heat injury, damping-off diseases, mechanical damage, lack of soil water, and predation by birds and insects. The first two of these are the least understood and warrant some discussion.

Damping-Off Diseases There are a large number of organisms that cause this disease, and the most important belong to the genera *Rhizoctonia* and *Pythium*. The attacks take place at the ground line or slightly below and quickly involve the upper rootlet and hypocotyl. The hypocotyl first appears water-soaked and discolored, then it loses its turgidity, and finally the seedling topples over and withers, resulting in a wasp-waisted appearance at ground line. The worst attacks take place under warm damp conditions and while the seedling is in the succulent stage. Indeed, the seedling may be attacked and killed before it even breaks ground.

The spores of several of these damping-off fungi exist in all soils, particularly rich organic soils, for the organisms are not obligate parasites. Usually they appear to be irregularly distributed, and thus in nurseries they will spread from foci of infection. Except in nurseries, losses from damping-off diseases are difficult to observe. Under field conditions, losses from damping-off diseases are combatted by treating seeds with fungicides before direct seeding operations. In nurseries, soils are commonly fumigated, seeds are treated with fungicide, and care is taken to avoid watering during the day when conditions are warm. The fumigation of nursery beds sterilizes the soil, eliminating both beneficial mycorrhizae and undesirable fungi. Both types of organisms rapidly reinvade the soil within a few weeks after fumigation. This is usually enough time for the seedlings to develop beyond the precarious succulent phase and thus not be susceptible to damping off and yet be able to take advantage of mycorrhizal infection.

Direct Heat Injury Heat injury is probably the most common cause of seedling mortality. This is especially true on exposed sunny aspects where the temperature of the soil surface can commonly exceed the threshhold level of 55°C (103°F) at which protoplasmic damage can occur. The succulent hypocotyl is very susceptible to heat injury, which develops at the ground line where the highest temperatures occur as a result of absorption of energy from the sun.

Heat injury becomes evident when the cortex collapses and the seedling falls over. The soft, dead cortex tissue becomes susceptible to decay organisms and becomes scorched and white in appearance. The spot is usually on one side but may become so large that it involves the entire hypocotyl and the plant withers at the ground line.

Plant factors which influence the likelihood of heat injury are the diameter of the hypocotyl, the shading effects of the cotyledons, the age of the seedling, and perhaps the cooling effect of the transpiration stream. Field conditions that are important and may be manipulated by the silviculturist are shade, soil color, and soil dryness. Dark or dry soils heat up much more readily than light-colored or moist soils. This may lead one to expect that duff and litter will reduce losses because of their mulching effects. However, higher temperatures are reached on the surface of the 0 horizon than on bare mineral soil, i.e. 70°C as compared with 66°C. This is so because of lower heat conductance and lower specific heat in organic matter. Shade can be provided by leaving harvesting slash or other debris on the ground and directing seeding on top of it. Germination percent will be lower, however, because of the fewer seeds making contact with mineral soil.

Mortality after the Succulent Stage After the seedling hypocotyl becomes hardened, it is relatively safe from the forms of injury that have been considered. It may be injured or killed by a large number of factors, the most important of which are:

1 Drought
2 Lack of sufficient light (This is most common even for tolerant species.)
3 High temperature.
4 Frost and frost heaving (physical lifting of the seedling and surrounding soil by the formation of ice crystals)
5 Smothering from litter fall or snow
6 Insects and diseases (These are rarely of outstanding importance to seedlings, except the brown-spot needle blight of longleaf pine and the pales weevil which attacks eastern white pine and other conifer species.)
7 Fire
8 Animal damage (trampling, browsing) and rodent damage

SITE PREPARATION

Definition in Ecological and Physiological Terms

The condition of the site can vary from favorable to unfavorable in terms of providing a suitable habitat for the establishment of a new forest. The silviculturist's aim is to provide the desired vegetation with a temporary advantage in becoming established by manipulating the environment. The site-preparation treatment must be selected after thorough consideration of ecological, physiological, managerial, and social factors. In particular, the treatment should be effective, cheap, and at the same time incur minimal environmental impacts.

Site preparation involves the deliberate manipulation of four site factors:

1 *Physical factors of the atmospheric microenvironment:* temperature, radiation, and evaporative stress.

2 *Forest floor and competing vegetation:* the condition and amounts of slash, litter, and weed species.

3 *Soil:* The physical factors of texture, porosity, bulkdensity, water-holding capacity, soil water availability, presence of hard pan, and the necessity for draining and the chemical factors of cation-exchange capacity, nutrition, pH, mycorrhizae, and soil microorganisms.

4 *Biotic factors:* the potential impacts of animals, insects, diseases, and birds.

To ensure rapid establishment of desired vegetation, the condition or levels of these factors must be favorable. Therefore, the silvicultural task is, first, to be able to evaluate the relative ecological and physiological requirements of plants, and second, to be able to manipulate the environment to enhance the growth of desired plants and to put undesired plants at an ecological disadvantage.

This basic concept of matching plants to environments, or of modifying environments (if possible) to suit the desired species, is the key to successful silviculture and predictable performance. It is often not necessary to match plants and environments exactly, but successful forest establishment can often occur if the situation is modified, even temporarily, in favor of the desired plants. This concept can be expressed in terms of the allocation of site resources. A given site has a definable set of resources (temperature, radiation, nutrients, soil, water availability, etc.) which is usually allocated by nature on a first-come first-served basis. In site preparation, the silviculturist can determine the extent to which these resources can be reallocated to favor the desired vegetation. In order to do this, some knowledge is required of the relative potential of species to dominate the site (Newton, 1973).

Figure 16-7 illustrates that the initial time T_1 is critical in determining which species are capturing the resources of the site. It also shows that any proposed treatment should be done as soon as possible after time T_0. If one can

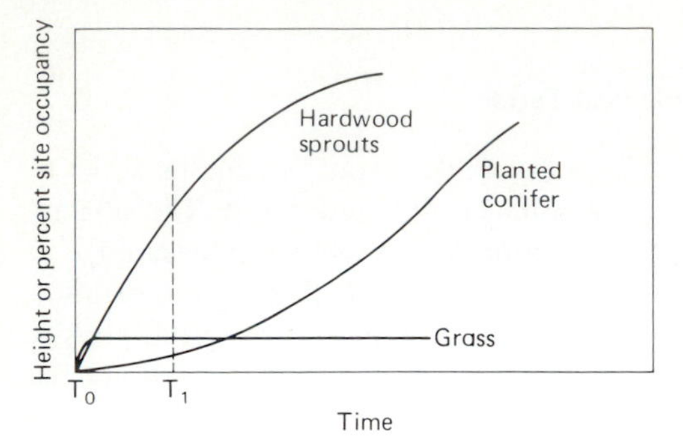

Figure 16-7 Diagrammatic representation of the concept of dominance potential.

predict the relative dominance potential of species on a given site, one can then predict the time available in which specific treatments can be effective and what kinds of treatments these must be.

The concept can be further developed if we consider a site which is currently fully occupied by brush or low-quality hardwoods (Fig. 16-8). If the forester determines that this vegetative type is inappropriate for the particular management objective, then at time T_1 this brush cover can be substantially reduced to occupy a much lower percentage of the area. At time T_1, then, the resources that were being consumed by the brush are now immediately available and will be taken up by invading vegetation unless the silviculturist directs the uptake of resources by further treatment. If allowed to revegetate naturally, the expression of dominance potential will result in the site being reoccupied in the manner shown in Fig. 16-8A. Obviously, by time T_2, which could be only 1 year after T_1, there is little point in planting the area to conifers, which have a particularly low dominance potential in this situation, since the site is already well on the way to being occupied. It can also be seen that it is simply impossible for the plantation to be successful until the brush occupies a lesser proportion of the site by becoming overtopped or through further treatment. The best time to introduce desired vegetation into the area was at time T_1, particularly if the existing brush had been brought to as low a level of site occupancy as feasible. Given the dominance potential of sprouting brush, it would probably be preferable to plant conifers at time T_1 and to plan a second brush treatment at time T_3 (Fig. 16–8B) to release the planted conifer from excessive competition. If this were to be done, a successful plantation largely occupied by desired conifer would be obtained by time T_4 rather than in the indefinite future in the case of the stand without the second release treatment.

These concepts of dominance potential, rate of site occupancy, and reallocation of site resources provide a useful framework within which the silviculturist can predict the likely relative behavior of plants in changing environments. Using

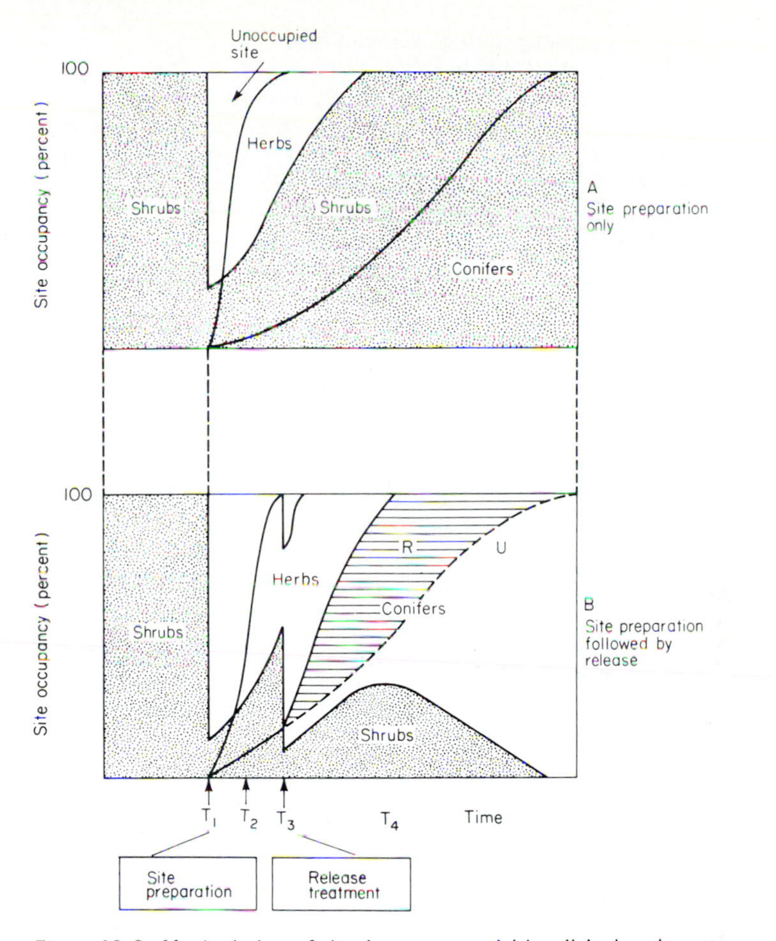

Figure 16-8 Manipulation of dominance potential by silvicultural treatment. After second treatment, shrubs display lower dominance potential when overtopped by conifer. Dominance potential of herbs is relatively unaffected in this example. R is path of released conifer, U is path of unreleased conifer, hatched area between paths R and U is the extent of increased conifer productivity resulting from the treatment. (*After Newton, 1976.*)

these concepts, the nature and timing of site-preparation treatments can be determined and the outcome of the regeneration strategy will be predictable.

Objectives and Methods of Site Preparation

The *objectives* of site preparation can be numerous, but they generally include one or more of the following:

1 Improve early soil moisture conditions by reducing competing vegetation or draining excess water

2 Improve survival by manipulating overhead shade
3 Make tree planting easier and cheaper by reducing slash
4 Increase wood production by improving early growth
5 Shorten the length of the rotation or harvest cycle since larger-dimension trees are obtained in a shorter time
6 Shorten the time for, and increase, financial returns on investments
7 Improve or reduce wildlife food and cover
8 Improve accessibility for future cultural and harvesting operations
9 Reduce fire hazard
10 Improve aesthetics

The methods of site preparation are commonly grouped under five categories of mechanical, chemical, fire, hand tools, and a combination of the previous four.

Mechanical The methods involve root-raking, bulldozing, shearing, ball-and-chain methods, brush crushing, wind-rowing, piling, terracing, ploughing, ripping, bedding, burying, and chipping. They all involve machinery and therefore are subject to problems of capital intensiveness, high energy consumption, limitations of slope, and potential soil damage from compaction or erosion. Each method should be evaluated for each specific site as to the extent it is likely to affect the operating plant environment both favorably and unfavorably and the likely overall ecological impact.

Chemical Only some broad generalization can be made and readers are directed to the following sources for further information:

General Information Barrons (1969), Presidents' Science Advisory Committee (1971), Weed Science Society of America (1974), Day (1975), University of California (1977).

Uses of Herbicides Bentley and Estes (1965), Newton (1973), Gratkowski and Lauterbach (1974), Newton and Norgren (1977).

Effect of Herbicide on Wildlife Habitat Krefting and Hansen (1969), U.S. Department of Interior (1964), Carter et al. (1975), Borrecco et al. (1972), Keith et al. (1959).

Fate of Herbicides in Environment Clark et al. (1964), Krammes and Willets (1964), Lutz et al. (1973), Reigner et al. (1968), Trichell et al. (1968), Norris (1967).

Toxicity and Hazard Condon (1968), Johnson (1971), Montgomery and Norris (1970), Norris (1971), Sheets and Harris (1965), Way (1969).

The objective of chemical treatments is to temporarily reduce the dominance potential of undesirable vegetation to enable the desired vegetation to more rapidly attain a higher site occupancy. Herbicides are seldom prescribed to kill all vegetation, rather, dosage rates and timing are chosen to selectively reduce the growth of one vegetative component.

The use of chemicals in forest management involves a complex of technical, biological, legal, and social problems. In most cases, social concerns regarding possible impacts of herbicide use, plus stringent federal and state restrictions on use, are constraining this method of site preparation despite a wealth of information on mode of action, toxicity, and hazard.

The situation is paradoxical in that the proponents of herbicide use point to the relatively low environmental impacts and costs compared with alternative methods involving mechanical treatments, hand removal, or fire. The adversaries decry the introduction of chemicals into the ecosystem, pointing to possible unknown hazards to humans and other biological systems. The controversy stems largely from societal and emotional concerns rather than from demonstrated and substantiated technical or biological problems. The evidence of low risk when herbicides are used according to specifications is well documented. However, the social reaction is a real effect and must be seriously considered along with any other operational constraints.

Before any chemical can be registered for use it must be fully tested. These tests include analyses of insecticidal, fungicidal, herbicidal, and growth regulatory activity, crop tolerance, persistence, duration of activity, toxicity (acute, subacute, and chronic), environmental impacts, and degradation products. These tests, which are rigorously reviewed by federal and state agencies, provide considerable information on mode of action and potential hazard. Specifications on rates and methods of application are designed to reduce risk to a minimal level. However, despite exhaustive documentation, the public in some areas has been so concerned regarding risk as to literally halt herbicide applications on public and private forest land.

The advantages and disadvantages of using herbicides are as follows (University of California, 1977; Newton, 1975).

Advantages

1 Economy and safety, compared to alternate control practices such as hand cutting or mechanical methods.

2 Time saving, large areas can be covered in a relatively short time.

3 Convenience, in application and labor requirements.

4 Fewer retreatments to control resprouting of herbaceous and woody perennials.

5 Some herbicides are sufficiently selective to minimize or avoid damage to conifers.

6 Can be used to either increase or decrease wildlife populations by direct manipulation of food supply or cover.

Disadvantages

1 Success of weed-control programs depends on selection of the proper type and formulation of herbicides or herbicide mixtures.

2 Application timing is often critical in maximizing conifer tolerance and in gaining maximum weed control.

3 Nontarget off-site plants may be adversely affected by drift or through root uptake of certain persistent herbicides.

4 Herbicide residues.

5 Possible negative social reaction.

Herbicides are described in terms of their method of application. Preemergence herbicides are applied to the soil to prevent grass or broad-leaved weed seeds from germinating. Herbicides of this type are atrazine and simazine. The amount moved into the soil depends on the content of organic matter and clays as well as rainfall. In general, these herbicides do not leach vertically in the soil beyond the top few inches.

Foliar-applied herbicides are designed to kill or reduce the growth of undesirable vegetation already existing at the time of treatment. Sensitivity of vegetation to such herbicides depends on the stage of phenological development. Vegetation is most sensitive to phenoxy herbicides in early summer when growth is active. Consequently, early summer is the best time to spray when the objective is to achieve maximum brush growth suppression without regard for susceptibility of any associated conifers. Where the object is to release pines from brush competition, spring or early summer spraying cannot be used since both species are susceptible. Spraying is therefore commonly done in the fall after the conifer buds have become dormant. At this time, the growth of brush is usually still sufficiently active to make it susceptible to the herbicide. Interestingly, California conifers decrease in sensitivity to 2,4-D and 2,4,5-T herbicides from pines, which are the most sensitive, Douglas-fir, which is intermediate, and true firs and incense-cedar, which are least sensitive. This ranking happens to be the same as that for increasing shade tolerance in California.

A new class of herbicides including the growth regulating materials glyphosate is extremely effective on certain weed species. They are also relatively harmless to conifers when applied during fall, winter, and early spring. Effects on brush are limited to deciduous species, and to late summer or fall applications. Glyphosate also controls herbs at almost any time that they are actively growing.

Herbicides can also be applied by frill injection, which involves directly applying the herbicide into the cambium by an injector or by cutting overlapping frills or girdles through the bark with an axe. Herbicides in this case must be xylem mobile and therefore water soluble, consequently the amine form of 2,4-D is commonly used. Picloram is often recommended since it is both xylem and phloem mobile and can therefore translocate into the root system. It can move into the soil and kill adjacent trees, but damage is usually slight. Injection is useful in treating certain hardwood trees such as tanoak and bigleaf maple, where aerial applications are often ineffective. It has the advantage that the treatment is absolutely effective, only the trees selected are affected, and all the herbicide stays in the tree and degrades as the tree decomposes. The major disadvantage in some steep areas is the high cost resulting from the intensive use of labor.

Table 16-3 presents a list of common herbicides and their uses in forestry. Phenoxy herbicides such as 2,4-D (2,4-dichlorophenoxyacetic acid) and 2,4,5-T (2,4,5-trichlorophenoxyacetic acid) are plant hormonelike materials. These herbicides are absorbed by the leaves and translocated along with photosynthate into the root system. Here the herbicide causes enhanced cell development in meristematic regions which probably causes excessive drain on the plant and results in plant dysfunction. Dosage and timing are determined by species sensitivity and the plant's stage of phenological development. The objective is to control herbicide penetration so that, initially, minimal foliar injury occurs and the material is translocated.

Toxicity *Toxicity* of a herbicide is a measure of the amount of the chemical needed to produce detrimental effects in test animals. Toxicity is usually

Table 16-3 Common Herbicides and Their Uses in Forestry

Registered herbicide	Use in forestry
Amitrole (Amitrole-T, Cytrol)	Site preparation for certain herbaceous and woody perennials. Conifers injured. Chlorophyll synthesis inhibited.
AMS (Ammate)	Site preparation for many woody species. Repeat applications necessary. Conifers injured.
Atrazine (Aatrex)	Preemergence spray for grasses and some broad-leaved weeds. Most conifers not injured.
Cacodylic acid (Phytar)	Conifer thinning and hardwood tree control by stem or frill injection.
Dalapon (Dowpon)	Annual and perennial grass control. Apply as foliar spray 6 weeks before planting since conifers are injured.
Dicamba (Banvel)	Site preparation and woody plant control. Conifers injured, grasses not controlled.
MSMA (Ansar)	Conifer thinning and hardwood tree control by stem or frill injection. Toxic to *Fomes* and pests.
Picloram (Tordon)	Site preparation, bark application, or stem injection. Conifers injured. Persists in soil.
Simazine (Princep)	Preemergence spray for annual grasses and some broad-leaved weeds. Conifers not injured.
Silvex (Kuron, Weedone)	Herbaceous and woody perennials, especially 2,4-D tolerant species. Grasses not injured. Conifer injury seasonally dependent. Has dioxin contaminant.
2,4,5-T (Esteron 2,4,5)	Herbaceous and woody perennials. Grass not injured. Conifer injury seasonally dependent. Has dioxin contaminant.
2,4-D	Herbaceous and woody perennials. Grasses not injured. Conifer injury seasonally dependent. Less selective to conifer than 2,4,5-T. Has no dioxin contaminant.
Asulam (Asulox)	Control of bracken fern.

Source: M. Newton, Oregon State University, Corvallis, and S. Radosevich, University of California, Davis, unpublished report.

expressed as being either acute, where test animals are actually killed, subacute, where toxicity is commonly based on a 90-day feeding test, or chronic, where test animals are fed for a lifetime to determine whether the chemical is mutagenic (causing mutations), carcinogenic (causing cancer), or teratogenic (causing birth defects).

Acute toxicity is usually expressed in terms of LD_{50}, which stands for *lethal dose 50*, or the amount of chemical (milligrams per kilogram of body weight of subject animal) required to kill 50 percent of a treated test animal population. Thus amitrole with an LD_{50} for rats of 24,600 is very safe compared with 2,4,5-T, which has an LD_{50} for rats of 350. This measure of LD_{50} is difficult to interpret, however, since toxicity is quite different from hazard. For example, the LD_{50} of aspirin is about the same as that for 2,4,5-T.

Chronic toxicity is of concern because of the delayed effects that might arise in a population treated with the chemical. Currently, chemicals that have chronic toxicity cannot be registered for use even if applications result in no detectable residues.

The potential for exposure of nontarget organisms to toxic doses of herbicides is due primarily to spillage, misapplication, drift, and volatility rather than direct application. Drift occurs in conjunction with aerial application and results when very small droplets (about 10 μm in size) are released. If droplet size is controlled at about 100 μm and if there is low wind movement, drift is minor. Droplet size is largely controlled by the kind of nozzles used in application. Volatility is the loss of herbicide from the target area because of vaporization. This can be reduced by using low-volatile esters or eliminated by using amines (salt-formulations) and by spraying in the morning when temperature is low and inversions are unlikely.

Hazard is a function of both the concentration to which the organism is exposed and the potential for exposure. Chemicals might be highly toxic, but if the concentrations used are low or their use is such that organisms are not exposed for any length of time, or if organisms are not exposed to toxic levels, then the material is not necessarily hazardous.

Environmental Aspects Some of the herbicide is decomposed by the action of light (photodecomposition). The major receptor for herbicides applied to a system, however, is the soil. Herbicides are washed from the foliage or deposited on the soil surface by the falling treated leaves. Herbicides are basically immobile and generally do not move substantially in the soil; they are tightly bound by adsorption to clays and organic particles and are rarely found more than 6 in below the surface. There they are broken down by microorganisms and chemical reactions. The half-life of phenoxy herbicides in soil is about 3 weeks; in three months, over 90 percent has degraded.

Adverse environmental impacts are due almost entirely to misuse of the herbicides and can largely be controlled. Examples of past misuse include spillage or direct application into rivers, streams, or lakes and drift damage to adjacent crops. These unfortunate misuses are now under strict control by federal, state, and local regulatory agencies. The likelihood of direct adverse effects of herbicides on wildlife is low. For a 1 kg ha^{-1} application, initial herbicide residue levels would generally be less than 100 parts per million (ppm) in vegetation, less than 3 ppm in the surface 2.5 cm of soil, and less than 0.05 ppm in streams unless extensive direct application is made to surface water (Newton and Norris, 1976). These quantities produce effects on sensitive plant species but not on animals exposed to the same applications. Oral ingestion of sufficient treated vegetation to cause an effect is unlikely. Ingested herbicides usually pass through the digestive system with little or no retention or accumulation, and the exposure period is short because of the degradation of the herbicide in the environment.

The use of chemicals as a site-preparation tool is thus a perplexing one. There are distinct technical, biological, and economic advantages to their use, but on the other hand, some people are concerned that there may develop some yet unsuspected biological problem that may be unsolvable and far-reaching in impact. These concerns have resulted in temporary restraints on the use of herbicides in forestry. Physical treatments such as mechanical site preparation and use of fire for removing vegetation may have a far greater immediate impact on the environment, habitat, wildlife, and soil erosion, but nevertheless may be chosen since their mode of action is physical and less emotional.

Fire Fire may be used in a broadcast fashion across an area where fuels are continuous, or it may be used to burn piles or windrows. Burning of slash piles facilitates future access, reduces the area of "unoccupied" land, and returns nutrients more rapidly to the site. In some cases, however, the burning of piles or windrows actually serves no real practical purpose and may simply be "cosmetic" and an unnecessary cost.

Broadcast burns are used to reduce hazard, and to reduce the bulk of unwanted vegetation on a site, which is a physical impediment to regeneration. The amount of fuels on a site after a harvesting operation depends on stand conditions. In the West, the slash residue usually amounts to about 62.5 metric tons ha^{-1} (25 tons per acre). The effect of the burn on the site depends on the size of the fuels, their quantity and arrangement, fuel moisture content, slope, and environmental conditions at the time of burning. These factors affect the *intensity* of the burn, i.e., the rate and burn-out time. Thus the important factor affecting the impact of fire on a site is the calories per second generated by the fire; that is, a very "hot" fire moving rapidly may have the same effect on a site as a slow-moving, low fire.

Because fire intensity and sites can vary so tremendously, it is difficult to generalize about the effects of burning on the forest system. Since most of the effects influence soil properties, these factors are discussed in Chap. 10. Further discussion on fire as an intermediate treatment is provided in Chap. 17.

Burning of slash also has an impact by influencing the composition of the weed species that become established after burning. Figure 16-9 shows two adjacent areas on the U.S. Forest Service Challenge Experimental Forest in California. The area to the left was broadcast burned and rapidly became revegetated with *Ceanothus*. The area to the right was windrowed and the unburned bays became revegetated with *Arctostaphylos*.

Another aspect of burning that is gaining increasing concern is the attendant problem of smoke. This is mainly a problem of aesthetics, and visual pollution has resulted in restrictions on the timing of burning. These restrictions are aimed not only at reducing the risk of escapes but also at ensuring that atmospheric conditions take the smoke away from populated areas.

There are four major components of smoke, of which the first is by far the most important (Darley et al., 1966).

1 Particulates, mainly carbon and ash. These provide the smoke with visibility, and their size and quantity are greatly influenced by temperature. The annoyance factor of smoke is frequently minimized in very intense fires which create a strong convection column that lifts the smoke into the upper atmosphere.

2 Carbon monoxide. This is produced at the rate of about 31 to 67 kg per metric ton (70 to 150 lb/ton) of slash. By comparison, the burning of gasoline produces 397 kg of CO per metric ton of fuel.

3 Hydrocarbons. These are produced at a rate of 2.2 to 13.5 kg per

Figure 16-9 The effects of two different kinds of site-preparation treatment are illustrated at the U.S. Forest Service Challenge Experimental Forest, California. The area on the left was broadcast burnt after harvesting and was rapidly invaded by *Ceanothus intergerrimus*. The area on the right was windrowed and the unburnt areas between the windrows was rapidly invaded by *Arctostaphylos viscida*.

metric ton (5 to 30 lb/ton), compared with 58.5 kg per metric ton (130 lb/ton) from burning gasoline.

4 Nitrous oxide. Very small quantities are produced, but these can be important since NO gives rise to smog.

An interesting potential effect of smoke in the forest is its inhibitory effect on the development of certain pathogens, including western gall rust and mistletoe (J. R. Parmeter, personal communication). Ongoing studies are determining the extent of this effect and the role that periodic fire plays in maintaining the health of the forest.

Combination of Methods It is commonly found that a combination of site-preparation treatments is more effective and causes a smaller ecological impact than a heavy application of one method. For example, in the West, the conversion of brush fields is commonly best achieved by mechanical crushing, burning to remove the organic matter, and one or two herbicide sprays to control sprouting vegetation. In flat lands in the South, a common treatment is ploughing, bedding, and fertilization. In the Pacific Northwest, a "brown and burn" treatment of combined herbicides plus fire has been used effectively in removing alder thickets with minimal ecological impact. The use of fertilizers at the time of site preparation is increasing, and in some locations, irrigation with industrial effluent has been used to solve the problem of waste disposal while simultaneously adding nutrients and water to the forest.

The silvicultural need to develop and prescribe sound, effective, and acceptable site-preparation treatments is indeed a major challenge. The approach to selecting the most appropriate site-preparation treatment requires an awareness of the controlling ecological, physiological, managerial, and social factors involved. This is often an extremely difficult task. Not only must a better job be done by silviculturists in attaining acceptable regeneration, but the techniques used must be visually and ecologically acceptable. Two examples will illustrate current concerns. One is the use of terracing in the Bitterroots of Montana. Here, regeneration was successful, essentially no erosion occurred, but the area was visually unacceptable. The second example is the passage of Public Law 92-500, which requires that there be zero discharge of sediments into streams by 1985. Silvicultural activities have been defined as potential causes of non-point-source pollution, and treatments must be developed and utilized which eliminate surface runoff and sedimentation.

SEEDING

Reproduction by seeding can be either natural or artificial.

Natural Seeding

Natural seeding is dependent on neighboring seed trees to produce the seed, and as we have seen in Chap. 8, this may occur annually or at intervals of up to

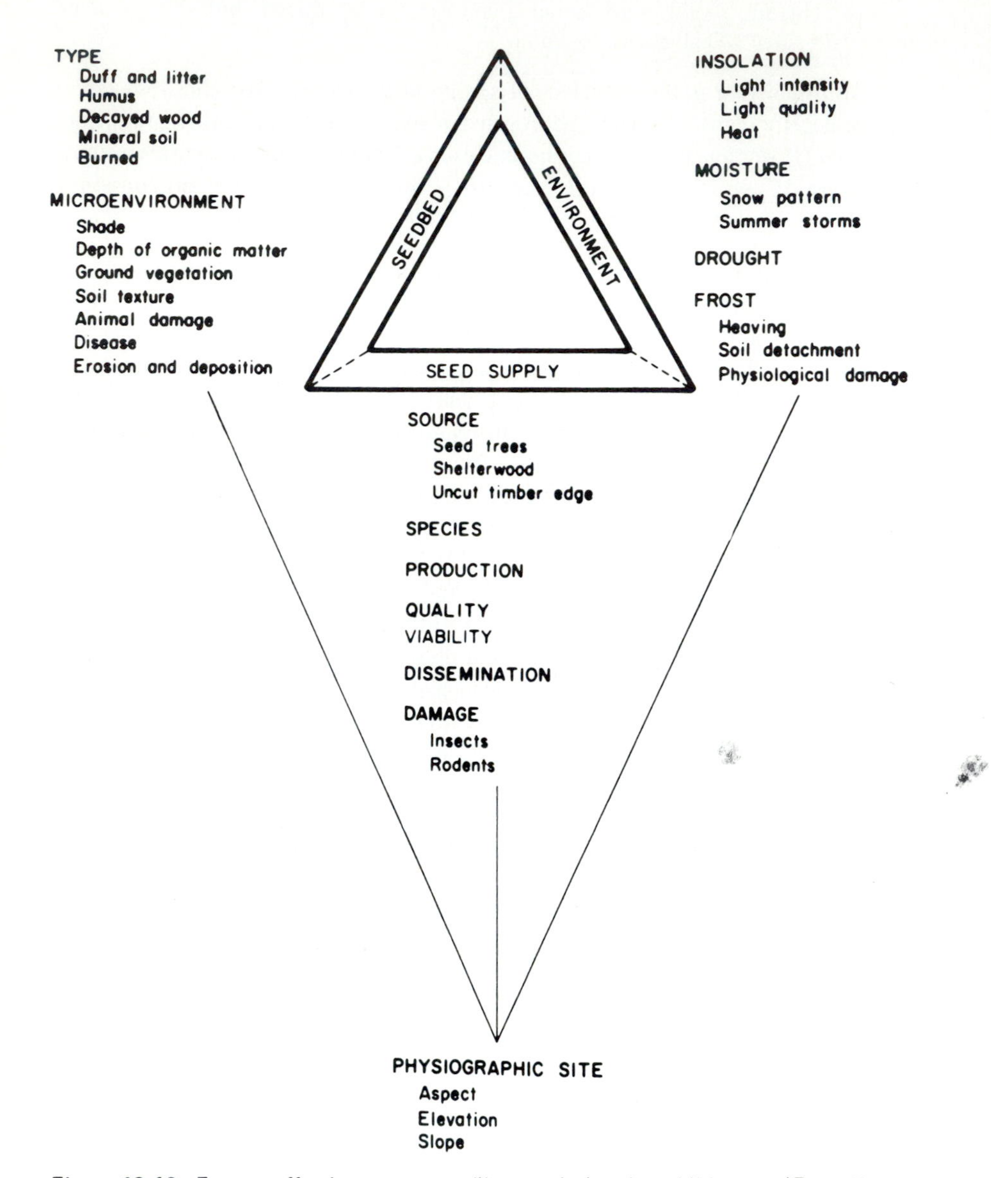

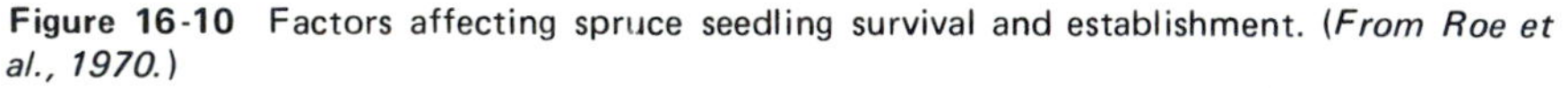

Figure 16-10 Factors affecting spruce seedling survival and establishment. (*From Roe et al., 1970.*)

7 years. Of course, harvesting cannot always be timed so that it immediately precedes an imminent seed fall. After logging, weed species commonly invade the area. Consequently, it is common practice to delay site preparation until the year in which seed fall is expected. If the delay is unacceptable, the area must be seeded artificially or planted.

The success of natural regeneration depends on (1) a supply of viable seed, (2) a suitable seedbed, and (3) an environment compatible with germination and seedling establishment (Roe et al., 1970). This concept can be presented as a triangle of factors in which incompatibility of one of the elements results in regeneration failure (Fig. 16-10). If the coincidence of all factors at a favorable level does not occur, the result is unacceptably low stocking. The reverse situation, which results in grossly overstocked areas, is also a problem in that it creates the necessity for precommercial thinning if one wishes to avoid growth loss to excessive competition.

Therefore, the principles involved in natural seeding involve:

1 The number and quality of parent trees that are needed to provide desirable levels of stocking

2 Timing the removal of the overstory or the site-preparation treatment such that an appropriate seedbed is created just prior to seed fall

3 Creating through shelterwood or seed tree, or by shape, size, and orientation of clearcut block, the desirable physical environment to promote germination and survival

4 Evaluating the need to modify treatments to accommodate potential pathogen, insect, or rodent problems

5 Modifying, if necessary, the physical environment, particularly the elimination of shade to ensure subsequent growth of established plants

The suitability of natural seeding depends very largely on management objectives and intensity. Where management is extensive, with rotations over 100 years, as in the management of public lands in the West, natural seeding is a desirable approach to regeneration. This is particularly true on lands that must meet such complex objectives as providing wood, habitat, and other diverse amenities. This is so because those multiple objectives are best met in forests with irregular stocking and stand structure. In forests where rotations are shorter and wood production becomes a more dominant goal, control of stocking becomes more critical. Evaluations must then be made as to whether initially higher costs of planting and control of stocking from the outset are acceptable. One other possible disadvantage of natural seeding is the lesser opportunity to upgrade the genetic quality of the regeneration.

Direct Seeding

The two main approaches to direct seeding are (1) sowing seeds in spots or rows and (2) broadcast sowing by hand or by using aircraft. Spot sowing involves placing 5 to 20 seeds in a "spot" which has been selected as suitable or especially prepared to aid germination. This method permits the filling of understocked areas, gives better control of spacing, and uses much less seed per acre than broadcast methods. Spot sowing is almost as expensive as planting and usually requires some precommercial thinning. Broadcast sowing can be done manually if the area is small, but has usually been done with either helicopters or fixed-wing aircraft. Helicopter application is preferred, mainly because slower air speeds permit flying at lower altitudes and result in a more precise application of seed.

Direct seeding permits the silviculturist to overcome many of the disadvantages of natural seeding:

1 Seed of desired species and quality can be used.
2 Stocking control can be somewhat achieved by controlling the number of sound seed applied per acre.
3 Timing of the operation can be made to coincide with desirable weather and site conditions.
4 Seed can be coated with pesticides to limit losses to insects, diseases, rodents, and birds.

Despite these advantages, the silviculturist still has relatively little control of spacing of regeneration. Using aerial methods, at the usual application rate of 25,000 to 40,000 viable seeds per hectare (10,000 to 15,000 seeds per acre) which is needed to guard against failure, if even a small percentage germinate and become established, the area will probably require precommercial thinning. Also, one should evaluate the climate since there are many areas, especially in the Southwest, where climatic conditions are not suitable. Fall seeding with unstratified seed may result in unacceptable over-winter losses as well as irregular stocking due to seed movement associated with surface-water runoff. The problem is magnified in areas with snow, and application either before snow fall or on top of the snow has proved to be largely unsuccessful. Stratified seed can be applied in the spring, but in some areas the soil rapidly becomes dry after snow melt. Seed germinating on the surface therefore has low survival.

The fate of sown seed in a forest environment is illustrated in Fig. 16-11, which shows the proportion of Douglas-fir seed which is lost to a variety of factors and the periods within germination and survival during which each factor operates. The magnitude of the predation on seed indicates that without an effective pesticide, the activity of rodents and pathogens can render a direct seeding program completely ineffective. A common pesticide used is an Endrin-Arasan mixture. However, some difficulty has been experienced in satisfactorily adhering the mixture to the seed without loss in seed viability (Mann et al., 1971; Radwan and Ellis, 1971). Also, since Endrin is a chlorinated hydrocarbon, there are restrictions on its use and therefore a need to develop a suitable substitute which is effective, biodegradable, and environmentally safe. These concerns have coincided with a desire to obtain greater control over stocking. Conse-

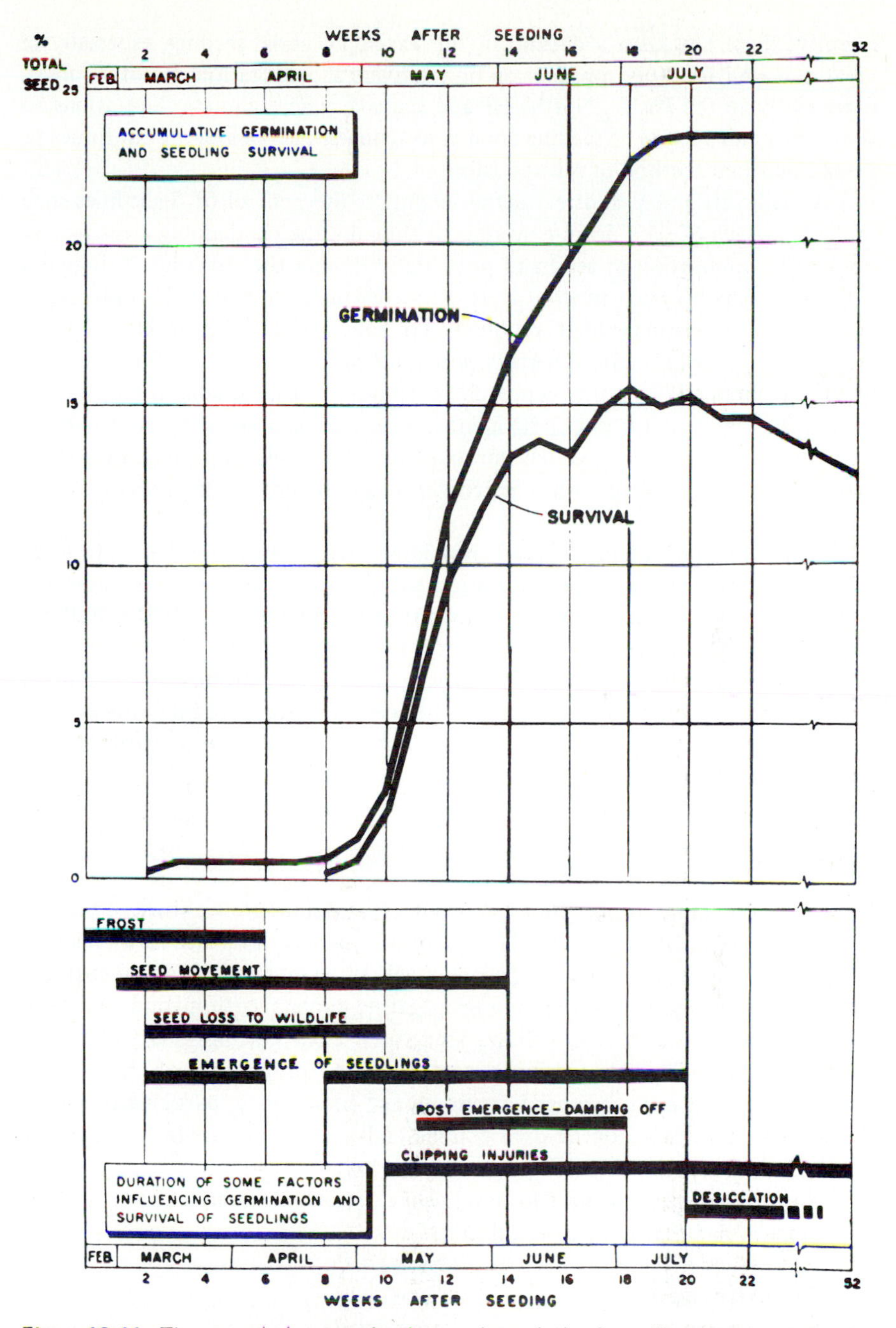

Figure 16-11 The cumulative germination and survival of seedlings produced by 440 Scandium^{46}-tagged Douglas-fir seeds with duration of some factors influencing fate of seeds. (*From Lawrence and Rediske, 1962.*)

quently, there has been a decline in the extent of aerial seeding, especially in areas managed for wood production on relatively short rotations. In other areas, particularly in the Pacific Northwest and Canada, where climatic limitations to the timing and success of seeding are not so stringent, aerial seeding continues to be an important method of reforestation.

A major difficulty in direct aerial seeding is the control of flight lines such that the swaths of seed do not overlap. If they do, the overlapping areas receive double the application of seed and potentially double the stocking. If there is a substantial gap between swaths, areas of no regeneration occur. This obviously compounds the management problem. To help avoid these problems, some attempts have been made to aerially sow multiple rows of seeds. Some success has been obtained using this approach by simultaneously sowing three rows of seeds with the rows 3.6 m (12 ft) apart. Seeds were coated with clay to obtain a uniform spherical shape, which improved aerodynamics, and precision sowing was obtained by seeding with a helicopter at an altitude of 25 ft (Barnett and Campbell, 1975).

Broadcast applications of seed over large areas require huge quantities of seed. With the increase in reforestation programs nationwide, it is becoming increasingly difficult to obtain large quantities of quality seed. If the available seed is used to raise seedlings in a nursery, much greater areas can be reforested with a given quantity of seed. Currently, seed produced from seed orchards is too limited in supply and too costly to be used for direct seeding; however, in the future it is conceivable that the cost of obtaining seed from orchards may be lower than the cost of collecting wild seed.

PLANTING

Planting is commonly done using bare-root or container-grown stock. These are the most common techniques used for planting conifers and many hardwoods.

Another method of raising planting stock is the growing of rooted cuttings. This is the standard method used in propagating poplars, cottonwood, and sycamore. Other species such as willows are grown from cuttings, particularly for stream-bank stabilization, erosion control, and the production of shade trees. The rooting of conifer cuttings has also gained importance, particularly on the West Coast, and may soon be used commercially, as is the case in New Zealand and parts of Europe.

A third approach to the raising of planting stock is research aimed at developing clonal material from tissue culture.

Bare-Root Planting

The Production of Stock and Initial Size This approach involves the raising of nursery stock usually in open nursery beds, followed by lifting, packing, storing, transporting to the nursery site, and out-planting.

The quality of nursery stock can be defined only in terms of those characteristics which enhance survival and growth in a particular area. Consequently, high-quality stock for one location may be defined in quite different terms from that to be out-planted in another area. In addition, the quality of stock required will depend on the out-planting environment and the kind and level of site preparation carried out.

Quality characteristics can be controlled by:

1 Selecting the most desirable provenance or genotype.

2 Controlling seedbed density to give desirable seedling size. This is normally 215 to 270 m^{-2} (20 to 25 ft^{-2}) of nursery bed, although in the case of producing large 3–0† Douglas-fir in the Pacific Northwest, densities of 86 to 107 m^{-2} (8 to 10 ft^{-2}) are best (Wilson and Campbell, 1972). In California it has been shown that two seasons after hand planting, field survival of ponderosa pine grown in the nursery at 538 m^{-2} (50 ft^{-2}) was 62 percent, whereas for seedlings grown at 107 m^{-2} (10 ft^{-2}) it was 83 percent. Results after machine planting the same stock were 79 and 92 percent, respectively (Baron and Schubert, 1963).

3 Controlling mycorrhizae, damping-off fungi, water, shade, and nutrition.

4 Incorporating root-wrenching treatments to control shoot/root ratios and to aid in hardening-off (Rook, 1971).

5 Age.

6 Transplanting in the nursery. However, this requires considerable extra nursery space and at least doubles the cost of seedlings.

7 Careful timing of lifting to ensure high root-growth capacity. Care must also be taken to minimize root exposure and damage to root systems in handling.

8 Appropriate storage such that temperatures inside the seedling bundles are maintained close to 0°C (Hermann et al., 1972; Brown, 1973). This is needed to keep plant respiration and fungal development as low as possible.

In general, it is best to plant the largest seedling that has a high probability of becoming established in a given environment. In locations such as British Columbia, Washington, and Oregon, where moisture is commonly not severely limiting, tall planting stock survives better and grows more rapidly than smaller stock. In these relatively favorable areas, each 2.5 cm of seedling height added in the nursery will result in a growth advantage of 7 to 10 cm by the seventh or eleventh year after out-planting (Smith, 1975; M. Newton, personal communication). The attainment of physical size as rapidly as possible is important from the standpoint of minimizing the effects of both weed competition and wildlife browsing (Newton and Black, 1965; Hartwell, 1973). If by planting larger seedlings plantations can reach breast height in 2 to 3 years rather than 4 to 6 years, the potential for rapid site occupancy is very significantly enhanced. On drier sites, such as in many of the southwestern states where spring droughts are com-

†The two digits refer to the number of years the seedling is grown in, respectively, the nursery bed and the transplant bed.

mon, planted seedlings are exposed to high transpirational stress when outplanted. In these areas, inability to handle adverse water relations is the most common cause of mortality. Consequently, greater survival is obtained using planting stock that is smaller, has a lower shoot/root ratio, and has a higher root-growth capacity. Small stock is exposed to competition from weeds, insects, and browsing animals for a longer period of time; thus the silviculturist must determine the optimum-sized stock that can handle the environment of the site in question.

The advantages of initial size are not always apparent, particularly on the drier sites. Engelmann-spruce survival does not seem to be well correlated with the total amount of food reserves (Ronco, 1973), and vigor of *Pinus radiata* on low-quality sites does not appear to be related to initial seedling size (Pawsey, 1972). With respect to hardwoods, trials in Wisconsin have shown that the use of large 1-1 stock with about 60 cm tops, particulary of *Fraxinus americana*, *Acer saccharum*, and *Liriodendron*, is best to overcome weed competition after clearcutting (Johnson, 1971).

The Planting Operation The planting of seedlings must be regarded as one integral step in the regenerative process between seed collection and the establishment of the new crop. Successful establishment can be obtained only if all links in the chain of events are compatible. Stock should be transported from the nursery and temporarily stored on the planting site under conditions which match as closely as possible the storage conditions at the nursery. The actual planting should be done by an experienced or trained crew since the best-quality stock can have low survival if poorly handled and planted. The need to avoid root exposure is particularly critical. Figure 16-12 shows how the effects of root exposure on survival of Douglas-fir is influenced by the lifting date of the seedling in the nursery and storage time. Generally, the closer the plants are to dormancy when lifted, the lesser are the effects of root exposure (Mullins, 1971). The effects of root exposure are not limited to first-year survival only, but in the case of white spruce, have continued to be a significant cause of reduced height growth and mortality 5 years after planting (Mullin, 1974).

Methods of planting vary, but the common objective is to place the seedling firmly in the ground without bending the root system and causing "J-roots." This takes considerable care, and the problem can be severe. In a study of loblolly pine, of some 2000 bar-planted seedlings investigated 3 years after planting, 35 percent were J-rooted, 11 percent were L-rooted, 21 percent had their root systems "balled," and only 33 percent were described as well-planted (Gruschow, 1959). To test the impact of planting on survival, an extensive test was carried out with loblolly pine (Hay and Woods, 1974). In this test, roots were deliberately planted in seven different ways, including J-shaped, U-shaped, slanted, and normal. Contrary to general belief, distortion of root systems at the time of planting in this particular situation produced negligible detrimental effects within the first 2 years after planting. In some cases, distortion of roots pro-

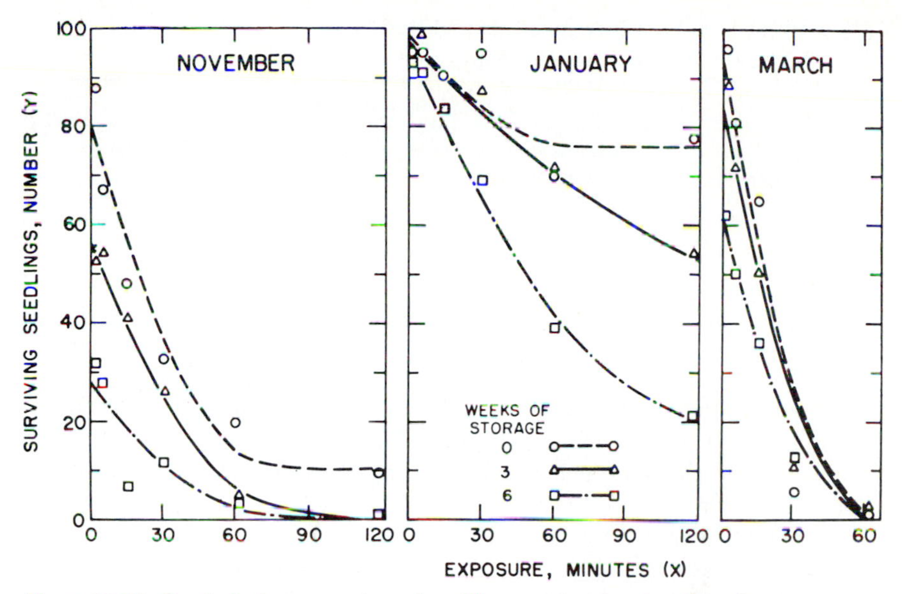

Figure 16-12 Survival of nursery-planted seedlings as related to duration of root exposure, by month of lifting and weeks of storage. (*From Hermann, 1967.*)

duced a beneficial effect which appeared to be due to the stimulation of development of lateral roots at the expense of the tap root.

Frequently, the intial deformation in root systems persists after many years of planting. For example, in an investigation of 13-year-old jack and red pine plantations that were established by bar planting, 65 percent of the 5000 trees evaluated were found to have predominantly one-plane root systems, and growth of these was 20 to 30 percent lower than those trees with well-balanced root systems (Rudolph, 1950). Similar kinds of problems can develop with seedlings that have spiralled or balled root systems from improper handling of container stock. The problem is complex, since in many other cases root systems either develop into a more balanced distribution or show no differences in survival and growth (Stiell, 1958).

In general, however, it is best to plant seedlings with normal, straight arrangement of roots in the planting hole. Care should be taken to ensure that dry soil, litter, or snow does not fall into the planting hole since this will create dry conditions or an air pocket next to the roots, which limits water uptake. Also, the soil should be firmly tamped around the root after planting to eliminate air spaces and minimize pore size, which facilitates capillary movement of water. The importance of care in planting is shown in Table 16-4, where it can be seen that the best planter was about twice as successful as the poorest in obtaining survival of red pine.

Planting should be done only when conditions are favorable for survival. In spring, it is critical that the soil be warm enough that, to avoid transpirational

Table 16-4 First-year Survival of Planted Red Pine
(Bars Enclose Values that Are Not Significantly Different)

Planter	First-year survival, %		
	Shallow-silt loam	Deep fine sand	Stony-silt loam
1	53	67	57
2	45	49	56
3	29	42	41
4	28	30	34

Source: Cooley, 1974.

stress, root initiation commence immediately. In the Southwest this temperature of the rooting zone should be 10 to 12°C and warming.

Timing of planting, including the choice between spring and fall, particularly in relation to environmental conditions, will therefore vary between geographic area, elevation, aspect, and type of site-preparation treatment. Choice of timing becomes critical in more harsh environments and may be nonlimiting in milder parts of the world.

Initial spacing is another critical decision, since it directly influences subsequent operations in the plantation. Initial wide spacing is used where (1) little mortality is expected, (2) the quality of trees is uniformly high, with therefore little need to thin out misshapen and forked trees, and (3) cost of labor and harvesting is greater than the values derived from early stand treatments. It has the advantage that less site preparation may be needed, fewer initial trees are planted (therefore costs of seedlings and planting are lower), and the need for precommercial thinning is reduced or avoided.

Closer spacing has consequences which are opposite to those just stated for wide spacing. In particular, closer spacing results in

1 Earlier canopy closure, which may be important in dominating competing vegetation.
2 More rapid suppression of lower branches.
3 The need for precommercial or commercial thinning to avoid merchantable growth losses due to competition.
4 Greater opportunity for thinning, which can be used to remove undesirable phenotypes and enhance the production of high-quality timber.
5 Greater total wood production. This volume will be spread over a larger number of smaller stems, particularly early in the rotation, and the costs of merchandising this volume may exceed its value.

The advantages and disadvantages associated with a particular level of initial stocking therefore depend on species, site quality, aspect, slope, nearness to market, access, pests, and management objectives and strategies. It should be expected, therefore, that particular levels of initial spacing will vary even within one forest. Choice of initial spacing directly influences the economics of growing wood, since it largely obligates the forest manager to a particular set of subsequent treatments, such as weed control and precommercial and commercial thinning, if satisfactory growth is to be maintained. In making decisions on spacing, the forester must appreciate that the objective is often not to maximize volume production but to maximize value production. It is often better to have a smaller volume per acre on fewer, larger, more valuable trees than to have a substantially larger volume per acre distributed over many small trees. The costs of falling, skidding, hauling, barking, and utilizing large numbers of small trees can be economically unattractive.

Container Planting

In the mid 1970s, there was a major shift in emphasis in reforestation in North America toward the growing of seedlings in containers. In 1973, the United States produced 963,105,000 trees for forest and windbarrier use, and about 26 million, or 3 percent, were grown in containers; in Canada, 90 percent of planting is of 2-2 or 3-0 bare-root stock, but about 34 million container-grown seedlings are produced (Stein et al., 1975; Armitage, 1974). The Pacific Northwest produces about 42 million container-grown seedlings, which is approximately one-fifth the total coniferous-seedling production in that region.

Of course, growing forest seedlings in some kind of container is not new and has been done for decades or even centuries in various parts of the world. What is new in the recent North American interest in containers is the scale of the operations, where millions of seedlings are raised annually in one location, mostly in specially constructed greenhouses (Fig. 16-13*a* and *b*). The seedlings are commonly grown for 6 months to a year depending on species and size of stock required. Out-planting of seedlings that are only a few weeks old has been largely discontinued because of excessive mortality. In some systems the container is planted together with the seedling, but in most cases the container is removed and only the "plug" is planted.

Interest in reforestation using container stock developed rapidly in the early 1970s, and considerable knowledge has been gained on container-seedling production for many species, container design, nursery operations, stock handling, and evaluation of field performance (Tinus et al., 1974). The reasons for this rapid swing toward reforestation with container-grown stock include (Stein et al., 1975):

1 Response by the profession to challenges for accelerated reforestation
2 Improved survival and growth of seedlings resulting from planting with undisturbed root systems
3 More positive response of some species
4 Extension of the planting season
5 Achievement of greater production and planting efficiency

Figure 16-13 a: The growing of 10 million container seedlings in a specially constructed greenhouse. b: The containers used vary considerably in size, shape, and material. The use of individual tubes facilitates the growing of seedlings at optimal densities since tubes with ungerminated seeds or with seedling mortality can be replaced with tubes containing healthy seedlings.

In addition, container growing allows industries to produce their own seedlings, since high-quality agricultural-type land is not required. It eliminates industry reliance on state-grown seedlings and involves industry foresters more directly in the total reforestation program.

There are several drawbacks, however, and these include (Stein et al., 1975):

1 The requirement of a higher level of technical knowledge

2 The rapid changes in seedlings due to variations in water and nutrient application resulting from the relatively small sizes of the containers and the small amounts of potting mix

3 Acceleration of the incidence and effects of disease and nutrient imbalance resulting from the conditions that support seedling growth

4 Current costs, which are about 1½ times that for growing bare-root stock

In addition, in some cases survival is not significantly enhanced, and there is the possibility of root restriction and/or spiraling. Thus there is still some controversy as to which method is best. However, the availability of both container stock and bare-root stock does increase the potential flexibility that a silviculturist has in prescribing reforestation procedures. Despite early expectations that container stock would facilitate survival on poorer sites and enable one to plant earlier and later in the planting season, this has not been generally true in actual practice. Container stock survives better on better sites than on poorer sites, stock from larger containers becomes established more readily than stock from smaller containers (Endean and Hocking, 1973), and "plugs" grow better than seedlings planted with their containers. Usually, the larger the seedling, the better is the field performance. However, the cost of growing large-container seedlings increases proportionately with seedling size. This is so because of reduced output per greenhouse and the increased length of time the seedlings must be tended. Thus the expected economic advantages associated with production of container stock will probably be obtained only with the small and less desirable containers.

Ideally, the container should adequately support the seedling's biological requirements, have all the necessary characteristics for full automation of both production and planting, and biodegrade within moments after planting (Stein et al., 1975). This is so because moisture held in the container is only adequate for a short time and there is consequently an immediate need for many roots to grow and make contact with the surrounding soil.

Use of container stock may be distinctly advantageous where small-sized planting stock is suitable and for some species such as hemlock or the true firs which are currently difficult to grow satisfactorily in bare-root nurseries. Planting small, cheap container stock is probably a more dependable alternative to aerial seeding, and container stock also may be preferable in the rehabilitation of strip-mined or eroded areas. However, for general reforestation purposes, container seedlings are not necessarily more desirable than bare-root seedlings where appropriate treatments for the production of stock with high root-growth capacity have been developed.

EVALUATION OF REFORESTATION SUCCESS

It is no easy matter to define when a stand is considered to be established (Stein, 1974*a*, *b*). The mere presence of desired species does not necessarily mean that the stand will rapidly develop to meet managerial objectives. An evaluation of reforestation or a comparison between alternative methods of regeneration is not possible until the seedlings have truly become "established" and the new forest is growing satisfactorily. Thus the definition of the term *established* is subjective and will vary with species, site, dominance potential of weed species, and management objectives.

The first concern in evaluating reforestation success is the definition of an acceptable tree. It presumably should be healthy, well formed, and free to grow relative to weed and pest competition. Each acceptable tree contributing to a definition of stocking should have a high dominance potential and desirable vigor characteristics. A tree remaining in a suppressed seedling stage for an indefinite period is not realistically contributing to any management objective.

Second, certain stand characteristics contribute to an evaluation of reforestation success. The stand should be adequately stocked in terms of numbers of trees per hectare, and these trees should be well distributed over the area. Adequate stocking is usually measured by means of a stocked-quadrat method. The actual minimum number per hectare that is acceptable depends on both management objectives and strategies. In some states the minimum acceptable stocking to be present on private land within a specific period after harvesting is prescribed by law.

Stocking standards or analyses of reforestation success are commonly expressed at the end of one or two growing seasons after reforestation. However, because of the potential impact of competition from weeds and pests, it is preferable to define standards of success in terms of either the attainment of an arbitrary height (such as breast height or 3 m) or the stature of the new forest crop relative to weed and pest competition.

Therefore, in evaluating costs of regeneration it is necessary to include the costs of all treatments needed to get the stand "established." On this basis, initial costs may be quite unimportant. Economic analyses of this kind have justified higher initial costs of site preparation and regeneration which have significantly reduced rotations and produced salable products at an earlier age. This approach is basic to the so-called high-yield or intensive forest practices. It has led to the trend toward planting as opposed to seeding, and toward the planting of larger rather than smaller stock. For example, in Ontario, estimates of costs related to probable height of plantations at 10 years of age suggested that use of transplants of *Picea glauca*, *Picea mariana*, *Pinus resinosa*, and *Pinus strobus* would be considerably more economic than the use of smaller stock (Mullin and Howard, 1973).

To enhance success in reforestation, each stage in the operation should be evaluated to determine where small gains can be made that will increase outplanting survival and subsequent growth. Selecting seed from the most desirable provenance, upgrading all treatments and operations in the nursery to enhance the production of seedlings with high root-growth capacity and potential to withstand water stress, using bigger stock, selecting the most appropriate kind and level of site preparation, ensuring a high-quality job of planting, and controlling weeds and pests are all necessary. If survival and capacity for growth can be increased by a few percentage points at each of these stages, then the additive effect will result in substantially higher success in the overall reforestation program. The key is to regard the operation from an integrated-systems standpoint and to involve all participating personnel in relating their work to the final goal of achieving successful reforestation.

Chapter 17

Intermediate Treatments

An *intermediate treatment* is any manipulation in a stand that occurs between two regeneration periods. The regeneration period provides the seed, seedbed, and growing conditions necessary for the germination of seed and the establishment of seedlings. Intermediate treatments are done to ensure the desired composition, stem quality, spacing, and growth performance in a "developing stand" as well as to recover values that might otherwise be lost. Release cuttings form the broadest grouping, including weeding, cleaning, liberation, improvement cuttings, and thinnings. The other treatments applied to stands, like salvage and sanitizing operations, have more of an emergency status. The desire to accelerate growth, to increase cone production, or to correct a site deficiency has promoted the use of fertilizers as an intermediate treatment. Prunings are aimed at enhancing log quality, and prescribed burning is applied to reduce fire hazard and to control understory competition or improve forage production.

The choice, frequency, and intensity of these treatments are determined after a thorough ecological appraisal and a translation of managerial objectives into desired stand structure and composition. Methodologies selected may be further constrained by societal and environmental considerations. In many

North American forests, intermediate treatments are used to modify some presently existing undesirable stand characteristics. In managed stands, intermediate treatments should be integral parts of the management plan and constitute part of the planned sequence of treatments. It is only by this approach of developing silvicultural prescriptions that managers can rationally evaluate the relative merits of alternative methods.

CLEANING, WEEDING, AND LIBERATION

Cleaning, weeding, and liberation are all applied early in the development of the stand and are designed to ensure control over composition and structure. As discussed in the previous chapter, the site resources of a particular area (i.e., nutrient base, water availability, temperature, light levels, and so on) can either be consumed in an uncontrolled manner by various species or can be enhanced by silvicultural treatments that favor particular components of the vegetation. Site resources can therefore be deliberately reallocated following the concept outlined in Fig. 2-1.

Cleaning

A *cleaning* takes place in a stand where the trees are saplings or smaller. Its sole purpose is to release one species from the dominance of another. A species is undesirable if it overtops or threatens to overtop the favored species. Cleaning is an operation that requires investment rather than giving an immediate cash return, so it is done in the cheapest, effective way, using spraying or injection of herbicides, severing of competitors, or even simply breaking of the tops. Only crop trees of the favored species are released, since eradication of the poorer species is not the objective. The purpose of cleaning is only to release enough of the favored species to ensure its rapid dominance of the site (Fig. 17-1). A single cleaning might suffice for releasing Douglas-fir saplings from competing red alder, but it has taken several cleanings to ensure the dominance of red spruce or eastern white pine in competition with sprouting hardwoods on good sites.

Figure 17-1 Cleaning; all trees are the same age. The competition of crop trees (A) is reduced, while the other individuals of the same species are allowed to develop without treatment.

Figure 17-2 A liberation cut removes the older overstory to release the sapling understory.

Weeding

Weeding is similar to cleaning, but it applies to the freeing of seedlings or saplings from competition with ground vegetation, vines, and shrubs. The competition is mowed or grubbed out from around the seedling or removed by herbicides.

Liberation

A *liberation cut* is also done when the individuals of the favored species are saplings or smaller, but it differs from a cleaning in that the trees removed form a much older age class. The trees to be removed in the older overstory may be of any species or form since the only objective is to release the understory (Fig. 17-2). If the overstory is merchantable, its removal is simplified, but the operation is usually done at a cost. These costs are sometimes reduced by use of a herbicide or by girdling rather than felling, in which case the treated overstory deteriorates slowly and the falling debris does little damage to the vigorous understory. The need for a liberation cut results from a treatment such as the under-planting of red pine under a hardwood canopy on the Kellogg Tract in Michigan or the invasion of a species like Engelmann spruce or subalpine fir into a stand of lodgepole pine in Utah.

IMPROVEMENT CUT

An *improvement cut* is done in either even- or uneven-aged stands where the released trees are of pole size or larger, i.e., the operation generally provides

Figure 17-3 An improvement cut in a stand where a combination of liberation and cleaning cuts had been postponed. This is the same stand as shown in Fig. 17-2 but at a later stage of development without treatment.

merchantable material. Its purpose is to release trees that will improve the composition, form, and/or growth of the residual stand (Fig. 17-3). The treatment is often prescribed where a cleaning or liberation cut may have been justified at an earlier age but the opportunity was neglected for financial or other reasons until the trees were big enough for an improvement cut. In fact, improvement cuts are often the initial cuts to bring previously unmanaged stands into a better condition for management. In uneven-aged stands where natural regeneration is an anticipated and desired consequence, the improvement cut is an adjunct to the selection method.

THINNINGS

A *thinning* is an operation performed in even-aged stands or even-aged groups in uneven-aged stands at any time prior to the beginning of the regeneration period, where the objective of harvesting trees is primarily to redistribute growth potential or to benefit the quality of the residual stand. In the past, incorporation of thinning schedules in forest management has been used as a hallmark of intensive forest practices. In some western European and Southern Hemisphere countries, about 50 to 60 percent of the total harvest has come from thinnings. In recent years, however, high wages and reduced workforce availability have encouraged heavier and more infrequent thinnings, as well as the adoption of wider initial spacing to keep the necessity of thinning to a minimum.

In order to differentiate thinnings from an improvement, liberation, or cleaning cut, the word *thinning* should be confined to use where cutting of the favored species, whether in pure or mixed stands, is meant. In a mixed, even-aged stand, a thinning deliberately favors quality individual trees without regard for species. Where species composition is deliberately changed, the treatment is one of the other intermediate cuts rather than a thinning.

In view of the importance attached to the potential use of thinning to control stocking, an understanding of thinning principles and methods is essential for a practicing forester. A site is capable of supporting thousands of seedlings per hectare, but relatively few (150 to 250) mature trees, so a fully stocked, developing stand experiences a continuing reduction in the number of trees with age. This natural attrition or mortality forms the main basis for thinning. This basic concept is illustrated in Figs. 14-24 and 14-25.

Thinning attempts to achieve a number of objectives, although no one method may accomplish them all. The following list summarizes the main characteristics and purposes of thinning:

1 The main purpose is to salvage and utilize material that would normally be lost due to natural stand mortality. This material amounts to approximately 25 to 35 percent of gross productivity.

2 Thinning increases merchantable yields by distributing volume growth on fewer larger stems.

3 Final volume production in thinned stands is usually less than in unthinned stands.

4 Thinned stands commonly have higher value production.

5 Thinning allows the forester to select the crop trees instead of nature's frequent choice of the biggest, roughest dominants.

6 With the exception of selection thinning, thinning will shorten the rotation if rotation age is determined by the attainment of a certain tree diameter.

7 Thinning will lengthen the rotation if rotation age is determined by the age at which growth rate of individual trees becomes unacceptably low or if based on the culmination of the mean annual increment.

8 Thinning provides an early return of invested capital and increases the rate of return on the investment.

Thinning Methods

There are five classical thinning methods: low, high (crown), selection, mechanical, and free. The application of alternative methods is discussed in detail in Smith (1962). A short description of each method is given here, but the main concern is to identify those silvicultural principles which provide the basis for sound choice and application of a particular method.

Low Thinning In low thinning the objective is to release the dominant and codominant trees by removing the lower crown classes (Fig. 17-4). Low thinnings can grade from light to heavy, and the choice depends on the management objectives and the stage of development of the stand. A formal description of these grades of thinning, which are called A (very light), B, C, D, and E (extra heavy), have been published (Hummel et al., 1959) and apply to Europe, but in

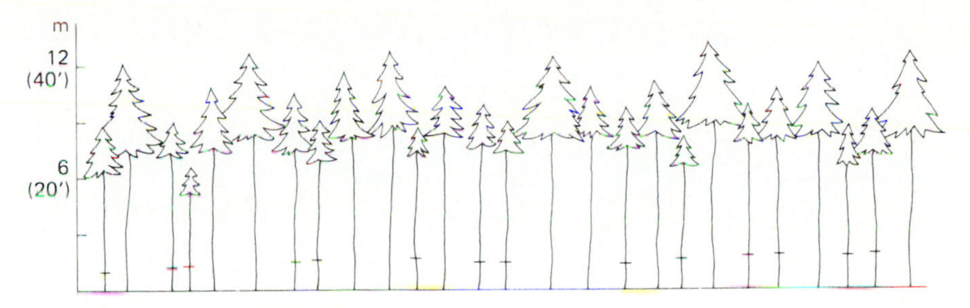

Figure 17-4 An initial heavy, low thinning.

the United States only three grades are recognized: light thinning removes suppressed, dying, and low-intermediate crown classes; medium thinning removes the same trees as light thinning plus most of the intermediates; heavy thinning removes the same trees as medium thinning plus the rest of the intermediates and some of the codominants.

The philosophy of low thinning is that the lower crown classes use significant amounts of water and nutrients and thus are detrimental to the growth of the upper crown classes. Low thinnings can be made throughout the rotation, and they are most generally used because a crew can be readily trained to apply a particular grade of cut uniformly over a stand.

High (Crown) Thinning The objective of a high (crown) thinning is to release the dominant crop trees and good codominant trees, if spacing requires it, by removing the competing codominants and dominants. If any tree in the lower crown classes is likely to die before the next thinning, it is also removed. A potential transition from low to high thinnings is illustrated in Fig. 17-5, in which the trees to be cut if low thinnings are to be continued are marked, as well as the trees to be cut if a high-thinning regimen is to be inaugurated.

There are two ways of applying a high thinning. The first is to select and usually mark the future crop trees and then to cut only part of those competing dominants or codominants which will release them. In subsequent thinning, the crop trees are further released. The second approach is to release any dominant or codominant tree that could be a candidate for becoming a crop tree. In subsequent thinnings, continual choices can be made regarding which trees are developing into the better crop trees.

The philosophy of high thinning is that the lower crown classes are using only insignificant amounts of nutrients and that the real competition is between the dominants and codominants for nutrients, light, and growing space.

A high thinning is more immediately profitable than a low thinning because the trees removed are much larger, but the potential damage to the residual stand due to felling the dominants can be much greater for the same reason. The presence of the lower crown classes also gives increased difficulty in maneuvering heavy equipment. By increasing the degree of release with each high thinning,

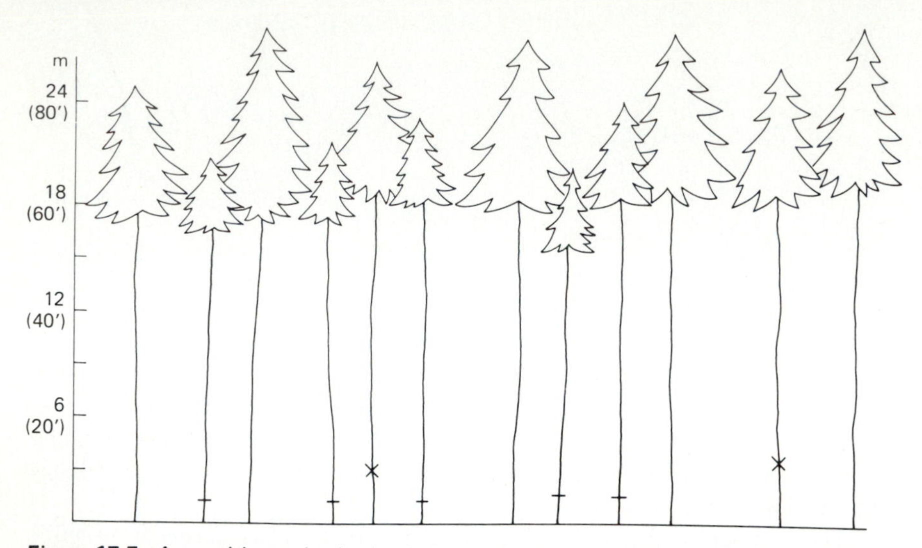

Figure 17-5 A transition point in the thinning of a stand is illustrated where the (–) trees would be cut if *low thinning were to be continued* and the (X) trees would be the *only* trees cut if a high thinning were initiated. (This diagram illustrates the same stand as Fig. 17-4 but at a later age).

a crop tree can maintain a more uniform radial growth in spite of the rapidly increasing diameter. High thinning gives the shortest rotation age on a diameter basis or the largest trees for a fixed rotation age.

Selection Thinning Selection thinning removes all trees in the dominant crown class in order to free trees in the codominant and intermediate crown classes which will become the future crop trees. It is a method that has three variations.

1 In young even-aged stands where a scattering of trees has outgrown the rest of the stand and threatens to develop into wolf trees for lack of side pressure (Fig. 17-6), the threatening trees are removed. Further thinnings in the stand can be low thinnings and eventually perhaps high thinnings. This variation is the only one in general use.

2 A stand is allowed to develop with low thinnings until the codominants

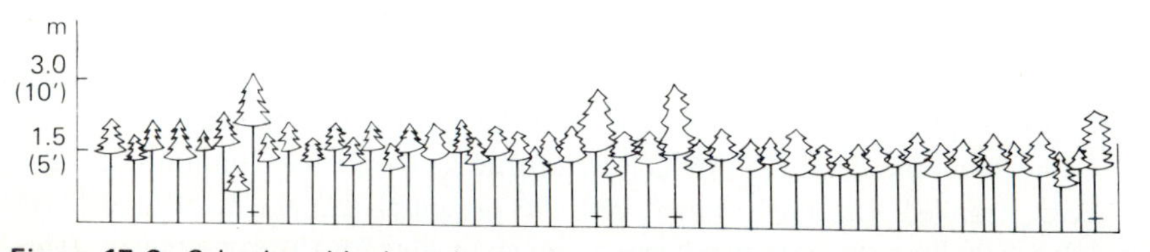

Figure 17-6 Selection thinning where a few trees in a young stand threaten to become wolf trees; their removal results in a more uniform development of the rest of the stand.

and some intermediates have developed high-quality boles, at which time the dominants are removed and the quality codominants become the crop trees. This variation is applicable only to species which are capable of responding well to release and which can reclaim the growth capacity of the site. It produces a large volume of thinnings from the largest trees in the stand at midrotation, and if the rotation criterion is the attainment of a particular diameter, lengthens the rotation.

3 Borggreve's (Dengler, 1944) variation is to allow the stand to develop until the dominants are merchantable, then dominants (10 to 20 percent of the basal area) are removed. After the codominants (now dominants) have developed to appropriate size, some of the old and new dominants (10 to 20 percent of the basal area) are removed. This promotion of the lower crown classes to dominants continues until further reduction in the number of trees will no longer control the site. When the residual trees are merchantable, they are clearcut and the area planted. This method requires a tolerant species whose lower crown classes will respond to the release.

Removing all the largest, fastest-growing trees from a stand is dysgenic selection, where the mean value of a characteristic, such as growth rate, in the second rotation is lower than that in the first. This could occur only if the stand is regenerated naturally and, of course, is irrelevant if the stand is regenerated artificially. In addition, superior growth of individual trees may result not from genotype but from some aspect of the environment. The first variation of the selection thinning is unlikely to be dysgenic.

An example where the three methods of low, high, and selection thinning have been applied is provided by four experimental plots of Sitka spruce set out in 1946 in Denmark. In 1959, thinnings began with Mr. Juncker demonstrating his practice of Borggreve selection thinning. The selection plot was thinned every year. Professor Möller thinned his plots by low- and high-thinning methods every 2 years. However, the control plot, while it was not thinned, had the mortality removed and sold. The average height of trees in 1959 was 10 m. The data accumulated on the plots include results from the 1965 thinnings, shown in Table 17-1. This table indicates that after 7 years of applying different thinning methods, the selection thinning provided the greatest return.

After each series of thinnings, when the trees on the selection thinning plot were in danger of losing control of the site because their growth response was inadequate, the remaining trees would be clearcut and the area replanted. By the repeated rotations of selection thinnings, greater returns were anticipated than would have been produced at the end of the one rotation for any of the other three plots.

As discussed in Chap. 14, gross growth and mean annual increment of stands are fairly constant. However, as shown in Table 17-1, the heavy cutting of dominants in selection thinning has depressed increment and total volume significantly.

Table 17-1 Effect of Different Thinning Methods on Growth, Yield, and Earnings for Sitka Spruce in Denmark

Plot	1959				After 1965 thinning		
	Average stand diameter, cm	Basal area, m^2	Volume, m^3	No. of stems	Volume, m^3	Average annual increment—7 years, m^3	Percent volume growth
Control	10.5	34.64	184	4,033	374	35	9
Low thin	10.4	34.32	178	4,060	325	38	12
High thin	10.5	35.15	185	4,085	258	38	15
Selection	11.1	37.82	206	4,019	173	30	18

Returns* from the Thinning Converted to Cubic Meters at 75 Danish Crowns per Cubic Meter

Plot	Danish Crowns net, 7 years	Volume, m^3	1965 volume	Total volume	
Control	400	5	374	379	
Low thin	4,000	53	325	378	Average height in
High thin	8,000	107	258	365	1965 was over 14 m
Selection	12,000	160	173	333	

*Returns are approximate since they represent rounded figures.

Source: C. M. Möller, personal communication.

Mechanical Thinning Mechanical thinning removes trees without regard for crown class, quality, or character of the trees removed. There are three basic approaches:

1 Spacing guides are used and all trees within a given radius around the selected tree are removed.
2 Row thinning is used in plantations where every second to seventh row is removed. A seventh-row thinning would be used to provide access for heavy equipment which could then thin between the access routes.
3 In excessively dense seedling or sapling stands, swaths of trees are removed by bulldozer in a checkerboard fashion. The small groups of seedlings left in a grid pattern may then require release by hand (Table 14-4).

Free Thinning In free thinning, as the name implies, trees are removed without conforming to any one of the preceding methods. Individual trees are removed depending on the forester's opinion as to what is best for stand development. Criteria used in selecting trees to be either released or cut include crown class, vigor, spacing, form, and branch characteristics.

Thinning Schedules

Whether or not to incorporate thinnings into a management plan is highly debated and depends on initial spacing, site, topography, product output, utilization standards, management objectives, availability of labor and equipment, nearness to market, and other factors. Commonly, however, managed, young-growth stands depend on a series of thinnings to maximize merchantable volume and value production. One of the most critical decisions is when to initiate the first thinning. The sequence and timing of future thinnings require managed-stand yield tables for effective planning. If thinnings are not to sacrifice future growth potential, some measure of stocking control is also necessary.

Timing of the First Thinning The timing of the first thinning depends on initial stocking, management considerations, and market considerations. From the biological viewpoint, stands should be thinned prior to the onset of serious between-tree competition if merchantable yields are to be enhanced. This, however, will be constrained if there is no market for the harvested material. If this is the case, management must consider the desirability of precommercial thinning. This is done if it can be justified that the precommercial thinning will enhance volume growth or economic returns earlier in the rotation than otherwise would be the case (see Table 14-4). The age at which a stand reaches merchantability is largely dependent on the average diameter of the trees. Consequently, thinning strategies are usually aimed at growing trees as rapidly as possible because tree size has a significant impact on log values.

Tree size also has a big impact on logging costs, as shown in Fig. 17-7. Early thinnings when diameters are small are therefore likely to be more costly than

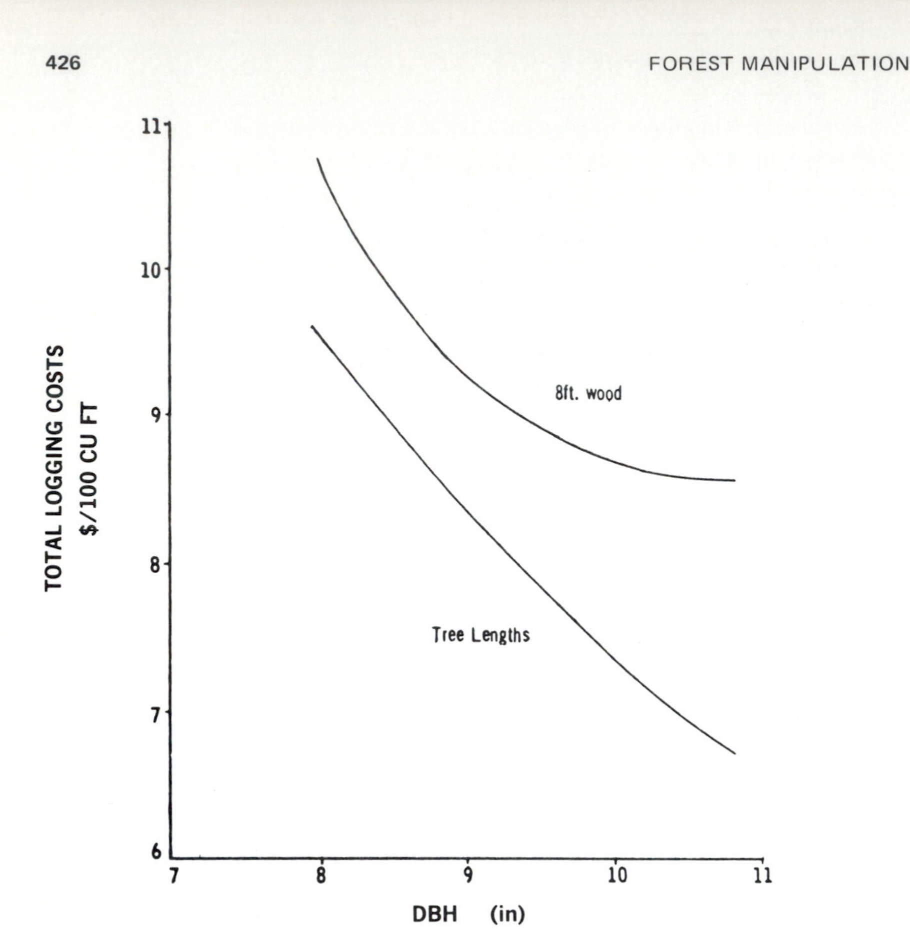

Figure 17-7 Logging costs in thinnings are a significant deterrent to doing thinnings, but the logging cost is reduced appreciably with increasing diameter of the trees cut. (*From Humphreys, 1970.*)

thinnings done later in life. The diagram also shows why it might be good strategy to make the first thinning a crown thinning rather than a low thinning.

Timing in thinning determines whether a stand responds well or poorly to a particular degree of stocking reduction. Early thinning is desirable at relatively high stocking levels to avoid competition and the risk of stagnation. Low stocking, on the other hand, postpones the first thinning. A higher site quality requires earlier thinning for a given initial density because competition begins earlier. Factors such as the tolerance of the species, the spreading habit of the crown, and the end product desired by management will affect the time of a thinning. However, the condition of the crown is the major factor in timing, since a crown that is allowed to deteriorate below 30 to 40 percent of the total height of the tree slows down the response that can be expected from the thinning. Age is also a significant factor, because as maturity is approached, height growth is negligible and response to thinning is dependent on the ability of the crown to expand laterally or become more efficient in photosynthate production or use of water. Pearson (1950) has found for ponderosa pine in the Southwest where moisture is limited, that the release of a 200-year-old tree with a small crown did give accelerated growth because of the increase in available moisture.

Sequence of Thinnings In developing thinning schedules, there is no particular merit in continuing to use the same method for each sequential thinning. In fact, the reverse is more commonly true. One may commence with a crown thinning and follow this with a series of low thinnings once the trees have reached merchantable size, or one may use a mechancial method followed by a series of free thinnings. The silviculturist's test is to develop a thinning strategy that is most suitable for the development of the particular stand while conforming to managerial and economic constraints. In planning a series of thinnings, the timing and nature of each treatment is strongly influenced by the nature of the earlier and later thinnings. Thinning intensity, which is a product of both the frequency and grade of a series of thinnings, determines the "heaviness" of the treatment.

Stocking Control The key to thinning is the control of stocking. Predictions must be made of the likely behavior and productivity of a given species on a particular site as the stand develops. An example of this is the data gathered from even-aged stands of upland hardwoods in the Central states (Gingrich, 1967). The data show a fairly constant growth over a wide range of average stand diameters and stocking (Fig. 14-22). In practice, stocking control is usually obtained by controlling stand basal area. However, a better grasp of the desirable range of stocking can be had by expressing the limits in terms of Reineke's stand density index.

In an unmanaged stand, basal area accumulates rapidly but the increment per tree is practically constant (Table 14-3); however, lowering the density of the stand within the limits of stocking where growth is constant would appreciably raise the increment per tree. Thus, to enhance merchantable yields, it is desirable to maintain stocking levels in the lower part of the optimal range because this results in larger material with perhaps some sacrifice of total volume. The levels of basal area corresponding to the upper and lower limits of the optimal range are remarkably similar for many moderately intolerant trees managed on short rotations for wood production. For species such as Douglas-fir, the southern pines, red pine, ponderosa pine, and Monterey pine, the upper and lower limits for stocking when managed on rotations of 30 to 50 years are approximately 130 and 90 ft^2 per acre. Managed-stand yield table data for a sequence of thinnings of red pine in the Lake states illustrate the application of stocking control that recognizes the optimal basal-area range (Fig. 14-26).

Determining optimal levels of stocking for each species is of critical importance in the intensive management of forests. Progress is being made in determining optimal levels through studies of growing stock levels such as have been done for Douglas-fir and ponderosa pine.

PRUNING

Pruning removes the branches from the bole of a tree at an early age in order to have knot-free wood laid down on the bole. In rotations less than 100 years, the only way in which one can obtain significant amounts of clear lumber is by artificial pruning. The branches of intolerant species will die when the trees are grown at close spacings, but the dead branches will normally not drop from the whole lower bole at a sufficiently early age. In addition, the close spacings required for death of all branches on the lower bole may result in a loss in merchantable volume production.

Some species in genera such as *Agathis* (New Zealand kauri) and *Eucalyptus* develop a natural abscission layer where the branches are attached to the trunk. When the lower branches lose vigor and become senescent, they are naturally shed in much the same way as leaves of deciduous trees are shed in the autumn. The branches of North American trees do not have abscission layers at their base, and therefore the trees do not develop branch-free boles except through death and decay of the branches.

The pruning of trees is commonly the most expensive intermediate treatment when expressed per individual treated tree. These costs vary greatly and depend on:

1. The number of trees pruned per acre
2. The diameter of the branches
3. The number of branches per whorl
4. The number of whorls to be pruned
5. The height of the pruning
6. Growth rate, and hence site quality
7. The interest rate charged on the investment
8. The waiting time prior to harvest

Pruning can be economically justified when the accumulated costs of the treatment are predicted to be less than the differential in value between pruned and nonpruned logs at the time of harvest. These predictions are notoriously difficult to make. Consequently, pruning normally will not be done in North America as long as lumber from old-growth trees remains on the market. Under some circumstances, pruning can be financially attractive. This has been shown to be the case for rapidly growing trees such as Monterey pine in New Zealand (Brown, 1965), as well as for some North American species. For example, Smith (1961) reported that a 5 percent return on investment can be obtained by pruning Douglas-fir, giving a good profit in 20 to 30 years. Red pine has been reported to yield between 3 and 8 percent over a 60-year period, and similarly,

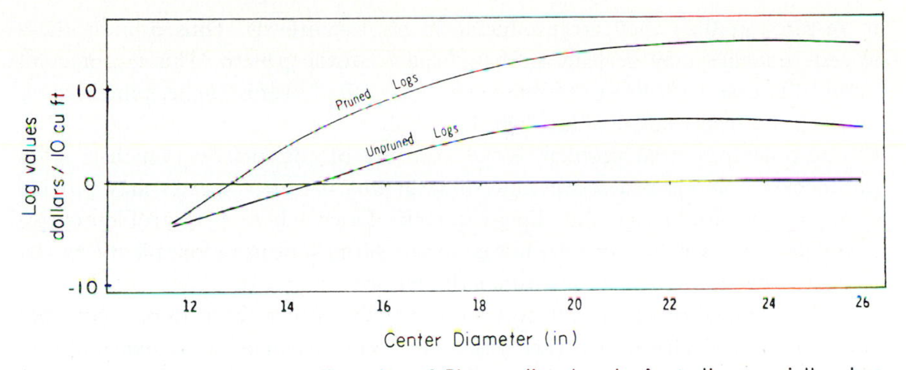

Figure 17-8 Pruning increases the value of *Pinus radiata* logs in Australia, especially when used to produce veneer. (*Adapted from Humphrey, 1970.*)

the pruning of white pine has been claimed to be profitable (Funk, 1961). While pruning increases the value of sawlogs, the benefits are not significant until the diameter is over 15 in. Pruning to produce knot-free plywood can give greatly increased returns (Fig. 17-8), and the increased value over unpruned logs occurred in this case at a diameter of 12 to 13 in. There is some question as to the future value of pruning for sawlogs because mill-run lumber may be cheaper to upgrade in the mill than pruning.

It can be seen from the listing of costs that pruning is least expensive when few trees are pruned per acre, the branches are small in diameter and few in number, and growth rates are fast. For these reasons, only crop trees may be pruned, pruning is done at a young age, and the preferred stands are on high quality sites and concurrently thinned and fertilized. In addition, commonly only the bottom log is pruned, since this is the highest-value log in the tree and the costs of pruning rise dramatically with height. In pruning young trees, the attempt is to prune the stem when its diameter is about 10 cm (4 in). If this is done, not only will the branches be small but the knotty core within the log will be confined to the central 10 cm. After pruning, the stand should be thinned, since the rate at which branch stubs grow over is a function of the length of the stub and the rate of growth of the tree.

In pruning trees, no more than 50 percent of the live crown should be removed at any one time. Reduction in live-crown ratio to below 50 percent is likely to put the tree at a competitive disadvantage since height growth might be temporarily reduced. If this happens, crop trees may subsequently become dominated by trees of poorer form which grew more rapidly because of their larger crowns. The effects of removal of live lower branches on growth depends on their contribution to total carbohydrate production. In closed stands, the lower branches of intolerant trees tend to be photosynthetically inefficient because of low light levels. Under conditions of high vapor-pressure deficits (see Chap. 5), the foliage on these lower branches can be utilizing more carbohydrate in respiration than they are producing in photosynthesis. Thus pruning these lowest branches may actually be beneficial to total growth. This is not commonly the case with more tolerant trees, where the lower branches remain more photosynthetically active at low light levels.

Another potential problem is the exposure of cut surfaces, which in pines produces a copious protective flow of resin but in other species may provide opportunities for insect and disease attack. Cases where pest problems have developed following pruning are not common. More serious perhaps is the extent to which the cut branches constitute a fire hazard.

The removal of significant portions of the live crown tends to produce trees with a more cylindrical bole (i.e., less taper). This effect of live-crown ratio on stem form is discussed in more detail in Chap. 6.

SITE AMELIORATION

Site amelioration has many connotations depending on management objectives. With timber production as the objective, fertilization has been a very productive way of enhancing growth. Irrigation would also be a stimulant to growth in many places and would provide a means for effluent disposal if the source of waste was close to commercial stands of trees. A different interpretation of site amelioration can be developed where objectives of management include recreation, aesthetics, and wildlife values. Here, sites can be ameliorated by appropriate plantings of shrubs or tree mixtures (Welch and Andrus, 1977).

Fertilization

As discussed in Chaps. 10 and 11, the productivity or site quality of an area is a function of biological factors (including species, genotype, density, and pests) and physical factors (including soil physical and chemical characteristics and topographic characteristics). The actual productivity of a site, however, can be modified through treatments which may either lower productivity, as in the case of erosion, or raise it by favorably modifying any of the biological or physical factors. The most common way to raise productivity is through fertilization, which raises the nutrient capital of a site by adding readily available sources of nutrients. Management of vegetation requires an understanding of nutrient supply, nutrient demands, and nutrient cycling since all silvicultural treatments interrupt this system–some in a relatively minor way, but others, such as harvesting and site preparation, can make dramatic changes in the nutrient balance.

Different species have different nutrient requirements, and consequently they can differ greatly in their capacity to respond to fertilizer treatments. Stone (1973) cited early work by Mitchell and Chandler (1939) who grouped species in the Northeast into classes according to their relative response to increasing nitrogen supply. Species regarded as nitrogen tolerant are red and white oak, red maple, and aspen, whereas nitrogen-demanding species are white ash, basswood, and tulip poplar. The nitrogen-tolerant species are superior competitors on sites with low nitrogen status compared with the nitrogen demanders, which are superior competitors on sites with abundant nitrogen.

Of all nutrient elements, nitrogen is by far the primary growth-limiting nutrient in North America. This is as true for the northern hardwood forests, where growth can be increased on the average by 80 percent (Auchmoody and Filip, 1973), as it is for the Douglas-fir forests of the Pacific Northwest, where the average growth rate following nitrogen fertilization is 18 percent (University of Washington, 1977). Not all species respond to nitrogen, however, as is the case with hemlock (*Tsuga heterophylla*), which in most cases shows no growth increase. In addition, the form of nitrogen can be important. For example, Engelmann spruce does not respond to nitrate nitrogen but responds well when the nitrogen is in the ammonium form (Bigg and Daniel, 1978). Response to nitrogen is common because of the high proportion of nitrogen which accumulates in organic matter. This is particularly true in northern latitudes and at high elevation where there is a large buildup of litter and organic matter on the forest floor.

Applications of fertilizer in forests are usually done to enhance tree growth. It must be remembered, however, that addition to the nutrient status of a forest ecosystem will tend to change the dynamics of all food chains and trophic levels in the system. The ecosystem is complex, and consequently the full impacts of fertilizer treatments are difficult to predict. After a brief discussion of the aims, methods, and results of forest fertilization from the timber-production standpoint, the ecological and environmental aspects of fertilization will be evaluated further.

There are five main objectives in using fertilization as a management tool (Stone, 1973):

1 The production of more and more useful wood

2 Rapid establishment of young forests, intensified culture, and wider choice of species

3 Repair, protection, and enhancement of vegetative cover, such as on exposed soil, burned or degraded sites, or heavily used areas

4 Improved food and cover for wildlife

5 Production of specialty crops such as Christmas trees and maple products

The most important incentive for forest fertilization is to produce more wood on a smaller land base. This incentive becomes increasingly more pronounced as commercial forest land is withdrawn for special purposes such as roadless areas and parks.

Forest fertilization has been standard practice in parts of Europe and the Southern Hemisphere for decades. Many millions of hectares are routinely fertilized annually. In North America, particularly in the South, the Pacific Northwest, the Northeast, and Canada, forest stands have also been fertilized for a long time, and the acreage treated annually is rapidly increasing despite the increasing costs of fertilizer. The most common fertilizer used is nitrogen—used in the form of urea pellets since this form has the highest proportion (46 percent) of nitrogen. Applications are almost always made with a helicopter in order to accurately control swath width and avoid gaps or overlaps which cause undesirable changes in site quality within the stand (Armson, 1972). Rates of application for the common elements N, P, and K are usually in the order of 50 to 300 kg of active element per hectare. Urea is usually applied at the rate of 200 kg ha^{-1}. This is usually regarded as adequate for at least a 5- to 7-year period and is sufficient to supplement the nutrients already present in the soil and supply the annual tree requirements, which for Douglas-fir are (Gessel et al., 1960):

Nutrient	Annual requirement, kg ha^{-1}
N	15–60
P	2–30
K	5–50
Ca	20–100

The results of fertilizer applications are extremely variable and range from actual negative responses to growth increases of 200 to 300 percent depending on species and sites. In general, a higher response is obtained on lower-quality sites, but increased growth on higher-quality sites is usually sufficient to warrant fertilizer applications.

A characteristic result of fertilizer application, particularly on lower-quality sites, is to increase the mortality of trees in the lower crown classes. This is so because fertilization increases growth rate and competition causing a faster expression of dominance. This effect is more pronounced in stands of intolerant trees than in stands of tolerant species. Another characteristic is that fertilized trees develop foliage which is distinctly larger in size and darker green in color. The foliar biomass increases also. For example, after fertilization with nitrogen, Douglas-fir on poor soils increased the amount of needles in the canopy from 20 to 53 percent (Heilmann and Gessel, 1963). Root growth is also commonly increased; fine-root biomass under a 90-year-old beech-birch-maple forest in New Hampshire doubled after an application of lime and N-P-K fertilizer (Safford, 1972).

As noted earlier, applications of fertilizer can affect the status of other components of the forest system than the crop trees. Understory plants and "weeds" respond to added nutrients, and this is particularly important when fertilizers are applied at the time of regeneration of even-aged stands. The incentive to ensure rapid growth of the desired tree seedlings is usually so great that other treatments such as herbicides or mechanical weeding are used to control competition. Fertilization early in the rotation is important because the time before canopy closure is when greatest demands are made on the available nutrient capital of the site (see Chap. 10).

The influence of fertilizer treatments on insect populations is complex since the treatment may either increase or decrease the availability of suitable food and space for population development (Weetman and Hill, 1973). Fertilizer treatments change sugar percentage, osmotic concentration, amino acid composition, lignification, and other internal factors that may be favorable or unfavorable to particular insects. Plant growth rate is increased, biomass is increased, and changes occur in growth of understory, rates of maturation, and within-stand microclimate which affect specific insects differently. Insects respond to those changes by changing their behavior, preferences, tolerance ranges, natality, mortality, longevity, rate of growth, and ultimate size (Singh, 1970; Weetman and Hill, 1973). As a consequence of this complexity, no good generalizations can be made as to whether fertilizer applications increase or decrease potential insect problems. More detailed reviews of this subject have been prepared by Mustanoja and Leaf (1965), Stark (1965), Lee (1968), and Foster (1968).

The effects of fertilizer on diseases in forests are similarly variable. Where fertilization corrects deficiencies, disease is usually reduced. But N and N-P-K fertilizers have been found to increase tree susceptibility to a number of fungi including *Cronartium fusiforme*, certain *Fugarium* and *Verticillum* species, and several rusts and fungi of poplars. Increases in diseases of seedlings are particularly common after fertilizing. On the other hand, N fertilizers usually decrease attack by *Lophodermium pinastri* (pine-needle cast), *Ceratocystis ulmi* (Dutch elm disease), and *C. fagarearum* (oak wilt) (Weetman and Hill, 1973).

Ectomycorrhizae are more prevalent in infertile soils, but depending on the fertilizer used, mycorrhizae may either increase or decrease (Shigo, 1973). Since ectomycorrhizae help reduce infection of pine roots by *Phytophthora cinnamomi* (Marx, 1971), any treatment that reduces mycorrhizal population could be important. The potential impact of fertilizer on nematodes and viruses is similarly variable and therefore difficult to predict. With regard to wildlife, it is well known that fertile soils support larger and more varied populations and that there is preferential feeding on fertilized browse.

Other impacts of fertilizer treatments include possible effects on quantity and quality of stream flow. Here it appears that the treatment of established stands is unlikely to cause significant impacts because of the effective filtering capacity of forest humus and soils. More significant impacts could be expected to occur when sites are fertilized at the time of planting when surface runoff and leaching are more likely. The greatest impact from fertilization occurs when applications are made directly into streams and lakes. Added nutrients increase the productivity and populations of algae, bottom vertebrates, and fish, but high accumulation of nutrients can lead to eutrophication. The impact of added nutrients to streams, however, is limited to just a short distance downstream from the point of entry (Werner, 1973). The quantity of stream flow after fertilization may tend to be reduced by the more rapid growth of vegetation and increased foliar biomass, which would transpire more water out of the soil profile.

All these considerations indicate that while fertilization may be well justified in terms of growing more wood faster on fewer acres, the potential impact on the forest system as a whole is not well understood and is highly variable.

Irrigation

Irrigation of forests is not a common practice but can be done, for example, in the growing of plantations of poplar on agricultural land in Italy. In recent years, the problem of disposal of effluents in an environmentally acceptable manner has awakened interest in disposing of this waste material on forests. The proposition seems doubly attractive since most wastes have a high nutritive value. Experiments have been conducted in various locations, notably in the South, where the objective is to determine the amounts of waste water and nutrients that forests can utilize without detrimental changes in stand structure.

The possibilities appear attractive except for the fact that the sources of most city and mill effluents are prohibitively long distances away from suitable productive forests. Further increases in the demands for fiber and energy may make the use of effluent wastes on forests more attractive in the future.

PRESCRIBED BURNING

Prescribed burning is a common practice in the South and is increasingly being used in parts of the West to control understory vegetation. In many forest stands where the desirable overstory is composed of intolerant species, there is a tendency for more tolerant species to become established and develop as an understory. This is clearly seen in many forests where a fire-exclusion policy over the past few decades has resulted in dramatic changes in forest composition and structure. In parts of the West, stands that used to be predominantly pine and Douglas-fir with little understory are now heavily stocked with a dense understory of the more tolerant white fir.

Prescribed burning is a most useful tool and requires considerable technical skill to ensure that there is enough fuel to carry a ground fire, but not so much that the fire will reach sufficient intensity to scorch the desired crop trees. By controlling the development of understory, the forester is also increasing or decreasing the development of particular species of wildlife as well as populations of insects and pathogens. As with all other intermediate treatments, prescribed burning can be regarded as having positive or negative impacts depending on the particular factor under consideration in the ecosystem and the overall management objectives. A further consideration in some localities is the visual impact of smoke.

SANITATION AND SALVAGE CUTTING

Sanitation cutting is the removal of insect-attacked or diseased trees in order to maintain the health of the stand. The removal of these trees is not necessarily expected to be economic in the direct sense. *Salvage cutting*, on the other hand, is the removal of trees in order to obtain an economic gain before their value is lost. These trees may be damaged by insects, disease, lightning, or fire. These two treatments are often done in conjunction with one another, and the general treatment is called a *sanitation-salvage cutting*.

The effect of this treatment on stand development is generally to remove trees in the oldest and largest size category. The trees are commonly low in vigor and "unhealthy," and their removal is important in maintaining the healthy development of the future crop trees. These old, large, unhealthy trees are, however, the source of food and habitat for species of wildlife and insects that are integral parts of the ecosystem. Insectivorous birds, for example, that nest in these trees often play an important role in keeping the forest free of the buildup of populations of insect pests. The trees may also provide the only habitat for endangered species. As a consequence, the forester must exercise care and caution before categorizing trees as "unhealthy" and removing them without careful consideration of their role in the total system. Trends in this direction are evidenced by the fact that some states have forest-practice legislation which includes limitations on the removal of snags or dead and dying trees.

Chapter 18

Reproduction Methods

The concern for immediate regeneration is the most significant difference between silviculture and exploitive logging. Through the hundreds of years that stands have been exploited on this continent with no concern for a future tree crop, nature through the often slow process of succession has reclaimed most cutover areas with trees. However, tremendous areas are nonproductive or poorly stocked. In pioneer times there was little reason for concern because the expanse of forest was apparently limitless, and the population pressure and demand for forest products were low. Wood production by the exploitation of old-growth forests is no longer appropriate to meet the demands of our economy for structural material, paper products, and a multitude of other uses. The difference between the wood production of previously unmanaged forests and the need for wood products must come from managed young-growth stands. We can no longer ignore the loss in production due to regeneration failures. Immediate regeneration is now possible with a reasonable expenditure of time and money because the silvicultural knowledge is available and stumpage prices have increased to provide sufficient margin to allow for good forestry practices—two conditions not generally present in the past even if a forest owner had wanted to practice forestry.

SILVICULTURAL SITUATIONS FOR REPRODUCTION

In order to discuss reproduction methods, the environmental situations that can be created in the harvesting of a stand must be examined. These are largely influenced by factors which are uncontrollable such as slope, aspect, elevation, latitude, climate, and soil. The effects of these factors, however, can to some extent be ameliorated by silvicultural treatments. In treating stands, the forester needs to recognize that there are four major situations that strongly influence regeneration survival and growth: open areas, closed areas, south-facing stand edges, and north-facing edges (Fig. 18-1).

Open Areas These are openings created in a stand by harvesting or windthrow that are free of tree root competition (except in droughty situations where roots often extend well beyond the margin of the trees) and free of the influence of the stand margin. The area usually receives the full intensity of direct solar radiation through the 4 h midday period. Wind influence on an open area depends on the size of area, orientation with respect to wind direction, and height of surrounding trees. Aspect and slope combine to modify the impact of openness—south and west slopes become increasingly severe with steepness. Increasing latitude above the Tropic of Cancer reduces the solar energy for a given slope, aspect and time of year (Frank and Lee, 1966). An open area has low interception losses, low transpiration losses, and high evaporation losses.

Closed Areas A completely closed condition in a stand has intense root competition, almost no direct solar radiation reaching the soil surface, reduced

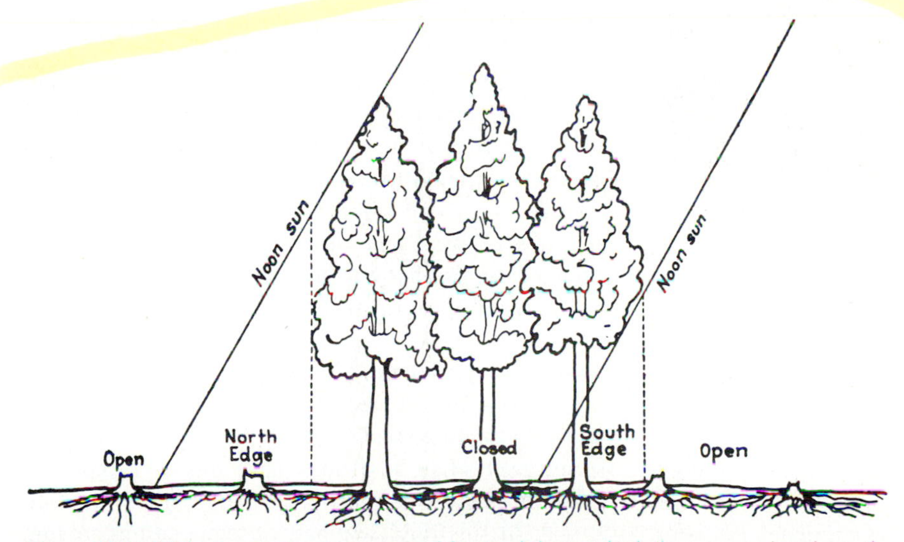

Figure 18-1 Silvicultural situations created in stands by manipulation: open, closed, north, or south edges.

wind movement, and high interception and evapotranspiration losses but low soil surface evaporation. Aspect, slope, and latitude influence the amount of solar radiation impinging on the canopy and the amount of long-wave radiation that reaches the soil surface. The closed condition is usually a matter of degree, except perhaps in the tropics where the several stories in a stand provide an almost completely closed condition at soil level. A given degree of openness in the canopy is a much more "closed" situation on north aspects and higher latitudes than on south aspects or lower latitudes. The effect of steepness of slope is to accentuate the effects of aspect and latitude, while the effect of tree height is to increase the comparative closure on north and south aspects.

South Edge The south-edge situation is limited to the distance that direct solar radiation can penetrate to areas under the tree canopy at the margin of the opening (Fig. 18-1). A south edge is an especially severe site because the area has normal root competition with high direct radiation to increase soil evaporation. In addition, a south edge produces a "heat trap" effect due to long-wave radiation emitted from ground, tree trunks, and canopy. Depth of stand penetration depends on slope, the sun's declination, and the height of the lower canopy. The steeper the south slope, the less is the depth of solar penetration. While the maximum penetration would be a north slope with a slope angle equal to the sun's declination, the effect would be minimal because the energy would be dispersed over the full depth of ray penetration and never impinge on the soil surface except as reflected or reradiated.

North Edge The north-edge situation has a width determined by the length of the shadow cast by the trees along the margin of an opening. With decreasing sun declination, the shadow lengthens as it does with increasing slope on north aspects, but increasing slope on south aspects decreases the length of shadow. Beyond the root zone of the border trees [i.e., the outer margin of the canopy (Smith, 1964)], the shaded area is free of tree root competition. Radiation is limited to diffused light except as sunflecks move fleetingly over a point. North edges are usually on the lee side of stands, since winds in the northern hemisphere are commonly from the southwest and west. Consequently, at higher elevations there is greater accumulation of snow which provides more moisture. Thus the north-edge situation in many locations is a protected site with a favorable environment for regeneration.

No effort is made to discuss east and west edges because they represent various modifications within the extremes represented by the north and south edges. Environmentally there is a significant difference between a southeast aspect and a southwest aspect, even when both may have the same input of solar energy. The southeast aspect receives its maximum energy in the morning when plants are normally turgid. At this time, soil and air temperatures are low and the plants are not likely to be stressed. However, an open or even closed situation on a southwest aspect receives maximum energy inputs in the afternoon when air and soil temperatures are high, and plants are more likely to develop stress conditions than those in comparable situations on other aspects. Any stand treatment on a southwest or south aspect has to be more expertly judged and requires more careful application than on other aspects where the margin of allowable error is greater.

PURPOSES OF REPRODUCTION METHODS

The choice of reproduction method is frequently a compromise between what is biologically ideal and what is economically and socially acceptable. A compromise is feasible because trees can normally reproduce under a range of environmental conditions. The choice depends on the best resolution of the conflicting needs of a stand and the purposes a particular method can serve.

The purposes of reproduction methods are (1) to remove the crop trees in such a manner as to ensure that regeneration occurs in a predictable period of time; (2) to provide a number of options for creating the specific environment most favorable for the establishment of a species; (3) to provide a means for selectively favoring one species over another; (4) to allow alternatives for meeting aesthetic, recreational, wildlife, watershed, and soil constraints in handling a stand; (5) to give an opportunity for maintaining or developing a certain stand structure; (6) to provide alternative methods of harvesting to meet managerial constraints in regulating forest production; (7) to adapt the reproduction method to the presence of wind, disease, insect, or fire hazards; and (9) to permit a choice among methods for efficient use of new technology in harvesting, utilization, and products.

REPRODUCTION METHODS

A distinction must be drawn between reproduction methods and reproduction or silvicultural systems because they are frequently incorrectly used as synonyms. A *reproduction method* describes the manner in which a stand will be cut to ensure regeneration, and there is no pretense of prescribing how the resulting stand will be treated after it is established. On the contrary, a *silvicultural system* is defined as "a process, following accepted silvicultural principles, whereby crops constituting forests are tended (weeding, cleaning, improvement cuts and thinnings), harvested and replaced. . . . Systems are conveniently classified according to the method of carrying out the fellings that remove the mature crop with a view to regeneration. . . ."[1] Thus, if the only concern is to get regeneration without commitment to any future handling of a stand, then the treatment is a reproduction method. If, on the other hand, a reasonably com-

[1] Unfortunately, the *Terminology* of Ford-Robinson (1971) ignores its own definition and calls the methods systems, i.e., clearcutting system when it is describing the method.

prehensive prescription involving a sequence of treatments is made for how the stand is to be treated after regeneration in anticipation of its needs to reach the objective of management for that stand at rotation age, then the reproduction method used for establishing the stand names the system, i.e., clearcut system. Probably a great portion of this country has been so concerned with problems of harvesting and regenerating old-growth stands that future stand treatments have not been identified. Consequently, there has been more concern with methods rather than systems.

The reproduction methods can be outlined as follows:

- I High forest methods: regeneration established from seed
  - A Even-aged stands
    - 1 Clearcutting methods
    - 2 Seed-tree method
    - 3 Shelterwood methods
  - B Uneven-aged stands
    - 1 Selection methods
- II Coppice forest methods: regeneration derived from sprouts and suckers
  - A Coppice method: all sprouts and suckers
  - B Coppice with standards: coppice with seed-origin older growth or older sprouts

These reproduction methods were first developed in Europe in the eighteenth century by the newly created profession of forestry to restore forests from centuries of local exploitive logging using simple coppice methods and often taking the best trees while leaving the worst within the limits of their transportation systems. The first method to be utilized was the clearcutting method since it enabled the forester to rapidly replace degraded stands with stands of higher value. In most cases, this involved replacing deteriorated hardwood stands with conifers. The development of the other even-aged methods followed, and subsequently the selection methods were introduced to permit the management of uneven-aged stands.

The applicability of these methods to North American forests can be questioned because they were developed in a different country, for different species, without the availability of equipment such as trucks and tractors, with lower labor costs, with different management objectives, and with a totally different relationship between the public and forest use. An analysis of the situation shows, however, that the classical reproduction methods remain highly relevant to today's situation because their application is based on a thorough understanding of ecological principles.

As illustrated in Fig. 2-1, the silviculturist is basically interpreting ecological trends and vegetation dynamics and using these as bases for predicting changes over time. Knowledge of relative plant tolerance and other silvical characteristics of trees (Fowells, 1965) is fundamental to the appropriate choice of reproduction method. This knowledge, plus information on microclimate (e.g., soil-water availability and evaporative stress), permits the silviculturist to determine, for a given site, the extent to which these microclimatic factors must be maintained or modified to support the growth of desired plants while ideally placing undesirable plants at a competitive disadvantage. The silviculturist then attempts as far as possible to create this desired microclimate through stand treatment and thus ensure regeneration of desired species. This results in the development of stands of distinctive structure and composition.

HIGH-FOREST REPRODUCTION METHODS

Reproduction methods in high forests have only four categories and there is the tendency to think of all reproduction methods as being, of necessity, one of these distinctly differentiated ways of securing reproduction. On the contrary, the methods form a continuum in the degree of exposure a site experiences in going from the selection method to the clearcutting method. The forester recognizes this continuum and chooses from it a treatment or treatments that satisfy the management objectives for the particular stand and its environment. A unique example illustrating the application of this concept is given below (p. 448).

For simplicity, the high-forest reproduction methods are illustrated as uniform regular strips or blocks, whereas in practice the areas conform to the topography. Boundaries follow contours and are oriented for protection from wind. Cut areas are perhaps best distributed over the range of site qualities so that each year's cut has a balance of material from good and poorer sites.

Even-Aged Methods

A stand is even-aged when the range in age classes within it is reasonably narrow. The acceptable range increases with rotation length. A stand with a range of 1 to 50 years would be uneven-aged, but the same stand at 250 years would be even-aged. A forest can be even-aged if all its component stands are even-aged, although each may be a different age.

Clearcutting Methods A reproduction method is a *clearcutting method* when the open situation dominates and edge situations are minimal, when all trees are removed before regeneration occurs, and when regeneration occurs without dependence on the protection of border trees. The prime objective of the clearcutting method is simply to reestablish an even-aged stand by removing the mature one.

A decision to clearcut may be dictated by a number of adverse considerations such as insect epidemic, disease, fire, or decadent stand conditions. Other reasons may be the desire to change species, to introduce a better genotype, or to meet the needs for regulating volume production.

Clearcutting has numerous constraints on its use, among which are:

1 Legal constraints. These may be laws such as the National Forest Management Act of 1976, state forest-practice acts, or county ordinances.

2 Silvical constraints. The seed must be capable of establishing in the open and of being adequately distributed over the area, and the seedbed and growing conditions must remain favorable until full stocking is accomplished. Planting circumvents some of the silvical constraints.

3 Economic constraints. These are few, other than the potential cost of regeneration failure with its loss of production and costs of correction.

4 Aesthetic and environmental constraints. In certain situations, potential aesthetic and environmental impacts are the major constraints on the use of clearcutting. This is particularly true in the harvesting of old-growth stands but may not be so severe in managing young-growth stands.

Clearcutting with Natural Reproduction A species' silvical characteristics are decisive in the choice of this method. If open situations are favored or required for successful establishment of a species in nature, as exemplified by stand histories, then clearcutting may warrant consideration. The extreme open situations created by clearcutting rarely occur in nature. The closest approximation is a fire-killed stand. But here, although root competition is removed, the dead trees cast considerable shade and erosion hazards are less. Assuming regeneration is favored by clearcutting, the width of the cut area depends on the dissemination characteristics of the seed that comes from the edge trees. Where species such as jack and lodgepole pine have serotinous cones which remain closed on the trees for many years, the slash after harvesting contains abundant seed which is released after ground temperature or light fire opens the cones. In this case, the width of the cut has no limit from the standpoint of ensuring regeneration. Tree seed that may be stored in the litter for more than a year by a few species is too unreliable to depend on for regeneration. On the contrary, very large numbers of seed from brush species such as *Ceanothus* and *Ribes* in the West remain viable for up to 100 years in the forest litter layer. The presence of long-lived seed of brush species can therefore strongly influence the rate of invasion and competition for seedlings and may deter the choice of clearcutting and prescribed burning.

Seedbed conditions influence the seeds-per-seedling ratio, which in turn influences the width at which adequate regeneration can be expected from a seed source. A favorable seedbed usually means a mineral soil with a sufficiently receptive surface to get seeds covered. Favorable growing conditions call for a site with as low a level of herbaceous or shrub competition as possible. Inasmuch as a clearcutting is subject to rapid deterioration of site conditions favorable to seedling survival, adequate stocking may depend on one or a few seed crops. To reduce this time factor, site preparation should be done immediately prior to seed fall rather than immediately after harvesting.

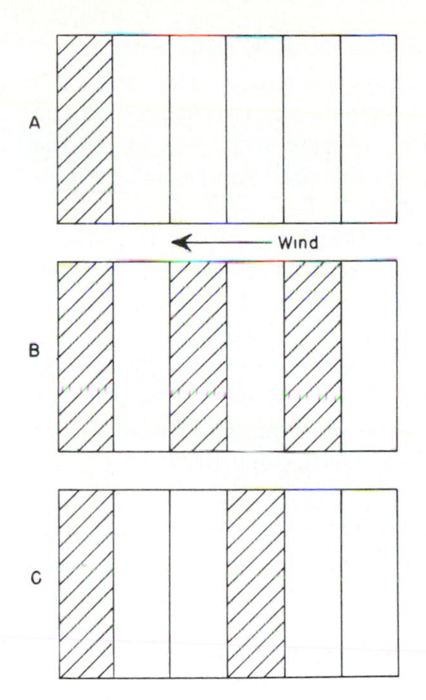

Figure 18-2 Clearcut areas by simple (A), alternate- (B), and progressive- (C) strip methods for the same species, with the assumption that the width of each strip is optimum for the natural establishment of seedlings. Cross-hatched areas are cut in one year in each method.

Clearcutting with natural regeneration has three common variations, and the choice of one over the others depends on the size of the area, optimum seeding distance from the edge, windthrow hazard in leave strips, cost of regenerating the area in leave strips, and the amount of volume needed from the area.

Simple Clearcutting Unit The area of a simple clearcut unit (Fig. 18-2) is controlled by a number of factors:

1 The dissemination distance of the seed
2 Economic considerations, such as compatibility of volume removed with mill capacity
3 Care to ensure that a particular age class is well distributed over the whole working circle rather than in one large area
4 A smaller area, which reduces the aesthetic and environmental impacts

Alternate-Strip Clearcutting If a stand to be managed on an even-aged basis is too large to be a single clearcut unit yet the timber needs to be harvested over a short period of time, then there are two alternatives for handling it under the clearcutting method: alternate- or progressive-strip cutting. In the alternate-strip method, alternating strips (Fig. 18-2) are cut in one given year and the method provides for the largest volume of cut with clearcutting. After regeneration has become established on the cut strips, the leave strips are cut. The problems of the leave strips are (1) they are susceptible to windthrow; (2) a large percentage of the total area may be in leave strips; (3) regeneration of the leave strips is a problem calling for planting or shelterwood method; and (4) the area may be aesthetically undesirable. The minimum width of leave strips depends on the risk of windthrow, an adequate seed supply for the clearcut area, and the economics of harvesting the strip and its regeneration.

Progressive-Strip Clearcutting In progressive-strip clearcutting (Fig. 18-2), a stand is cut progressively in a series of adjacent strips. The number of strips in a series depends on the length of time a strip takes to regenerate satisfactorily and the maximum age difference between strips that is compatible with maintaining an even-aged stand. Progressive strips reduce the potential losses from windthrow since only the last strip is dangerously susceptible. The smaller area in the final leave strips reduces the regeneration problem. The rate of timber flow is lowered from that in alternate-strip clearcutting and the aesthetics are improved.

The diagrammatic representation of the methods of strip clearcutting (Fig. 18-2) assumes that the same total area, species, and site conditions are being compared. The width of the cut strips varies with the species and is determined by the silvics of the species. For Douglas-fir, the width could be 400 m (1320 ft) on north or northeast slopes (Isaac, 1943) but less than 60 m (200 ft) on a south slope. If 5 years are allowed for regeneration of cut strips and the leave strips are planted, then block A (simple clearcutting) takes 26 years to cut over, block B (alternate-strip clearcutting) takes 6 years, and block C (progressive-strip clearcutting takes 11 years.

The advantages of the clearcutting method can be listed as follows: (1) lower logging costs are possible; (2) no logging damage to standing timber occurs; (3) less skill is demanded than in other reproduction methods; (4) the introduction of improved stock is allowed; (5) sanitation problems such as dwarf mistletoe can be solved; (6) it is the only method for serotinous-coned species; (7) grazing is possible after seedlings are beyond damage in size; (8) decadent overmature stands can be dealt with; and (9) narrow clearcuts favor big game.

The disadvantages of the clearcutting method can be listed as follows: (1) species must survive in open situations; (2) species control is poor; (3) seedling distribution is uneven and parts of an area may become understocked or overstocked; (4) erosion and site-deterioration hazard may be high; (5) regeneration may be delayed and growth conditions may deteriorate rapidly; (6) seeds must be light enough to have a satisfactory seeding distance; (7) unprofitable trees may have to be cut; and (8) clearcutting rates the lowest aesthetically of any of the high-forest reproduction methods but remains unaesthetic only long enough for seedlings to reach the sapling stage.

Clearcutting with Planting Clearcutting with planting has been widely adopted mainly because of the need to obtain prompt regeneration and control spacing. As with most simple answers to complex problems, planting failures may also be common because species cannot survive in the open. However, most planting failures can be traced to improperly prepared planting sites, poor stock, and poor planting techniques (see Chap. 16).

The advantages of clearcutting with planting can be listed as follows: (1) planting after logging should give immediate control of the site; (2) rotation is shortened; (3) species composition is controlled and a change in species or introduction of a superior genotype is possible; (4) spacing is uniform and reduces the need for a noncommercial thinning; (5) the size of cut area can be unlimited; (6) management plans are simplified; and (7) planting may be more profitable in the long run, especially if natural reproduction is difficult or slow.

The disadvantages of clearcutting with planting can be listed as follows: (1) planting costs are high; (2) planting costs must be capitalized rather than carried as an operating cost; (3) failure is a risk; (4) fewer trees per acre are possible compared with natural regeneration; (5) a seasonal labor supply is needed; (6) with uniform vigor and spacing, there is danger of stagnation unless the stand is thinned; and (7) plantations are usually of one species so they have the advantages and disadvantages of pure stands.

Clearcutting with Direct Seeding Since Mann (1960) demonstrated that direct-seeding success depended on excellent site preparation and seed protection, direct seeding has been successfully used in the South, Pacific Northwest, and Canada. When successful, the method is initially cheaper than seedling planting, but the risks are higher. It is a reproduction method for general consideration in extraordinary situations such as fresh-burned inaccessible areas or very large fresh burns.

Seed-Tree Method As a result of the Morris Act of 1902, the seed-tree method became the first reproduction method prescribed by law, except as the Organic Act of 1897 laid down what trees could be cut. Its application mostly failed its purpose because delayed seed crops gave brush the control of the site (Eyre and Zehngraf, 1948). The results reflect what can be expected of legislated silvicultural methods out of context with the environment.

The seed-tree method consists of leaving enough good seed-producing trees scattered over an area to ensure adequate stocking in a predicted period of time. The method ensures an even distribution of seed over the area and allows a larger area to be cut than clearcutting with natural regeneration. The method is suited to intolerant species, but the seed trees do provide some degree of protection. An added skill is necessary in selecting seed trees for their seed production, good phenotype, and most important their ability to resist windthrow and breakage, as well as to decide on the proper number of trees needed.

Usually the seed trees will be harvested after regeneration is established, and care must be taken to avoid excessive damage to the new crop.

The number of seed trees required on an area depends on: (1) the number of seeds per tree; (2) the dissemination distance; (3) the survival time of seed trees in terms of windthrow and other losses; (4) frequency of seed crops; (5) seeds per seedling ratio; (6) the desired number of seedlings per acre; (7) the cost of trees if abandoned; and (8) the length of time the seedbed is receptive and growing conditions favorable.

The advantages of the seed-tree method can be listed as follows: (1) more uniform distribution of seed occurs as compared with clearcutting; (2) better composition control is allowed; (3) there is no limitation on area cut; (4) logging costs are low; and (5) better aesthetics than clearcutting are produced.

The disadvantages of the seed-tree method can be listed as follows: (1) control of spacing and timing of the new crop is difficult; (2) it is limited to windfirm trees; (3) it is commonly limited to light-seeded species; (4) soil protection is difficult; (5) rodents, seedbed conditions, and growing conditions are difficult to control; and (6) if seed trees are not harvested, there is a monetary loss; if they are harvested, seedlings are damaged.

Shelterwood Methods The essential purpose of the shelterwood method is to accomplish the regeneration of the site under the shade and protection of the final crop trees. In contrast to the relatively rigid conditions created by the clearcutting and seed-tree methods for the establishment of even-aged stands, the shelterwood method is capable of producing any degree of site protection in a stand. This capability of manipulating stand density to provide the environment required for the regeneration of one or several species on a variety of sites makes the shelterwood method the most flexible way of reproducing even-aged stands. Another distinction is the shelterwood method's capacity for producing an abundance of uniformly distributed seed.

The shelterwood method is characterized by a series of cuts made toward the end of the rotation (reproduction period). Three kinds of cuts can be made.

1 Preparatory cut. This is a light, partial cut to correct some unfavorable condition in the stand such as poor crown development (seed production would be delayed if such trees were depended on for immediate seeding), to improve windfirmness, or to correct poor seedbed conditions (litter accumulation). There can be several preparatory cuts. In old-growth stands that have begun to open up, preparatory cuts are usually unnecessary, but in dense, unmanaged, young-growth stands, they may be needed.

2 Seed cut. There is only one seed cut. The canopy is opened up to the degree that is judged to favor the regeneration of the desired species and to exclude undesirable species.

3 Removal cuts or final cut. One or more removal cuts occur after the seed cut has provided adequate regeneration and this has grown sufficiently in height. The number of removal cuts depends upon how "sensitive" the seedling stand is to sudden exposure. In addition to reducing environmental shock for the seedling, the overstory may be removed slowly if there is danger of the seedlings being overtopped by vegetation that the overstory can control. There is also the economic value accumulating on the residual trees as well as a restocking potential for areas damaged in earlier removal cuts. A less intensive application of the method may reduce the number of cuts to a seed cut and one removal or final cut if species, site, and economics make the simplification feasible. When for any reason a stand has acceptable established advanced growth, the function of the seed cut has been performed, so the shelterwood method can be reduced to a final removal cut, provided it is done in a manner that ensures the survival of the established regeneration.

Uniform Shelterwood Method The uniform shelterwood method (Fig. 18-3) applies each cut over the whole stand. It ensures the most even-aged stand.

The advantages of the uniform shelterwood method can be listed as follows: (1) it allows ultimate control of site conditions for the regeneration of even-aged stands; (2) it is the best method for heavy-seeded species; (3) it provides the best control over regeneration composition, amount, and distribution; (4) it is the most flexible method since it is applicable to tolerant and intolerant species; (5) good soil protection is provided; (6) high aesthetic qualities are produced; and (7) there are no biological constraints on its application to large areas.

The disadvantages of the uniform shelterwood method can be listed as follows: (1) logging costs are increased because of the returns to the same area for smaller volumes and the care exercised to prevent excessive damage; (2) high technical skill is needed; (3) it requires a fairly wind-firm species; (4) unavoidable damage to residual stand and reproduction occurs during logging; and (5) very intolerant trees like longleaf pine need rapid removal of overstory after seedling establishment.

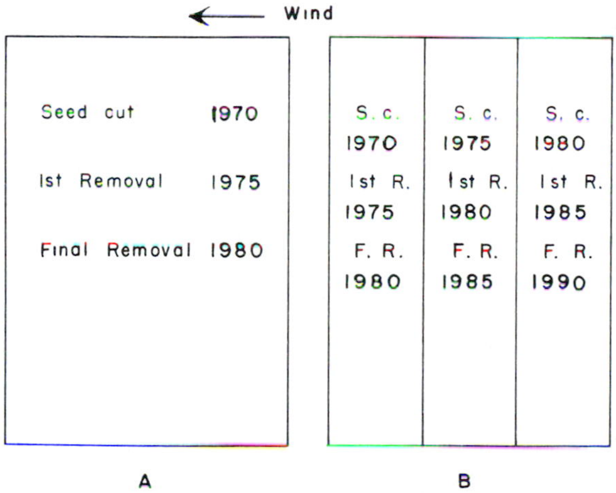

Wind

Seed cut 1970
Ist Removal 1975
Final Removal 1980

S. c. 1970 | S. c. 1975 | S. c. 1980
Ist R. 1975 | Ist R. 1980 | Ist R. 1985
F. R. 1980 | F. R. 1985 | F. R. 1990

A B

Figure 18-3 Shelterwood methods commonly used are uniform shelterwood (A) and strip shelterwood (B).

Strip Shelterwood Method In this method a stand is broken into strips (Fig. 18-3). The number and width of the strips depend on the reasons why the strip shelterwood method was used rather than the uniform shelterwood method. Strip shelterwood may be chosen to distribute the volume of a large stand over a longer period, to reduce the windthrow hazard by leaving uncut strips to the windward, to provide better seedbed preparation in the seed-cut strip and less logging damage to seedlings in the removal-cut strip by taking logs through seed-cut strip, to provide a source of seed from the adjacent strip after the removal cut, and to improve the aesthetic values of the operation.

Shelterwood Method Variations The number of variations in the shelterwood method is quite large. A few of the more interesting ones are the following:

1 Group shelterwood. Small groups of reproduction may serve as centers around which concentric shelterwood strips progress until the whole area is regenerated.

2 Wagner's shelterwood method. A tapered density of overstory is developed on the north edge of a stand, which ensures creation of the right conditions for reproduction. As reproduction is established, the tapered edge is moved toward the south.

3 Strip-clearcut shelterwood. The clearcut strip is no wider than the north-edge effect will protect. When the strip is regenerated, the next strip is clearcut.

Regenerating an Even-Aged Mixed Stand

The following unique example[2] illustrates the concept that a forester can choose the conditions needed to fulfill the management objectives from a continuum of potential reproduction methods. The original mixed stand at Saas consisted of silver fir, beech, Norway spruce, and larch, and the same mixture of even-aged regeneration was required.

The natural regeneration of the stand had already begun and can be diagrammed as in Fig. 18-4, in which A is a line of small openings made by harvesting trees in which very tolerant silver fir was capable of being established; B and C are portions of the tapered edge, with B conditions favorable only for tolerant beech regeneration and C conditions favorable for the less tolerant Norway spruce; and D is a scattering of intolerant larch seed trees. The whole pattern moves west after the individual species have developed to the point where they can compete on fairly equal terms. Silver fir requires protection and considerable time to develop but can compete very well if 20 to 30 cm tall. Beech and Norway spruce can regenerate under the same conditions, but spruce will gain dominance over beech and eliminate it. Beech can establish under lower light than spruce, so beech is given time at B to get 20 to 40 cm of growth. Then

[2] Saas Revier, Steyr Forstverwaltung, Austria in 1951 under Forstmeister W. Streinz and Oberförster Liftinger.

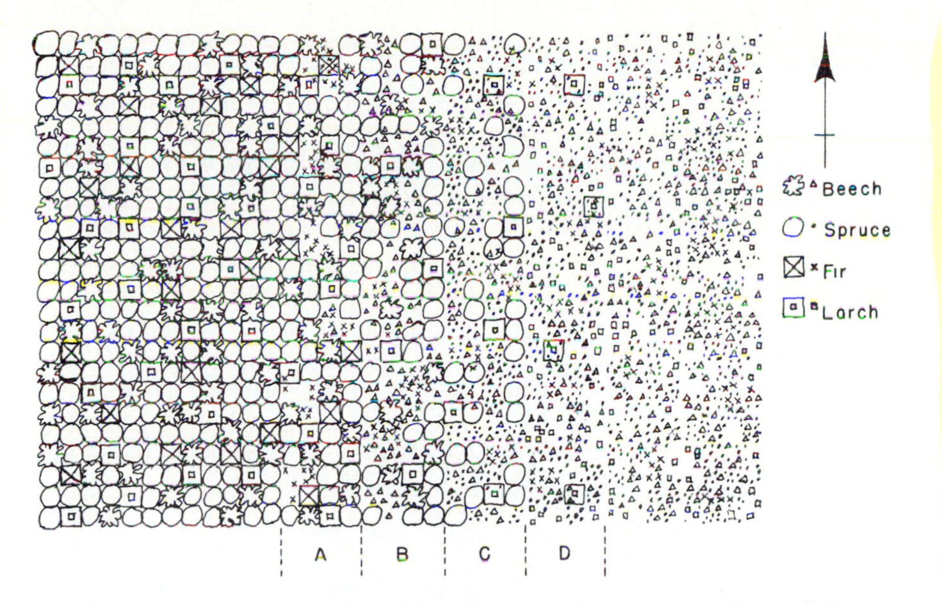

Figure 18-4 Regenerating a mixed even-aged stand with (A) silver fir (*Abies alba*), very tolerant; (B) beech (*Fagus sylvatica*), tolerant; (C) Norway spruce (*Picea abies*), tolerant; and (D) larch (*Larix decidua*), very intolerant. Small openings (A) are made well ahead of tapered stand density (B–C), with a final scattering of larch seed trees (D). Stand in illustration has regenerated and is ready for the next progressive cut to the west.

the tapered edge moves west (with another series of small openings for silver fir being cut to the west also) and spruce establishes among the beech and fir. In another 3 to 5 years, the spruce is established on C and the tapered edge moves west again, leaving only seed trees of larch. The larch seed regenerates in the gaps of the other species, and its rapid height growth brings it quickly into competition with the others. With regeneration of everything established, the composition is controlled by thinnings to attain the desired 65 to 75 percent spruce, 15 to 20 percent beech, 5 to 10 percent silver fir, and 5 percent larch. It is a labor-intensive procedure suitable to a time when trees were valuable, labor costs low, and labor for woods work readily available. Times have changed.

Uneven-Aged Stands

The nature and structure of uneven-aged stands have already been discussed in Chap. 14. It remains to clarify how such stands can be maintained.

Selection Methods The application of the selection method is characterized by two conditions: the stands are uneven-aged, and the regeneration never loses the protection (or competition) of the older age classes around it.

Single-Tree Selection In the single-tree selection method, individual trees (or a small group) are removed and new regeneration occurs in their place. It requires a very tolerant species to become established and survive under such small openings in the canopy. Cutting procedures under the selection method call for frequent partial cuts in a stand. The interval between cuts in the same stand is called the *cutting cycle*. There is no rotation age at which the mature crop is harvested, as in even-aged management. However, the crop trees cut at each cutting cycle have reached maturity, and the age of maturity and the length of the cutting cycle determine the number of age classes present in each stand. In the discussion of stand growth (Chap. 14), the shape of the inverted-J form could be maintained only by balancing the harvest in each diameter class.

An example of a stand managed under either the single-tree or group selection is given in Fig. 18-5. Assume a unit of a working circle consists of 400 ha with an annual growth of 4.2 m^3 ha^{-1} yr^{-1} (300 board feet per acre per year). Maturity age for crop trees is 100 years, and the cutting cycle is 10 years. Each stand would be 40 ha, so each year 4.2 X 10 X 40 = 1680 m^3 would be cut. Every stand would have 10 age classes because on the eleventh year the cut would be in stand 1 again.

Year Mature Trees Cut	Age in 1980	Year Mature Trees Cut	Age in 1980	Year Mature Trees Cut	Age in 1980	Year Mature Trees Cut	Age in 1980	Year Mature Trees Cut	Age in 1980
1980	100	1981	99	1982	98	1983	97	1984	96
1990	90								
2000	80								
2010	70								
2020	60								
2030	50								
2040	40								
2050	30								
2060	20								
2070	10	2071	9	2072	8	2073	7	2074	6
1		2		3		4		5	

								Year Mature Trees Cut	Age in 1980
1985	95	1986	94	1987	93	1988	92	1989	91
								1999	81
								2009	71
								2019	61
								2029	51
								2039	41
								2049	31
								2059	21
								2069	11
2075	5	2076	4	2077	3	2078	2	2079	1
6		7		8		9		10	

Figure 18-5 Diagrammatic representation of the selection method for a unit of a working circle of 10 stands illustrating a management plan with a 10-year cutting cycle which would be applicable to single-tree or group selection. Each stand has 10 age classes. A balanced stand is maintained by reducing tree numbers in *each* diameter class in line with the *q*-ratio desired.

Figure 18-6 Group selection in the northern hardwood type on the Upper Peninsula Experimental Forest of Michigan, with F. H. Eyre (1948).

The actual structure of a stand may be made to fit any management objective, such as maximizing production of sawtimber or value appreciation, by use of de Liocourt's law or a derivitive of it: the *q* factor (see Chap. 14). Structure manipulation is further clarified in Chap. 14.

The advantages of the single-tree selection method can be listed as follows: (1) only a selection method is capable of maintaining an uneven-aged stand; (2) reproduction of tolerant species is easily obtained; (3) site protection is excellent with little or no exposure to insolation and wind; (4) stands can be readily adapted to fluctuating market conditions; (5) a fire is less likely to be as disastrous as in a young, even-aged stand; (6) capital returns come at short intervals even from small woodlots; and (7) uneven-aged stands usually have the highest aesthetic rating, especially single-tree selection.

The disadvantages of the single-tree selection method can be listed as follows: (1) a highly skilled forester is required to practice it; (2) logging costs are higher because of the small volume per acre, the frequent returns to each stand required, and the care necessary to hold damage to an acceptable limit; (3) stem quality especially with hardwoods is lower than a comparable site with even-aged stand; (4) crop trees are scattered throughout the stand; (5) the necessary intermediate cuts in the younger age classes in order to maintain the age class distribution have a tendency to be neglected; (6) inventory examinations and growth and yield projections are difficult and time-consuming; (7) grazing under most conditions is not permissible because young age classes may be damaged; and (8) conditions are unfavorable for the establishment and growth of intolerant species.

Group-Selection Method Group selection creates larger openings in an uneven-aged stand, but the opening must not be so large as to lose the site protection by the surrounding trees. The opening should be no larger than one to two tree heights (Fig. 18-6), but aspect and slope influence the actual size. There is some loss in site protection and in aesthetics, otherwise the method reduces to a degree the disadvantages of the single-tree selection method.

The advantages of the group-selection method can be listed as follows: (1) the regeneration in the small groups grows up under even-aged conditions which gives better stem form; (2) the larger openings permit establishment of intolerant species; (3) harvesting is more concentrated so logging costs are lower; (4) harvesting in groups lowers damage to residual stand; (5) the intermediate cuts may be made less frequently without sacrificing diameter class distribution, although composition may suffer; and (6) inventorying is somewhat easier.

COPPICE FOREST REPRODUCTION METHODS

The term *coppice* describes all regeneration that is derived from vegetative sprouting of dormant or adventitious buds. Most sprouting trees are hardwoods with a few exceptions, of which redwood is outstanding as a vigorous sprouter.

Coppice methods of reproducing stands are perhaps the oldest and historically the most widespread of all methods. They were described in hieroglyphics in ancient Egypt and have been used for centuries in Europe, India, Africa, and the Americas. A stand of sprouts will have double the volume of seedlings at 20 years; however, the seedlings will equal the sprouts at 40 years and outgrow them beyond 40 years. The reason for this early phenomenal growth rate is usually attributed to the fact that the sprout has the advantage of the parent root system and to the stimulation of wound hormones. A sprout loses its advantage over a seedling because the seedling builds its own root system while the sprout is confined to part of the parent's system and loses efficiency with age.

Coppice growth was primarily for fuel and was the principle source of fuel in Europe until after World War II. In the Sihlwald, the community forest of Zurich, Switzerland, under management since the fifteenth century, firewood brought prices equivalent to structural wood. Today, the Sihlwald is totally converted to high forest because the market for firewood has almost disappeared, since gas and oil have taken over the heating and cooking market.

Fuel wood was abundant in the United States, so there was no necessity for managing stands for maximum volume. However, the recent geometric growth in the demand for pulp-wood products has stimulated a keen interest in coppice's potential. Union Camp Corporation has 20,000 hectares earmarked for coppice production. With an initial investment of $375 to $500 per hectare, it hopes to recover 11.2 metric tons of wood per hectare per year (5 tons per acre per year) on a 12-year rotation. Sycamore is being experimented with in the Southeast and grown on a 3- to 6-year rotation for about the same production per hectare at harvest (Steinbeck et al., 1970; Dutrou and Saucier, 1976).

Coppice Reproduction Method

The coppice method is the simplest of all methods for regenerating a stand of trees. It depends on vigorous sprouts from small stumps since stumps of larger, older trees sprout too weakly to grow vigorously. While all hardwoods sprout, few have stumps with enough resistance to rot to allow several rotations of sprouts to develop.

Sprouts develop on the top, side, and root collar of stumps. Root-collar sprouts minimize the hazard of rot infection from the parent stump, so trees are cut off close to the ground to force the production of root collar sprouts. Sprouts develop best when stands are harvested during fall or winter when the tree is dormant. If cut in summer, the sprouts are weak, the stumps bleed, and the growing season is shorter.

Suckers develop from adventitious buds on roots of a few species such as beech and aspen. A sucker has the initial rapid growth of a sprout, but its final independence from the parent allows a steady growth to maturity comparable to that of a tree of seedling origin. Trembling aspen is the best example of the habit, since beautiful stands originate from suckers. In fact, it is the only means of regenerating aspen in the West.

A coppice stand is strictly even-aged, since all sprouts start at the same time and they are more intolerant than seedlings of the same species. A coppice stand has no slash problems, since utilization is generally intensive and the slash decomposes rapidly. No site preparation is necessary and the composition is automatic. Some stumps lose their ability to sprout and need to be replaced because sprouts produce little or no seed; the replacements are planted seedlings.

Coppice growth requirements can be listed as follows: (1) the composition of the stand must be of species that are able to sprout, to have sprouts persist to merchantable size, to retain the ability to sprout again at merchantable size, and to be durable enough to produce several rotations; (2) the site must be very fertile and have a high moisture supply so frequent cuts do not cause site deterioration; and (3) a frost-free site is necessary since sprouts are very susceptible.

The advantages of the coppice method can be listed as follows: (1) rapid volume growth occurs on short rotations; (2) low investment in growing stock produces a high return; (3) regeneration is simple; (4) there are no area limits on harvesting; (5) high vigor reduces hazard from injurious agents; and (6) it is adaptable to mechanical harvesting.

The disadvantages of the coppice method can be listed as follows: (1) there is a need for a market for small material such as pulpwood; (2) it is restricted as

Figure 18-7 Coppice with standards after the coppice material had been harvested on the Forêt de Conches, France (1952).

to the number of species that can be used; (3) site deterioration may be severe; (4) frost hazard limits site selection; (5) the frequent clearcutting probably makes it aesthetically the poorest reproduction method; (6) some planting must be done; and (7) replacing a coppice forest with improved strains would be difficult because of the continued sprouting of the old stand.

Coppice with Standards Method

The principal purpose of the coppice with standards method is to modify the major weakness of the coppice method, which produces only small and often poor-formed trees, by growing some larger trees (Fig. 18-7). By allowing a part of the coppice stand to grow for another one or more coppice rotations, trees of any size can be produced. Sprouts can be used for standards, but standards are frequently planted seedlings and the seedlings are often conifers. Standards adversely affect the growth of coppice trees around them because the standards are free to expand their crowns with no lateral pressure. As the proportion of standards increases, the volume of growth per year decreases at a more rapid rate. By increasing the number of standards, a coppice forest can be gradually converted into a high forest.

Epilogue

There are those in forestry who sway with the new and deprecate the old wisdom, yet the recognition of good and bad practices in Heinrich Cotta's time could be a quote from today. Cotta, a German forester, in the preface of his book *Anweisung zum Waldbau* (*Advice on Silviculture*, 1816, translated by B. E. Fernow in *Forest Quarterly*, 1902) said

> If the inhabitants of Germany should leave their country it would be all grown up with woods within a century. Since there would be nobody to use them, the soil would be enriched and the woods would not only increase in size, but in productive power. If, however, the people returned again and made just as large drafts as before for wood, litter and pasturage, the woodlands, even with the best forest management, would again not only be reduced in size, but also become less fertile.
>
> Forests form and thrive best where there are no people—and hence no forestry, and those are perfectly justified who say: Formerly we had no forestry science and enough wood; now we have that science, but no wood.
>
> One could say with the same justice: Those people are healthier who do not need a physician than those who do. But it would not follow that the

physicians are to be blamed for the diseases. There would be no physicians if there were no diseases, and no forestry science without deficiency in wood supplies. This science is only a child of necessity or need, and need is therefore its natural concomitant; hence the phrase should be: We have now a forestry science because we have a dearth of wood.

Forestry, however, does not offer any nostrums and can do nothing against the course of nature. The celebrated physician Verdey said: "The good physician lets people die; the poor one kills them." With the same right one can say the good forester allows the most perfect forests to become less so; the poor one spoils them. That is to say, just as the good physician cannot hinder that men die because that is the course of nature, so the best forester cannot hinder that the forests, which came to us from past times, become less now they are being utilized.

Germany formerly contained immense, perfect, most fertile forests. But the large forests have become small, the fertile have become sterile. Each generation of man has seen a smaller generation of wood. Here and there we admire still the giant oaks and firs, which grew up without any care, while we are perfectly persuaded that we shall never in the same place be able, with any art or care, to reproduce similar trees. The grandsons of those giant trees show the signs of threatening death before they have attained one-quarter of the volume which the old ones contained, and no art nor science can produce on the forest soil which has become less fertile, such forests as are here and there still being cut down.

The good forester then, also, allows the forest to become less, but only where it cannot be helped; the poor forester, on the other hand, spoils them everywhere.

Without utilization, the forest soil improves constantly; if used in orderly manner it remains in a natural equilibrium; if used faultily it becomes poorer. The good forester takes the highest yield from the forest without deteriorating the soil, the poor one neither obtains this yield nor preserves the fertility of the soil

It is hardly credible how much one can benefit or damage by the kind of management; the true forestry science contains, therefore, much more than those think, who know only its generalities.

Thirty years ago, I prided myself on knowing forestry science well. Had I not grown up with it and in addition had learned it in the universities! Since then I have not lacked the opportunity for increasing my knowledge in many directions, but during this long period I have come to see very clearly how little I know of the depths of the science, and to learn that this science has by no means reached that point which many believe to have been passed.

Many perhaps may be in the condition in which I was thirty years ago; may they in the same manner be cured of their conceit! Forestry is based on the knowledge of nature; the deeper we penetrate its secrets, the deeper the depths before us. What the light of an oil lamp makes visible is easily overlooked; many more things we can see by torchlight, more unknown things become apparent, and it is a sure sign of shallowness, if anybody believes he knows it all.

Our foresters can still be divided into empiricists and scientists; rarely are both united.

What the former considers sufficient in a forest management is easily learned, and the systematic teachings of the other are soon memorized. But in practice the art of the first stands to a thorough forestry science in the same relation as the quack medicine to the true pharmacopoeia; and the other often does not know the forest for the many trees. Things look very differently in the forest from what they do in books; the learned man stands therefore, frequently, left by his learning and at the same time without the bold decision of the empiricist.

Three principal causes exist why forestry is still so backward; first the long time which wood needs for its development; second, the great variety of sites on which it grows; third, the fact that the forester who practices much writes but little, and he who writes much practices but little.

The long development period causes that something is considered good and prescribed as such which is good only for a time, and later becomes detrimental to the forest management. The second fact causes that what many declare good or bad proves good or bad only in certain places. The third fact brings it about that the best experiences die with the man who made them, and that many entirely one-sided experiences are copied by the merely literary forester so often that they finally stand as articles of faith which nobody dares to gainsay, no matter how one-sided or in error they may be.

Bibliography

Aaltonen, V. T.: Soil and forest with special reference to northern European silviculture, Paul Parey, Berlin, 1948.

——: Forest soil research in Finland, *Unasylva*, 4(3):116–121 (1950).

Adams, M. W., A. H. Ellingboe, and E. C. Rossman: Biological uniformity and disease epidemics, *Bio. Science*, 21:1067–1070 (1971).

Ahlgren, I. F., and C. E. Ahlgren: Ecological effects of forest fires, *Bot. Rev.*, **26**:483–533 (1960).

Alban, D. H.: An improved growth intercept method for estimating site index of red pine, *USDA Forest Ser. Res. Pap. NC-80*, 1972.

Alexander, R. R.: Crown competition factor (CCF) for Englemann spruce in the Central Rocky Mountains, *USDA Forest Ser. Res. Note RM-188*, 1971.

——, D. Tackle, and W. G. Dahms: Site indexes for lodgepole pine, with corrections for stand density: methodology, *Rocky Mtn. Forest and Range Exp. Sta., U.S. Forest Ser. Res. Pap. RM-29*, 1967.

Allen, G. S., and J. N. Owens: The life history of Douglas-fir. Environment Canada, Forestry Service, Ottawa, 1972.

Armitage, F. B.: Statement for Canada on tree planting and direct seeding. *In* Symposium on stand establishment, IUFRO Joint Meeting Divisions 1 and 3, Wageningen, The Netherlands, Oct. 15–19, 1974, pp. 410–419.

Armson, K. A.: Fertilizer distribution and sampling techniques in the aerial fertilization of forests, *Univ. of Toronto, Fac. Tech. Rep. 1*, 1972.

——: Forest soils: Properties and processes, Univ. of Toronto Press, Toronto, Canada, 1977.

—— and R. van den Driessche: Natural root grafts in red pine (*Pinus resinosa* Ait.), *Forest Chron.*, **35**:232–241 (1959).

Art, H. W., and P. L. Marks: A summary table of biomass and net annual primary production in forest ecosystems of the world. *In* H. E. Young (ed.), Forest biomass studies, Proc. of XVth IUFRO Congress, sec. 25, Univ. of Florida, Gainesville, FL, Life Sciences and Agriculture Exp. Sta., Univ. of Miami, 1971, pp. 3–32.

Assmann, E.: Waldertragskunde, BLV Verlagsgesellschaft mbH, Munich, 1961.

Auchmoody, L. R., and S. M. Filip: Forest fertilization in the eastern United States: hardwoods. *In* Forest fertilization symp. proc., *USDA Forest Ser. Gen. Tech. Rep. NE-3*, 1973, pp. 211–225.

Averell, J. H.: Rules of thumb for thinning loblolly pine, *J. For.*, **43**:649–651 (1945).

Bailey, R. G.: Ecoregions of the United States, U.S. Forest Ser., Ogden, Utah, 1976.

Baker, F. S.: The principles of silviculture. McGraw-Hill, New York, 1950.

Bakuzis, E. V., and H. L. Hansen: Distribution of balsam fir reproduction and basal area in the edaphic field of forest communities in the central pine section of Minnesota, *Minn. For. Note 120*, 1962.

Balmer, W. E.: Operational aspects in the production of forest tree planting material. *In* Symposium on stand establishment, IUFRO Joint Meeting Divisions 1 and 3, Wageningen, The Netherlands, Oct. 15–19, 1974, pp. 91–102.

Banks, C. H.: Spiral grain and its effect on the quality of South African timber, *Bosb. Suid-Afr.*, **10**:27–33 (1969).

Bannan, M. W.: The vascular cambium and tree ring developments. *In* T. T. Kozlowski (ed.), Tree growth, Ronald Press, New York, 1962, pp. 3–21.

Barber, S. A.: A diffusion and mass flow concept of soil nutrient availability, *Soil Science*, **93**:39–49 (1962).

Barnes, G. H.: Yield of even-aged stands of western hemlock, *USDA Forest Ser. Tech. Bul. 2373*, 1962, p. 52.

Barnes, R. L., and G. W. Bengtson: Effect of fertilization, irrigation, and cover cropping on flowering and on nitrogen and soluble sugar composition of slash pine, *For. Science*, **14**:172–180 (1968).

Barnett, J. P., and T. W. Campbell: Viability of seeds sown with an aerial multiple-row seeder, *Tree Plant. Notes*, **26**(2):1–2 (1975).

Barney, W. C.: Effect of soil temperature and light intensity on root growth of loblolly pine seedlings, *Plant Physiol.*, **20**:146–163 (1951).

Baron, F. J., and G. H. Schubert: Seedbed density and pine seedling grades in California nurseries, *U.S. Forest Ser. Res. Note PSW-31*, 1963.

Barrons, K. C.: Some ecological benefits of woody plant control with herbicides, *Science*, **165**:465–468 (1969).

Baskerville, G. L.: Some interrelationships of silviculture and logging, *Pulp and Paper Magazine of Canada*, June 1965.

Batchelor, L. D., and H. S. Reed: The seasonal variation of the soil moisture in a walnut grove in relation to the hygroscopic coefficient, *Calif. Agric. Exp. Sta. Tech. Pap. 10*, 1923.

Baule, H., and C. Fricker: The fertilizer treatment of forest trees, BLV Verlagsgesellschaft mbH, Munich, Germany, 1970.

Baver, L. D., W. H. Gardner, and W. R. Gardner: Soil physics, 4th ed., Wiley, New York, 1972.

Beard, J. S.: Climax vegetation in tropical America, *Ecology*, **25**:127–158 (1944).

Beck, D. E.: Polymorphic site index curves for white pine in the southern Appalachians, *U.S. Forest Ser. Res. Pap. SE-80*, 1971*a*.

——: Growth intercept as an indicator of site index in natural stands of white pine in the southern Appalachians, *U.S. Forest Ser. Res. Note SE-154*, 1971*b*.

Bednall, B. H.: The problem of lower volumes associated with second rotations in *Pinus radiata* plantations in South Australia, Bul. 17, Woods and Forests Dept., South Australia, 1968.

Bengston, G. W.: Fertilizing southern pine forests: Does it pay? *In* Proc. symp. on planted southern pines, U.S. Forest Ser., Southeast Area, State and Private Forestry, Atlanta, GA, 1968, pp. 49–72.

Bentley, J. R., and K. M. Estes: Uses of herbicides on timber plantation, USDA Forest Ser., Region 5 and Pacific Southwest Forest and Range Experiment Station, San Francisco, CA, 1965.

Bethel, J. S.: The effect of position within the bole upon fiber length of loblolly pine, *J. For.*, **39**:(1):30–33 (1941).

Bickford, C. A., F. S. Baker, and F. G. Wilson: Stocking, normality and measurement of stand density, *J. For.*, **55**:99–104 (1957).

Bigg, W. L., and T. W. Daniel: Effects of nitrate, ammonium, and pH on the growth of conifer seedlings and their production of nitrate reductase, *Plant and Soil* (in press).

Bilan, M. V.: Stimulation of cone and seed production in pole-size loblolly pine, *For. Science*, **6**:207–220 (1960).

Bingham, R. T., and A. E. Squillace: Phenology and other features of the flowering of pines, with special reference to *Pinus monticola*, Dougl., *U.S. Forest Ser. Northern Region and Intermountain Forest and Range Exp. Sta. Res. Pap. 53*, 1957.

—— and J. W. Wright: Breeding blister rust resistant western white pine. II. First results of progeny tests including preliminary estimates of heritability and rate of improvement, *Silvae Genet.*, **9**:33–41 (1960).

Bissett, I. J. W., and H. E. Dadswell: The variation in cell length within one growth ring of certain angiosperms and gymnosperms, *Aust. For.*, **14**:17–29 (1950).

—— and A. W. Wardrop: Factors influencing tracheid length in conifer stems, *Aust. For.*, **15**:17–30 (1951).

Black, C. C.: Ecological implications of dividing plants into groups with distinct photosynthetic productive capacity. *In* J. B. Cragg (ed.), Advances in ecological research, Academic, New York, 1971.

Blackman, F. F.: Optima and limiting factors, *Ann. Bot.*, **19**:381–395 (1905).

Bollen, W. B., C.-S. Chen, K. C. Lu, and R. F. Tarrant: Influence of red alder on fertility of a forest soil, Res. Bul. 12, Forest Res. Lab., Ore. State Univ., Corvallis, 1967.

Bormann, F. H.: The relationship of ontogenetic development and environmental modification to photosynthesis in *Pinus taeda* seedlings, in K. V. Thimann (ed.), The physiology of forest trees, Ronald, New York, 1957, pp. 197–215.

——, G. E. Likens, D. W. Fisher, and R. S. Pierce: Nutrient loss accelerated by clear cutting of a forest ecosystem, *Science*, **159**(3817):882–884 (1968).

Borrecco, J. E., H. C. Black, and E. F. Hooven: Response of blacktail deer to herbicide-induced habitat changes, in Proc. west. assoc. state game and fish comm., Portland, 1972, pp. 437–451.

Bowen, G. D., C. Theodorou, and M. F. Skinner: Towards a mycorrhizal inoculation programme, *Proc. American-Australian Forest Nutr. Conf., Canberra 1971*, 1973.

Bower, D. R., and J. L. Smith: Partial girdling multiplies shortleaf cones. Southern For. Expt. Sta., For. Notes No. 132, 1961.

Boyle, J. R.: Nutrients in relation to intensive culture of forest crops, *Iowa State J. Res.*, **49**:297–303 (1975).

Braun, E. L.: Deciduous forests of Eastern North America. McGraw-Hill, New York, 1950, p. 596.

Brazier, J. D.: An assessment of the incidence and significance of spiral grain in young conifer trees, *Forest Prod. J.*, **15**(8):308–312 (1965).

Brender, E. V., and W. H. McNab: Loblolly pine seed production in the lower Piedmont under various harvesting methods, *J. For.*, **70**(6):345–347 (1972).

Briegleb, P. A.: An approach to density measurement in Douglas-fir, *J. For.*, **50**:529–536 (1952).

Brix, H.: An analysis of dry matter production of Douglas-fir in relation to temperature and light intensity, *Can. J. Bot.*, **45**(11):2063–2072 (1967).

——: Effects of nitrogen fertilization on photosynthesis and respiration in Douglas-fir, *For. Science*, **17**(4):407–414 (1971).

Brown, C. L.: Growth and form. *In* Trees: structure and function. M. H. Zimmermann and C. L. Brown. Springer-Verlag, Vienna, pp. 125–167, 1971.

——: Physiology of wood formation in conifers, *Wood Science*, **3**(1):8–22 (1970).

—— and R. E. Goddard: Silvical considerations in the selection of plus genotypes, *J. For.*, **59**:420–426 (1961).

Brown, G. S.: The yield of clear wood from pruning: some results with radiata pine, *Comm. For. Rev.*, **44**:197–221 (1965).

Brown, G. W., A. R. Gahler, and R. B. Marston: Nutrient losses after clearcut logging and slash burning in the Oregon coast range, *Water Resources Res.*, **9**(5):1450–1453 (1973).

Brown, K. M., and C. Merritt: A shadow pattern simulation model for forest openings, Res. Bul. 868, Purdue Univ., 1970.

Brown, R. M.: Cold storage of forest plants, Forest Record 88, Forestry Commission, England, 1973.

Brown, W. H., and D. M. Matthews: Philippine dipterocarp forests, *Phil. J. Sciences*, **9**:413–561 (1914).

Bruce, David: Comparisons of measures of stand density, Lodgepole Pine Management Work Conference, Bend, OR, 1965.

——: Derivation of stocking guides—Douglas-fir and hemlock-spruce, unpublished data, 1972.

Bruce, D., D. DeMars, and D. Reukema: Douglas-fir managed yield simulator—DFIT user's guide, *U.S. Forest Ser., Pacific Northwest For. Range Exp. Sta. Gen. Tech. Rep. PNW-57*, 1977.

Buck, J. M., R. S. Adams, J. Cone, M. C. Conkle, W. J. Libby, C. J. Eden, and M. J. Knight: California tree seed zones, California region, U.S. Forest Serv., San Francisco, CA, 1970.

Buckman, R. E.: Growth and yield of red pine in Minnesota, *USDA Tech. Bul. 1272*, 1962.

Burckhardt, H.: Säen und Pflanzen nach forestlicher Praxis, Ein Handbuck der Holzerziehung, 1854 [from Dengler, 1944].

Burger, H.: Rein und gemischte Bestande, *Zeitsch. For. Jagdwes*, **60**:100–108 (1928).

Burns, F. Y.: Southern forest soils, Proc. 8th For. Symp., La. School of Forestry, 1959.

Burroughs, E. R., Jr., and B. R. Thomas: Declining root strength in Douglas-fir after felling as a factor in slope stability, *USDA Forest Ser. Res. Pap. INT-190*, 1977.

Burtt-Davey, J.: The classification of tropical woody vegetation types, *Oxford Univ. Imp. For. Inst. Pap. 13*, 1938.

Buttrick, P. L. Forest economics and finance. Wiley, New York, 1943.

Cajander, A. K.: The theory of forest types, *Acta Forestalia Fennica*, **29**:1–108 (1926).

Campbell, T. E., and W. F. Mann, Jr.: Regenerating loblolly pine by direct seeding, natural seeding, and planting, *U.S. Forest Ser. Res. Pap. SO-85*, 1973.

Cannell, M. G. R., S. Thompson, and R. Lines: An analysis of inherent differences in shoot growth within some north temperature conifers. *In* M. G. R. Cannell and F. T. Last (eds.), Tree physiology and yield improvement, Academic, New York, 1976, pp. 173–205.

Canny, M. J., B. Nairn, and M. Harvey: The velocity of translocation in trees, *Aust. J. Bot.*, **16**(3):479–485 (1968).

Carbonnier, C.: Ett gallringsförsö i planterad granskog, *Statens skogsforskningsinstitut Rep. 55*, 1957.

Carmean, W. H.: Suggested modifications of standard Douglas-fir site curves for certain soils in southwestern Washington, *For. Science*, **2**:242–250 (1956).

——: Tree height-growth patterns in relation to soil and site. *In* Tree growth and forest soils, Proc. 3rd North Am. Forest Soils Conference, Oregon State Univ. Press, 1970, pp. 499–511.

—— and A. Vasilevsky: Site-index comparison for tree species in northern Minnesota, *North Central Forest Exp. Sta., U.S. Forest Ser. Research Paper NC-65*, 1971.

Carter, M. C., J. W. Martin, J. E. Kennamer, and M. K. Causey: Impact of chemical and mechanical site preparation on wildlife habitat. *In* B. Bernier and C. H. Winget (eds.), Forest soils and forest land management, Les Presses de l'Université Laval, Quebec, Canada, 1975, pp. 323–331.

CAST: The phenoxy herbicides. A task force report to the council for agricultural science and technology, CAST Report 39, Dept. of Agronomy, Iowa State Univ., Ames, IA, 1975.

Champion, H. G.: A preliminary survey of the forest types of India and Burma, *Indian Forest Records*, **1**:1–286 (1936).

Chase, A. J., and H. E. Young: The potential of softwood thinnings and standing dead softwoods as a source of wood pulp, University of Maine, *Life Sciences and Agric. Exp. Sta. Tech. Bul. 82*, 1976.

Chisman, H. H., and F. X. Schumacher: On the tree-area ratio and certain of its applications, *J. For.*, **38**:311–317 (1940).

Choong, E. T., B. H. Box, and P. J. Fogg: Effects of intensive cultural management on growth and certain wood properties of young loblolly pine, *Wood and Fiber*, **2**(2):105–112 (1970).

Cieslar, A.: Licht- und Schattenholzarten, Lichtgenuss und Bodenfeuchtigkeit, *Centralblatt ges. Forstwesen*, **34**:4–22 (1909).

Clark, D. E., J. E. Young, R. L. Younger, L. M. Hunt, and J. K. McLaren: Animal metabolism of herbicides, fate of 2.4-dichlorophenoxyacetic acid in sheep, *J. Agric. and Food Chem.*, **12**(10):43–45 (1964).

Clark, J.: Photosynthesis and respiration in white spruce and balsam fir, *N.Y. State Univ. Coll. For. Tech. Publ. 85*, 1961.

Cleary, B.: The effect of plant moisture stress on the physiology and establishment of planted Douglas-fir and ponderosa pine seedlings, Ph.D. thesis, Oregon State Univ., 1970.

Clements, F. E.: Plant succession, *Carnegie Inst. Publ. 242*, Washington, D.C., 1916.

Clements, J. R.: The effect of late summer irrigation on red pine buds and shoot lengths, *Tree Plant. Notes*, **22**(2):4–5 (1971).

Coile, T. S.: Forest soil problems in the Piedmont Plateau, *J. For.*, **35**(4):344–349 (1937).

——: Forest classification: Classification of forest sites with special reference to ground vegetation, *J. For.*, **36**:1062–1066 (1938).

—— and F. X. Schumacher: Relation of soil properties to site index of loblolly and shortleaf pines in the Piedmont region of the Carolinas, Georgia and Alabama, *J. For.*, **51**:739–744 (1953).

Cole, D. W., and S. P. Gessel: Movement of elements through a forest soil as influenced by tree removal and fertilizer applications. *In* C. T. Youngberg (ed.), Forest-soil relationships in North America, Oregon State Univ. Press, Corvallis, 1965, pp. 95–104.

—— and S. P. Gessel: Cedar River research—A program for studying pathways, rates and processes of elemental cycling in a forest ecosystem, *For. Resources Monogr. 4*, Univ. of Washington, Seattle, WA, 1968.

——, S. P. Gessel, and S. F. Dice: Distribution and cycling of nitrogen, phosphorus, potassium and calcium in a second-growth Douglas-fir ecosytem. *In* H. E. Young (ed.), Symp. on Primary Productivity and Mineral Cycling on Natural Ecosystems, Univ. of Maine, Orono, 1967, pp. 197–232.

——, J. Turner, and S. P. Gessel: Elemental cycling in Douglas-fir ecosystems of the Pacific Northwest, Proc. XIII International Botanical Congr., Leningrad, U.S.S.R., in press.

Condon, P. A.: The toxicity of herbicides to mammals, aquatic life, soil micro-organisms, beneficial insects and cultivated plants, 1950–1965. A list of selected references. National Agricultural Library, Library List 87, USDA, Beltsville, MD, 1968.

Cooley, J. H.: Planting technique and care of stock affect survival of planted red pine, *U.S. Forest Ser. Res. Note NC-159*, 1974.

Covell, R. R., and D. C. McClurkin: Site index of loblolly pine on Ruston soils in the southern coastal plain, *J. For.*, **62**:263–264 (1967).

Cown, D. J.: The effects of artificial fertilization on the wood properties of conifers: A literature review and problem analysis, For. Prod. Rep. 402 (unpub.), N.Z. For. Res. Inst., Rotorua, 1972.

Craib, J. J.: The place of thinning in wattle silviculture and its bearing upon management of exotic conifers, *Zeitsch. Weltforstweiss*, **1**:77–108 (1934).

Cremer, K. W.: Seasonal patterns of shoot development in *Pinus radiata* near Canberra, *Aust. For. Res.*, **6**(2):31–52 (1973).

Cromer, R. N., M. Raupach, and A. R. P. Clarke: An examination of ashbed soils growing radiata pine in Gippsland, Victoria, *Aust. For.*, **34**(1):1–10 (1970).

Curtis, J. D.: Silvicultural limitations of shallow soils, *U.S. Forest Ser. Intermtn. For. Range. Ex. Sta., Misc. Publ. 24*, 1961.

Curtis, R. O.: Stand density measures: An interpretation, *For. Science*, **16**:402–414 (1970).

Dadswell, H. E., J. M. Fielding, J. W. P. Nicholls, and A. G. Brown: Tree-to-tree variations and the gross heritability of wood characteristics of *Pinus radiata, Tappi*, **44**:174–179 (1961).

—— and A. W. Wardrop: Growing trees with wood properties desirable for paper manufacture, *CSIRO Div. For. Prod. Rep. 357*, 1959.

Daft, M. J., and E. Hacskaylo: Arbuscular mycorrhizas in the anthracite and bituminous coal wastes of Pennsylvania, *J. Appl. Ecol.*, **13**:523–531 (1976).

Daniel, T. W., R. L. Meyn, and R. R. Moore: Reineke's stand density index in tabular form in English and metric units with its application, Utah Agric. Exp. Sta. Res. Paper 37, 1979.

Darley, E. F., F. R. Burleson, E. H. Mateer, J. T. Middleton, and V. P. Osterli: Contribution of burning of agricultural wastes to photochemical air pollution, *Air Pollution Cont. J.*, **16**(12):685–690 (1966).

Daubenmire, R.: Forest vegetation of northern Idaho and adjacent Washington and its bearing on concepts of vegetation classification, *Ecol. Monogr.*, **22**:301–330 (1952).

——: Vegetative indicators of rate of height growth in ponderosa pine, *For. Science*, **7**:24–34 (1961).

—— and J. B. Daubenmire: Forest vegetation of eastern Washington and northern Idaho, *Wash. Agric. Exp. Sta. Tech. Bul. 60*, 1968.

Davis, L. S.: Investments in loblolly pine clonal seed orchards, *J. For.*, **65**:882–887 (1967).

—— and J. A. Henderson: Ecosym—progress report 1—The conceptual framework, Dept. For. Out. Recrea., Utah State Univ., 1976.

Davis, W. J., and T. T. Kozlowski: Stomatal responses to change in light intensity as influenced by plant water stress, *For. Science*, **21**(2):129–133 (1975).

Day, B. E. The phenoxy herbicides. Council for Agricultural Science and Technology Report No. 39. CAST, Ames, Iowa, 1975.

Day, W. R.: The soil conditions which determine windthrow in forests, *Forestry*, **23**(2):90–95 (1950).

Day, M. W., and V. J. Rudolph: Thinning red pine by percent of height, *Mich. St. Univ. Res. Rep. 125*, 1971.

——, D. P. White, and J. W. Wright: Fertilizer application can improve red pine seed production, *Tree Plant. Notes*, **23**(1):25–27 (1972).

De Bell, D. S., and C. W. Ralston: Release of nitrogen by burning light forest fuels, *Soil Science Soc. Am. Proc.*, **34**(6):936–938 (1970).

Decker, J. P.: Early history of photorespiration, *Eng. Res. Center Bioeng. Bul. 10*, Ariz. State Univ., Tempe, 1970.

Della-Bianca, L., and D. L. Olson, Jr.: Soil-site studies in Piedmont hardwood and pine hardwood upland forests, *For. Science*, **7**:320–329 (1961).

Dengler, A.: Waldbau auf ökologischer Grundlage, Springer-Verlag, Vienna, 1944.

Dewers, R. S., and D. M. Moehring: Effect of soil water stress on initiation of ovulate primordia in loblolly pine, *For. Science*, **16**:219–221 (1970).

Dickinson, C. H., and G. Pugh: Biology of plant litter decomposition, 2 vol., Academic, New York, 1974.

Dickmann, D. I.: Carbohydrate relations, in Tree physiology colloquium, Coop. Extension, Univ. Wisconsin, pp. 111–140.

Dickson, B. A., and R. L. Crocker: A chronosequence of soils and vegetation near Mt. Shasta, California, *J. Soil Science*, **4**(2):123–154 (1953).

Dickson, R. E., J. F. Hosner, and N. W. Hosley: The effects of four water regimes upon the growth of four bottomland tree species, *For. Science*, **11**:299–305 (1965).

Dindal, D. L.: Proc. first soil microcommunities conf., Syracuse, N.Y., CONF.-711076, U. S. Atomic Energy Comm. Tech. Info. Center, 1973.

Dittmer, H. J.: A quantitative study of the subterranean pedology of three field grasses, *Am. J. Bot.*, **25**:654–657 (1938).

Doak, C. C.: Evolution of foliar types, dwarf shoots, and cone scales of *Pinus*, *Illinois Biol. Monogr.*, **13**:1–106 (1935).

Doran, J. C.: Pattern in height growth in *Radiata pine* progenies in Gippsland, Victoria, *Aust. For. Res.*, **6**(3):21–26 (1974).

Dorman, K. W.: Forest tree improvement for Georgia: A problem analysis of research needs in forest tree breeding and forest genetics, *Georgia For. Res. Coun. Rep. 9*, 1962.

Duchaufour, P.: Forest soil dynamics in an Atlantic climate. Laval Univ. Press, Quebec, 1959.

——: Précis de pedologie. Masson et Cie., Paris, 1969.

Duff, G. H., and N. J. Nolan. Growth and morphogenesis in the Canadian forest species. 1. The controls of cambial and apical activity in *Pinus resinosa* Ait. *Canad. Jour. Bot.* **31**(4):471–513 (1953).

Duffield, J. W., and E. G. Snyder: Benefits from hybridizing American forest tree species, *J. For.*, **56**:809–814 (1958).

Dunning, D.: A tree classification forests of the Sierra Nevada, *J. Agric. Res.*, **36**:755–771 (1928).

Durzan, D. J.: Nutrition and water relations of forest trees: A biochemical approach, Proc. 3rd. North Am. For. Biol. Workshop, C.S.U., Colorado, 1974, pp. 15–63.

Dutrow, G. F., and J. R. Saucier: Economics of short rotation sycamore, *USDA Forest Ser. Res. Rep. SO-114*, 1976.

Duvigneaud, P., and S. Denaeyer-DeSmet: Biological cycling of minerals in temperate deciduous forests. *In* D. E. Reichle (ed.), Ecological Studies 1, Analysis of Temperate Forest Ecosystems, Springer-Verlag, Vienna, 1971, pp. 199–225.

Ebell, L. F.: Girdling: its effect on carbohydrate status and on reproductive bud and cone development in Douglas-fir, *Can. J. Bot.*, **44**(3):453–466 (1966).

——: Cone-production and stem growth response of Douglas-fir to rate and frequency of nitrogen fertilization, *Can. J. For. Res.*, **2**(2):327–338 (1972*a*).

——: Cone induction response of Douglas-fir to form of nitrogen fertilizer and time of treatment, *Can. J. For. Res.*, **2**(2):317–326 (1972*b*).

—— and E. E. McMullan: Nitrogenous substances associated with differential cone production responses of Douglas-fir to ammonium and nitrate fertilization, *Can. J. Bot.*, **48**:2169–2177 (1970).

Echols, R. M.: Variation in tracheid length and wood density in races of Scotch pine, *Yale Univ. Bul. 64*, 1958.

——: Patterns of wood density distribution and growth rate in ponderosa pine. *In* Proc. Symp. on the Effect of Growth Acceleration on the Prop. of Wood, For. Prod. Lab., Madison, WI, 1972, pp. H1–H16.

—— and M. T. Conkle: The influence of plantation and seed source elevation on wood specific gravity of 29-year-old ponderosa pines, *For. Science*, **17**(3):388–394 (1971).

Eddington, C.: Genetics: Key to the South's super forest of the future, *Forests and People* (La. For. Assoc.), **19**:6–10 (1969).

Edlin, H. L.: The summerwood-springwood ratio in conifers: How it arises, varies, and affects end use for timber, board and paper, *Forestry*, **38**(1):91–112 (1965).

Einspahr, D. W., M. K. Benson, and M. L. Hardiner: Influence of irrigation and fertilization on growth and wood properties of quaking aspen. *In* Proc. Symp. on the Effect of Growth Acceleration on the Prop. of Wood, For. Prod. Lab., Madison, WI, 1972.

——, R. E. Goddard, and H. S. Gardner: Slash pine, wood and fiber property heritability study, *Silvae Genet.*, **13**(4):103–108 (1964).

Eis, S.: Root grafts and their silvicultural implications, *Can. J. For. Res.*, **2**:111–120 (1972).

——: Cone production of Douglas-fir and grand fir and its climatic requirements, *Can. J. For. Res.*, **3**:61–70 (1973).

Elliott, G. K.: Wood density in conifers, *Tech. Commun. For. Bur., Oxf. 8*, 1970.

Endean, F., and D. Hocking: Performance after planting of four types of container-grown lodgepole pine seedlings, *Can. J. For. Res.*, **3**(2):185–195 (1973).

Erickson, H. D., and G. M. B. Lambert: Effects of fertilization and thinning on chemical composition, growth and specific gravity of young Douglas-fir, *For. Science*, **4**(4):307–315 (1958).

Eyre, F. H., and P. Zehngraf: Red pine management in Minnesota, *USDA Circular 778*, 1948.

Fabricius, L.: Einfluss von Lichtentzug und Wurzelkonkurrenz bei der Kiefer, *Forstwissen. Centralbl.*, **51**:477–506 (1929).

Farrar, J. L.: The use of factor gradients in evaluating site, Proc. Fifth World For. Cong. 1960, 1962, pp. 524–529.

Ferree, M. J., T. D. Shearer, and E. J. Stone, Jr.: A method of evaluating site quality in young red pine plantations, *J. For.*, **56**:328–332 (1958).

Ferrell, William K.: Preliminary report on some of the forest soils of the western white pine region of Idaho, *N.W. Science*, **24**:1–35 (1950).

Fiedler, H. J., and W. Hoffman: R. L. Krutzsch and the beginnings of teaching soil science in Tharandt 150 years ago, *Wiss. Z. Tech. Univ. Dresden*, **17**(6):1715–1725 (1968).

Fielding, J. M.: The seasonal course of height growth and development of *Pinus radiata*, *Aust. For. Res.*, **2**(1):46–50 (1966).

——: The influence of silvicultural practices on wood properties. *In* J. A. Romberger and P. Mikola (eds.), International Review of Forest Research, vol. 2, Academic, New York, 1967, pp. 95–126.

——: Branching and flowering characteristics of Monterey pine, *Aust. For. Timb. Bur. Bul. 37*, 1970.

Fillon, Alphonse: La sylviculture pratique mise en valeur des sols pauvres. Berger-Levrault, Paris, 1890.

Flint, F. F.: Development of the megagametophyte in *Liquidambar styraciflua* L. Madrono, **15**:25–29 (1959).

Florence, R. G., and J. R. McWilliam: The influence of spacing on seed production, its application to forest tree improvement, *Queensland For. Serv. Res. Note 2*, 1954.

Ford-Robinson, F. C. (ed.): Terminology of Forest Science, Technology, Practice and Products, Soc. Am. Foresters, Wash. D.C., 1971.

Forward, D. F., and N. J. Nolan: Growth and morphogenesis in the Canadian forest species. VII Progress and control of longitudinal growth of branches. *In Pinus resinosa Ait, Cana. J. Bot.*, **42**(7):923–950, 1964.

Foster, A. A.: Damage to forests by fungi and insects as affected by fertilizers, in Forest fertilization–theory and practice, T.V.A. Muscle Shoals, AL, 1968, pp. 42–46.

Fowells, H. A.: Silvics of forest trees of the United States, *USDA Agric. Handbook 271*, 1965.

——, and B. M. Kirk: Availability of soil moisture to ponderosa pine, *J. For.*, **43**:601–605 (1945).

——: Forest seed collection zones in California, *Calif. For. and Range Exp. Sta. Note 51*, 1946.

Frank, A. B.: Über die auf Wurzelsymbiose beruhende Ernährung gewisser Bäume durch untererdische Pilze, *Ber. dtsch. bot. Ges.*, **3**:128–145 (1885).

Frank, E. C., and R. Lee: Potential solar beam irradiation on slopes, *Rocky Mt. For. and Range. Exp. Sta. Res. Pap. RM-18*, 1966.

Frederickson, R. L.: Comparative chemical water quality—Natural and disturbed streams following logging and slash burning. *In* J. T. Krygier and J. D. Hall (eds.), Forest land uses and stream environment, Oregon State Univ., Corvallis, 1971, pp. 125–157.

Fricke, K.: Licht und Schattenholzarten ein wissenschaftlich nicht begründetes Dogma, *Clb. g. Forstwesen*, **30**:315–325 (1904).

Froehlich, H. A.: The influence of different thinning systems on damage to soil and trees. *In* G. J. Hamilton (ed.), Aspects of thinning, Proc. IUFRO conf. on economics and harvesting of thinnings, Edinburgh, 1974; *Forestry Comm. Bul. 55*, pp. 102–105, 1976.

Funk, D. T.: Pruning white pine (*Pinus strobus*): a literature review, *U.S. For. Ser., Central St. For. Exp. Sta. Tech. Pap. 185*, 1961.

Funsch, R. W., R. H. Mattson, and G. R. Mowry: CO_2 supplemented atmosphere increases growth of *Pinus strobus* seedlings, *For. Science*, **16**(4):459–460 (1970).

Gates, D. M.: Transpiration and leaf temperature, *Ann. Rev. Plant Physiol.*, **19**:211–239 (1968).

Gessel, S. P., and W. J. Lloyd: Effect of some physical soil properties on Douglas-fir site quality, *J. For.*, **48**:405–410 (1950).

——, K. J. Turnbull, and F. T. Tremblay: How to fertilize trees and measure response, Nat. Pl. Food Inst., Washington, D.C., 1960.

Gevorkiantz, S. R., P. O. Rudolf, and P. J. Zehngraff: A tree classification for aspen, jack pine and second-growth red pine, *J. For.*, **41**:268–274 (1943).

——: Measuring stand normality, *J. For.*, **42**:503–598 (1944).

Gingrich, S. F.: Measuring and evaluating stocking and stand density in upland hardwood forest in the Central States, *For. Science*, **13**:38–53 (1967).

Goodall, D. W. Vegetational classification and vegetational continua. *Angewante Pflanzensoziologie. Veroff. Karntner Landes. Klagenfort*, **1**:168–182 (1954).

Goodman, R. M.: Red pine cone production stimulated by heavy thinning, *U.S. Forest Ser. Lake States For. Exp. Sta. Tech. Notes 628*, 1962.

Grah, R.: 1977, personal communication.

Graham, B. V., Jr., and F. H. Bormann: Natural root grafts, *Bot. Rev.*, **32**(3): 255–292 (1966).

Grano, C.: Precommercial thinning of loblolly pine, *J. For.*, **67**:825–827 (1969).

Gratkowski, H. J., and J. Lauterbach: Releasing Douglas-firs from varnishleaf ceanothus, *J. For.*, **72**(3):150–152 (1974).

Grier, C. C.: Wildfire effects on nutrient distribution and leaching in a coniferous ecosystem, *Can. J. For. Res.*, **5**:599–607 (1975).

Griffing, C. G., and W. W. Elam: Height growth patterns of loblolly pine saplings, *For. Science*, **17**(1):52–54 (1971).

Griffith, B. G., E. W. Hartwell, and T. E. Shaw: The evolution of soils as affected by the old field white pine-mixed hardwood succession in central New England, *Harvard Forest Bul. 15*, 1930.

Gruschow, D. F.: Observations on root systems of planted loblolly pine, *J. For.*, **57**:894–896 (1959).

Hacskaylo, E.: Mycorrhizae, Proc. 1st North Am. Conf. on Mycorrhizae, *U.S. Forest Ser. USDA Misc. Publ. 1189*, 1971.

——: Mycorrhiza: The ultimate in reciprocal parasitism? *BioScience*, **22**(10): 577–583 (1972).

—— and C. M. Tompkins: World literature on mycorrhizae, *Contribution of Reed Herbarium XXII*, 1973.

Hacskaylo, J.: Inorganic deficiency symptoms in white pine (*Pinus strobus*), Ohio Agric. Res. and Dev. Center, Worster, 1962.

Haddock, P. G.: Forestry and soils in British Columbia, paper at Canadian Soc. of Soil Sci. Agric. Inst. of Canada meeting at U.B.C., Vancouver, B.C., June 1965, mimeographed, 1965.

Haig, I. T.: Second growth yield, stand and volume tables for the western white pine type. *USDA For. Ser. Tech. Bul. 323*, 1932.

Hakkila, P.: Branches, stumps and roots as a future raw material source in Finland. *In* H. E. Young (ed.), Forest biomass studies, Proc. of XVth IUFRO Congress, sec. 25, Univ. of Florida, Gainesville, FL, Life Sciences and Agric. Exp. Sta., Univ. of Maine, 1971, pp. 51–62.

Hall, G. S.: Age distribution of needles in red pine crowns, *For. Science*, **12**(3): 369–370 (1966).

Hall, O. F.: Where does thinning fit into the management of Lake states pulpwood stands, *Proc. Soc. Am. For. 1954*, 1955, pp. 70–74.

Halverson, H. G., and J. L. Smith: Controlling solar light and heat in a forest by managing shadow resources, *U.S. Forest Ser. Pacific Southwest For. and Range Exp. Sta. Res. Pap. PSW-102*, 1974.

Hartmann, F.: Forstökologie. Verlag G. Fromme & Co., Vienna, 1952.

Hartwell, H. D.: A comparison of large and small Douglas-fir nursery stock outplanted in potential wildlife damage areas, *Washington State Dept. Nat. Res. Note 6, Forest Land Management Division Contrib. 173*, 1973.

Hastings, W. G.: Revolutionizing nursery practice, *J. For.*, **21**(2):180–182 (1923).

Hatch, M. D., C. B. Osmond, and R. O. Slatyer: Photosynthesis and photorespiration. Wiley, New York, 1971.

Hatchell, G. H., C. W. Ralston, and R. R. Foil: Soil disturbance in logging, *J. For.*, **68**(12):772–775 (1970).

Hay, R. L., and F. W. Woods: Shape of root systems influences survival and growth of loblolly seedlings, *Tree Plant. Notes*, **25**(3):1–2 (1974).

Heiberg, S. O.: Forest soil in relation to silviculture, *J. For.*, **37**(1):42–46 (1939).

Heilmann, P. E., and S. P. Gessel: Nitrogen requirements and the biological cycling of nitrogen in Douglas-fir stands in relationship to the effects of nitrogen fertilization, *Plant and Soil*, **18**:386–402 (1963).

Helms, J. A.: Factors influencing net photosynthesis in trees: an ecological viewpoint. *In* M. G. R. Cannell and F. T. Last (eds.), Tree physiology and yield improvement, Academic, New York, 1976, pp. 55–78.

Henderson, J. A., R. L. Mauk, D. L. Anderson, T. A. Davis, and T. J. Keck: Preliminary forest habitat types of the Uinta Mountains, Dept. of For. and Out. Recreation, Utah State Univ., 1977.

——, R. L. Mauk, D. L. Anderson, R. Ketchie, P. Lawton, S. Simon, R. H. Sperger, R. W. Young, and A. Youngblood: Preliminary forest habitat types of northwestern Utah and adjacent Idaho, Dept. of For. and Out. Recreation, Utah State Univ., 1976.

—— and L. S. Davis: ECOSYM–a classification and information system for wildland resource management, USDA For. Ser., SEAM Programs, Billings, MT, 1977.

Henry, Edmond: Les sols forestiers. Berger-Levrault, Paris, 1908.

Hermann, R. K.: Seasonal variation in sensitivity of Douglas-fir seedlings to exposure of roots, *For. Science*, **13**(2):140–149 (1967).

——, D. P. Lavender, and J. B. Zaerr: Lifting and storing western conifer seedlings, *Oregon State Univ. For. Res. Lab. Res. Pap. 17*, 1972.

Hewlett, J. D., and W. L. Nutter: An outline of forest hydrology. Univ. of Georgia Press, Athens, 1969.

Heyer, G.: Das Verhalten der Waldbäume gegen Licht und Schatten, Erlangen, 1852.

Heyward, F., and R. M. Barnette: Effect of frequent fires on chemical composition of forest soils in the longleaf pine region, *Univ. of Fla. Agric. Exp. Station 265*, 1934.

Hilgard, E. W.: Soils. Macmillan, New York, 1914.

Hills, G. A.: The use of site in forest management, *For. Chron.*, **29**:128–136 (1953).

——: Soil-forest relationships in the site regions of Ontario, *First N.A. For. Soils Conf. 1958*, 1958, pp. 190–212.

——: Regional site research, *For. Chron.*, **36**:401–423 (1960).

——: The classification of forest productivity systems, *Proc. 5th World Forestry Congress*, **1**:572–576 (1962).

Hodges, J. D.: Patterns of photosynthesis under natural conditions, *Ecology*, **48**:234–242 (1967).

Hodgkins, E. J.: Testing soil-site index tables in southwest Alabama, *J. For.*, **54**:261–266 (1956).

——: Forest site classification in the Southeast: An evaluation. Southern forest soils 8th ann. symp., Louisiana State Univ., 1959.

——: Estimating site index for longleaf pine through quantitative evaluation of associated vegetation, *Proc. Soc. Am. For.*, **1960**:28–32 (1960).

——: Productivity estimation by means of plant indicators in the longleaf pine forests of Alabama. Tree growth and forest soils, Third N.A. for. soils conf. 1970, Ore. State Univ. Press, 1970, pp. 461–474.

Holdridge, L. R.: Life zone ecology, Trop. Sci. Center, San Jose, Costa Rica, 1967.

——: Determinination of world plant formations from simple climatic data, *Science*, **105**:367–368 (1947).

——, W. C. Grenke, W. H. Hatheway, T. Liang, and J. A. Tosi, Jr.: Forest environments in tropical life zones–A pilot study. Pergamon, New York, 1971.

Hoover, M. D.: Effect of removal of forest vegetation upon water yields, *Am. Geophys. Union Trans.*, **25**:969–977 (1944).

Horn, H. S.: The adaptive geometry of trees. Princeton U. Press, Princeton, 1971.

Howe, J. P.: Effect of irrigation on wood formed at various heights in ponderosa pine, *Wood Science*, **3**(2):126–218 (1970).

Hoyer, G. E.: Provisonal optimum thinning intensity tables: A basis for thinning yield control of Douglas-fir, *DNR Rep. 6*, State of Washington, Dept. Nat. Res., 1966.

Hoyle, M. C.: Nature and properties of some forest soils in the White Mountains of New Hampshire, *USDA Forest Ser. Res Pap. NE-260*, 1973.

Huber, B., and H. Polster: Zur Frage der physiologischen Ursachen der unterschiedlichen Stofferzeugung von Pappelklonen, *Biol. Zbl.*, **74**(7/8):370–420 (1955).

—— and J. Rüsch: Über den Anteil von Assimilation und Atmung bei Pappelblattern, *Ber. deut. bot. Ges.*, **74**(2):55–62 (1961).

Hughes, F. E.: Tension wood: A review of literature. *For. Abstr.*, **26**(1):2–9; **26**(2):179–186 (1965).

Hughes, J.: Regeneration opportunities and alternatives, in Southern pine management—Today and tomorrow, 20th Annual For. Symp., Louisiana State Univ., Baton Rouge, LA, 1971, pp. 21–31.

Huhta, V., E. Karppinen, M. Nurminen, and A. Valpas: Effect of silvicultural practices upon arthropod, annelid and nematode populations in coniferous forest soil, *Ann. Zool. Fennici.*, **4**(2):87–143 (1967).

Hummel, F. C., G. M. L. Locke, J. N. R. Jeffers, and J. M. Christie: Code of sample plot procedure. *U.S. For. Comm. Bul. 31*, 1959.

Humphreys, N.: Value resulting from cultural techniques of thinning and pruning, Proc. symp. on *P. radiata*, Aust. Nat. Univ., 1971.

Hyland, Fay: Fiber analysis and distribution in the leaves, juvenile stems, and roots of ten Maine trees and shrubs, *Life Sci. and Agric. Exp. Sta. Tech. Bul. 71*, 1974.

Hyun, S. K.: Developing advanced generation breeding population for a hybrid breeding program, Inter. union for. res. org., sec. 22, Working group on quantitative genetics, Gainesville, FL, 1971.

Ilyer, J. G.: Renovation of fertility of biocide-treated soils, Proc. int. symp. on soil fertility evaluation, New Delhi, vol. 1. 1971, pp. 613–618.

—— and S. A. Wilde: Fertility of forest soils: Its concept and reality, Proc. 8th int. cong. soil sci., Bucharest, Romania, 1964, pp. 1003–1007.

Isaac, L. A.: Reproductive habits of Douglas-fir, Charles L. Pack Forestry Foundation, 1943.

Isakov, I. P.: Estimating the amount of moisture transpired by stands of *Pinus sibirica*, *Leoovedenie*, **1**:29–36 (1974).

Ivanov, A., A. Siina, and Y. G. Tselniker: Determination of the transpiration capacity of forest cover, *Bot Zhur. U.S.S.R.*, **36**:5–20 (1952); *Bot. Abst.*, **26**:7125 (1951).

Jackson, D. I., and G. B. Sweet: Flower initiation in temperate woody plants, *Hortic. Abstr.*, **42**:9–24 (1972).

Jackson, L. W. R.: Relation of leaf structure to shade tolerance of dicotyledonous tree species, *For. Science*, **13**(3):321–323 (1967).

Jarvis, P. G., and M. S. Jarvis: Growth rates of woody plants, *Physiol. Plant.*, **17**:654–666 (1964).

Jenkinson, J. L.: Seasonal patterns of root growth capacity in western yellow pines, in Proc. natl. conv. soc. Am. for., Washington, D.C., Sept. 28–Oct. 2, 1975, pp. 445–453.

Jenny, H.: Factors of soil formation. McGraw-Hill, New York, 1941.

—— and R. Overstreet: Surface migration of ions and contact exchange, *J. Phys. Chem.*, **43**:1185–1196 (1939).

Jensen, L. A., and Noll, E.: Experience in germination testing of Pacific Northwest Douglas-fir seed, *Oregon Agr. Exp. Sta. Tech. Pap. 1236*, 1959.

Johnson, J. E.: The public health implications of widespread use of the phenoxy herbicides and picloram, *BioScience*, **21**(17):899–905 (1971).

Johnson, P. S.: Growth and survival of interplanted hardwoods in southern Wisconsin oak clear cuttings, *U.S. Forest Ser. Res. Note NC-118*, 1971.

Johnsson, H.: Forest tree breeding by selection: Clonal seed orchards, seedling seed orchards, progeny tests, *Silvae Genet.*, **13**:41–49 (1964).

Jones, E. P., Jr., and O. O. Wells: Ice damage in a Georgia planting of loblolly pine from different seed sources, *U.S. Forest Ser. Res. Note SE-126*, 1969.

Jones, R.: Review and comparison of site evaluation methods, *USDA Forest Ser. Res. Pap. RM-51*, 1969.

Jorgensen, J. R., C. G. Wells, and L. J. Metz: The nutrient cycle: Key to continuous forest production, *J. For.*, **73**(7):400–403 (1975).

Karizumi, N.: The mechanism and function of tree root in the process of forest production. I. Method of investigation and estimation of the root biomass, *Bul. Gov. For. Exp. Sta. 259*, Meguro, Japan, 1974, pp. 1–97.

Karlsson, I.: Delayed incompatibility in grafted clones of Douglas-fir, *B.C. For. Ser. Res. Note 52*, 1970.

Keays, J. L.: Complete-tree utilization, resumé of a literature review. *In* Forest Biomass Studies. Section 25, Yield and Growth. XVth Int. Union For. Res. Org. Conf., Univ. of Florida, Gainesville, Fla. Published by Life Sci. and Agric. Expt. Sta., Univ. Maine, Orono. pp 93–102. 1971.

—— and J. V. Hatton: Complete tree utilization studies: II Yield and quality of kraft pulp from the components of *Picea glauca, Tappi*, **54**(10):1721–1724 (1971).

Keen, F. P.: Relative susceptibility of ponderosa pine to bark-beetle attack, *J. For.*, **34**:919–927 (1936).

Keeves, A.: Some evidence of loss of productivity with successive rotation of *P. radiata* in the South-East of South Australia, *Aust. For.*, **30**(1):51–63 (1966).

Keith, J. O., R. M. Hansen, and A. L. Ward: Effect of 2,4-D on abundance and foods of pocket gophers, *J. Wildlife Mgt.*, **23**:137–145 (1959).

Kellman, M. C.: The viable seed content of some forest soils in coastal British Columbia, *Can. J. Bot.*, **48**(7):1383–1385 (1970).

Kessel, S. L.: Soil organisms. The dependence of certain pine species on a biological soil factor, *Emp. For. J.*, **5**:70–74 (1927).

Kimmins, J. P.: Evaluation of the consequences for future tree productivity of the loss of nutrients in whole-tree harvesting, *For. Ecol. and Mgt.*, **1**:169–183 (1977).

Kinlock, B. B., Jr., G. K. Parks, and C. W. Fowler: White pine blister rust: Simply inherited resistance in sugar pines, *Science*, **167**:193–195 (1970).

Kiss, G., and O. Sziklai: Development of Douglas-fir seed and pollen cones, *Proc. West. For. Genet. Assoc.*, 1965, pp. 55–59.

Kittridge, J.: Forest Influences. McGraw-Hill, New York, 1948.

Kitzmiller, J. H.: The tree improvement master plan for the California region, USDA Forest Service, 1976.

Klem, G. S.: Quality of wood from fertilized forests, *Tappi*, **51**(11):99A–103A (1968).

Klemmedson, J. O., A. M. Schultz, H. Jenny, and H. H. Biswell: Effect of prescribed burning of forest litter on total soil nitrogen, *Soil Sci. Soc. Am. Proc.*, **26**(2):200–202 (1962).

Klock, G. O.: Forest and range soils research in Oregon and Washington. A bibliography with abstracts 1964–1968, *USDA Forest Ser. Pap. PNW-90*, 1969.

Knight, H.: Loss of nitrogen from the forest floor by burning, *For. Chron.*, **42**(2):149–152 (1966).

Koch, P.: Utilization of the southern pines. *USDA Handbook 420*, 1972.

Kojima, S., and V. J. Krajina: Vegetation and environment of the coastal western hemlock zone in Strathcona Provincial Park, *Syesis*, **8**:1–123 (1975).

Köppen, W.: Die Klimate die Erde. Bruytero, Berlin, 1923.

Köstler, J.: Waldbau. Parey Verlag, Berlin, 1950.

Kozak, A., O. Sziklai, B. G. Griffith, and J. H. G. Smith: Variation in cone and seed yield from young open-grown Douglas-firs on the U.B.C. Research Forest, *Univ. British Columbia For. Res. Pap. 57*, 1963.

Kozlowski, T. T.: Growth and development of trees, vol. II, Cambial growth, root growth, and reproductive growth. Academic, New York, 1971.

—— and C. E. Ahlgren (eds.): Fire and ecosystems. Academic, New York, 1974.

—— and T. Keller: Food relations of woody plants, *Bot. Rev.*, **12**(4):293–382 (1966).

——, J. H. Torrie, and P. E. Marshall: Predictability of shoot length from bud size in *Pinus resinosa Ait.*, *Can. J. For. Res.*, **3**(1):34–38 (1973).

—— and C. H. Winget: Patterns of water movement in forest trees, *Bot. Gaz.*, **124**:301–311 (1963).

Kraft, G.: Zur Lehre von den Durchforstungen, Schlagstellungen und Lichtungshieben, Hannover, 1884 [from Denger, 1944].

Krajicek, J. E., K. A. Brinkman, and S. F. Gingrich: Crown competition–A measure of density, *For. Science*, **7**:35–42 (1961).

Krajina, V. J.: Ecology of forest trees in British Columbia. *Ecol. West. N. Amer.*, **2**:1–146 (1969).

——: Ecosystem perspectives in forestry, M. R. MacMillan Lectureship in Forestry, Univ. of British Columbia, 1972.

Kramer, P. J.: Plant and soil water relationships: A modern synthesis. McGraw-Hill, New York, 1969.

—— and W. S. Clark: A comparison of photosynthesis in individual pine needles and entire seedling at various light intensities, *Plant Physiol.*, **22**:51–57 (1947).

—— and J. P. Decker: Relation between light intensity and rate of photosynthesis of loblolly pine and certain hardwoods, *Plant Physiol.*, **19**:350–358 (1944).

—— and T. T. Kozlowski: Physiology of trees. McGraw-Hill, New York, 1960.

Krammes, J. S., and D. B. Willets: Effect of 2,4-D and 2,4,5-T on water quality after a spraying treatment, *USDA Forest Ser. Res. Note. PSW-52*, 1964.

Krefting, L. W., and H. L. Hansen: Increasing browse for deer by aerial applications of 2,4-D, *J. Wildlife Mgt.*, **33**(4):784–790 (1969).

Krueger, K. W., and R. H. Ruth: Comparative photosynthesis of red alder, Douglas-fir, Sitka spruce, and western hemlock seedlings, *Can. J. Bot.*, 47(4):519–527 (1969).

Krugman, S. L., W. I. Stein, and D. M. Schmitt: Seed biology. *In* C. S. Schopmeyer (ed.), Seeds of woody plants in the United States, *USDA Handbook 450*, 1974, pp. 5–40.

Küchler, A. W.: Manual for use with potential natural vegetation of the conterminous United States, *Amer. Geog. Soc. Spec. Publ. 36*, 1964.

Kuntz, J. E.: Root growth, in Tree physiology colloquium, Univ. of Wisc. Extension, 1973, pp. 71–84.

Laatsch, W.: The formation of fertile forest soils, *Mitt. St. Forstverw. Bayerns*, **34**:274–286 (1964).

Lane, C.: Forest stand conversion from hardwoods to pines—Effects on soil nutrients, micro-organisms and forest floor weights during the first seven years, *For. Science*, **21**(2):155–163 (1975).

Langsaeter, A.: Om tynning i enaldret gran- og furuskog (About thinning in even-aged stands of spruce, fir and pine), *Meddel. f. d. Norske Skogforsøksvesen*, **8**:131–216 (1941).

Lanner, R. M.: Living stumps in the Sierra Nevada, *Ecology*, **42**:170–173 (1961).

——: The pine shoot primary growth system, *Dissert. Abstr.*, **30B**(1):13 (1969).

——: Shoot growth patterns of loblolly pine, *For. Science*, **17**(4):486–487 (1971).

——: Patterns of shoot development in *Pinus* and their relationship to growth potential. *In* M. G. R. Cannell and F. T. Last (eds.), Tree physiology and yield improvement, Academic, New York, 1976, pp. 223–243.

Larcher, W.: Physiological plant ecology. Springer-Verlag, Vienna, 1975.

Larson, P. R.: A physiological consideration of the springwood-summerwood transition in red pine, *For. Science*, **6**:110–122 (1960).

——: Auxin gradients and regulation of cambial activity, in T. T. Kozlowski (ed.), Tree Growth, Ronald, New York, 1962, pp. 97–117.

——: The indirect effect of drought on tracheid diameter in red pine, *For. Science*, **9**(1):52–62 (1963).

——: Some indirect effects of environment on wood formation. *In* M. H. Zimmermann (ed.), The formation of wood in forest trees, Academic, New York, 1964, pp. 345–365.

——: Wood formation and the concept of wood quality, *Bul. Yale Sch. For.*, **74**:54 (1969).

—— and J. C. Gordon: Photosynthesis and wood yield, *Agric. Science Rev.*, **7**:7–14, 1969.

Lassen, L. E., and E. A. Okkonen: Effect of rainfall and elevation on specific gravity of coast Douglas-fir, *Wood and Fiber*, **1**(3):227–235 (1969).

Lavender, D. P., and G. H. Atherton: Sonic treatment of Douglas-fir and ponderosa pine seeds and seedlings, *Oregon State Univ. For. Res. Lab. Res Note. 48*, 1964.

Lawrence, W. H., and J. H. Rediske: Fate of sown Douglas-fir seed, *For. Science*, 8(3):210–218 (1962).

Leaf, A. L.: Plant analysis as an aid in fertilizing forests. *In* L. M. Walsh and J. D. Beaton (eds.): Soil testing and plant analysis, Soil Sci. Soc. Am., Madison, WI, 1973.

——, C. B. Davey, and G. K. Voigt: Forest soil organic matter and nutrient element dynamics, Proc. first soil microcommunities conf., Syracuse, N.Y., CONF.-711076, U.S. Atomic Energy Comm. Tech. Info. Center, 1973, pp. 33–46.

Leaphart, C. D., and M. A. Grismer: Extent of roots in the forest soil mantle, *J. For.*, 72(6):358–359 (1974).

Ledig, F. T., and T. O. Perry: Net assimilation rate and growth in loblolly pine seedlings, *For. Science*, **15**(4):431–438 (1969).

Lee, Y.: A review of research literature on forest fertilization, *For. Res. Lab. Inform. Rep. BC-X-18*, 1968.

Leith, H. von, and M. Vogt: Der Lichtkompensationspunkt einiger Waldschattenpflanzen im Sommer und im Frühjahr, *Abstr. in Proc. 9th Int. Bot. Congr. Montreal*, **2**:227–228 (1959).

Lester, D. T.: Variation in cone production of red pine in relation to weather, *Can. J. Bot.*, **45**:1683–1691 (1967).

Lexen, B.: Bole area as an expression of growing stock, *J. For.*, **41**:883–885 (1943).

Libby, W. J.: The backcross hybrid Jeffrey X (Jeffrey X Coulter) pine, *J. For.*, **56**:840–842 (1958).

——, W. J. Brener, D. Cole, L. Pugsley, W. I. Stein, and C. E. Farnsworth: Biological methods of securing genetic gains in forest trees, Soc. Am. For., Tree Seed Committee Report, 1968.

——, R. F. Stettler, and F. W. Seitz: Forest genetics and forest tree breeding, *Ann. Rev. Genet.*, **3**:469–494 (1969).

Liebig, J. von: Die Grundsatze der Agriculturchemie. Pub. F. Vieweg und Sohn, Braunschweig, 1855.

Likens, G. E., F. H. Bormann, N. M. Johnson, D. W. Fischer, and R. S. Pierce: Effects of forest cutting and herbicide treatment on nutrient budgets in Hubbard Brook watershed-ecosystem, *Ecol. Mono.*, **40**:23–47 (1970).

Lindsey, A. A.: Success of the Holdridge life zone model in yielding potential evapotranspiration estimates for U.S. weather stations, unpublished, 1978.

—— and J. O. Sawyer, Jr.: Vegetation-climate relationships in Eastern United States, *Proc. Indiana Acad. Sci.*, **80**:210–214 (1971).

Liocourt, F. de: De l'aménagement des sapinières, *Société Forestière de Franche-Comté et Belfort Bull.*, **6**:396–405 (1898).

Logan, K. T.: Monthly variations in photosynthetic rate of jack pine provenances in relation to their height, *Can. J. For. Res.*, **1**:256–261 (1971).

Lohrey, R. E.: Planted pines grow better than seeded pines on hardwood dominated site, *Tree Plant. Notes*, **24**(2):12–13 (1973).

Lovengreen, J. A.: (Analysis of a completed sample plot in Norway spruce), *Det. Forst. Forsogs i Denmark*, **20**:355–372 (1951). (from O. F. Hall, 1970).

Low, A. J.: A study of compression wood in Scots pine (*Pinus sylvestris*), *Forestry*, **37**(2):179–201 (1964).

—— and G. van Tol: Initial spacing in relation to stand establishment. *In* Symposium on stand establishment, IUFRO Joint Meeting Divisions 1 and 3 (Wageningen, The Netherlands, Oct. 15–19, 1974), pp. 296–314.

Lowery, D. P.: Spiral grain patterns in Douglas-fir, *Proc. Montana Acad. Sci., Missoula*, 25:62–67 (1965).

Lowry, W.: Apparent meteorological requirements for abundant cone crop in Douglas-fir, *For. Science*, **12**:185–192 (1966).

Ludlow, M. M., and P. G. Jarvis: Methods for measuring photorespiration in leaves. *In* Z. Sestak, J. Catsky, and P. G. Jarvis (eds.), Plant photosynthetic production: manual of methods, pp. 294–315. Dr. W. Junk N. V. The Hague, 1971.

Lunt, H. A.: Forest soil problems in New England, *Ecology*, **19**(1):50–56 (1938).

——: The forest soils of Connecticut, *Conn. Agric. Exp. Sta. Bul. 523*, Storrs, 1948.

Lutz, H. J.: Disturbance of forest soil resulting from the uprooting of trees, *Yale Univ. School of Forestry Bul. 45*, 1940.

—— and R. F. Chandler: Forest soils. Wiley, New York, 1946.

——, G. E. Beyers, and T. J. Sheets: The persistence and movement of picloram and 2,4,5-T in soils, *J. Environ. Qual.*, 2(4):485 (1973).

Lyford, W. H.: Forest soil microtopography, Proc. first soil microcommunities conf., Syracuse, N.Y., CONF. -711076, U.S. Atomic Energy Comm. Tech. Info. Center, 1973, pp. 47–58.

—— and D. W. MacLean: Mount and pit microrelief in relation to soil disturbance and tree distribution in New Brunswick, Canada, *Harvard Forest Pap. 15*, 1966.

Lynch, D. W.: Effects of stocking on site measurement and yield of second-growth ponderosa pine in the Inland Empire, *Int. Mtn. For. and Range Exp. Sta., U.S. Forest Ser. Res. Pap. 56*, 1958.

Lyr, H., and G. Hoffmann: Growth rates and growth periodicity of tree roots, *Int. Rev. For. Res. N.Y.*, 2:181–236 (1967).

McArdle, R. E., W. H. Meyer, and D. Bruce: The yield of Douglas-fir in the Pacific Northwest, *USDA Forest Ser. Tech. Bul. 201*, 1949.

McCree, K. J., and J. H. Troughton: Prediction of growth rate at different light levels from measured photosynthesis and respiration rates, *Plant Physiol.*, **41**:559-566 (1966).

McDougall, G. A.: Radial growth studies, *For. Ent. Path. Br. Dept. For. Canada Prog. Rept. 19*, 1963.

McLintock, T. F.: Hardpans in the spruce-fir forests of Maine, *Soc. Am. Foresters Proc. 1957*, 1957, pp. 65–66.

—— and C. A. Bickford: A proposed site index for red spruce in the Northeast, *U.S. Forest Ser. Northeastern Exp. Sta. Pap. 93*, 1957.

Mackenzie, A. M.: The Bowmont Norway spruce sampling plots, *Forestry*, **35**:129–138 (1962).

Maki, T. E.: Need for fertilizers in wood production, *Unasylva*, **20**(3)(82):49–54 (1966).

Mancini, F.: The soils of the forest of Paneveggio, *Ann. Acad. Ital. Science For.*, 8:373–454 (1959).

Mann, W. F., Jr.: Guidelines for direct seedling loblolly pine, *U.S. Forest Ser. Southern For. Exp. Sta. Occas. Pap. 188*, 1961.

——, T. E. Campbell, and J. P. Barnett: Loss of endrin from repellent-coated pine seed, *U.S. Forest Ser. Res Note SO-135*, 1971.

Marquis, D. A.: Scarify soil during logging to increase birch reproduction, *Northern Logger*, **14**:424–42 (1965).

Marx, D. H.: Ectomycorrhizae as biological deterrents to pathogenic root infection. *In* Mycorrhizae, North Am. conf. mycorrhizae proc., 1, *USDA Forest Ser. Misc. Publ.*, **1189**:81–96 (1971).

——: Mycorrhizae and establishment of trees on strip-mined land, *Ohio J. Science*, **75**(6):288–297 (1975).

——: Mycorrhizae of exotic trees in the Peruvian Andes and synthesis of ectomycorrhizae on Mexican pines, *For. Science*, **21**(4):353–358 (1975).

—— and C. B. Davey: The influence of ectotrophic mycorrhizal fungi on the resistance of pine roots to pathogenic infections. III. Resistance of aseptically formed mycorrhizae to infection by *Phytophthora cinnamomi*, *Phytopathology*, **59**:549–558 (1969).

Mason, D. T.: The life history of lodgepole pine in the Rocky Mountains, *USDA Forest Ser. Bul. 154*, 1915.

Matthews, D. M.: Ipil-ipil, a firewood and reforestation crop [*Leucaena glauca* (*L.*) Benth.], *Philippine Isl. Bur. For. Bul. 13*, 1914.

Matthews, J. D.: Factors affecting the production of seed by forest trees, *For. Abstr.*, **24**:i–xiii (1973).

Mayr, H.: Waldbau auf naturgesetzlicher Grundlage. Parey, Berlin, 1909.

Megraw, R. A., and W. T. Nearn: Detailed DBH density profiles of several trees from Douglas-fir fertilizer/thinning plots. *In* Proc. of the symp. on the effect of growth acceleration on the prop. of wood, For. Prod. Lab., Madison, WI, 1972, pp. G1–G24.

Meinzer, O. E.: Hydrology. Dover, New York, 1942, pp. 277–278 dealing with transpiration ratios.

Melin, E., 1955, quoted by Hacskaylo, E.: Metabolic exchanges in ectomycorrhizae, in Mycorrhizae, *U.S. Forest Service Misc. Publ.* **1189**:175–182 (1971).

—— and H. Nilson: Transfer of radioactive phosphorus to pine seedlings by means of mycorrhizae hyphae, *Physiol. Plant.*, **3**:88–92 (1950).

Merriam, C. H.: Life zones and crop zones of the United States, *USDA Biol. Sur. Bul. 10*, 1898.

Merritt, C.: Effect of environment and heredity on the root-growth pattern of red pine, *Ecology*, **49**(1):34–40 (1968).

Meyer, A. C.: The elements of hydrology. Wiley, New York, 1928.

Meyer, H. A.: Structure, growth and drain in balanced uneven-aged forests, *J. Forestry*, **50**:85–92 (1952).

Meyer, W. H.: Diameter distribution series in even-aged forest stands, *Yale Univ. School For. Bull. 28*, 1930.

——: Yields of second growth spruce and fir in the Northeast, *USDA Forest Ser. Tech. Bul. 142*, 1929.

——: Yield of even-aged stands of Sitka spruce and western hemlock, *USDA Forest Ser. Tech. Bul. 544*, 1937.

——: Yield of even-aged stands of ponderosa pine, *USDA Forest Ser. Tech. Bul. 630*, 1938.

Mikola, P.: Application of mycorrhizal symbiosis in forestry practice. *In* G. C. Marks and T. T. Kozlowski (eds.), Ectomycorrhizae, Academic, New York, 1973, pp. 383–411.

——, O. Laiho, J. Erikäinen, and K. Kuvaja: The effects of slash burning on the commencement of mycorrhizal association, *Acta. Forest. Fenn.*, **77**(3):3–12 (1964).

Mirov, N. T., and F. S. Baker: Physiology of forest tree seed and seedlings. Unpubl. manuscript. Univ. Calif., Berkeley. 1942.

Mitchell, H. C.: Regulation of farm woodlands by rule of thumb, *J. For.*, **41**:243–248 (1943).

Mitchell, H. L.: Patterns of specific gravity variation in North American conifers, *Proc. Soc. Am. For.*, **1964**:169–179 (1965).

——: Effect of nitrogen fertilizers in the growth rate and certain quality characteristics of sawlog size red oak, yellow poplar and white oak. *In* Proc. of symp on the effect of growth acceleration on the prop. of wood, For. Prod. Lab., Madison, WI, 1972, pp. J1–J15.

—— and P. R. Wheeler: Specific gravity—A measure of intrinsic wood quality, *Proc. Soc. Am. For.*, **1959**:53–57 (1959).

Mitscherlich, E. A.: Das Wirkungsgesetz der Wachstumfaktoren, *Landwirtschaft Jahrbuch Bog.*, **11**:15 (1921).

Moir, W. H., and E. P. Bachelard: Distribution of fine roots in three *Pinus radiata* plantations near Canberra, Australia, *Ecology*, **50**:658–662 (1969).

Möller, C. M.: Untersuchungen über Laubmenge, Stoffverlust und Stoffproduktion des Waldes, *Forst. Forsogs. Danm.* **17**:1–287, (1945).

——· The effect of thinning, age and site on foliage, increment and loss of dry matter, *J. For.*, **45**:393-404, (1947).

Monteith, M. L.: Dew: facts and fallacies, in The water relations of plants, Blackwell Science, London, 1963.

Montgomery, M. M., and L. A. Norris: A preliminary evaluation of the hazards of 2,4,5-T in the forest environment, *USDA Forest Ser. Res. Note PNW-116*, 1970.

Moore, R. R.: A study of stand structure in the uneven-aged stands in the Engelmann spruce–subalpine fir type of the Utah State University forest, Ph.D. dissertation, Univ. of Washington, Seattle, 1964.

Mullin, R. E.: Some effects of root dipping, root exposure and extended planting dates with white spruce, *For. Chron.*, **47**(2):90–93 (1971).

——: Effects of root exposure on establishment and growth of outplanted trees, Int. symp. on ecology and physiology of root growth, Biologischen Gesselschaft der Deutschen Demokratischen Republik, Sec. Physiologie, G. Hoffman, (ed.), pp. 202–229, Akademie-Verlag, Berlin, 1974.

—— and C. P. Howard: Transplants do better than seedlings and . . . *For. Chron.*, **49**(5):213–218 (1973).

Mustanoja, K. J., and A. L. Leaf: Forest fertilization research 1957–1964, *Bot. Rev.*, **31**(2):151-246 (1965).

Myers, C. A.: Yield tables for managed stands—with special reference to the

Black Hills," *U.S. Forest Ser., Rocky Mt. For. Range Exp. Sta. Res. Paper RM-21*, 1966.

Myers, C. A.: Yield tables for managed stands of lodgepole pine in Colorado and Wyoming, *U.S. Forest Ser., Rocky Mt. For. Range Exp. Sta. Res. Paper RM-26*, 1967.

Myers, C. A., C. B. Edminster, and F. G. Hawksworth: SWYLD2: Yield tables for even-aged and two-storied stands of southwestern ponderosa pine, including effects of dwarf mistletoe, *U.S. Forest Ser., Rocky Mt. For. Range Exp. Sta. Res. Paper RM-163*, 1976.

Namkoong, G.: Nonoptimality of local races, Proc. 10th S. conf. on for. tree improvement, 1969, pp. 149–153.

Naskoweak, A. F.: Spiral grain in trees—a review, *For. Prod. J.*, **13**(7):266–275 (1963).

Nelson, T. C., and F. A. Bennett: A critical look at the normality concept, *J. For.*, **63**:107–109 (1965).

Newton, M.: Forest rehabilitation in North America: Some simplifications, *J. For.*, **71**:159–162, (1973).

——: Constructive use of herbicides in forest resource management, *J. For.*, **73**(6):329–336 (1975).

—— and H. C. Black: Large planting stock of Douglas-fir helps evade damage by animals and sprouting brush on favorable sites, *Res. Progress Rep. Western Weed Control Conf.*, 1965.

—— and J. D. Norgren: Silvicultural chemicals and protection of water quality. U.S. EPA, Seattle, Report EPA 910/9-77-036, 1977.

—— and L. A. Norris: Evaluating short- and long-term effects of herbicides on non-target forest and range biota, *Down to Earth*, **32**(3):18–26 (1976).

Nicholls, J. W. P.: The effect of environmental factors on wood characteristics. I. The influence of irrigation on *Pinus radiata* from South Australia, *Silvae Genet.*, **20**(1/2):26–33 (1971*a*).

——: The effect of environmental factors on wood characteristics, *Silvae Genet.*, **20**(3):67–73 (1971*b*).

Norris, L. A.: Chemical brush control and herbicide residues in the forest environment. *In* M. Newton (ed.), Herbicides and vegetation management in forests, ranges and noncrop lands, Oregon State Univ. School of Forestry, Corvallis, OR, 1967.

——: Chemical brush control, assessing the hazard, *J. For.*, **69**:715–720 (1971).

Northcott, P. L.: Is spiral grain the normal growth pattern? *For. Chron.*, **33**(4): 335–352 (1957).

Odum, E. P.: The strategy of ecosystem development, *Science*, **164**:262–270 (1969).

Owens, J. N.: The relative importance of initiation and early development on cone production in Douglas-fir, *Can. J. Bot.*, **47**:1039–1049 (1969).

—— and F. H. Smith: The initiation and early development of the seed cone of Douglas-fir, *Can. J. Bot.*, **42**:1031–1047 (1964).

Ovington, J. D.: Some aspects of energy flow in plantations of *Pinus sylvestris*, *Ann. Bot. Lond.*, **25**:15–20 (1961).

Panshin, A. J., and C. de Zeeuw: Textbook of wood technology, 3rd ed. McGraw-Hill, New York, 1970.

Parmeter, J. R.: Ecological considerations in even-age management: Microbiology and pathology. *In* R. K. Hermann and D. P. Lavender (eds.), Even-age management, Proc. symp. Oregon State Univ., 1973, pp. 113–135.

——— and B. Uhrenholdt: Some effects of pine-needle or grass smoke on fungi, *Phytopath.*, **65**(1):28–31 (1975).

Pawsey, C. K.: Height and diameter growth cycles in *Pinus radiata*, *Aust. For. Res.*, **1**(1):3–8 (1964).

———: Survival and early development of *P. radiata* as influenced by size of planting stock, *Aust. For. Res.*, **5**(4):13–24 (1972).

Pearson, G. A.: Management of ponderosa pine in the southwest, *USDA Forest Ser. Agric. Mono. 6*, 1950.

Pelisek, J.: Classification and characteristics of the dynamics of soil moisture in the forest regions of Czechoslovakia, *Lesnictri*, **31**(8):689–712 (1958).

Perry, T. V., and C. Wang: The value of genetically superior seed, *J. For.*, **56**: 843–845 (1958).

Pfister, R. D., B. L. Kovalchik, S. F. Arno, and R. D. Presby: Forest habitat types of Montana, *Interm. For. and Range Exp. Sta., U.S. Forest Ser.*, 1974.

Pflugbeil, Ernst: Proposed site indices for Englemann spruce on the College Forest of the Utah State University, M.S. thesis, Utah State Univ., 1960.

Pharis, R. P.: The roles of gibberellin and other phytohormones in strobilis induction, sexuality and development, IUFRO Section 22, Varparanto, Finland, 1970, pp. 1–9.

———, S. D. Ross, and R. L. Wample: Promotion of flowering in seedling Douglas-fir and lodgepole pine by certain of the gibberellins, alone, in combination with each other, and with an auxin. *In* C. P. P. Reid and G. H. Fechner (eds.), Proc. 3rd, N. Am. for. biol. workshop, Colorado State Univ., Fort Collins, 1974.

——— and W. Morf: Physiology of gibberellin-induced flowering in conifers. *In* E. F. Wightman and G. S. Setterfield (eds.), Biochemistry and physiology of plant growth substances, pp. 1341–1356, Runge Press, Ottawa, 1968.

——— and J. N. Owens: Hormonal induction of flowering in conifers, *Yale Sci. Mag.*, **41**:10–19 (1966).

Pillow, M. Y.: Specific gravity relative to characteristics of annual rings in loblolly pine, *USDA Forest Prod. Lab. Tech. Bul. 1989*, 1954.

Pollard, D. F., and K. T. Logan: The role of free growth in the differentiation of provenances of black spruce, *Picea mariana* (Mill.) B.S.P., *Can. J. For. Res.*, **4**(3):308–311 (1974).

Polster, H.: Die physiologischen Grundlagen der Stofferzeugung in Walde-Untersuchungen über Assimilation, Respiration und Transpiration unser Hauptholzarten, Bayerische Landwirtschaftsverlag GmbH, Munich, 1950.

Popescu-Zeletin, J.: Die Kontrollmethode, Beitrage zur Auffassung ihrer rechnerischen Grundlagen, *Allg. Forst-u. Jagd-Zeit'g*, **112**:135–147, 196–211, 233–254, *U.S. Forest Serv. Div. Silvic. Trans. 300*, 1936.

Porterfield, R. L.: A goal programming model to guide and evaluate tree improvement programs, *For. Science*, **22**(4):417–430 (1976).

———, B. J. Zobel, and F. T. Ledig: Evaluating the efficiency of tree improvement programs, *Silvae Genet.*, **24**(2–3):33–44 (1975).

Posey, C. E.: The effects of fertilization upon wood properties of loblolly pine (*Pinus taeda L.*), *Sch. For. N.C. State Univ. Tech. Rep. 22*, 1964.

Power, D. R., and J. L. Smith: Partial girdling multiplies short-leaf cones. *U.S. Forest Ser. Southern For. Exp. Sta. For. Notes 132*, 1961.

Pratt, P. F.: Effect of fertilizers and organic materials on the cation-exchange capacity of an irrigated soil, *Soil Science*, **83**:85–89 (1957).

President's Science Advisory Committee: A report of the panel on herbicides, Exec. Office of the President, Office of Science and Technology, Washington, D.C., 1971.

Puritch, G. S.: Cone production in conifers—A review of the literature and evaluation of research needs, *Can. For. Serv. Information Pap. BC-X-65*, 1972.

——: Effect of waters stress on photosynthesis, respiration, and transpiration of four *Abies* species, *Can. J. For. Res.*, **3**:293–298 (1973).

Radwan, M. A., and W. D. Ellis: Factors affecting endrin content of endrin-coated Douglas-fir seed, *Northwest Science*, **45**(3):188–192 (1971).

Read, R. A.: Tree species occurrence as influenced by geology and soil on the Ozark north slope, *Ecology*, **33**:239–246 (1952).

Reigner, I. C., W. Supper, and R. Johnson: Will the use of 2,4,5-T to control streamside vegetation contaminate public water supplies? *J. For.*, **66**:914–918 (1968).

Reineke, L. H.: Perfecting a stand-density index for even-aged forests, *J. Agric. Res.*, **46**:627–38 (1933).

Remezov, N. P., and P. S. Pgrebnyak: Forest soil science, Israel Prog. for Scientific Translation, Jerusalem, 1969.

Renbuss, M. A., G. A. Chilvers, and L. D. Pryor: Microbiology of an ashbed, *Proc. Linnean Soc. of NSW*, **97**(4):302–310 (1973).

Reukema, D. L.: Seed production of Douglas-fir increased by thinning, *PNW For. Range Exp. Sta. Res. Notes 210*, 1961.

Reukema, D. and D. Bruce: Effects of thinning on yield of Douglas-fir: Concepts and some estimates obtained by simulation. *U.S. Forest Ser., Pacific Northwest For. Range Exp. Sta. Gen. Tech. Rep. PNW-58*, 1977.

Reynolds, R. R.: Twenty-nine years of selection timber management on the Crossett experimental forest, *U.S. Forest Ser. Res. Paper S0–40*, 1969.

Riekerk, H., and S. G. Gessel: The movement of DDT in forest soil solutions, *Soil Sci. Soc. Am. Proc.*, **32**(4):595–596 (1968).

Roberts, B. R.: Effects of water stress on the translocation of photosynthetically assimilated carbon-14 in yellow poplar. *In* M. H. Zimmerman (ed.), The formation of wood in forest trees, pp. 273–288, Academic, New York, 1964.

Roberts, F. L.: A study of the absorbing surfaces of the roots of loblolly pine, M.A. thesis, Duke University, 1948.

Robinson, R. K.: Mycorrhizae and the "second rotation decline" of *Pinus patula* in Swaziland, *S. Afr. For. J.*, **84**:16–19 (1973).

Roe, A. L., R. R. Alexander, and M. D. Andrews: Englemann spruce regeneration practices in the Rocky Mountains, *USDA Prod. Res. Pap. 115*, 1970.

Romberger, J. A., and R. G. Gregory: Analytical morphogenesis and the physiology of flowering in trees. *In* C. P. P. Reid and G. H. Fechner (eds.), Proc. 3rd N. Amer. for. biol. workshop, Fort Collins, CO, 1974, pp. 132–147.

Romell, L. G.: Mull and duff as biotic equilibria, *Soil Science*, **34**:161–188 (1932).

Ronco, F.: Food reserve of Englemann spruce planting stock, *For. Science*, **19**(3):213–219 (1973).

——: Diagnosis: "Sunburned" trees, *J. For.*, **73**(1):31–35 (1975).

Rook, D. A.: Effect of undercutting and wrenching on growth of *P. radiata* D. Don seedlings, *J. Appl. Ecol.*, 8(2):477–490 (1971).

Rowan, S. J.: Hardwood seedling production trends in southern nurseries 1965–1970, *J. For.*, **70**(3):162–163 (1972).

Rudinsky, J. A., and J. P. Vité: Certain ecological and phylogenetic aspects of the pattern of water conduction in conifers, *For. Science*, **5**:259–266 (1959).

Rudolf, P. O.: Forest plantings in the Lake states, *USDA Tech. Bul. 1010*, 1950.

——: Local red pine seed develops best plantations, *U.S. For. Ser. Lake States Exp. Sta. Tech. Note 296*, 1948.

Russell, R. J.: Climates of California, *Univ. Calif. Publ. Geog.*, **2**:73–84 (1926).

Safford, L. O.: Effect of fertilization on the biomass of fine roots in a beech-birch-maple stand (abstr.), North Am. forest biol. workshop 2, Corvallis, OR, 1972.

—— and S. Bell: Biomass of fine roots in a white spruce plantation, *Can. J. For. Res.*, **2**:169–172 (1972).

Salman, K. A., and J. W. Bongberg: Logging high-risk trees to control insects in the pine stands of northeastern California, *J. For.*, **40**:533–539 (1942).

Sargent, C. S.: Report on the forests of North America (exclusive of Mexico), Dept. Interior 10th Census, vol. 9, 1884.

Saucier, J. R., and A. F. Ike: Response in growth and wood properties of American sycamore to fertilization and thinning. *In* Proc. symp. on the effect of growth acceleration on the properties of wood, For. Prod. Lab., Madison, WI, 1972, pp. F1–F12.

Sawyer, J. O., Jr., and A. A. Lindsey: The Holdridge bioclimatic formations of Eastern and Central United States, *Proc. Indiana Acad. Sci.*, *1963*, **72**:105–112 (1964).

Schimper, A. F. W.: Plant geography upon a physiological basis, Clarendon Press, Oxford, 1903.

Schmidtling, R. D.: Cultivating and fertilizing stimulate precocious flowering in loblolly pines, *Silvae Genet.*, **20**(5):220–221 (1971).

——: Fruitfulness in conifers: Nitrogen, carbohydrate and genetic control. *In* C. P. P. Reid and G. H. Fechner (eds.), Proc. 3rd. North Am. for. biol. workshop, Fort Collins, CO, 1974, pp. 148–164.

Schnur, G. L.: Yield, stand and volume tables for even-aged upland oak forests, *USDA Tech. Bul. 560*, 1937.

Schramm, J. E.: Plant colonization studies on black wastes from anthracite mining in Pennsylvania, *Am. Philo. Soc.*, **56**:1–194 (1966).

Schubert, G. H.: Silviculture of southwestern ponderosa pine, *USDA Forest Ser. Res. Pap. RM-123*, 1974.

—— and R. S. Adams: Reforestation practices for conifers in California, State of California Resources Agency, State Division of Forestry, Sacramento, CA, 1971.

Schultz, R. P.: Root development of intensively cultivated slash pine, *Proc. Soil Sci. Soc. Am.*, **36**(1):158-162 (1972).

—— and F. W. Woods: The frequency and implications of intraspecific root-grafting in loblolly pine, *For. Science*, **13**(3):226–239 (1967).

Schultze, E. D.: Der CO_2-Gaswechsel der Buche (*Fagus silvatica*) in Abhängigkeit von den Klimafactoren im Freiland, *Flora*, **159**:177–232 (1970).

Schumacher, F. X.: Yield, stand and volume tables for red fir in California, *U.C. Agric. Exp. Sta. Bul. 456*, 1928.

——: Yield, stand and volume tables for Douglas-fir in California, *U.C. Agric. Exp. Sta. Bul. 491*, 1930.

Schwappach, A.: Die Samenproduction der wichtigsten Waldholzarten in Preussen, *Zeitschrift f. Forst-u. Jagdwes*, **26**:147-174 (1895).

Shantz, H. L., and R. Zon: Atlas of American agriculture, USDA, 1924.

Sheets, T. J., and C. I. Harris: Herbicide residues in soils and their phytotoxicity to crops grown in rotations, *Residue Rev.*, **11**:119–140 (1965).

Shigo, A. L.: Insect and disease controls: Forest fertilization relations. *In* Forest fertilization symp. proc., *USDA Forest Serv. Gen. Tech. Rep. NE-3*, 1973, pp. 117–121.

Shiroya, T., G. R. Lister, V. Slankis, G. Krotkov, and C. D. Nelson: Translocation of the products of photosynthesis to roots of pine seedlings, *Can. J. Bot.*, **40**:1125–1135 (1962).

Shoulders, E.: Root pruning southern pine in the nursery, *U.S. Forest Ser. Southeastern For. and Range Exp. Sta. Res. Pap. 5*, 1963.

——: Rainfall influences female flowering of slash pine, *U.S. Forest Ser. Res. Note S0-159*, 1973.

Silen, R. R.: A simple progressive tree improvement program for Douglas-fir, *U.S. Forest Ser. PNW For. Exp. Sta. Res. Note 45*, 1966.

——: Earlier forecasting of Douglas-fir cone crop using male buds, *J. For.*, **65**: 888–892 (1967).

—— and K. E. Row: Inheritance of stockiness in ponderosa pine families, *U.S. Forest Ser. PNW For. Exp. Sta. Res. Note. 166*, 1971.

Singh, P.: Host plant nutrition and composition effects on agriculture pests, *Res. Inst. Can. Dep. Agric. Inform. Bul. 6*, 1970.

Slankis, V., V. C. Runeckles, and G. Krotkov: Metabolites liberated by roots of white pine (*Pinus strobus*) seedlings, *Physiol. Plant.*, **17**:301–313 (1964).

Slatyer, R. O. Plant-water relationships. Academic, New York, 1967.

Smith, A. H. Taxonomy of ectomycorrhiza-forming fungi. *In* Mycorrhizae, *U.S. Forest Ser. Misc. Pub. 1189*, 1971.

Smith, D. M.: The practice of silviculture, 7th ed. Wiley, New York, 1962.

Smith, J. H. G.: Root spread can be estimated from crown width of Douglas-fir, lodgepole pine and other British Columbia tree species, *For. Chron.*, **40**: 456–473 (1964).

——: Comments on "A critical look at the normality concept," *J. For.*, **63**:706–707 (1965).

——: Big stock vs. small stock, Western Reforestation Coord. Comm., Vancouver, B.C., mimeographed, 1975.

—— and J. Walters: Prune large immature Douglas-fir now, *Fac. For. U. of B.C. Res. Note 30*, 1961.

Smith, W. H., L. E. Nelson, and G. L. Switzer: Development of the shoot system of young loblolly pine. II. Dry matter and nitrogen accumulation, *For. Science*, **17**(1):55–62 (1971).

Soc. Am. Foresters: Forest cover types of North America (exclusive of Mexico), Soc. Am. Foresters, Washington, D.C., 1954.

Soil Survey Staff: Soil taxonomy, *USDA Soil Cons. Ser. Agric. Handbook 436*, 1975.

Solomon, D. S.: The influence of stand density and structure on growth of northern hardwoods in New England, *USDA Forest Ser. Res. Pap. NE-362*, 1977.

Spilsbury, R. H., and D. S. Smith: Forest site types of the Pacific Northwest, Dept. Lands and Forests, *B.C. Forest Tech. Publ. T. 30*, 1947.

Spurr, S. H.: Forest inventory. Ronald, New York, 1952.

—— and W. Hsiung: Growth rate and specific gravity in conifers, *J. For.*, **52**(3): 191–200 (1954).

Staebler, G. R.: What are the economic implications of using off-source seed? Symposium on movement of seed from source in the Douglas-fir region. Proc. west. for. genet. assoc., Olympia, WA, 1966.

Stage, A. R.: A mathematical approach to polymorphic site curves for grand fir, *For. Science*, **9**:167–180 (1963).

——: A tree-by-tree measure of site utilization for grand fir related to stand density index, *USDA Forest Ser. Res. Note INT-77*, 1968.

——: A growth definition for stocking: units, sampling and interpretation, *For. Science*, **15**:255–265 (1969).

Stahelin, R.: Thinning even-aged loblolly and slash pine stands to specific densities, *J. For.*, **47**:538–549 (1949).

Stahl, E.: Der Sinn der Mycorrhizenbildung, *Jahrb. Wiss. Bot.*, **34**:534–668 (1900).

Stark, N.: Nutrient cycling in a Jeffrey pine ecosystem, Montana Forest and Conserv. Exp. Sta., University of Montana, Missoula, 1973.

Stark, R. W.: Taxonomic groups of seed destroying insects. *In* Symp. on cone and seed insects, Ann. Meeting Ent. Soc. Am. Pac. Br. For. Biol. Lab., Victoria, B.C., 1960, pp. 11–15.

——: Recent trends in forest entomology, *Ann. Rev. Ent.*, **10**:303–324 (1965).

Steele, R., D. Ondov, S. Cooper, and R. D. Pfister: Preliminary forest habitat types of eastern Idaho and western Wyoming, U.S. Forest Ser. Intermt. For. and Range Exp. Sta., 1977.

——, R. D. Pfister, R. A. Ryker, and J. A. Kittams: Preliminary habitat types of the Challis, Salmon and Sawtooth National Forests, U.S. Forest. Ser. Intermt. For. and Range Exp. Sta., 1974.

——: Forest habitat types of central Idaho, U.S. Forest Ser. Intermtn. For. and Range Exp. Sta., 1975.

Steila, D.: An evaluation of the Thornthwaite and Holdridge classifications as applied to the Mediterranean Borderland, *Prof. Geographer*, **18**:358–364 (1966).

Stein, W. I.: What do we mean by stocking? Western For. and Conserv. Assoc.

Permanent Assoc. Committees Proc. 1973, Portland, Oregon, 1974*a*, pp. 105–109.

——: When are stands considered established? in Symposium on stand establishment, IUFRO Joint Meeting, Divisions 1 and 3, Wageningen, The Netherlands, Oct. 15–19, 1974, 1974*b*, pp. 422–427.

——, J. L. Edwards, and R. W. Tinus: Outlook for container-grown seedling use in reforestation, *J. For.*, **73**(6):337–341 (1975).

Steinbeck, K., R. G. McAlpine, and J. T. May: Short rotation culture of sycamore: A status report, *J. For.*, **70**(4):210–213 (1972).

Steinbrenner, E. C., J. W. Duffield, and R. K. Campbell: Increased cone production of young Douglas-fir following nitrogen and phosphorus fertilization, *J. For.*, **58**:105–110 (1960).

Stephens, G. R., Jr.: Stimulation of flowering in eastern white pine, *For. Science*, **10**(1):28–34 (1964).

Stephenson, G. K., and E. B. Snyder: Genetic variation–key to superior trees, Southern For. Exp. Sta., U.S. Forest Ser., 1969.

Stiell, W. M.: A planting method experiment with white spruce, *Pulp. Pap. Mag. Can*, **59**(6):207–208 (1958).

——: Comparative cone production in young red pine planted at different spacings, *Can. J. For. Serv. Publ. 1306*, 1971.

Stoate, T. N., I. Mahood, and E. C. Crossin: Cone production in Douglas-fir (*Pseudotsuga menziesii*), *Emp. For. Rev.*, **40**:104–110 (1961).

Stocker, O.: Die photosynthetischen Leistungen der Steppen und Wüstenpflanzen, in W. Ruhland (ed.), *Handbuch der Pflanzenphysiologie*, **5**(1):460–491 (1960).

Stoeckeler, J. H.: Soil factors affecting the growth of quaking aspen forests in the Lake states, *Univ. Minn. Agric. Exp. Sta. Tech. Bul. 233*, 1960.

——, and G. W. Jones: Forest nursery practice in the Lake states, *USDA Handbook 110*, 1957.

Stoehr, H. A.: Stem taper of trembling aspen, *For. Science*, **1**:222–226 (1955).

Stone, E. C.: The ecological importance of dew, *Q. Rev. Biol.*, **38**(4):328–341 (1963).

—— and J. L. Jenkinson: Physiological grades for ponderosa pine nursery stock based on predicted root growth capacity, *J. For.*, **69**(1):31–33 (1971).

—— and G. H. Schubert: Root regeneration by ponderosa pine seedlings lifted at different times of the year, *For. Science*, **5**:322–332 (1959).

——, R. W. Benselor, F. J. Baron, and S. L. Krugman: Variation in the root regenerating potential of ponderosa pine from four California nurseries, *For. Science*, **9**(2):217–225 (1963).

Stone, E. L.: Objectives in forest fertilization: Current and potential. *In* Forest fertilization, *Symp. Proc. USDA Forest Ser. Gen. Tech. Rep. NE-3*, 1973.

—— and P. E. Lemmon: Soil and the growth of forests. *In Soil, Yearbook Agric. USDA*, pp. 721–732, 1957.

——, J. E. Stone, and R. C. McKittrick: Root grafting in pine trees, *N. Y Food and Life Sciences*, **6**(2):19–21 (1973).

Sucoff, E.: Timing and rate of bud formation in *Pinus resinosa*, *Can. J. Bot.*, **49**(10):1821–1832 (1971).

—— and S. G. Hong: Effects of thinning on needle water potential in red pine, *For. Science*, **20**:25–29 (1974).

Sukachev, V. N., and N. Dylis (eds.): Fundamentals of Forest Biogeocoenology. Oliver and Boyd, London, 1964.

Sutton, R. F.: Form and development of conifer root systems, *Tech. Comm. Commonwealth For. Bur. 7*, 1969.

Sweeney, J. R.: Responses of vegetation to fire, *Univ. Calif. Publ. in Botany*, **28**(4):143–250 (1956).

Sweet, G. B., and K. J. Thulin: Graft incompatibility in radiata pine in New Zealand, *N.Z. J. For. Science*, **3**:83–90 (1973).

Tanaka, Y., A. J. Long, and J. K. Winjum: Effects of undercutting and wrenching on 2 + 0 Douglas-fir seedlings. *In* Proc. Western Forest Nursery Council Meeting, Portland, OR, 1974, pp. 60–71.

Tappeiner, J. C.: Effect of cone production on branch needle and xylem ring growth of Sierra Nevada Douglas-fir, *For. Science*, **15**:171–174 (1969).

Tarrant, R. F.: Bibliography of Pacific N.W. Forest Soils Publications, *Pacific Northwest For. Range Exp. Sta. Res. Note 169*, 1958.

—— and R. E. Miller: Accumulation of organic matter and soil nitrogen beneath a plantation of red alder and Douglas-fir, *Soil Sci. Soc. Am. Proc.*, **27**(2): 231–234 (1963).

Tatum, L. A.: The southern corn leaf blight epidemic, *Science*, **171**:113–116 (1971).

Taylor, R. F.: The application of a tree classification in marking logdepole pine for selection cutting, *J. For.*, **37**:777–782 (1939).

Teeguarden, D. T., and D. T. Gordon: A precommercial thinning on ponderosa and Jeffrey pine, *J. For.*, **57**:900–904 (1959).

Tepper, H. B.: Leader growth of young pitch and shortleaf pines, *For. Science*, **9**(3):344–353 (1963).

Thompson, P. T.: A test of the Holdridge model in midlatitude mountains, *Prof. Geographer*, **18**:286–292 (1966).

Thornthwaite, C. W.: The climates of North America according to a new classification, *Geogr. Rev.*, **21**:633–655 (1931).

——: An approach to a rational classification of climate, *Geogr. Rev.*, **38**:55–94 (1948).

Tinus, R. W., W. I. Stein, and W. E. Balmer (eds.): Proc. North Am. containerized forest tree seedling symp., Denver, CO, Aug. 26–29, 1974, *Great Plains Agric. Council Publ. 68*, 1974.

Toda, R.: A brief review and conclusions of the discussion on seed orchards, *Silvae Genet.*, **13**:104 (1964).

Tosi, J. A., Jr.: Climatic control of terrestrial ecosystems: A report on the Holdridge model, *Econ. Geography*, **40**:173–181 (1964).

Toumey, J. W. The vegetation of the forest floor; light versus soil moisture, *Proc. Int. Cong. Pl. Sci.*, **1**:575–590 (1929). Ithaca, N.Y.

Trappe, J. M.: Selection of fungi for ectomycorrhizal inoculation in nurseries, *Ann. Rev. Phytopath.*, **15**:203–222 (1977).

Trichell, D. W., H. C. Morton, and M. G. Merkle: Loss of herbicides in run-off water, *Weed Science*, **16**(4):447–449 (1968).

Troup, R. S.: Silvicultural systems. Clarendon Press, Oxford, 1928.

Turkel, H. S., A. L. Rebuck, and A. R. Grove, Jr.: Floral morphology of white oak, *Penn. State Univ. Agric. Exp. Sta. Bul. 593*, 1955.

U.K. Bur. Soils: Bibliography on effect of silvicultural practices on soil properties, 1935–1958, *Biblio. Bur. Soils Harpenden 190*, 1959.

——: Bibliography on physical properties of forest soils, 1941–1958, *Biblio. Bur. Soils Harpenden 185*, 1959.

——: Bibliography on soil requirements of forest trees, 1940–1958, *Biblio. Bur. Soils Harpenden 186*, 1959.

——: Bibliography on the effects of clear felling and forest thinning on soil properties, 1930–1968, *Biblio. Bur. Soils. 1292*, 1969.

Ulrich, A., and F. J. Hills: Plant analysis as an aid in fertilizing sugar crops. Part 1. Sugar beets. *In* L. M. Walsh and J. P. Beaton (eds.), Soil testing and plant analysis, no. 17, Soil Sci. Soc. Am., Madison, WI, 1973, pp. 271–288.

University of California: Forest pest control, *U.C. Div. of Agric. Sciences Leaflet 2962*, 1977.

Urie, D. H.: Soil-site relationships on second growth Douglas-fir stands in the central Oregon Coast Range, M.S. thesis, Oregon State Univ., Corvallis, 1959.

U.S. Department of Agriculture: Woody plant seed manual, *Misc. Publ. 654*, 1948.

——: Seeds of woody plants in the United States, *Agriculture Handbook 450*, 1974.

U.S. Department of Interior: Effects of pesticides on fish and wildlife, *Fish and Wildlife Service Circular 226*, 1964.

U.S. Forest Service: Volume, yield, and stand tables for second-growth southern pines, *USDA Forest Ser. Misc. Publ. 50*, 1929.

——: Forest tree seed orchards: A directory of industry, state and federal forest tree seed orchards in the United States, USDA Forest Ser., November 1971.

——: Forest fertilization symposium, *Proc. N.E. For. Exp. Sta. Tech. Rep. NE-3*, 1973.

Van Cleve, K., L. A. Viereck, and R. L. Schlentner: Accumulation of nitrogen in alder (*Alnus*) ecosystems near Fairbanks, Alaska, *Arctic Alp. Res.*, **3**(2): 101–114 (1971).

Van Den Berg, D. A., and R. M. Lanner: Bud development in lodgepole pine, *For. Science*, **17**(4):479–486 (1971).

Van Dorsser, J. C., and D. A. Rook: Conditioning of radiata pine seedlings by undercutting and wrenching: Description of methods, equipment and seedling response, *N.Z. J. For.*, **17**(1):61–73 (1972).

Van Vredenburch, C. L. H., and J. G. A. La Bastide: The influence of meteorological factors on the cone crop of Douglas-fir in the Netherlands, *Silvae Genet.*, **18**:182–186 (1969).

Varnell, R. J., A. E. Squillace, and G. W. Bengtson: Variation and heritability of fruitfulness in slash pine, *Silvae Genet.*, **16**:125–128 (1967).

Vazzo, J. A., and E. Hacskaylo: Inoculation of *Pinus caribaea* with Ectomycorrhizal Fungi in Puerto Rico, *For. Sci.* **17**:239–245 (1971).

Viro, P. J.: Prescribed burning in forestry, *Comm. Inst. Forestalis Fennica 67.7*, 1969.

Waggoner, P. E., and N. C. Turner: Transpiration and its control by stomata in a pine forest, *Con. Agric. Exp. Sta. Bul. 736*, 1971.

Wagle, R. F., and J. H. Kitchen: Influence of fire on soil nutrients in a ponderosa pine type, *Ecology*, **53**(1):118–125 (1972).

Wakeley, P. C.: Geographic source of southern pine seed, *J. For.*, **42**:23–32 (1944).

—— and J. Marrero: Five-year intercept as site index in southern pine plantations, *J. For.*, **56**:332–336 (1958).

Walker, C., and H. F. Perkins: Forest soils and silviculture in Georgia, *Report 4 Georgia For. Res. Council*, 1958.

Walters, J., and J. Soos: Shoot growth patterns of some British Columbia conifers, *For. Sci.*, **9**:73–85 (1968).

——: The vertical and horizontal organization of growth in some conifers of British Columbia. Res. Pap. Fac. Forestry, University of British Columbia, Canada.

Way, J. M.: Toxicity and hazards to man, domestic animals and wildlife from some commonly used auxin herbicides, *Residues Rev.*, **26**:37–62 (1969).

Weed Science Society of America: Herbicide handbook, 1974.

Weetman, G. F., and B. Webber: The influence of wood harvesting on the nutrient status of two spruce stands, *Can. J. For. Res.*, **2**:351–369 (1972).

—— and S. B. Hill: General environmental and biological concerns in relation to forest fertilization. *In* Forest fertilization, *Symp. Proc. USDA Forest Ser. Gen. Tech. Rep. NE-3*, 1973, pp. 19–35.

—— and A. L. Leaf: Forest fertilization—A bibliography with abstracts, *World Forestry Series Bul. 2*, 1956.

Weidman, R. H.: Evidences of racial influence on a 25-year test of ponderosa pine, *J. Agric. Res.*, **59**:855–887 (1939).

Welch, B. L., and Dean Andrus: Rose hips—A possible high energy food for wintering mule deer, *USDA Forest Ser. Res. Note INT-218*, 1977.

Wenger, K. F.: The stimulation of loblolly pine seed trees by pre-harvest release, *J. For.*, **52**:115–118 (1954).

——: Annual variation in the seed crops of loblolly pine, *J. For.*, **55**:567–569 (1957).

Werner, R. A.: Clockwise spiral ascent of dye in southern pine, *For. Science*, **17**(1):44–45 (1971).

Werner, R. G.: Water quality—Limnological concerns about forest fertilization. *In* Forest fertilization, *Symp. Proc. USDA Forest Ser. Gen. Tech. Rep. NE-3*, 1973, pp. 72–78.

Westing, A. H.: Formation and function of compression wood in gymnosperms, *Bot. Rev.*, **31**(3):381–480 (1965).

Westveld, M.: A budworm vigor-resistance classification for spruce and balsam fir, *J. For.*, **52**:11–24 (1954).

White, D. P.: Available water: The key to forest site evaluation, First North Am. Forest Soils Conf., Michigan State Univ., 1958.

White, E. H., W. L. Pritchett, and W. K. Robertson: Slash pine root biomass and nutrient concentration, Forest biomass studies, 15th IUFRO Conf., Gainesville, FL, 1971, pp. 164–176, *Life Sci. and Agric. Expt. Sta. Misc. Publ. 132*, 1971.

Whitt, D. M., and L. D. Baver: Particle size in relation to base exchange capacity and hydration properties of Putnam clay, *J. Am. Soc. Agron.*, **29**:703–708 (1930).

Wilcox, H.: Cambial growth characteristics. *In* T. T. Kozlowski, (ed.), Tree growth, Ronald, New York, 1962, pp. 57–88.

——: Growth studies of the root of incense cedar, *Libocedrus decurrens*. II. Morphological features of the root system and growth behavior, *Am. J. Bot.*, **49**:237–245 (1962).

Wilcox, H. E.: Morphology of ectendomycorrhizae in *Pinus resinosa*. *In* E. Hacskaylo, (ed.), Mycorrhizae, Proc. 1st North Am. Conf. on Mycorrhizae, April 1969, *USDA Forest Ser. Misc. Publ. 1189*, 1971, pp. 54–68.

Wilde, S. A.: Soil science and semantics, *J. Soil Science*, **4**:1–4 (1953).

——: Forest soils, their properties and relation to silviculture. Ronald, New York, 1958.

——: Growth of Wisconsin coniferous plantations in relation to soils, *Univ. of Wisconsin Res. Bul. 262*, 1965.

——: Mycorrhizae: Their role in tree nutrition and timber production, *Univ. of Wisconsin Res. Bul. 272*, 1968.

——: Forest humus: Its classification on a genetic basis. *Soil Science*, **111**(1): 1–12 (1971).

——, A. Giordano, B. H. Shaw, A. W. Fedkenhever, and W. T. Seip: Tree spacing in forest plantations as related to soils and revenue, *Univ. of Wisconsin Res. Bul. 589*, 1968.

—— and J. G. Ilyer: Effect of natural subirrigation on the uptake of nutrients by forest plantations, *Acta Forestalia Fennica*, **76**:3–9 (1963).

——, F. G. Wilson, and D. P. White: Soils of Wisconsin in relation to silviculture, *Wisconsin Consv. Dept. Publ. 525*, 1949.

Wilson, B. C., and R. K. Campbell: Seedbed density influences height, diameter, and dry weight of 3-0 Douglas-fir, *Tree Plant. Notes*, **23**(2):1–4 (1972).

Wilson, F. G.: Numerical expression of stocking in terms of height, *J. For.*, **44**: 758–761 (1946).

——: Evaluation of three thinnings at Star Lake, *For. Science*, **1**:227–231 (1955).

Winstead, J. E.: Fiber tracheid length and wood specific gravity of seedlings as ecotypic characters in *Liquidambar styraciflua*, *Ecology*, **53**(1):165–172 (1972).

Wittich, W.: Classification, mapping and interpretation of soils for forestry purposes, Paper GP/109/1/F, 5th World For. Congress, Seattle, WA, 1960.

Wood, J. P., and E. P. Bachelard: Root grafting in radiata pine stands in the Australian Capital Territory, *Aust. J. Bot.*, **18**(3);251–259 (1970).

Woodman, J. N.: Variation of net photosynthesis within the crown of a large forest-grown conifer, *Photosynthetica*, **5**(2):50–54 (1971).

Woods, D. B., and N. C. Turner: Stomatal response to changing light by four tree species of varying shade tolerances, *New Phytol.*, **70**(1):77–84 (1971).

Wort, D. S.: Physiology of cambial activity. *In* T. T. Kozlowski (ed.), Tree growth, Ronald, New York, 1962, pp. 89–95.

Wright, J. W.: Species hybridization in the white pine, *For. Science*, **5**:210–222 (1959).

——: Genetics of forest tree improvement. *FAO For. and For. Products Studies 16*, 1962.

——, R. T. Bingham, and K. W. Dorman: Genetic variation within geographic ecotypes of forest trees and its role in forest tree improvement, *J. For.*, **56**:803–808 (1958).

Wuenscher, J. E., and T. T. Kozlowski: Relationship of gas exchange resistance to tree seedling ecology, *Ecology*, **52**(6):1016–1023 (1971).

Young, H. E.: The complete tree concept—A challenge and an opportunity, *Proc. Soc. Am. For.*, **1964**:231–233 (1965).

——: Challenge of complete tree utilization, *For. Prod. J.*, **18**:83–86 (1968).

—— (ed.): Forest biomas studies, sec. 25, 15th IUFRO Congress, Univ. of Florida, Gainesville, FL, 1971.

——: Complete-tree concept: 1964–1974. *For. Prod. J.*, **24**:13–16 (1974).

——: The enormous potential of the forests: A positive rebuttal to Grantham and Ellis, *J. For.*, **73**:99–102 (1975).

—— and P. M. Carpenter: Weight, nutrient element and productivity studies of seedlings and saplings of eight tree species in natural ecosystems, *Maine Agric. Exp. Sta. Tech. Bul. 28*, 1967.

Youngberg, C. T.: Forest-soil relationships in North America, Proc. 2d. North Am. forest soils conf., Oregon State Univ. Press, Corvallis, 1965.

—— and C. B. Davey: Tree growth and forest soils, Proc. 3rd North Am. forest soils conf., N. Carolina State Univ. Press, Raleigh, 1970.

Zahner, R.: Site quality relationships of pine forests in southern Arkansas and northern Louisiana, *For. Science*, **4**:163–176 (1958).

——: Internal moisture stress and wood formation in conifers, *For. Prod. J.*, **13**(6):240–247 (1963).

——: Site quality and wood quality in upland hardwoods: Theoretical considerations of wood density. *In* C. T. Youngberg and C. B. Davey (eds.), Tree growth and forest soils, Oregon State Univ. Press., Corvallis, 1970, pp. 447–497.

—— and W. W. Oliver: The influence of thinning and pruning on the date of summerwood initiation in red and jack pines, *For. Science*, **8**(1):51–63 (1962).

——, J. E. Lotan, and W. D. Baughman: Earlywood-latewood features of red pine grown under simulated drought and irrigation, *For. Science*, **10**(3): 361–370 (1964).

Zak, B.: Role of mycorrhizae in root disease, *Ann. Rev. Phytopathol.*, **2**:377–392 (1964).

——: Characterization and identification of Douglas-fir mycorrhizae. *In* E. Hacskaylo, (ed.), Mycorrhizae, Proc. 1st North Am. conf. on mycorrhizae, April 1969, *USDA Forest Ser. Misc. Publ. 1189*, 1971, pp. 38–53.

Zelitch, I.: Stomatal control, *Ann. Rev. Plant Physiol.*, **20**:329–350 (1969).

——: Photosynthesis, photorespiration, and plant productivity. Academic, New York, 1971.

Zimmermann, M. H., and C. L. Brown: Trees: Structure and function, Springer-Verlag, New York, 1971.

Zinke, P. J.: Site quality for Douglas-fir and ponderosa pine in northwestern California as related to climate, topography, and soil, *Proc. Soc. Am. For.*, **1958**:167–171 (1959).

——: Forest site quality as related to soil nitrogen content, *Int. Soc. of Soil Sci., 7th Int. Congress*, Amsterdam, 1950, *Transactions Comm. 4*, pp. 411–418.

——: The pattern of influence of individual forest trees on soil properties, *Ecology*, **43**(1):130–133 (1962).

——: The depletion of soil moisture storage in forest soils. *In* Proc. 4th North Am. forest soils conf., Laval Univ. Press, Quebec, 1974, pp. 27–36.

—— and W. L. Colwell: Some general relationships among California forest soils. *In* C. Youngberg (ed.), Proc. 2d North Am. forest soils conf., Oregon State Univ. Press, Corvallis, 1963, pp. 353–365, 1964.

Zobel, B. J.: The genetic improvement of southern pines, *Sci. Amer.*, **225**(5): 94–103 (1971).

——, R. C. Kellison, and D. B. Kirk: Wood properties of young loblolly and slash pines. *In* Proc. Symp. on the Effect of Growth Acceleration on the Properties of Wood, Forest Prod. Lab., USDA, Madison, WI, 1972, pp. M1–M20.

——, R. Kellison, and M. Mathias: Genetic improvement in forest trees—Growth rate and wood characteristics in young loblolly pine. *In* Proc. 10th Southern Conf. on For. Tree Impr., Texas For. Ser., Houston, 1969, pp. 59–75.

Zon, R., and W. N. Sparhawk: Forest resources of the world. McGraw-Hill, New York, 1923.

INDEX